BEAM INSTRUMENTATION WORKSHOP 2002

BEAM INSTRUMENTATION WORKSHOP 2002

Tenth Workshop

Upton, New York   6–9 May 2002

EDITORS
Gary A. Smith
Thomas Russo
Brookhaven National Laboratory
Upton, New York

SPONSORING ORGANIZATION
Collider-Accelerator Department,
Brookhaven National Laboratory

CD-ROM INCLUDED

Melville, New York, 2002
AIP CONFERENCE PROCEEDINGS ■ VOLUME 648

Editors:

Gary A. Smith
Thomas Russo

Building 911
Brookhaven National Laboratory
Upton, NY 11973
USA

E-mail: gasmith@bnl.gov
         trusso@bnl.gov

The articles on pp. 162-173, 212-219, 237-247, 368-375, 376-383, 417-424, and 483-490 were authored by U.S. Government employees and are not covered by the below mentioned copyright.

Authorization to photocopy items for internal or personal use, beyond the free copying permitted under the 1978 U.S. Copyright Law (see statement below), is granted by the American Institute of Physics for users registered with the Copyright Clearance Center (CCC) Transactional Reporting Service, provided that the base fee of $18.00 per copy is paid directly to CCC, 222 Rosewood Drive, Danvers, MA 01923. For those organizations that have been granted a photocopy license by CCC, a separate system of payment has been arranged. The fee code for users of the Transactional Reporting Service is: 0-7354-0103-9/02/$19.00.

© 2002 American Institute of Physics

Individual readers of this volume and nonprofit libraries, acting for them, are permitted to make fair use of the material in it, such as copying an article for use in teaching or research. Permission is granted to quote from this volume in scientific work with the customary acknowledgment of the source. To reprint a figure, table, or other excerpt requires the consent of one of the original authors and notification to AIP. Republication or systematic or multiple reproduction of any material in this volume is permitted only under license from AIP. Address inquiries to Office of Rights and Permissions, Suite 1NO1, 2 Huntington Quadrangle, Melville, N.Y. 11747-4502; phone: 516-576-2268; fax: 516-576-2450; e-mail: rights@aip.org.

L.C. Catalog Card No. 2002114218
ISBN 0-7354-0103-9
ISSN 0094-243X
Printed in the United States of America

Contents

Preface ix
Organizing Committees x
Conference Photo xi
Schedule xii
Faraday Cup Award xiii

FARADAY CUP AWARD – INVITED TALK

The Magnetic Quadrupole Pick-Ups in the CERN PS 3
    A. Jansson

TUTORIALS

The Physics and Properties of Free-Electron Lasers 23
    S. Krinsky
A Tutorial on Beam Loss Monitoring 44
    R. E. Shafer
Optical System Design for High-Energy Particle Beam Diagnostics 59
    B. Yang
Fast Digitization and Digital Receiver Technology 79
    R. C. Kimball

INVITED TALKS

FNAL Instrumentation: Lessons Learned 103
    J. L. Crisp
Diagnostics for Recirculating and Energy Recovered Linacs 118
    G. A. Krafft and J.-C. Denard
Tune Measurement in RHIC 134
    M. Brennan, P. Cameron, P. Cerniglia, R. Connolly, J. Cupolo, W. Dawson,
    C. Degen, A. DellaPenna, J. DeLong, A. Drees, D. Gassner, M. Kesselman,
    R. Lee, A. Marusic, J. Mead, R. Michnoff, C. Schultheiss, R. Sikora,
    and J. Van Zeijts
Laser Beam-Profile Monitor Development at BNL for SNS 150
    R. Connolly, P. Cameron, J. Cupolo, D. Gassner, M. Grau, M. Kesselman,
    S. Peng, and R. Sikora
Short Bunch Beam Diagnostics 162
    P. Krejcik

CONTRIBUTED TALKS

Reconfigurable Instrumentation Technologies, Architectures and Trends..179
    R. Uršič

Optical Method for Mapping the Transverse Phase Space of a Charged Particle Beam..187
    R. B. Fiorito, A. G. Shkvarunets, and P. G. O'Shea

Beam Diagnostics in the SNS Linac......................................195
    M. A. Plum, L. Day, S. Ellis, R. Hardekopf, R. Meyer Sr., J. O'Hara, J. Power, C. Rose, R. Shafer, M. Stettler, and J. Stovall

Streak Camera Characterization Using a Femtosecond Ti:Sapphire Laser..203
    M. Ferianis and M. Danailov

Advanced Intraundulator Electron-Beam Diagnostics Using COTR Techniques...212
    A. H. Lumpkin, W. J. Berg, S. Biedron, M. Borland, Y. C. Chae, R. Dejus, M. Erdmann, Z. Huang, K.-J. Kim, J. Lewellen, Y. Li, S. V. Milton, E. Moog, D. W. Rule, V. Sajaev, and B. X. Yang

The LHC 450 GeV to 7 TeV Synchrotron Radiation Profile Monitor Using a Superconducting Undulator...........................220
    R. Jung, P. Komorowski, L. Ponce, and D. Tommasini

LHC Beam Loss Monitor System Design..............................229
    B. Dehning, G. Ferioli, W. Friesenbichler, E. Gschwendtner, and J. Koopman

A Very High Resolution Optical Transition Radiation Beam Profile Monitor...237
    M. Ross, S. Anderson, J. Frisch, K. Jobe, D. McCormick, B. McKee, J. Nelson, T. Smith, H. Hayano, T. Naito, and N. Terunuma

ELETTRA Photon Source Beam Stabilization..........................248
    S. N. Thanos and R. H. Hosking

POSTERS

Progress of Turn-By-Turn System for HLS.............................259
    J. H. Wang, J. H. Liu, B. G. Sun, W. M. Li, Z. P. Liu, Z. Z. Zhang, and P. Lu

Upgrades to PEP-II Tune Measurements..............................267
    A. S. Fisher, M. Petree, U. Wienands, S. Allison, M. Laznovsky, M. Seeman, and J. Robin

Proposed Profile Monitor Designs for the Advanced Hydrodynamic Facility (AHF).....................................275
    W. C. Sellyey and J. F. O'Hara

Design and Experiment of 200MeV Energy Spectrum Analysis System of Linac in NSRL......................................283
    P. Lu, Y. J. Pei, B. Sun, W. Li, H. Xu, Z. Liu, J. Hong, J. Wang, and D. He

Enhancements to the Digital Transverse Dampers at the Brookhaven AGS 289
    M. Wilinski, A. Drees, R. Michnoff, T. Roser, and G. A. Smith

Biasing Wire Scanners and Halo Scrapers for Measuring 6.7-MeV Proton-Beam Halo 297
    J. D. Gilpatrick, M. Gruchalla, J. Kamperschroer, and J. O'Hara

The Mechanical Design and Preliminary Testing Results of Beam Position Monitors for the LANSCE Isotope Production Facility and Switchyard Kicker Projects 305
    J. F. O'Hara, J. D. Gilpatrick, J. E. Ledford, R. B. Shurter, R. J. Roybal, and B. E. Bentley

Electron Beam Diagnostics at the Radiation Source ELBE 313
    P. Evtushenko, U. Lehnert, P. Michel, C. Schneider, R. Schurig, and J. Teichert

Cavity BPMs for the NLC 321
    R. Johnson, Z. Li, T. Naito, J. Rifkin, S. Smith, and V. Smith

A Fast VME Data Acquisition System for Spill Analysis and Beam Loss Measurement 329
    T. Hoffmann, D. A. Liakin, and P. Forck

Design of an Improved Ion Chamber for the SNS 337
    R. L. Witkover and D. Gassner

Preliminary Design of the Beam Loss Monitoring System for the SNS 345
    R. Witkover and D. Gassner

Booster Applications Facility Instrumentation 353
    D. Gassner, S. Bellavia, K. A. Brown, I. H. Chiang, P. Pile, and R. Prigl

Profile Measurement of Scanning Proton Beam for LiSoR Using Carbon Fibre Harps 361
    R. Dölling, L. Rezzonico, U. Frei, S. Benz, P.-A. Duperrex, and M. Humbel

A Digital Signal Receiver VXI Module for BPM and Phase Detection Processing 368
    B. E. Chase and K. G. Meisner

BPM System for the SNS Ring and Transfer Lines 376
    W. C. Dawson, P. Cameron, P. Cerniglia, J. Cupolo, C. Degen, A. DellaPenna, A. Huhn, M. Kesselman, J. Mead, and R. Sikora

The Log-Ratio Beam Position Monitor 384
    A. Kalinin

Design and Upgrade of a Compact Imaging System for the APS Linac Bunch Compressor 393
    B. Yang, E. Rotela, S. Kim, R. Lill, and S. Sharma

New Beam Position Monitor System Design for the APS Injector 401
    R. M. Lill, O. Singh, and N. Arnold

A Transverse Injection Damper at RHIC 409
    A. Drees, M. Brennan, P. Cameron, R. Connolly, R. Michnoff, and C. Montag

Spallation Neutron Source Beam Current Monitor Electronics 417
    M. Kesselman and W. C. Dawson

Beam Based Calibration of BPM Position Sensitivity at SPring-8 Storage Ring 425
    S. Sasaki, K. Soutome, and H. Tanaka

Nonintercepting Imaging Diagnostics for the APS Injector during Storage Ring Top-Up Operations 433
    A. H. Lumpkin, W. J. Berg, and B. X. Yang

Simple "Package Design" Ion Chamber Monitors for TRIUMF's Proton Beamlines 439
    D. Gray and B. Minato

An Update of the Diagnostic Systems Proposed for the New Third Generation UK Light Source, DIAMOND 447
    S. R. Buckley, M. J. Dufau, and R. J. Smith

Study and Design of a New Over-damped Cavity Kicker for the PEP II Longitudinal Feedback System 458
    F. Marcellini, M. Tobiyama, P. MacIntosh, J. Fox, H. Schwarz, D. Teytelman, and A. Young

In-Situ Calibration: Migrating Control System IP Module Calibration from the Bench to the Storage Ring 467
    J. M. Weber and M. J. Chin

Set-up of PEP-II Longitudinal Feedback Systems for Even/Odd Bunch Spacings 474
    D. Teytelman and J. Fox

Operation of the Beam Diagnostics System for Tevatron Electron Lens 483
    X. Zhang, K. Bishofberger, J. Fitzgerald, G. Kuznetsov, M. Olson, A. Semenov, V. Shiltsev, and N. Solyak

Techniques for Electro-Optic Bunch Length Measurement at the Femtosecond Level 491
    P. Bolton, D. Dowell, P. Krejcik, and J. Rifkin

Laser-Compton Scattering as a Potential Electron Beam Monitor 497
    K. Chouffani, D. Wells, F. Harmon, G. Lancaster, and J. Jones

Design of a Multi-Bunch BPM for the Next Linear Collider 508
    A. Young, D. McCormick, M. Ross, S. R. Smith, H. Hayano, T. Naito, N. Terunuma, and S. Araki

Global Orbit Feedback in SRRC 516
    C. H. Kuo, J. Chen, K. H. Hu, and K. T. Hsu

Simple Amplitude and Phase Detector for Accelerator Instrumentation 523
    K. H. Hu, J. Chen, C. H. Kuo, D. Lee, and K. T. Hsu

APPENDICES

Workshop Participants 533
Participating Vendors 545
Author Index 547

Preface

The Tenth Beam Instrumentation Workshop was held May 6-9, 2002 at Brookhaven National Laboratory in Upton, New York. Dr. Derek Lowenstein, Chairman of the Collider-Accelerator Department, delivered the Workshop opening remarks. This marks a return to the beginning for this meeting, the first of these was held at Brookhaven thirteen years ago in the fall of 1989.

The workshop was sponsored by the Collider-Accelerator Department and was attended by 120 participants from 10 countries. There were four tutorial sessions, six invited talks, 11 contributed talks and four discussion sessions. In addition, there was a poster session with 42 poster presentations as well as a vendor display with 10 participating vendors. The week was marked by the award of the 2002 Faraday Cup Award for innovative work in beam instrumentation to Andreas Jansson of CERN for his work on Magnetic Quadruple Beam Pickups. This Award is given by the Workshop Committee and is donated by Julien Bergoz of Bergoz, Ltd.

The workshop banquet was held at Danfords on the Sound, an inn with great harbor views at Port Jefferson located on Long Island Sound. The food was excellent; the company superb, and a fine time was had by all.

On the final afternoon, tours were available of the Relativistic Heavy Ion Collider, the National Synchrotron Light Source and the Free Electron Laser Facilities. Two tour busses carried over seventy people to each of these facilities where tour guides waited to show them around. This was a lot to pack into one afternoon, and we are indebted to Elaine Lowenstein and all the tour guide volunteers who made this possible.

The success of a workshop of this kind is the result of the hard work of many people. The organizing committee for this workshop worked hard in the shadow of the events of September 11, to produce a viable and meaningful program as well as choosing the Faraday Cup winner from so many fine submissions. It is remarkable that the workshop came together on time. The members of the small local committee labored to get all the details together, including the additional approval required for the first time from DOE. Special thanks go to the conference secretary, Anna Petway for her work in keeping things pulled together and on track. Thanks to Nick Franco and Ila Campbell for handling the web site and the registration information. We are grateful to Marty Kesselman for his handling of the vendor exhibits. We thank Roger of graphic arts for photos of the banquet, Melanie Covitz and the staff of the conference group for their help and advice, and Patricia Yalden of graphic arts for help with the poster.

Of course the ultimate success of the workshop rests with the participants themselves. In the talks, posters, and discussions sessions, truly rests the heart of the work that is done. So as a community of instrumentation specialists we owe our thanks to all who labored to present and then to write for these proceedings their contributions.

The Spallation Neutron Source will be the host of the next workshop in 2004. We look forward to this and hope to see many of you again.

Thomas Russo
Gary Smith
BIW 2002 Co-Chairmen

Organizing Committee

Thomas Russo (Co-Chair), BNL
Gary A. Smith (Co-Chair), BNL
Walter C. Barry, LBNL
Glenn Decker, ANL
Jean-Claude Denard - Soleil
Robert O. Hettel, SLAC
Kenneth Jacobs, UWSRC
Roland Jung, CERN
Geoffrey Krafft, TJNAF
Ralph S. Pasquinelli, FNAL
Michael A. Plum, LANL
Thomas Shea, ORNL
R. Coles Sibley, ORNL
Stephen R. Smith, SLAC
Gregory D. Stover, LBNL
Robert C. Webber, FNAL
James R. Zagel, FNAL

Local Arrangements Committee

Thomas Russo (Co-Chair), BNL
Gary A. Smith (Co-Chair), BNL
Anna Petway, BNL
Nick Franco, BNL
Angelika Drees, BNL
Martin Kesselman, BNL

10th Beam Instrumentation Workshop
Hosted by Brookhaven National Laboratory

	Sunday	Monday	Tuesday	Wednesday	Thursday
8:00		Registration			
		Welcome/Opening	Tutorial: Beam Loss Monitors Robert E. Shafer	Tutorial: Digital Technology Ralph Kimball	Tutorial: The Physcis and Properties of FEL's Samuel Krinsky
9:00		Tutorial: Optics for Emittance Measurements Bingxin Yang			
10:00			Coffee Break	Coffee Break	Coffee Break
		Coffee Break	Electron Linac Instrumentation Geoffrey A. Krafft	Laser Diagnostics Roger Connolly	FNAL Instrumentation James L. Crisp
11:00		Short Bunch Diagnostics Patrick Krejcik			
12:00		RF and Beam Control J. Michael Brennan	Contributed Talks #4, #5, #6, #7	Contributed Talks #8, #9, #10, #11	Faraday Cup Talk
					Closing
1:00					
		Lunch	Lunch	Lunch	Lunch
2:00					
3:00		Contributed Talks #1, #2, #3	Poster Session	RHIC Tune Measurement Peter Cameron	Tours of RHIC NSLS DUV-FEL
		Break		Break	
4:00		Discussion Groups #1, #2	Vendors	Discussion Groups #3, #4	
5:00	Registration and Social Gathering at Danfords (until 8 PM)			Buses Leave For Banquet At Danfords	

Andreas Jansson (3rd to the right) receiving the Faraday Cup Award from Julien Bergoz (2nd to the right), Gary A. Smith (left) BIW02 co-chair and Thomas Russo (far right) BIW02 co-chairman.

2002 FARADAY CUP AWARD

The 2002 Faraday Cup Award was presented to Andreas Jansson from European Organization for Nuclear Research (CERN), Geneva, Switzerland, for his work on the Magnetic Quadrupole Pick-Ups in the CERN PS.

FARADAY CUP AWARD - INFORMATION AND RULES

Purpose

The Faraday Cup Award, donated by Bergoz Inc., Crozet, France, is intended to recognize and encourage innovative achievements in the field of accelerator beam instrumentation.

Award

The award consists of a US $5000 prize and a certificate to be presented at the next U.S. Beam Instrumentation Workshop (BIW). Winners participating in the BIW will share a

$1000 travel allowance. The selection of recipients is the responsibility of the BIW Organizing Committee.

Award Criteria

The Faraday Cup Award shall be presented for an outstanding contribution to the development of an innovative beam diagnostic instrument of proven workability. The prize is only awarded for demonstrated device performance and published contribution.

Criteria Interpretation

Beam Diagnostic Instrument:
A device to measure the properties of charged elementary particle, atomic, or simple molecular beams during or after acceleration, or the properties of neutral particle beams produced in an intermediate stage of charged particle acceleration. The device may operate by detecting secondary beams of charged, neutral, massive or massless particles, but its purpose should be to diagnose the primary charged particle beam. The mass of primary beam particles shall be no greater than the order of 1000 atomic mass units.

Delivered Performance:
The performance of the device should have been evaluated using a charged particle beam, rather than in a "bench top" demonstration.

Publication:
A description of the device, its operating principle, and its performance should have been published in a journal or in the proceedings of a conference or workshop that is in the public domain. Laboratory design notes, internal technical notes, etc., do not qualify but may be submitted to support other publications. Full and open disclosure is necessary to the extent that a potential user could design a similar device. More than one article may be submitted (together) to satisfy this requirement; for example, an article describing the principle plus another article describing the performance.

Eligibility:
Nominations are open to candidates of any nationality for work done at any geographical location. There are no restrictions for candidates; however, in the event of deciding between work of similar quality, preference will be given to candidates in an early stage of their beam instrumentation career. The award may be shared between persons contributing to the same accomplishment. Once accepted by the Award Committee a nomination shall remain eligible for three successive competitions unless withdrawn by a candidate.

Disclosure

The Award Committee may release the names of entrants and a list of publications related to an entry if requested by a third party. Unpublished supporting material will not be disclosed nor will the names of persons supporting a nomination. Discussion regarding individual entries, scoring, etc. is regarded as confidential and will not be disclosed.

Nominations

The nomination package shall include the name of the candidate, relevant publications, a statement outlining his/her personal contribution and that of others, letters from two professional accelerator physicists, engineers or laboratory administrative personnel who are familiar with the device and its development. Two master copies of this package, suitable for copying, must be submitted not later than October 1st, the year prior to the workshop, to:

Conference Chairman
Beam Instrumentation Workshop

PREVIOUS WINNERS

1992 A. V. Feschenko – Bunch Shape Monitors Using Secondary Electron Emission
1993 R. B. Fiorito and D. W. Rule – Optical Transition Radiation
1994 E. Rossa – Real Time Single Shot Three Dimensional Measurement of Picosecond Photon Bunches
1996 W. Barry and H.-C. Lihn – Measurement of Subpicosecond Electron Bunch Length and Profiles Using Coherent Transition Radiation
1998 A. Peters – Cryogenic Current Comparator for nA beams
2000 K. Wittenburg – The PIN-Diode Beam Loss Monitor System at HERA

Faraday Cup Award Invited Talk

10th Beam Instrumentation Workshop
May 6-9, 2002

The Magnetic Quadrupole Pick-Ups in the CERN PS

Andreas Jansson

*European Organization for Nuclear Research (CERN),
CH-1211 Geneva 23, Switzerland*

Abstract. The idea of using the non-linearities of beam position monitors to measure the second moment $\sigma_x^2 - \sigma_y^2$ of the transverse beam distribution is almost as old as the synchrotron. However, although a few successful experiments have been reported, the method has not become widely accepted. One reason for this has been that little or no effort was put into optimizing the pick-ups that were used for the new purpose. In a standard beam position pick-up, the signal from the second moment is extremely weak and embedded in a strong common-mode background. Separating the signal from the background has therefore been a major stumbling block. Driven by the need for a non-destructive measurement of injection matching to preserve the small emittance of the LHC beam, a dedicated quadrupole pick-up has been developed for the CERN PS. The design employs magnetic coupling in a special pick-up geometry to remove the otherwise dominating background signal, thereby reducing the common-mode rejection requirements by about 60 dB. Two pick-ups have been installed in the machine. When the data from these pick-ups is combined, it is possible to measure both the matching parameters and the emittance of each bunch in the injected beam. This paper gives an overview of the pick-up design, describes the methods used to analyze the data, and presents some measurement results, including comparisons with other instruments in the machine.

INTRODUCTION

In any accelerator or storage ring, position pick-ups are standard diagnostic devices. A typical pick-up of the electrostatic type is shown in Figure 1 and consists of four electrodes around the beam. The passing beam induces a signal on each of these electrodes,

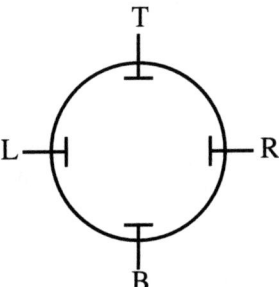

FIGURE 1. A typical position pick-up (electrostatic). The beam passes perpendicular to the plane of the drawing.

and the individual electrode signals can be combined in four (linearly independent) ways

$$\Sigma = T+R+B+L \tag{1}$$
$$\Delta_H = R - L \tag{2}$$
$$\Delta_V = T - B \tag{3}$$
$$\Xi = T-R+B-L \tag{4}$$

where to lowest order

$$\Sigma = Z_\Sigma I, \quad \Delta_H = Z_\Delta I \bar{x}, \quad \Delta_V = Z_\Delta I \bar{y}, \quad \Xi = Z_\Xi I \kappa. \tag{5}$$

Here, I is the beam current and $(\bar{x}, \bar{y})$ the beam position relative to the pick-up centre in the horizontal and vertical plane, respectively. The Z coefficients are the so-called transfer impedances, and depend on the pick-up geometry. From this, the beam position can be determined as

$$\bar{x} = \frac{Z_\Sigma}{Z_\Delta} \frac{\Delta_H}{\Sigma}, \quad \bar{y} = \frac{Z_\Sigma}{Z_\Delta} \frac{\Delta_V}{\Sigma}. \tag{6}$$

The fourth possible signal combination Ξ, called the quadrupole signal, is not used in a position pick-up. However, it can be shown that

$$\kappa = \iint (x^2 - y^2) \rho(x,y) \, dx \, dy = \sigma_x^2 - \sigma_y^2 + \bar{x}^2 - \bar{y}^2. \tag{7}$$

The so-called quadrupole moment κ (in some references it is given the somewhat incorrect name 'second moment') is of interest since it provides information on the beam size. A pick-up that measures κ is often referred to as a quadrupole pick-up.

The idea of a quadrupole pick-up was first introduced by Gol'din[1] in a 1966 theoretical paper written in Russian. His ideas were taken up by Nassibian at CERN, who generalized the theory to higher-order field components[2] and pointed out the basic limitations of the method. A prototype was even built for the CERN Booster, but was never used.

The first experimental use of quadrupole pick-ups was reported in 1983, when Miller et al[3] used six strip-line BPMs distributed along the SLC linac to measure the emittance of the passing beam. To calculate the emittance from the pick-up readings, they solved a matrix equation derived from the linac optics. More recently, it was realized that this type of equations is often numerically unstable[4], and that a stable implementation of the method may require multiple measurements with different optics[5].

In rings, the use of quadrupole pick-ups has largely focused on the frequency content of the raw signal. Beam width oscillations produce sidebands to the revolution frequency harmonics in the quadrupole signal, at a distance of twice the betatron frequency, and this can be used to detect higher order instabilities as well as injection mismatch. This was done at the CERN Anti-proton Accumulator[6], where the phase and amplitude of the detected signals were also used to find a proper correction to the injection mismatch, using an empirical response matrix[7]. Similar measurements were later performed at the Fermilab Anti-proton Accumulator[8] and in the Low Energy Anti-proton Ring at

CERN[9]. However, measurements based on frequency analysis are complicated by the fact that the interesting sidebands can also be produced by position oscillations, which demands a very good injection steering.

Although quite a few papers have been published over the years, very little work has been done to adapt and optimize the design of the pick-up itself for measuring beam size. The typical approach has been to use existing beam position pick-ups, and to extract the quadrupole signal using sophisticated electronics. This lack of detector development may be one of the reasons why quadrupole pick-ups have remained 'exotic' instruments.

WHY MAGNETIC COUPLING?

There are two main problems associated with quadrupole pick-ups. The first is that the quadrupole moment κ is not only dependent on the beam dimensions, but also on the beam position. This is a fundamental problem which cannot be solved by optimizing the pick-up design. However, the problem can be circumvented by measuring the beam position and subtracting its contribution from the quadrupole moment. This is possible with reasonable accuracy as long as the beam displacement from the centre is small compared to the beam dimensions.

The second and more serious problem is the range in signal levels involved. In a typical position pick-up, the Σ signal is very much stronger than the Ξ signal, and this difference grows rapidly with the aperture to beam-size ratio. Since in reality only a finite accuracy can be obtained when combining the four electrode signals, this sets a limit to the measurement of the quadrupole moment.

At first glance, it may seem that this second problem is also fundamental. However, it can be circumvented by using a magnetic detector. The magnetic field created by an infinitely long beam of constant cross-section is, when written in cylindrical (ρ, θ) coordinates[10],

$$\vec{B}(\rho,\theta) = -I\frac{\mu_0}{2\pi}\left[\frac{1}{\rho}\hat{\theta} + \bar{x}\left(\frac{\cos\theta}{\rho^2}\hat{\theta} - \frac{\sin\theta}{\rho^2}\hat{\rho}\right) + \bar{y}\left(\frac{\sin\theta}{\rho^2}\hat{\theta} + \frac{\cos\theta}{\rho^2}\hat{\rho}\right) + \right.$$
$$\left. + (\sigma_x^2 - \sigma_y^2 + \bar{x}^2 - \bar{y}^2)\left(\frac{\cos 2\theta}{\rho^3}\hat{\theta} - \frac{\sin 2\theta}{\rho^3}\hat{\rho}\right) + \ldots\right] \quad (8)$$

The first term, corresponding to the monopole mode, is responsible for the Σ signal. This field mode has a field component only in the $\hat{\theta}$ direction. Thus, by designing a detector that measures only the radial component, the dominating signal can be entirely suppressed. Such a coupling is achieved by shaping the antenna loops as a cylinder around the beam. To maximize the sensitivity to the quadrupole field mode, the four antenna loops should then be positioned where its radial component is strongest, i.e. at 45° angle to the horizontal plane (see Figure 2).

If the loops measuring the radial field component have an azimuthal opening angle of ϕ, the transfer impedances can be expected to scale as[10]

$$Z_\Delta \propto \left(1 - \frac{a^2}{d^2}\right)\frac{\sin\frac{\phi}{2}}{a^2}, \quad Z_\Xi \propto \left(1 - \frac{a^4}{d^4}\right)\frac{\sin\phi}{a^3}, \quad (9)$$

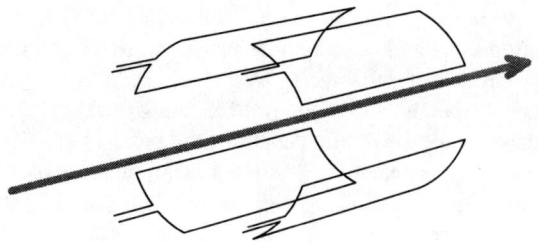

FIGURE 2. Arrangement of antenna loops to couple to the radial component of the magnetic field. The arrow symbolizes the beam.

where d is the radial position of the loops. Here, the effect of a conducting boundary at radial position a has also been taken into account.

PRACTICAL IMPLEMENTATION

To maximize the signal, the antenna loops should be placed as close as possible to the beam, and their preferred position is therefore given by the machine aperture.

When designing a magnetic quadrupole pick-up for the CERN PS, however, the antenna loops were placed a little further away, to make space for a ceramic vacuum tube inside the loops. The main reason was to avoid vacuum feed-throughs that could reduce the low-frequency response of the pick-up by introducing a parasitic series impedance in the antenna loops circuit. The ceramic was made circular for symmetry reasons, and was coated with a thin layer of titanium on the inside. The resistive layer acts as a shunt, significantly reducing the high frequency impedance of the pick-up, while being transparent in the passband of the pick-up.

A cylindrical cavity encloses the loops, providing continuity for the low frequency image currents. The space between ceramic and the cavity walls is divided longitudinally

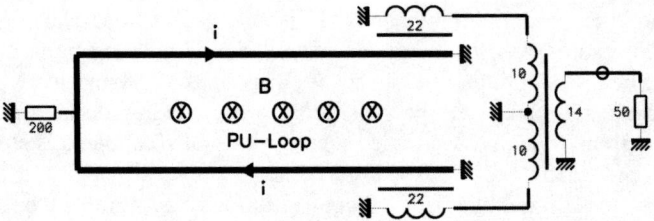

FIGURE 3. Schematic layout of one antenna loop, showing the transformer arrangement, the winding ratios and the position and value of the termination resistance.

in four sections by thin metal vanes. These shorten the path of the image currents, thereby further reducing the inductance seen by the beam. However, since they are placed in the symmetry planes of the quadrupole field, they do not affect the Ξ transfer impedance of the pick-up.

The four antenna loops were made of parallel, inter-connected rods, with an azimuthal opening angle ϕ of 45°. The current induced in each loop is read out with a series of transformers (see Figure 3). These also act as impedance transformers, to improve the low frequency response by reducing the load seen by the loop. The transformer end of the loop is connected to ground, to reduce capacitive coupling between windings. At the other end of the loop, a second ground point is connected via a resistor, with the purpose of damping out loop resonances.

The four antenna loop signals from the pick-up are combined in a passive 50 Ω hybrid. Due to the pick-up geometry, the signal combination is different from the case of an electrostatic pick-up

$$\Delta_H = TR - BR - BL + TL \tag{10}$$
$$\Delta_V = TR + BR - BL - TL \tag{11}$$
$$\Xi = TR - BR + BL - TL \tag{12}$$

where TR is the top right loop, BL the bottom left etc. The Σ signal is zero by design, and is not used.

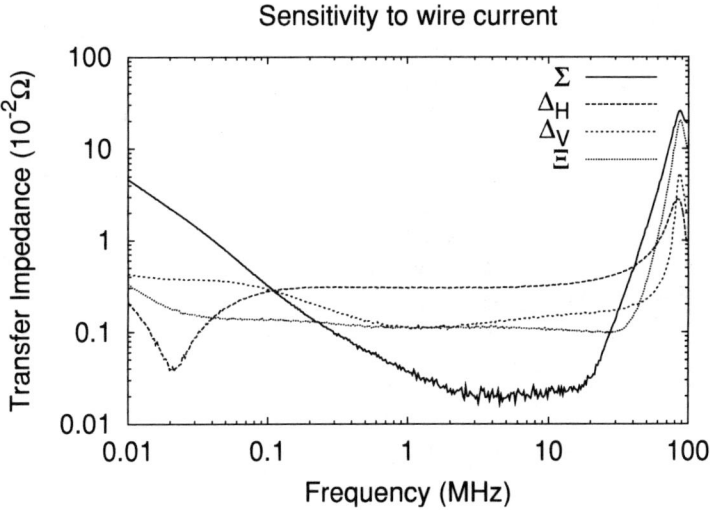

FIGURE 4. Measurement of the common-mode coupling (component independent of position) using a wire movable along the x axis. Ideally, all signals should be zero. The common-mode rejection of the Σ signal is very good up to about 20 MHz, where the tail of the loop resonance begins. The other signal levels are affected by a small (less than 0.5 mm) offset between the electrical and geometrical centre (this is within the error of the absolute wire positioning accuracy). The rise of the Σ signal at low frequencies is an effect of the measurement instrument, that also influences the other measurements slightly. Figure reprinted from [10] with permission from Elsevier Science.

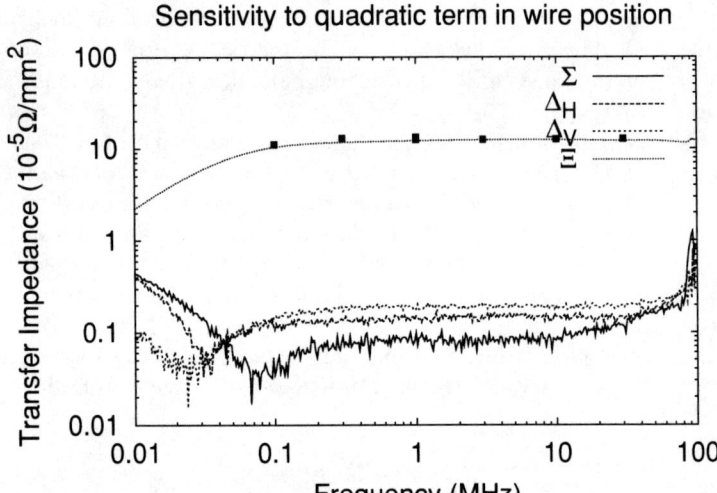

FIGURE 5. Measurement of the quadrupole mode coupling using a wire movable along the x axis. This is the coupling that is used to measure the quadrupole moment κ. The Ξ response is flat well above 20 MHz (the dots are simulated values for Ξ). Figure reprinted from [10] with permission from Elsevier Science.

The transfer impedances of the pick-up were both simulated and measured in the laboratory (see Figures 4 and 5) with good agreement. As expected, the measured sum transfer impedance is essentially zero in the passband.

Two pick-ups have been installed in consecutive straight sections of the machine. The optical parameters at their locations are given in Table 1. As shown later, it is crucial that the pick-ups are installed at locations with different ratios between horizontal and vertical beta value. The pick-ups were also installed as close as possible to each other to minimize the dependence of their relative betatron phase on the programmed machine tunes and the beam intensity (space charge detuning).

TABLE 1. Optical parameters at pick-up locations. The pick-ups are installed in consecutive straight sections of the PS machine.

Name	β_x	β_y	D_x	$\Delta\mu_x$	$\Delta\mu_y$
QPU 03	22.0 m	12.5 m	3.04 m	0.365	0.368
QPU 04	12.6 m	21.9 m	2.30 m		

SIGNAL ACQUISITION AND TREATMENT

After the hybrids, the composite signals from the two pick-ups are amplified and cabled to Digital Oscilloscopes in an adjacent building. At the maximum sampling rate of 500

Ms/s, 200 turns can be acquired. For normalization purposes, the signal from a wall current monitor is sampled simultaneously, as the pick-ups themselves do not give a measurement of the total beam current. All data treatment is then done on the digitized data. The analysis of the data is made in a LabView program. In order to resolve single bunches, the data is treated in the time domain, considering each bunch passage separately. This is in contrast to earlier measurements in rings, where frequency analysis was always used.

The first step in the analysis is to rid the signal of its intensity dependence, by normalizing to the measured beam current. The analysis is performed in two different ways, depending on whether the position and quadrupole moment are expected to be constant or varying along the bunch.

If there is no variation in position and size along the bunch, and one assumes that the quadrupole pick-up and the wall current monitor have the same frequency response, then the shape of a given pulse must be exactly the same in all signals (apart from a baseline offset and noise effects). The normalization problem then consists in determining the scaling factor between a pulse in the beam current signal and the corresponding pulse on the pick-up outputs.

To do this, time slices of about one RF period centred on the bunch are selected. Each selected slice is a vector of N samples and, under the above assumption, corresponding slices are proportional to each other. The quadrupole moment can therefore be found as the least squares solution to an overdetermined matrix equation, which in the case of the quadrupole signal has the form

$$\begin{pmatrix} \Sigma_1 & 1 \\ \Sigma_2 & 1 \\ \vdots & \vdots \\ \Sigma_N & 1 \end{pmatrix} \cdot \begin{pmatrix} \kappa \\ c \end{pmatrix} = \frac{Z_\Sigma}{Z_\kappa} \begin{pmatrix} \Xi_1 \\ \Xi_2 \\ \vdots \\ \Xi_N \end{pmatrix}. \tag{13}$$

The constant c depends on the base line difference and is not used. The same calculation is performed for the position signals, and the position contribution to the quadrupole moment is then subtracted.

An attractive feature of this method, apart from noise suppression, is that the base line is automatically, and unambiguously, corrected.

However, the assumption that beam size and position is does not vary along the bunch is not always correct. For example, sometimes, only parts of the bunch may oscillate due to a bad injection. Moreover, there is no fundamental reason why the beam size should be constant along the bunch, although this is true for a Gaussian bunch.

If the position varies along the bunch, one can not use the average bunch position to correct the quadrupole moment for its position dependence, since

$$<x^2> \neq <x>^2 \tag{14}$$

The correction must be done point-by-point along the bunch. For this purpose, a second normalization algorithm is used, which first establishes and subtracts the base line, and then calculates the position as well as the quadrupole moment in each point. After this correction, an average beam quadrupole moment can be calculated, but it is also possible to study variations of the beam size along the bunch.

BEAM-BASED CALIBRATION

Apart from calibration measurements in the lab, a number of tests can be performed using the beam, to verify that the pick-ups work as expected.

For example, one can take advantage of the position dependence of the quadrupole moment to make a consistency check between the position and quadrupole moment measurement of the pick-up, using data with large beam position oscillations but stable beam size. Such data can easily be obtained at injection by an appropriate trigger delay, since the beam size oscillations damp away much faster than beam position oscillations. A plot of expected versus measured variation of the quadrupole moment with beam position is shown in Figure 6, showing a good agreement. This test can easily be automated, and is a good indicator of whether the beam position correction works well.

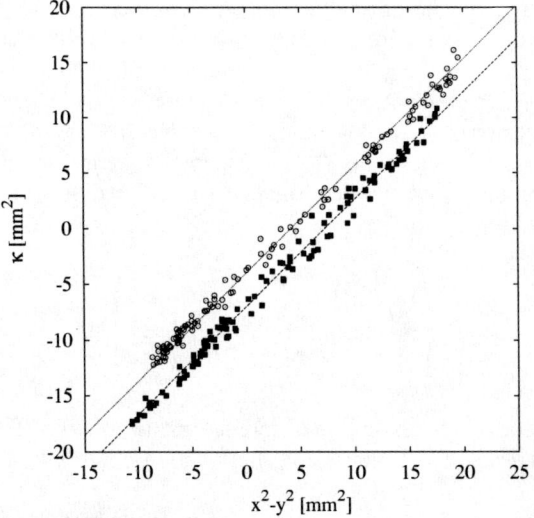

FIGURE 6. Quadrupole moment (uncorrected) versus expected beam position contribution. The squares and circles represent measurements made with the same pick-up on two different beams. The slope of the line is the same in both cases, and is very close to one (0.983). Figure from [11].

Another test is to compare the quadrupole moment measured by the beam with what is expected from the beam emittances measured by other instruments. The standard method for emittance measurement on a circulating beam in the PS is the fast wire-scanner. In order to test the calibration of the pick-ups, measurements were done on several different stable beams, approximately 15 ms after injection. The quadrupole pick-up signal was acquired over 200 machine turns, at the same time as the wire traversed the beam. The comparative measurement was performed on all the operational beams available in the machine, with the exception of the very high intensity beams that saturate the pick-up amplifiers. Thus there was a significant difference in both beam and machine parameters between the different measurements. This was done in an attempt to randomize any systematic errors. The beam parameters are given in Table 2, where the different beams

TABLE 2. Parameters of beams used for comparative measurements. Emittances and momentum spread are 2σ values.

Name	ε_x	ε_y	σ_p	I_{bunch}
SFTPRO	19 μm	12 μm	2.7×10^{-3}	2.7×10^{12}
AD	25 μm	9 μm	2.7×10^{-3}	3.3×10^{12}
LHC	3 μm	2.5 μm	2.2×10^{-3}	6.9×10^{11}
EASTA	8 μm	1.4 μm	2.5×10^{-3}	1.4×10^{11}
EASTB	7.5 μm	1.4 μm	1.6×10^{-3}	8.6×10^{10}
EASTC	12 μm	3 μm	2.4×10^{-3}	4.2×10^{11}

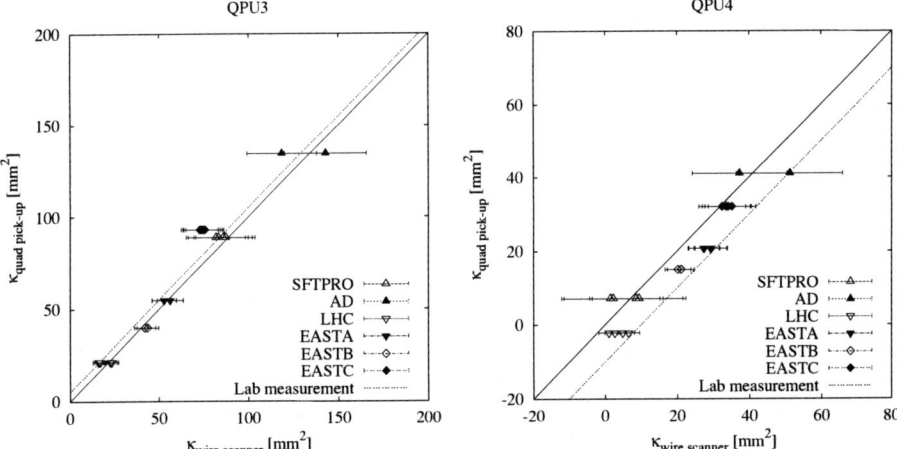

FIGURE 7. Comparison between the measured value from the two quadrupole pick-ups and the expected results calculated from the emittances measured with the wire-scanners. The solid line is the ideal case, and the dotted line includes pick-up offsets measured in the lab prior to installation. All possible ways of combining the wire-scanner measurements are displayed. Note that the cases where the two wire-scanner results are inconsistent are cases with large estimated systematic error. Figure from [11].

have been tagged with their operational names.

The r.m.s. variation in the measured quadrupole moments from turn to turn was of the order of 0.2-0.5 mm^2, depending on the beam intensity. Assuming that the beam size was perfectly stable, this gives an estimate of the single-turn resolution of the pick-up measurement. Also the wire-scanner measurements were stable, although for some beams there was a systematic disagreement between the two wire-scanners measuring in the same plane.

To compare the two instruments, the emittances measured with the wire-scanners were used to calculate the expected quadrupole moment at the locations of the pick-ups. The momentum spread required for both the wire scanner measurement and the subsequent calculation was obtained by a tomographic analysis of the bunch shape[12]. The propagated systematic error in the comparison was estimated on the assumption that the wire-scanner accuracy is 5% in emittance, the beta function at the pick-ups is known to 5%, the dispersion to 10% and the momentum spread to 3% accuracy. These estimates are rather optimistic, but give considerable propagated errors for certain measurement points. For simplicity, possible correlations between errors (e.g. beta function errors at different locations in the machine) were ignored, and all different error sources were added in quadrature. To accentuate the cases with wire-scanner disagreement, each of the four different ways of combining the two horizontal and two vertical wire-scanners was calculated separately and displayed as separate points. The result is shown in Figure 7.

Overall, the measured data seem to indicate that the offsets are slightly smaller than measured in the lab, which could be explained by the fact that the pick-ups were dismantled in the lab to be moved to the machine. However, the effect is within the error-bar, and no strong conclusion can therefore be made. Moreover, the pick-ups have been dismantled and rebuilt in the lab, without effect on the measured offsets.

The point corresponding to the EASTC beam appears to disagree somewhat in both planes, although the effect is just within the error-bar. There are a number of possible explanations for this that are currently under investigation. The general conclusion from the measurement series is that the wire-scanner and quadrupole pick-up agree within the measurement accuracy. The systematic errors due to optics parameters make it impossible to detect with certainty any difference in pick-up behaviour between the laboratory measurements with a simulated beam, and the measurements on a real beam in the machine. In order to calibrate the pick-ups more accurately using the beam, the wire-scanners and the pick-up should be situated in the same straight section, which is excluded in the PS due to space limitations.

Comparative measurements of injection matching have also been done using a secondary emission (SEM) grid with a fast acquisition system[13], that can measure beam profiles turn-by-turn for a single bunch. This is a destructive device and can only be used in rare dedicated machine development sessions. It is also limited both in bandwidth and

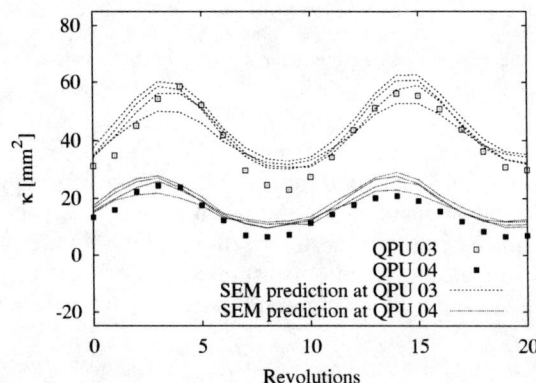

FIGURE 8. Beam size oscillations at injection measured with the quadrupole pick-ups and a turn-by-turn SEM grid. The SEM-grid beam size data were used to calculate the expected quadrupole moment at the pick-up locations. Beam position contributions and known pick-up offsets have been subtracted from the quadrupole moments. Figure from [11].

maximum beam intensity, and therefore it has not been possible to make a full systematic study on beams with different characteristics. Instead, a special beam was prepared, with low intensity to spare the grid, and long bunches due to the bandwidth limitations.

The SEM grid data was used to calculate the expected value of the quadrupole moment at the pick-up locations, using the beta values, dispersion, and relative phase advance in Table 1. The results are shown in Figure 8, and show a rather good agreement with what was actually measured with the pick-ups. The small differences can be accounted for by systematic error sources, i.e. the optical parameters used in the comparison.

EMITTANCE MEASUREMENT

When the circulating beam is stable, the quadrupole moments of a given bunch, as measured by the two pick-ups, are constant and given by

$$\begin{aligned} \kappa_1 &= \varepsilon_x \bar{\beta}_{x1} - \varepsilon_y \bar{\beta}_{y1} + \bar{D}_{x1}^2 \sigma_p^2 \\ \kappa_2 &= \varepsilon_x \bar{\beta}_{x2} - \varepsilon_y \bar{\beta}_{y2} + \bar{D}_{x2}^2 \sigma_p^2 \end{aligned} \quad (15)$$

When the momentum spread is known, the system of equations can be solved for the emittances, if the ratio between horizontal and vertical beta function is significantly different at the two locations. Thus, measuring the emittance of a stable circulating beam with quadrupole pick-ups is in fact rather straightforward. Such a measurement made in the PS is shown in Figure 9, and compares well with wire-scanner results.

Statistical errors due to random fluctuations in the measurement of κ can, although they are usually small, be reduced by averaging over many consecutive beam passages. The dominant errors are therefore systematic, coming from offsets in the pick-ups and errors in the beta functions, lattice dispersion and momentum spread. The pick-up offsets

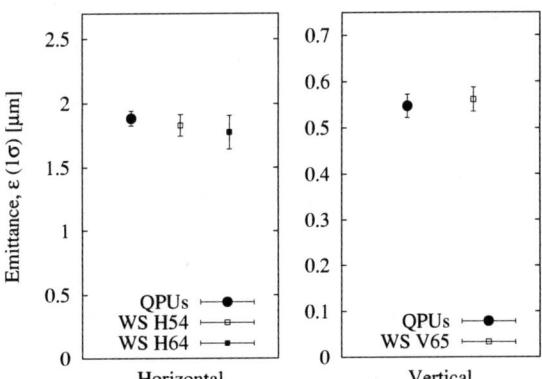

FIGURE 9. Filamented emittance of a proton beam measured with quadrupole pick-ups (QPU) and wire-scanner (WS). There is a good agreement. The error bar is the standard deviation for 10 measurements. Figure reprinted from [14]. ©2001 IEEE.

are, however, known from test bench measurements. Furthermore, by comparing the amplitude and phase of position oscillations as measured by the two pick-ups, the beta ratios and relative betatron phases can be determined.

The main uncertainty is thus the absolute value of the beta function, as for almost any other emittance measurement (e.g. wire-scanner). The accuracy can therefore be expected to be comparable to that of a wire-scanner.

Note that with three pick-ups, suitably located, the momentum spread could also be measured.

MATCHING MEASUREMENT

Even though quadrupole pick-ups can be used to measure filamented emittance, the main reason for installing such instruments in the machine is to be able to measure betatron and dispersion matching at injection, as no other instrument (apart from the destructive SEM-grid) is able to do this. One would like not only to detect mismatch, but also to quantify the injection error in order to be able to correct it.

In a ring, the turn by turn evolution of the beam envelope, and therefore the quadrupole moment, can be expressed in a rather simple analytical formula. If the beam is initially mismatched in terms of Twiss functions or dispersion, the value of κ will vary with the number of revolutions n performed as[15]

$$\begin{aligned}\kappa_n = &\ \bar{\beta}_x(\varepsilon_x + \Delta\varepsilon_x) - \bar{\beta}_y(\varepsilon_y + \Delta\varepsilon_y) + \bar{D}_x^2 \sigma_p^2 \\ &+ \bar{\beta}_x \varepsilon_x \delta_{\beta_x} \cos(2v_x n - \phi_{\beta_x}) + \bar{\beta}_x \sigma_p^2 \delta_{D_x}^2 \cos(2v_x n - 2\phi_{D_x}) \\ &- \bar{\beta}_y \varepsilon_y \delta_{\beta_y} \cos(2v_y n - \phi_{\beta_y}) - \bar{\beta}_y \sigma_p^2 \delta_{D_y}^2 \cos(2v_y n - 2\phi_{D_y}) \\ &+ \sqrt{\bar{\beta}_x \sigma_p^2 \bar{D}_x \delta_{D_x}} \cos(v_x n - \phi_{D_x}) \end{aligned} \quad (16)$$

assuming linear optics with no coupling between planes. Here, barred parameters refer to properties of the lattice, and $v_{x,y} = 2\pi q_{x,y}$.

The two middle lines of Eq. (16) are signal components at twice the horizontal and vertical betatron frequencies. They arise from both dispersion and betatron mismatch. The betatron mismatch is parametrized by the mismatch vector

$$\vec{\delta}_{\beta_x} = \begin{pmatrix} \frac{\beta_x}{\bar{\beta}_x} - \frac{\bar{\beta}_x \gamma_x + \bar{\gamma}_x \beta_x - 2\bar{\alpha}_x \alpha_x}{2} \\ \frac{\bar{\alpha}_x \beta_x - \alpha_x \bar{\beta}_x}{\bar{\beta}_x} \end{pmatrix} \approx \begin{pmatrix} \frac{\Delta\beta_x}{\bar{\beta}_x} \\ \bar{\alpha}_x \frac{\Delta\beta_x}{\bar{\beta}_x} - \Delta\alpha_x \end{pmatrix} \quad (17)$$

where, again, the last approximation is valid for small mismatch. The fourth line of Eq. (16) is a signal at the horizontal betatron frequency, which is due to horizontal dispersion mismatch. This is parametrized by the vector

$$\vec{\delta}_{D_x} = \begin{pmatrix} \frac{\Delta D_x}{\sqrt{\bar{\beta}_x}} \\ \sqrt{\bar{\beta}_x}\Delta D'_x + \bar{\alpha}_x \frac{\Delta D_x}{\sqrt{\bar{\beta}_x}} \end{pmatrix} \quad (18)$$

There is no corresponding signal at the vertical betatron frequency due to the absence of vertical lattice dispersion. Therefore, it is not possible to distinguish vertical dispersion mismatch from vertical betatron mismatch by studying the quadrupole signal. However, one does not usually expect a large vertical dispersion mismatch.

The first line contains constant terms, and also gives the steady state value that will be reached when the oscillating components have damped away. The steady state (filamented) emittance is given by

$$\varepsilon_x + \Delta\varepsilon_x = \varepsilon_x \frac{1}{2}\left(\bar{\beta}_x \gamma_x + \bar{\gamma}_x \beta_x - 2\bar{\alpha}_x \alpha_x\right) + \sigma_p^2 \frac{(\Delta D_x)^2 + (\bar{\beta}_x \Delta D_x' + \bar{\alpha}_x \Delta D_x)^2}{\bar{\beta}_x} \approx$$

$$\approx \varepsilon_x + \varepsilon_x \frac{|\vec{\delta}_{\beta_x}|^2}{2} + \sigma_p^2 \frac{|\vec{\delta}_{D_x}|^2}{2} \quad (19)$$

where the last approximation is valid for small betatron mismatch. The emittance increase from mis-steering is not included here, since beam position oscillations filament much slower than beam width oscillations and therefore do not affect the emittance over the first few turns.

By fitting the above function to the data, the injected emittances, the betatron mismatches in both planes, and the horizontal dispersion mismatch are directly obtained. The tunes can also be free parameters in the fit, which automatically estimates and corrects for space charge detuning. An example of a fit to measured data is shown in Figure 10. A requirement for a good fit convergence is, as when measuring filamented emittance, that the ratio between beta functions should be different at the pick-up locations. Also, the tunes must be such that enough independent data points are obtained. In the PS, this means that the working point $Q_h = Q_v = 6.25$, which is close to the bare

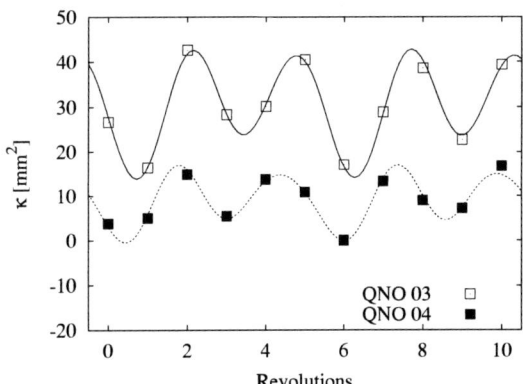

FIGURE 10. Theoretical expression for the quadrupole moment fitted to measured data. Here, seven turns (14 data points) were used to determine 10 free parameters (emittances, betatron and dispersion mismatches, and the tunes), but there is a relatively good match also for the subsequent turns. The measured detuning of the beam width oscillation frequencies were quite significant, $\Delta Q_h = 0.01$ and $\Delta Q_v = 0.05$ (as compared to the tunes measured from position oscillations). Figure from [11].

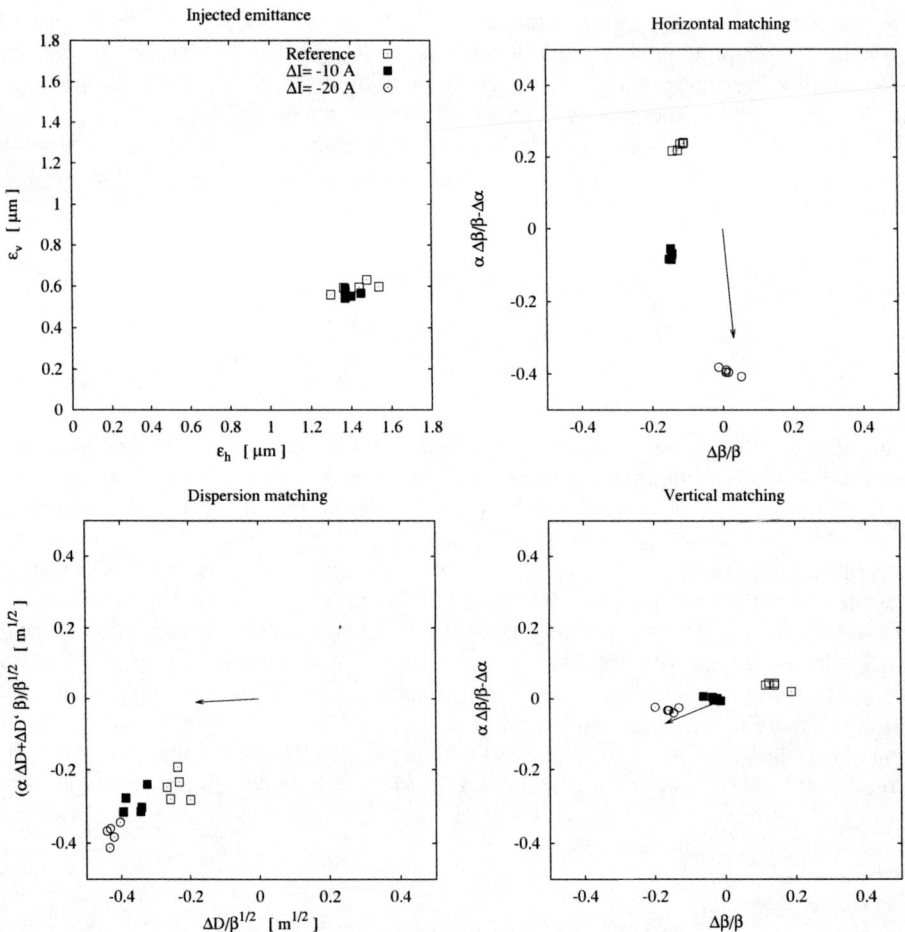

FIGURE 11. Injected emittance, betatron and dispersion mismatch vectors for three different settings of a transfer line quadrupole. Note the large dispersion mismatch. The vectors illustrate the variation in mismatch that is expected for a correction of -10A (calculated from beam optics theory). There is a good agreement between expected and measured behavior, indicating that the measurement works well. Figure from [11].

tune, should be avoided. With two pick-ups, at least five machine turns (10 data points) are required for the fit, if the tunes are also free parameters. Some more turns can be used to check the error, but the maximum number of turns is limited by decoherence.

Note that since the beam size oscillations due to dispersion mismatch are also detuned by space charge, measuring the dispersion component separately (by changing the energy of the beam and measuring the coherent response) would result in an accumulated phase error in the dispersion term.

To test the injection matching measurement, a series of measurements was done with different settings of some focusing elements of the PS injection line. An example of such a measurement is shown in Figure 11, where a quadrupole was changed in steps of 10 A, and the resulting variation of the fit parameters recorded. The variation of the different error vectors expected from beam optics theory is also shown, and there is a rather good agreement, both in direction and magnitude of the changes. The injected emittances are unchanged, as expected.

By using the theoretical response matrix for dispersion and betatron matching, a proper correction to the measured error can be calculated[16]. So far, actual corrections of the measured mismatches have not been made, since the dominant error (the dispersion mismatch) can not be corrected without a complete change of optics of the entire line. Studies for a new dispersion-matched optics are underway.

CONCLUSIONS

A new type of quadrupole pick-up has been developed for the CERN PS, and two such instruments are now installed in the machine. The design uses induction loops, coupling to the radial magnetic field component, to separate the small quadrupole signal from the strong common-mode (intensity) signal.

Comparison with other instruments in the machine show good agreement. All observed deviations are within the estimated systematic error bars. The systematic errors come mainly from the imperfect knowledge of beta value and dispersion needed to evaluate the data. Systematic errors are indeed expected to dominate the total error in the quadrupole pick-up measurement, as is the case for most emittance measurement devices.

For matching applications, the pick-ups can be used to determine phase and amplitude of horizontal and vertical betatron mismatch, as well as horizontal dispersion mismatch. This analysis can be done individually on each injected bunch. Since the mismatch is detected as an oscillation, the effect of systematic errors (e.g. pick-up offsets) is not very important.

As emittance measurement devices, the pick-ups have some interesting properties. The single turn resolution makes it possible to measure and follow the evolution of the emittance over many turns (limited only by acquisition memory). When measuring filamented emittance, it possible to reduce the effect of noise by averaging over many turns.

The pick-ups are non-destructive and have no moving parts that wear out, as is the case for a wire-scanner. The can therefore be used on every machine cycle. This makes it possible to create a watchdog application to monitor the evolution of the emittances over a long period, and detect any injection mismatch.

ACKNOWLEDGMENTS

I would like to thank David J. Williams for his enthusiastic support and experienced help, without which the pick-up would probably not have left the propsal stage. At Davids retirement, Lars Søby took over his role and made important contributions, including designing the pick-up amplifiers.

Jean-Mary Roux helped with the mechanical design and made the CAD drawings; Erk Jensen helped with HFFS simulations; Jeroen Belleman gave useful comments on the pick-up design; Uli Raich and Christian Dutriat made sure the wire-scanners and the turn-by-turn SEM-grid were functioning; Michael Benedikt, Christian Carli, Mats Lindroos and the operations team helped set up the beams and participated in some of the measurements. Many others colleagues not mentioned here have also contributed to the project in different ways.

Finally, I would like to thank the organizers of the Beam Instrumentation Workshop for selecting this work for the 2002 Faraday Cup Award.

REFERENCES

1. L. L. Gol'din, *Instruments and Experimental Techniques*, pp. 780–784 (1966).
2. G. Nassibian, The measurement of the multipole coefficients of a cylindrical charge distribution (1970), CERN internal note SI/Note EL/70-13.
3. R. H. Miller, et al., "Non-Invasive Emittance Monitor," in *Proc. 12th Int. Conf. on High Energy Accelerators*, Batavia, IL, 1983.
4. S. J. Russell, *Nucl. Instr. and Meth. in Phys. Res. A*, **430**, 498–506 (1999).
5. S. J. Russell, *Review of Scientific Instruments*, **70**, 1362 (1999).
6. G. Carron, et al., "Measurement of Coherent Quadrupole Oscillations at Injection into the Antiproton Accumulator," in *Proc. 15th Int. Conf. on High Energy Accelerators*, Chicago, IL, 1989.
7. V. Chohan, et al., "Measurement of Coherent Quadrupole Oscillations at Injection into the Antiproton Accumulator," in *Proc. 2nd European Particle Accelerator Conf.*, Nice, France, 1990.
8. F. M. Bieniosek, and K. Fullett, "Measurement and Reduction of Quadrupole Injection Oscillations in the Fermilab Antiproton Accumulator," in *Proc. 16th Particle Accelerator Conf.*, Dallas, TX, 1995, vol. 3, pp. 1942–1944.
9. M. Chanel, "Study of Beam Envelope Oscillations by Measuring the beam transfer function with a quadrupolar pick-up and kicker," in *Proc. 5th European Particle Accelerator Conf.*, Sitges, Spain, 1996, pp. 1015–1017.
10. A. Jansson, and D. J. Williams, *Nucl. Inst. and Meth. in Phys. Res. A*, **479**, 233–242 (2002).
11. A. Jansson, A non-invasive single bunch matching and emittance monitor (2002), physics/0202057, accepted for publication in Phys. Rev. ST-AB.
12. S. Hancock, et al., "Tomographic measurements of longitudinal phase space density," in *Proc. Conf. on Computational Physics*, Granada, Spain, 1998, publ. in: Computer Physics Communications, 118 (1999) 61-70.
13. M. Benedikt, et al., "Injection Matching Studies using Turn by Turn Beam Profile Measurement in the CERN PS," in *Proc. 5th European Workshop on Beam Diagnostics and Instrumentation for Particle Accelerators*, Grenoble, France, 2001.
14. A. Jansson, and L. Søby, "A non-invasive single bunch matching and emittance monitor for the CERN PS," in *Proc. 19th IEEE Particle Accelerator Conf.*, Chicago, IL, 2001.
15. A. Jansson, *Non-Invasive Measurement of Emittance and Optical Parameters for High-Brightness Hadron Beams in a Synchrotron*, Ph.D. thesis, Stockholm University (2001).
16. M. Giovannozzi, A. Jansson, and M. Martini, "Simultaneous Matching of Dispersion and Twiss Parameters in a Transfer Line," in *Proc. Workshop on Automatic Beam Steering and Shaping*, Geneva, Switzerland, 1998, CERN Yellow Report 99-07.

Tutorials

10th Beam Instrumentation Workshop
May 6-9, 2002

The Physics and Properties of Free-Electron Lasers

Samuel Krinsky

Brookhaven National Laboratory, Upton, NY 11973

Abstract. We present an introduction to the operating principles of free-electron lasers, discussing the amplification process, and the requirements on the electron beam necessary to achieve desired performance.

INTRODUCTION

In the storage rings of synchrotron radiation facilities, the electrons are radiating incoherently [1,2]. Since there is no multi-particle coherence, the radiated intensity is linearly proportional to the number of electrons N_e. If the electrons in a beam are spatially bunched on the scale of the radiation wavelength [3,4], coherent radiation with intensity proportional to N_e^2 will be emitted. Since N_e can be very large, coherent emission offers the potential of greatly enhancing the intensity. The technical challenge is to produce the required bunching on the scale of the radiation wavelength. A task that increases in difficulty as the wavelength is decreased.

To obtain coherent emission at short wavelengths, one must develop methods to bunch the electron beam utilizing the radiation. One approach that has already been successfully applied down to the vacuum ultraviolet is the free electron laser (FEL). The FEL [5] is based on a resonant interaction between an electromagnetic wave and an electron beam traveling along the axis of an undulator magnet. The periodic undulator magnetic field produces a transverse component of the electron velocity that couples the energy of the electron to that of the wave. Under general conditions this coupling will merely result in a shifting of energy back and forth between the electron beam and the radiation. However, in resonance, there can be sustained energy transfer from the electrons to the wave. FELs are reviewed in refs. [6-8].

In designing an FEL, one must decide on the type of electron accelerator to be used: e.g. storage ring [9], room temperature linac [10], or superconducting linac [11]. Storage rings provide very high stability and continuous operation; however, the FEL action perturbs the electron beam, thus limiting performance. The development of photocathode RF electron guns [12,13] has made linacs attractive as drivers for FELs. They can produce high peak current and small normalized emittance. The microbunch pulse length in photo-injectors is typically on the order of 10 ps, and bunch compression can be used to reduce the pulse length down to the vicinity of 100 fs. The macropulse structure in room temperature linacs consists of pulse trains separated

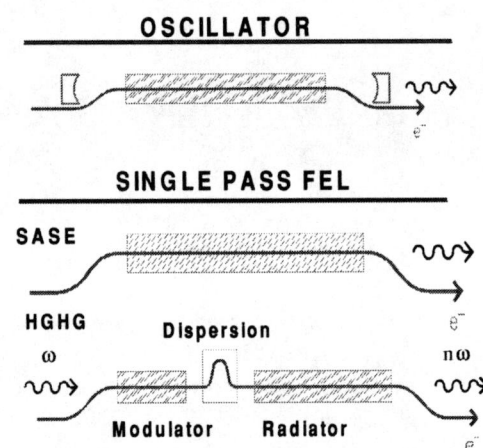

FIGURE 1. FEL configurations: oscillator; self-amplified spontaneous-emission; high-gain harmonic generation.

by dead time. Superconducting linacs can provide continuous-wave beams and very high stability.

A fundamental consideration in FEL design is whether to use a high-Q optical cavity, or to operate the FEL as a high-gain single-pass amplifier (see Fig. 1). An optical cavity has many advantages: it requires less gain per pass, simplifying the undulator, and it facilitates the production of narrow bandwidth output radiation. However, it is difficult to utilize optical cavities at short wavelengths because one requires high quality mirrors resistant to radiation damage. For this reason, present effort in the design of short wavelength FELs, from the VUV down to hard x.-rays, is predominantly focused on using single pass FEL amplifiers employing long undulators [10].

UNDULATOR RADIATION

Let us begin our discussion by considering an electron traversing an undulator magnet [1,2,14]. For the purposes of illustrating the basic principles, it is convenient for us to consider the undulator to be helical, resulting in a constant longitudinal velocity (along the undulator axis, z-direction)

$$v^* = c\left(1 - \frac{1+K^2}{2\gamma^2}\right). \tag{1}$$

The transverse velocity is

$$\vec{v}_T = -\frac{K}{\gamma}c\left(\hat{x}\cos k_w z + \hat{y}\sin k_w z\right). \quad (2)$$

Here, γmc^2 is the electron energy and K is the magnetic strength parameter,

$$K \cong 0.93 B_w(T)\lambda_w(cm), \quad (3)$$

where B_w is the amplitude of the helical undulator field and $\lambda_w = 2\pi/k_w$ is the undulator period length.

Consider a wave front radiated in the forward direction. After a time interval λ_w/v^*, the electron has passed through one additional undulator period, and a second wavefront emitted at this time follows the first by a time interval

$$T_s = \frac{\lambda_w}{v^*} - \frac{\lambda_w}{c}, \quad (4)$$

as illustrated in Fig. 2. In the forward direction the radiation spectrum is peaked at wavelength $\lambda_s = cT_s$, i.e.

$$\lambda_s = \frac{\lambda_w}{2\gamma^2}(1+K^2). \quad (5)$$

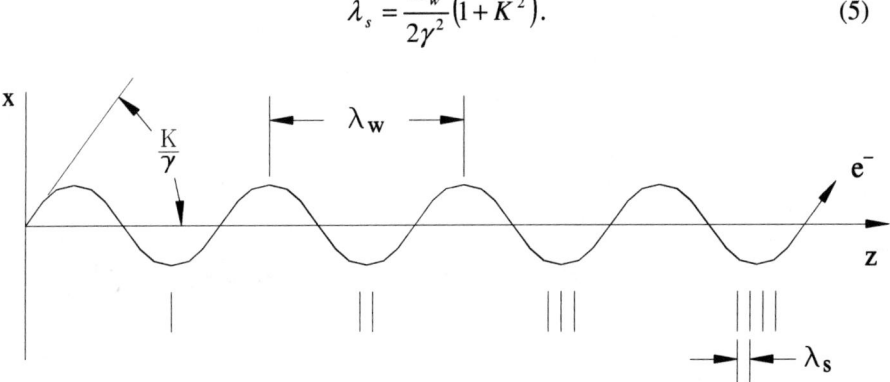

FIGURE 2. As the electron traverses successive undulator periods, additional radiation wavefronts are emitted. Since the electron is traveling almost at the speed of light, the electron slips only slightly behind the wavefronts emitted previously. Therefore, the radiation wavelength is much shorter than the undulator period.

For radiation propagating at a polar angle ϑ relative to the z-axis, the spatial separation of wavefronts emitted before and after an electron has traversed one period of the undulator is

$$\lambda_s(\vartheta) = \frac{\lambda_w}{2\gamma^2}\left(1 + K^2 + \gamma^2\vartheta^2\right). \tag{6}$$

In an undulator with N_w periods, the radiated pulse from one electron has a time duration of $N_w T_s$. Consequently, the linewidth at fixed observation angle is

$$\frac{\Delta\lambda}{\lambda_s} \cong \frac{1}{N_w}. \tag{7}$$

From Eq. (6), we see that the spectral broadening due to accepting radiation in a cone of half-angle $\Delta\vartheta$ about the forward direction is $\Delta\lambda/\lambda_s = \gamma^2(\Delta\vartheta)^2/(1+K^2)$. This broadening will be small only if $\Delta\vartheta < \vartheta_w$, where we define

$$\vartheta_w \equiv \sqrt{\frac{1+K^2}{2N_w\gamma^2}} = \sqrt{\frac{\lambda_s}{L_w}}, \tag{8}$$

with $L_w = N_w \lambda_w$. Angle ϑ_w characterizes the central cone of the undulator radiation. The power per unit solid angle per unit frequency emitted in the forward direction at $\omega = \omega_s = 2\pi c/\lambda_s$ by an electron beam of energy γ and current I_e traversing N_w periods is given by [1] (mks units, Alfven current $I_A = 4\pi\varepsilon_o mc^3/e = 17,000 Amp$)

$$\left[\frac{dP}{d\omega d\Omega}\right]^{SPONT}_{N_w} = mc^2 N_w^2 \frac{\lambda_w}{\lambda_s} J, \qquad J = \frac{K^2}{1+K^2}\frac{I_e}{I_A} \tag{9}$$

FREE-ELECTRON LASER: LOW-GAIN REGIME

Let us now consider an electron passing through an undulator with longitudinal and transverse velocities as specified in Eqs. (1) and (2). Suppose the electron to be interacting with a co-propagating electromagnetic wave with electric field $(\omega_s = ck_s)$

$$\vec{E} = E_o\left[\hat{x}\cos(k_s z - \omega_s t + \phi) - \hat{y}\sin(k_s z - \omega_s t + \phi)\right]. \tag{10}$$

The energy transfer between the electron and wave is described by [6-8]

$$\frac{d\gamma}{dz} = \frac{\vec{e}}{\gamma mc^3}\vec{v}\circ\vec{E} = -\frac{eKE_o}{mc^2\gamma}\cos(\zeta(z)+\phi), \tag{11}$$

where the ponderomotive phase is

$$\zeta(z) \equiv (k_s + k_w)z - \omega_s t(z). \tag{12}$$

Here, t(z) is the time of arrival of the electron at position z along the undulator axis. Differentiating Eq. (12), we find

$$\frac{d\zeta}{dz} = k_s + k_w - \frac{\omega_s}{v^*} \tag{13}$$

The condition for resonance is

$$\frac{d\zeta}{dz} = 0 \rightarrow v^* = \frac{\omega_s}{k_s + k_w} \rightarrow (c - v^*)\frac{\lambda_w}{v^*} = \lambda_s. \tag{14}$$

The resonance condition requires that while traveling through one period of the undulator, an electron falls one radiation wavelength behind the wave, as illustrated in Fig. 3. In resonance, the phase relation that exists between the electron and wave in one period is repeated in subsequent periods making possible a sustained energy transfer.

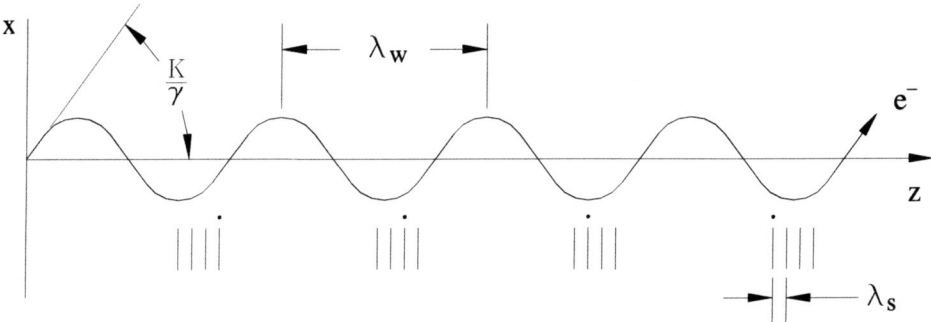

FIGURE 3. Resonance condition: as the electron and wave propagate along the axis of the undulator, the electron slips one radiation wavelength behind the wave for each undulator period traversed.

The resonant energy γ_r is determined by

$$\frac{1+K^2}{2\gamma_r^2}\lambda_w = \lambda_s. \tag{15}$$

When $\left|\frac{\gamma - \gamma_r}{\gamma_r}\right| \ll 1$, the equations describing the electron motion become

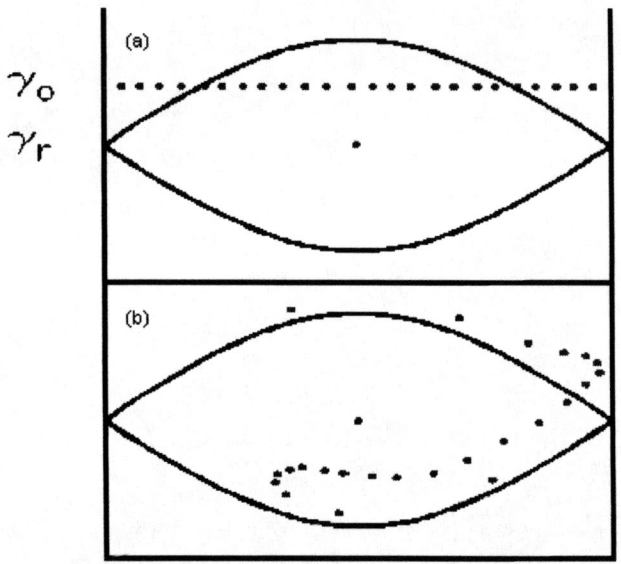

FIGURE 4. We show the longitudinal ζ, γ-phase space for an electron beam (a) initially monoenergetic and uniformly distributed in phase, and (b) after experiencing a small net energy loss to a co-propogating electromagnetic wave.

$$\frac{d\zeta}{dz} = 2k_w \frac{\gamma - \gamma_r}{\gamma_r} \tag{16}$$

$$\frac{d\gamma}{dz} = -\frac{eKE_o}{mc^2 \gamma_r} \cos(\zeta(z) + \phi), \tag{17}$$

If the undulator parameters K and k_w are constant, and we ignore the variation of E_o and ϕ, then the ponderomotive phase is determined by the pendulum equation

$$\frac{d^2\zeta}{dz^2} + \Omega_p^2 \cos(\zeta + \phi) = 0, \tag{18}$$

with

$$\Omega_p^2 = \frac{2ek_w KE_0}{mc^2 \gamma_r^2}. \tag{19}$$

This is an appropriate description in the low-gain regime, when amplification of the wave in a single pass through the undulator is small. For an electron beam which is initially monoenergetic and uniformly distributed in phase, an illustration of net energy loss by the electron beam to the radiation is presented in Fig. 4.

When the gain per pass is small, a large total gain can be achieved by placing the undulator between the mirrors of an optical resonator. If the electron beam is comprised of a long train of bunches, with neighboring bunches space by twice the length of the optical cavity, the radiation can be repeatedly amplified as it interacts with successive bunches.

FREE-ELECTRON LASER: HIGH-GAIN REGIME

It is difficult to use optical cavities at short wavelengths because of the need for high-quality mirrors resistant to radiation damage. Therefore, to generate short wavelength radiation, high-gain single pass amplifiers employing long undulators are of interest. The mathematical description of high-gain amplifiers [6-8, 15-17] must

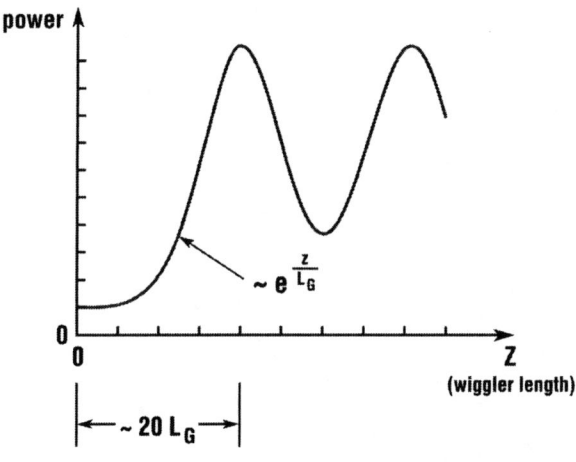

FIGURE 5. The radiated power increases exponentially until the instability saturates. In the case of self-amplified spontaneous-emission, the amplifier starts-up from shot noise, and it takes about 20 power gain-lengths to reach saturation.

take into account the variation of the radiation field. High gain results from a collective instability leading to exponential growth of the radiated power,

$$P \propto \exp(z/L_G), \qquad (20)$$

where L_G is the e-folding or gain length (see Fig. 5).

The mechanism leading to exponential growth is as follows: Electrons gain or lose energy depending upon their phase ζ with respect to the electromagnetic wave. The resulting energy modulation of the electron beam gives rise to a spatial bunching due to the dispersion in the undulator (i.e. v^* larger for higher-energy electrons). The density modulation at the radiation wavelength then produces enhanced coherent emission, amplifying the radiation intensity. The positive feedback loop is closed since the increase in radiation intensity enhances the energy modulation of the electron beam.

Suppose the electron beam entering the undulator is initially monoenergetic with energy γ_o, and uniformly distributed in phase, and there exists a small coherent laser seed. An illustration [7] of the evolution of the electron beam's longitudinal phase space distribution and the corresponding increase of the radiation field amplitude is presented in Fig. 6.

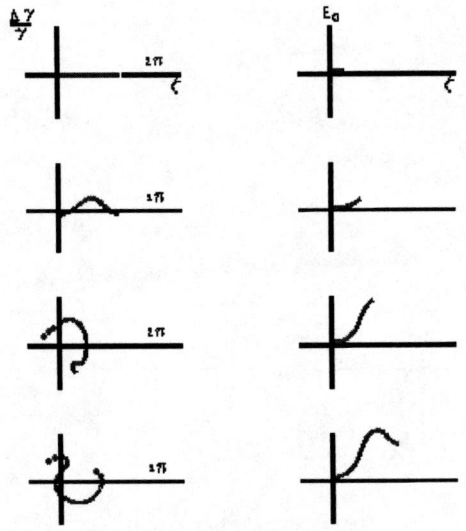

FIGURE 6. Evolution of the longitudinal phase space of the electron beam, initially monoenergetic and uniformly distributed in phase, and of the amplitude of the radiation field which starts at a small non-vanishing value.

As the radiation intensity increases, the "pendulum frequency," Ω_p [Eq. (19)], becomes larger. The exponential process saturates when the frequency of rotation in the phase space bucket becomes comparable to the growth rate:

$$\Omega_p \cong \frac{1}{L_G}. \qquad (21)$$

Within the one-dimensional approximation, in which dependence on transverse coordinates is ignored, the growth rate can be written in the form,

$$\frac{1}{L_G} = \sqrt{3}\, 2\rho\, k_w, \qquad (22)$$

where the dimensionless Pierce parameter ρ is given by [17]

$$(2\rho)^3 = \frac{\lambda_s \lambda_w}{\pi A} \frac{J}{\gamma_o}. \qquad (23)$$

The scaled current J was defined in Eq. (9) and A=cross-sectional area of electron beam. Using Eqs. (19), (22) and (23) in Eq. (21), one finds that the saturated radiation power is approximately given by

$$P_{sat} \cong \rho\, E_e(GeV)\, I_e(Amp), \qquad (24)$$

where $E_e(GeV)$ is the electron energy in GeV and $I_e(Amp)$ is the electron current in Amperes. We see that the Pierce parameter determines the gain length via Eq. (22) and the efficiency with which the exponential gain process can extract energy from the electron beam and give it to the radiation via Eq. (24).

One-Dimensional Theory

We shall present a brief introduction to the one-dimensional theory (neglecting dependence on transverse coordinates) of the FEL in the linear regime before saturation [15-17]. Consider the electron beam at the undulator entrance to be monoenergetic, $\gamma = \gamma_o$, and uniformly distributed in phase, with n_1 electrons per unit length. Define,

$$\tau = k_w z, \qquad \zeta = (k_s + k_w)z - \omega_s t, \qquad (25)$$

where $k_s = \omega_s/c = 2\gamma_0^2 k_w/(1+K^2)$. Introduce the line density $\Lambda(\zeta,\tau) \equiv$ the number of electrons per unit length, and the energy deviation $p(\zeta,\tau) \equiv 2(\gamma(\zeta,\tau)-\gamma_o)/\gamma_o$, and write the radiated electric field in the form

$$\vec{E} = \text{Re}\left[(\hat{x}+i\hat{y})E\, e^{i(k_s z - \omega_s t)}\right]. \tag{26}$$

The slowly varying complex amplitude $E \equiv E_o e^{i\phi}$ [see Eq. (10)].

Let us suppose the line density, energy deviation and the electric field amplitude have small sinusoidal perturbations:

$$\Lambda(\zeta,\tau) \cong n_1 + \Lambda_q(\tau)e^{i(1+q)\zeta}, \tag{27}$$

$$p(\zeta,\tau) \cong p_q(\zeta,\tau)e^{i(1+q)\zeta}, \tag{28}$$

$$E(\zeta,\tau) \cong E_q(\tau)e^{iq\zeta}, \tag{29}$$

where $q = (\omega - \omega_s)/\omega_s$ is the frequency detuning. We keep only terms linear in the perturbations. The variation in the electron energy deviation due to the radiated electric field is described by Eq. (11). Neglecting the non-resonant E_q^* term, one finds

$$\frac{\partial p_q}{\partial \tau} = -\frac{2d_2}{\gamma_o^2} E_q. \tag{30}$$

The change in the density modulation resulting from the energy modulation of the electron beam is given by the equation of continuity $[\partial \Lambda/\partial \tau + \partial(p\Lambda)/\partial \zeta = 0]$. Linearizing, one derives

$$\frac{\partial \Lambda_q}{\partial \tau} = -i(1+q)n_1 p_q. \tag{31}$$

The increment in the electric field driven by the density modulation is determined by the one-dimensional paraxial wave equation

$$\left(\frac{\partial}{\partial \tau} + iq\right)E_q = \frac{d_1}{\gamma_o}\Lambda_q, \tag{32}$$

where (mks units)

$$d_1 = \frac{eK}{2k_w A\varepsilon_o}, \quad d_2 = \frac{eK}{2k_w mc^2}, \tag{33}$$

and A is the cross-sectional area of the electron beam. Eqs. (30), (31) and (32) imply

$$\frac{\partial^2}{\partial \tau^2}\left(\frac{\partial}{\partial \tau}+iq\right)E_q = i(1+q)(2\rho)^3 E_q, \qquad (34)$$

with $(2\rho)^3 = 2d_1 d_2 n_1 / \gamma_o^3$, the Pierce parameter introduced in Eq. (23).

The solution to Eq. (34) has the form

$$E_q(\tau) = a_1 e^{s_1 \tau} + a_2 e^{s_2 \tau} + a_3 e^{s_3 \tau} \qquad (35)$$

where the coefficients a_1, a_2, a_3 are determined from the initial conditions $E_q(0), \Lambda_q(0),$ and $p_q(0) = 0$. The Laplace transform parameters s_1, s_2, s_3 are the solutions of the cubic dispersion relation

$$s^2(s + iq) = i(2\rho)^3. \qquad (36)$$

A useful approximation is

$$s \cong 2\rho\left[\mu - \frac{i}{3}\left(\frac{q}{2\rho}\right) - \frac{1}{9\mu}\left(\frac{q}{2\rho}\right)^2\right]. \qquad (37)$$

There are three modes: growing; decaying and oscillating; corresponding to $\mu = \frac{\sqrt{3}}{2} + \frac{i}{2}, -\frac{\sqrt{3}}{2} + \frac{i}{2}, -i$. For the exponentially growing mode,

$$\operatorname{Re} s_1 \cong (2\rho)\frac{\sqrt{3}}{2}\left[1 - \frac{1}{9}\left(\frac{q}{2\rho}\right)^2\right]. \qquad (38)$$

It can now be shown that the evolution of the electric field is determined by [18-19,23]

$$E_q(\tau) = H_q^{(2)}(\tau)E_q(0) + H_q^{(1)}(\tau)\frac{d_1}{\gamma_o}\Lambda_q(0), \qquad (39)$$

where the transfer functions are given by (m=1,2)

$$H_q^{(m)}(\tau) = \frac{s_1^m e^{s_1 \tau}}{(s_1 - s_2)(s_1 - s_3)} + \frac{s_2^m e^{s_2 \tau}}{(s_2 - s_3)(s_2 - s_1)} + \frac{s_3^m e^{s_3 \tau}}{(s_3 - s_1)(s_3 - s_2)}. \qquad (40)$$

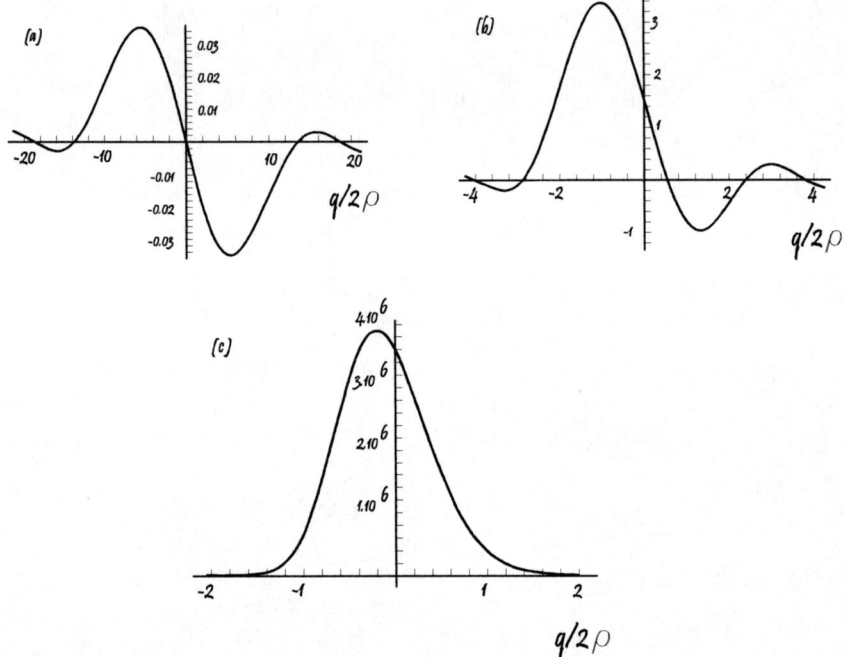

FIGURE 7. Plot of $\left|H_q^{(2)}(k_w z)\right|^2 - 1$ versus $q/2\rho$, for: (a) $2\rho k_w z = 0.5$; (b) $2\rho k_w z = 1.0$; (c) $2\rho k_w z = 10$. For low gain (a), the interference of the three modes is important. In the high-gain regime (c), the growing mode dominates.

In the exponential-gain regime the growing mode dominates and

$$\left|H_q^{(m)}(\tau)\right|^2 \cong \frac{1}{9}(2\rho)^{2m-4} e^{-q^2 \omega_s^2 / 2\sigma_\omega^2} e^{z/L_G}. \tag{41}$$

The power gain length, L_G, has the value given in Eq. (22), and the gain bandwidth is

$$\sigma_\omega = \omega_s \sqrt{\frac{3\sqrt{3}\,\rho}{\tau}}. \tag{42}$$

Self-Amplified Spontaneous-Emission (SASE)

In the absence of an external seed laser, $E_q(0) = 0$, so the FEL amplifier starts up from the shot noise in the electron beam [18-23],

$$\Lambda_q(0) = \frac{k_s}{2\pi} \sum_j e^{i(1+q)\omega_s t_j(0)}. \qquad (43)$$

Here, the sum is over the electrons comprising the beam and $t_j(0)$ is the arrival time of the jth electron at the undulator entrance. It follows from Eq. (39), that

$$E_q(\tau) = H_q^{(1)}(\tau) \frac{d_1 k_s}{\gamma_o} \frac{1}{2\pi} \sum_j e^{i(1+q)\omega_s t_j(0)}. \qquad (44)$$

We can treat the arrival times $t_j(0)$ as independent random variables. Therefore, at a fixed position, $\tau = k_w z$, along the undulator, $E_q(\tau)$ and its Fourier transform $E(\tau,\zeta)$ are sums of independent random terms. It follows from the Central Limit Theorem [24] that the probability distribution describing the spectral intensity $I \propto |E_q|^2$, or the time-domain intensity $I \propto |E|^2$, is the negative exponential distribution

$$p_I(I) = \frac{1}{\langle I \rangle} e^{-I/\langle I \rangle}. \qquad (45)$$

The intensity fluctuation is 100%.

The output intensity as a function of time exhibits spiking [25] (see Fig. 8), and the width of the intensity peaks are characterized by the coherence time [24,26] $T_{coh} = \sqrt{\pi}/\sigma_\omega$. The spectral intensity also exhibits spikes (Fig. 9), and their widths are inversely proportional to the electron bunch duration T_b.

Let us consider the energy in a single SASE pulse.

$$W(\tau) \propto \int_0^{\omega_s T_b} |E(\zeta,\tau)|^2 \, d\zeta \qquad (46)$$

For fixed τ, the pulse can be divided up into M statistically independent time-intervals of width T_{coh}. The fluctuation within a single coherent region is 100%, but the fluctuation σ_W of the energy in the entire pulse is reduced and given by [24,26]

$$\sigma_W^2 = \frac{\langle (W - \langle W \rangle)^2 \rangle}{\langle W \rangle^2} = \frac{1}{M} = \frac{T_{coh}}{T_b}. \qquad (47)$$

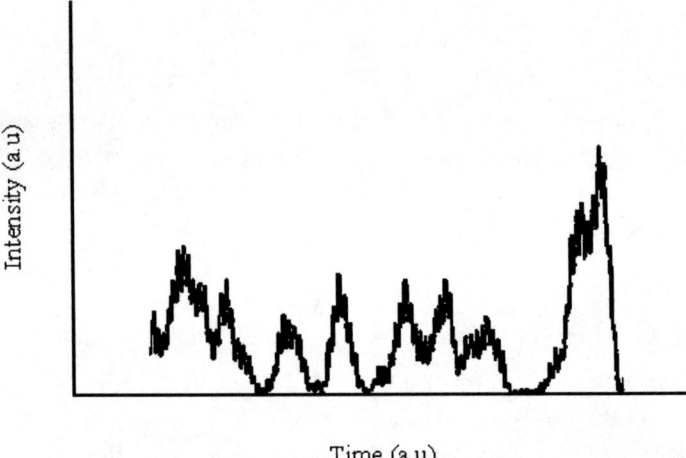

FIGURE 8. Intensity spiking in the time-domain. The width of the peaks is characterized by the SASE coherence time $T_{coh} = \sqrt{\pi}/\sigma_\omega$.

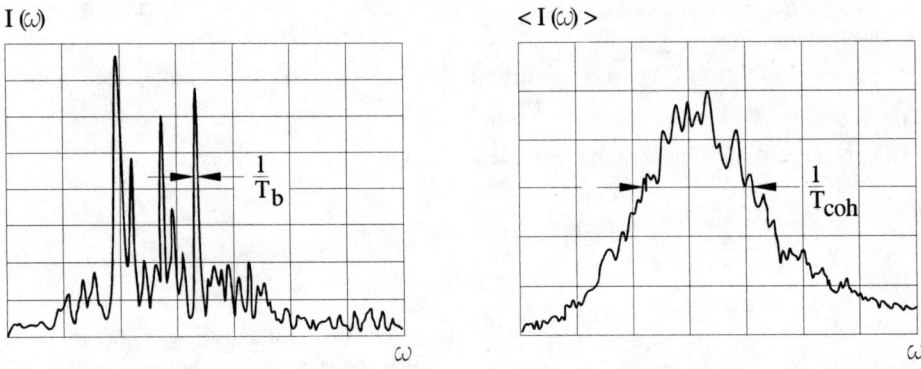

FIGURE 9. Intensity spiking in the frequency-domain. In the single-shot spectrum shown on the left, the width of the peaks is inversely proportional to the electron bunch duration T_b. The average of many SASE pulses is illustrated on the right, and in this case the width is proportional to the gain bandwidth $\sigma_\omega = \sqrt{\pi}/T_{coh}$.

The energy per pulse is described by the gamma distribution [24],

$$p_W(W) = \frac{M^M}{\Gamma(M)} \left(\frac{W}{\langle W \rangle}\right)^{M-1} \frac{1}{\langle W \rangle} \exp\left(-M \frac{W}{\langle W \rangle}\right). \quad (48)$$

It follows from Eqs. (41) and (44) that the average output SASE power spectrum can be expressed in the form [20]

$$\frac{dP}{d\omega} = \left[\left(\frac{dP}{d\omega}\right)_{Input} + \left(\frac{dP}{d\omega}\right)_{Noise}^{1-D}\right] \frac{1}{9} e^{-(\omega-\omega_s)^2/2\sigma_\omega^2} e^{z/L_G}, \quad (49)$$

where

$$\left(\frac{dP}{d\omega}\right)_{Noise}^{1-D} = \frac{\rho \gamma_o mc^2}{2\pi}, \quad (50)$$

arises from the shot noise in the electron beam and $(dP/d\omega)_{Input}$ represents an input laser seed. In the case of SASE (no seed), the average output power is [18,19]

$$P = \frac{1}{9} \frac{P_{sat}}{\sqrt{2N_{coh}}} e^{z/L_G}. \quad (51)$$

P_{sat} is the saturation power defined in Eq. (24) and $N_{coh} = n_1 c T_{coh}$ is the number of electrons in a coherence length. Typically, N_{coh} is large, and it takes about 20 SASE gain lengths to reach saturation (Fig. 5).

The noise (within the one-dimensional approximation) amplified in SASE is the spontaneous undulator radiation with frequency $\omega = \omega_s$ emitted in the first two power gain lengths into a cone of solid angle λ_s^2/A about the forward direction (A being the cross-sectional area of the electron beam). This follows from observing that [22]

$$\left(\frac{dP}{d\omega}\right)_{Noise}^{1-D} = \frac{3}{4} \frac{\lambda_s^2}{A} \left(\frac{dP}{d\omega d\Omega}\right)_{N_w=2L_G/\lambda_w}^{SPONT}, \quad (52)$$

which can be verified from Eqs. (9), (22) and (23).

Constraints On Electron Beam Quality

To achieve high gain, the resonant condition must hold for most electrons. This imposes tolerances on the energy spread and emittance. On should note that, as illustrated in Fig. 2, since the electrons are traveling almost at the speed of light, the radiation emitted by an electron moves ahead of it by only one radiation wavelength for each undulator period traversed. Therefore, an electron can influence only those electrons less than one slippage distance, $N_W \lambda_s$ in front of it. Tolerances to assure desired gain restrict the properties of the electron beam within a slice shorter than one slippage distance. To assure the entire electron beam contributes, tolerances must be imposed on the entire beam.

Recall from Eq. (15), that the resonance condition is

$$\frac{1+K^2}{2\gamma_r^2} \lambda_w = \lambda_s. \tag{53}$$

Suppose an electron has energy $\gamma = \gamma_r + \delta\gamma$. In this case, while traversing one undulator period, the electron slips a distance $\lambda_s + \delta\ell$ behind the wave:

$$\frac{1+K^2}{2\gamma^2} \lambda_w = \lambda_s + \delta\ell, \tag{54}$$

where

$$\delta\ell \cong -2\lambda_s \frac{\delta\gamma}{\gamma_r}. \tag{55}$$

Let us assume that in order to maintain coherent energy transfer over a section of undulator of length $2L_G$, it is necessary that

$$|\delta\ell| < \frac{\lambda_s}{4N_G}, \tag{56}$$

where $N_G = 2L_G / \lambda_w$ is the number of undulator periods in two power gain lengths. The significance of the tolerance imposed in Eq. (53) is that no electron will fall more than 90° out of phase with the wave while traversing two power gain lengths. From Eqs. (55) and (56), we see that the tolerance on the energy spread is

$$\left|\frac{\delta\gamma}{\gamma_r}\right| < \frac{1}{8N_G}. \tag{57}$$

Now consider an electron that has the resonant energy γ_r but which is traveling at a small angle ϑ relative to the z-direction. Since such an electron must travel a longer distance to traverse one undulator period, it slips a distance $\lambda_s + \delta\ell$ behind the wave:

$$\left(\frac{1+K^2}{2\gamma_r^2} + \frac{\vartheta^2}{2}\right)\lambda_w \cong \lambda_s + \delta\ell, \qquad (58)$$

where

$$\delta\ell = \frac{\vartheta^2}{2}\lambda_w. \qquad (59)$$

To assure coherent transfer of energy from the electron beam to the wave over two power gain lengths, we again impose the constraint of Eq. (53), resulting in the tolerance

$$\vartheta < \sqrt{\frac{\lambda_s}{4L_G}}. \qquad (60)$$

In the special case when the horizontal and vertical emittances are equal, $\varepsilon_x = \varepsilon_y = \varepsilon$, and the horizontal and vertical betatron functions are equal, $\beta_x = \beta_y = \beta$, we can write $\vartheta^2 = 2\varepsilon/\beta$ so the inequality (57) can be interpreted as a tolerance on the emittance [27]:

$$\varepsilon < \frac{\lambda_s}{4}\frac{\beta}{2L_G}. \qquad (61)$$

Diffraction of radiation out of the electron beam can result in a reduction of gain. The loss of gain [28,21] should be small as long as the Rayleigh range corresponding to the transverse dimensions of the electron beam is long compared to the gain length. For a cylindrical electron beam of radius $R = \sqrt{\beta\varepsilon}$, we can write this condition as

$$k_s \beta\varepsilon > 2L_G. \qquad (62)$$

In order to satisfy both Eqs. (61) and (62), we should have $\beta > 2L_G$. Since the gain is larger for higher electron density, the optimum will be near $\beta \cong 2L_G$.

In the linear regime before saturation, the coupled Vlasov-Maxwell equations have been used to derive a dispersion relation incorporating the effects of the energy spread, emittance, and focusing of the electron beam and the diffraction and guiding of the radiation. The gain length was expressed in the scaled form[1] [29]:

$$\frac{1}{2L_G} = \frac{2\pi}{\lambda_w} D\, G\left(\frac{4\pi\varepsilon}{\lambda_s}, \frac{\sigma_\gamma}{D}, \frac{\lambda_w}{D\lambda_\beta}, \frac{\omega-\omega_s}{\omega_s D}\right). \tag{63}$$

with

$$D = 4\left(\frac{J}{\gamma_o}\right)^{1/2}. \tag{64}$$

Here, the scaled current J was defined in Eq. (9), σ_γ is the fractional energy spread and $\lambda_\beta = 2\pi\beta$ the betatron wavelength. The scaling function G can be calculated very accurately [29] and the results agree with computer simulations to within 5-10%. This work was later extended in refs. [30-32]. The analytic determination of the gain length makes possible rapid computation, and facilitates FEL design optimization.

Let us now briefly discuss the dependence of the FEL parameters on output wavelength λ_s. As we reduce λ_s the required transverse emittance ε as estimated in Eq. (61) decreases proportionally. Thus, it follows that for a given normalized emittance, $\varepsilon_n \equiv \gamma\varepsilon$, the required energy increases. As the energy is increased, the current must also be increased to prevent the scaling parameter D [Eq. (64)] from becoming too small. Once the energy and wavelength are determined, the undulator period λ_w and field strength parameter K must be chosen to satisfy the resonance condition of Eq. (5), as well as certain practical and economic constraints. In particular, as one decreases the period, one must also decrease the magnet gap to prevent the field strength parameter from becoming too small. Therefore, the vertical aperture required for the electron beam, and perhaps electron beam diagnostics, limits

---

[1] When the electron beam size is large enough and the angular spread and energy spread are small, the gain length given in Eq. (63) approaches the result of 1-D theory, Eq. (22). In this regime,

$$G\left(\frac{4\pi\varepsilon}{\lambda_s}, 0, \frac{\lambda_w}{D\lambda_\beta}, 0\right) \cong \kappa \left(\frac{\lambda_s}{4\pi\varepsilon}\frac{\lambda_w}{D\lambda_\beta}\right)^{1/3},$$

with

$$\kappa = \frac{\sqrt{3}}{2}\left(\frac{\pi\beta\varepsilon}{2A}\right)^{1/3}.$$

The area of the electron beam $A = n_o/n_1$, where n_o is the peak value (at R=0) of the number of electrons per unit volume and n_1 is the number of electrons per unit length. For the parabolic transverse distribution, considered in ref. [29], $A = 3\pi\beta\varepsilon$.

how small the period can be made. Also, the magnet period must not be made too long, or else the size and cost of the system will become unnecessarily large.

In this paper, we have confined our attention to the fundamental frequency. However, near saturation, significant intensity is produced at the low harmonics of the fundamental. We refer the reader to the literature to pursue this interesting subject [33-36].

Some Recent Experiments on FEL Amplifiers

In Table 1, we present some of the fundamental parameters for four proof-of-principle SASE experiments which have been carried out over the last few years in the visible and vacuum ultraviolet: TTF1 at DESY [37]; LEUTL at ANL [38]; VISA at BNL [39]; and DUV-FEL at BNL [40]. The parameters correspond to reported experiments and do not necessarily represent the best or shortest wavelength performance achieved to-date. In the last column, we give the parameters for the design of the LCLS at SLAC [10]. Saturation of the SASE process has been observed at 95 nm at TTF1, at 130 nm at LEUTL and 800 nm at VISA. Agreement obtained between the experimental results and theory provides a firm foundation for the development of future x-ray facilities based on FEL amplifiers. The key challenge is to produce and transport the required high-brightness electron beams.

Table. 1. Parameters for FEL Projects

	TTF1	LEUTL	VISA	DUV-FEL	LCLS
λ_s (nm)	95	530	800	400	0.15
E_e (MeV)	250	217	72.6	140	14.3
λ_w (mm)	27	33	18	39	30
K	1.2	3.1	1.26	1.1	3.7
$\gamma\varepsilon$ (μm)	6	8.5	2	6	1.2
β_{av} (m)	1.2	1.4	0.27	3.2	7
I_e (Amp)	1300	266	200	500	3400
L_G (m)	0.67	0.57	0.17	0.68	4.7

The SASE FEL can produce high-intensity radiation with good transverse coherence but limited temporal coherence. Consideration has been given to seeding the FEL amplifier to produce improved temporal coherence. If a low power laser exists at the wavelength of interest, then the FEL can be used to amplify the signal to high power. In the cases of greatest interest, no such seed is available. However, one can use a seed at a longer wavelength and carry out harmonic generation in the FEL amplifier to generate temporally coherent short wavelength output. A proof-of-principle experiment of such a high-gain harmonic-generation (HGHG) FEL was successfully carried out in the infrared [41]. The DUV-FEL is designed to continue investigation of HGHG in the visible and vacuum ultraviolet. Another approach to seeding a short wavelength FEL consists of installing a monochromator after an initial section of undulator, and then amplifying the output in a second undulator [42]. A variant of this approach is the regenerative amplifier [43], in which the undulator is placed in a low-Q optical cavity whose reflectors consist of a mirror and a grating. The radiation from earlier bunches provide the monochromatized seed for the later bunches.

In SASE, the output pulse duration is determined by the density profile of the electron bunch. There is a strong desire to produce radiation pulses of femtosecond duration, which is generally shorter than the electron bunch. One possibility is to provide a short seed in an HGHG FEL. Another approach is to put an energy chirp on the electron bunch - producing a frequency chirp of the output radiation. A monochromator can then be used to select a short slice of the pulse [44]. Improvement of the output of an FEL amplifier is currently an active area of research, and many other schemes are currently under investigation.

ACKNOWLEDGMENTS

I wish to thank Dr. Li-Hua Yu for many illuminating discussions on the physics of free-electron lasers. This work was supported by the Department of Energy, Office of Basic Energy Sciences, under contract No. DE-AC02-98CH10886.

REFERENCES

1. Krinsky, S., Perlman, M.L., and Watson, R.E., "Characteristics of Synchrotron Radiation and of its Sources" in *Handbook on Synchrotron Radiation*, edited by E.E. Koch, Amsterdam: North-Holland, 1983, pp. 65-171.
2. Kim, K.J., "Characteristics of Synchrotron Radiation" in *Proceedings U.S. Particle Accelerator School*-1987, edited by M. Month et al, AIP Conference Proceedings 184, New York: American Institute of Physics, 1989, pp. 565-632.
3. Nakazato, T. et al, *Phys. Rev. Lett.* **63**, 1245 (1989).
4. Blum, E.B., Happek, U., and Sievers, A.J., *Nucl. Instrum. Methods* **A307**, 568 (1991).
5. Madey, J.M.J., *J. Appl. Phys.* **42**, 1906 (1971).
6. Colson, W.B., in *Laser Handbook*, vol. 6, edited by Colson, W.B. et al, Amsterdam: North-Holland, 1990.
7. Murphy, J.B. and Pellegrini, C., in *Laser Handbook*, vol. 6, edited by Colson, W.B. et al, Amsterdam: North-Holland, 1990.
8. Saldin, E.L., Schneidmiller, E.A., and Yurkov, M.V., *The Physics of Free Electron Lasers*, Berlin: Springer, 2000.
9. Litvinenko, V.N., Park, S.H., Pinayev, I.V., and Wu, Y., *Nucl. Instrum. Methods* **A475**, 195 (2001).
10. Galayda, J., ed., *Linac Coherent Light Source Conceptual Design Report*, SLAC-R-593 (2002).
11. Neil, G.R. et al, *Phys. Rev. Lett.* **84**, 662 (2000).
12. Fraser, J. and Sheffield, R., *Nucl. Instrum. Methods* **A250**, 71 (1986).
13. Wang X., Qiu, X., and Ben-Zvi, I., Phys. Bev. **E54**, R3121 (1996).
14. Krinsky, S., *IEEE Trans. Nucl. Sci.* **NS-30**, 3078 (1983).
15. Kroll, N.M. and McMullin, *Phys. Rev .***A17**, 300 (1978).
16. Debenev, Y.S., Kondratenko, A.M. and Saldin, E.L., *Nucl. Instrum. Methods* **A193**, 415 (1982).
17. Bonifacio, R., Pellegrini, C., and Narducci, L.M., *Opt. Commun.* **50**, 373 (1984).
18. Kim, K.J., *Nucl. Instrum. Methods* **A250**, 396 (1986).
19. Wang, J.M. and Yu, L.H., *Nucl. Instrum. Methods* **A250**, 484 (1986).
20. Kim, K.J., *Phys. Rev. Lett.* **57**, 1871 (1986)

21. Krinsky, S. and Yu, L.H., *Phys. Rev.* **A35**, 3406 (1987).
22. Yu, L.H. and Krinsky, S., *Nucl. Instrum. Methods* **A285**, 119 (1989).
23. Krinsky, S., *Phys. Rev.* **E59**, 1171 (1999).
24. Saldin, E.L., Schneidmiller, E.L. and Yurkov, M.V., *Nucl. Instrum. Methods* **A407**, 291 (1998)
25. Bonifacio, R., De Salvo, L., Pierini. P., Piovella, N., and Pellegrini, C., *Phys. Rev. Lett.* **73**, 70 (1994).
26. Yu, L.H. and Krinsky, S., *Nucl. Instrum. Methods* **A407**, 261 (1998).
27. Yu, L.H. and Krinsky, S., *Nucl. Instrum. Methods* **A272**, 436 (1988).
28. Moore, G.T., *Nucl. Instrum. Methods* **A239**, 19 (1985).
29. Yu, L.H., Krinsky, S., and Gluckstern, R.L., *Phys. Rev. Lett.* **64**, 3011 (1990).
30. Chin, Y.H., Kim, K.J. and Xie, M., *Phys. Rev.* **A46**, 6662 (1992).
31. Xie, M., *Nucl. Instrum. Methods* **A445**, 59 (2000).
32. Huang, Z. and Kim, K.J., *Nucl. Instrum. Methods* **A475**, 59 (2001)
33. Bonifacio, R., De Salvo, L., and Pierini, P., *Nucl. Instrum. Methods* **A293**, 627 (1990).
34. Schmidt, M.J. and Elliott, C.J., *Phys. Rev.* **A34**, 4843 (1986); **A41**, 3853 (1990).
35. Freund, H.P., Biedron, S.G., and Milton, S.V., *IEEE J. Quantum Electron.* **QE-36**, 275 (2000); *Nucl. Instrum. Methods* **A445**, 53 (2000).
36. Huang, Z. and Kim, K.J., *Nucl. Insstrum. Methods* **A475**, 112 (2001)..
37. Ayvazyan et al, *Phys. Rev. Lett.* **88**, 104802 (2002).
38. Milton, S.V. et al, *Science* **292**, 2037 (2001).
39. A. Tremaine et al, "Characterization of an 800 nm SASE FEL at Saturation," Proc. FEL2001.
40. A. Doyuran et al, Proc. EPAC 2002.
41. Yu, L.H. et al, *Science* **289**, 932 (2000).
42. Saldin, E.L., Schneidmiller, E.A. and Yurkov, M.V., *Nucl. Instrum. Methods* **A445**, 178 (2000).
43. B. Faatz et al, *Nucl. Instrum. Methods* **A429**, 424 (1999).
44. Schroeder, C.,B. et al, "Chirped Beam Two-Stage SASE-FEL for High-Power Femtosecond Pulse Generation," *Proc. FEL 2001*.

A Tutorial on Beam Loss Monitoring

Robert E. Shafer

*TechSource, Inc.
Santa Fe, NM*

Abstract. The beam loss monitoring system is one of the two most widely distributed beam diagnostic systems at most particle accelerator facilities. This tutorial reviews the characteristics of the ionizing radiation from beam losses, and the properties of beam loss radiation detectors.

INTRODUCTION

The beam loss monitoring system is one of the two most widely distributed beam diagnostic systems at most particle accelerator facilities. In addition to being a beam-tuning device, beam loss monitors (BLMs) are the front-line devices for protecting the beam line components from damage due to beam loss. In addition, the BLMs monitor losses that lead to long-term activation and radiation damage, as well as provide alarms when the radiation from beam losses may lead to excessive radiation levels outside the radiation enclosures.

The Effects of Ionizing Radiation

The effects of ionizing radiation can be categorized in the following table.

Material damage	overheating, thermal stress, radiation damage.
Cryogenic systems	excessive heat load, magnet quenching.
Optics	darkening (optical transmission).
Solid-state electronics	single event upset, long-term damage (dislocations).
Activation	personnel hazard (exposure).
Prompt radiation	backgrounds in experiments.
	personnel hazard (neutrons).

Sources of Ionizing Radiation

Ionizing radiation can come from both beam and non-beam sources:

Beam halo	Residual gas scattering
Residual gas stripping (H- beams)	Magnetic stripping (H- beams)
Focus and steering errors	Intercepting beam diagnostics
Foreign objects in the beam	Synchrotron radiation
X-rays from rf cavities	

Types of Ionizing Radiation

Types of ionizing radiation from high-energy particle beams include protons, electrons, pions, muons, gammas (including x-rays), and neutrons. Lost protons, if they are over a few GeV, will produce secondaries via hadronic showers, which includes pions, neutrons, and muons. Pi-zeros produce high-energy gammas. Lost electrons produce electromagnetic showers via bremsstrahlung. Gamma rays produced in electromagnetic showers convert back to electrons via Compton scattering and pair production.

DETECTION OF IONIZING RADIATION

Eventually, the primary mechanism by which a beam loss monitor detects beam loss is by ionization or by fluorescence. Both ionization and fluorescence represent a transfer of energy from the incoming charged particle to the *atomic electrons*. Interactions between the incident particle and *nuclei* are far less likely, and usually transfer momentum, rather than energy, resulting in multiple Coulomb scattering and beam divergence growth.

Energy loss of incident charged particles scattering on atomic electrons is described by the Bethe-Bloch equation, found in most textbooks on nuclear and particle physics.

$$dE/dx = -\frac{4\pi e^4 N_A}{\beta^2 mc^2} \frac{Z}{A} \left[\ln\left(\frac{2\beta^2 mc^2}{I}\right) - \ln(1-\beta^2) - \beta^2 \right] \text{ eV per gram/cm}^2 \quad (1)$$

where mc^2 is the electron rest mass, and β refers to the incident charged particle velocity, with charge $z=1$. A complete discussion of this equation can be found in the relevant textbooks. A plot of *dE/dx* for protons in aluminum and lead are shown in Figure 1. A complete set of *dE/dx* and range tables for protons in most elements can be found at the NIST website[1].

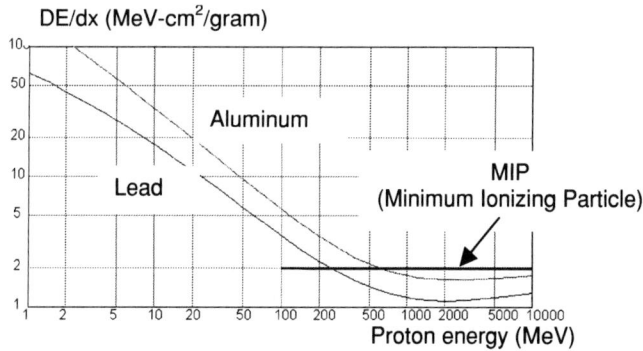

Figure 1. Plot of energy loss dE/dx vs. energy of incident proton.

The minimum in *dE/dx*, generally between 1 and 5 GeV for protons, is between 1 and 2 MeV-cm^2/gram. It is convenient to define a mythical Minimum Ionizing

Particle (*MIP*) as a particle that has an energy loss of 2 MeV-cm²/gram (shown as a line in Figure 1).

Using the definition of a rad of radiation dose as 100 ergs per gram leads to another definition, in terms of *MIP*s.

$$1\, rad = \frac{100\, ergs}{gram} \cdot \frac{MeV}{1.6 \cdot 10^{-6}\, ergs} \cdot \frac{MIP \cdot gram}{2\, MeV \cdot cm^2} = 3.1 \cdot 10^7\, MIPs\, per\, cm^2 \qquad (2)$$

So now we can describe the rad response of a beam loss monitor in terms of either energy deposition (100 ergs/gram), or in terms of a charged particle flux ($3.1 \cdot 10^7$ *MIP*s/cm²).

Radiation Detection Methods

The most common method for detecting ionizing radiation is to observe the interaction of charged particles with the atomic electrons in the detector, by measuring the ionization charge (ionization chambers), the fluorescence (phototube-scintillator combinations), or the secondary emission current (SEM chambers). Other methods of detecting the ionizing radiation include measuring Cerenkov light (from relativistic charged particles) or Compton electrons (from high energy gammas). Other detection methods (e.g., Smith Purcell radiation) have not been found to be useful.

Considerations in selecting a beam loss monitor

There are many factors that must be considered in selecting a beam loss monitor design. Some relate to the type of radiation, some relate to the expected dose rates and peak pulsed doses, and some relate to other factors such as reliability, physical space, calibration issues, cost, etc. Another consideration is whether to use an integrating type, whose output is measured in Coulombs per rad, or a pulse type detector, whose output is measured in counts per rad. A few factors are listed below.

- Detector output signal (current-integrating or pulse-type outputs)
- Sensitivity (Coulombs or pulses per rad)
- Detector dynamic range (rads per sec and instantaneous rad doses)
- Saturation characteristics for high radiation doses
- Sensitivity to backgrounds (e.g., RF cavity x-rays, synchrotron radiation)
- Sensitivity to magnetic fields
- Sensitivity to high voltage drift (e.g., photomultipliers)
- Uniformity of calibration (unit to unit)
- RAMI analysis (reliability, availability, maintainability, inspectability)
- On-line system testability
- Periodic calibration requirements
- Radiation hardness of materials used in construction
- Bandwidth (time resolution)
- Robustness (suitability for use in an accelerator enclosure environment)
- Physical size
- Cost

BEAM LOSS MONITORS USING IONIZATION DETECTION

When a charged particle passes through a gas, the gas is ionized, producing ion electron pairs. The amount of energy loss in creating an ion-electron pair is called the ionization constant. The following list shows the ionization constant for some common gases[2].

Gas	Ionization constant	Gas	Ionization Constant
Air	34 eV	hydrogen	36 eV
Helium	41	nitrogen	35
Neon	36	argon	26
Krypton	23	xenon	21

We can use these numbers to estimate the ionization yield per MIP in a cm of argon gas at STP:

$$N = \frac{1\,ion\,pair}{26\,eV} \frac{2 \cdot 10^6\,eV \cdot cm^2}{MIP\,gram} \frac{40\,grams}{22,414\,cm^3} = \frac{140\,ion\,pairs}{cm}\,per\,MIP \quad (3)$$

We can also estimate the number of Coulombs per rad in argon:

$$N = \frac{140\,ion\,pairs}{MIP\,cm} \cdot \frac{3.1 \cdot 10^7\,MIPs}{cm^2\,rad} \cdot \frac{1.6 \cdot 10^{-19}\,C}{ion\,pair} = 700\,pC/cm^3\,per\,rad \quad (4)$$

We can also make the same estimate more directly from the definition of a rad:

$$1\,rad = \frac{100\,ergs}{gram} \frac{1\,eV}{1.6 \cdot 10^{-12}\,ergs} \frac{1\,ion\,pair}{26\,eV} \frac{1.6 \cdot 10^{-19}\,C}{ion\,pair} \frac{40\,grams}{22,414\,cm^3} = 700\,pC/cm^3 \quad (5)$$

We can also calculate the cross section for creating an ion pair in argon, to compare to nuclear interaction rates:

$$\sigma = \frac{1\,pair}{26\,eV} \frac{2 \cdot 10^6\,eV\,cm^2}{gram} \frac{40\,grams}{6 \cdot 10^{23}\,atoms} = 5 \cdot 10^{-18}\,cm^2\,per\,atom \quad (6)$$

This is roughly 6 orders of magnitude larger than typical nuclear cross sections.

Because we will also discuss solid-state "ionization chambers" (silicon PIN diodes), the number of electron-hole pairs per cm in silicon per MIP is

$$N = \frac{1\,pair}{3.6\,eV} \frac{2 \cdot 10^6\,eV\,cm^2}{MIP\,gram} \frac{2.3\,grams}{cm^3} = 1.4 \cdot 10^6 \frac{electron-hole\,pairs}{cm}\,per\,MIP \quad (7)$$

So the charge production in solid-state ion chambers is much larger than in gas ion chambers.

Finally, we calculate the response of a 100 cm², 21-foil secondary-electron-emission monitor (SEM) to MIPs:

$$1\,rad = \frac{3.1 \cdot 10^7\,MIPs}{cm^2} \cdot 100\,cm^2 \cdot \frac{0.01\,electrons}{surface} \cdot 20\,surfaces \cdot \frac{1.6 \cdot 10^{-19}\,C}{electron} = 100\,pC \quad (8)$$

So a SEM detector is a very inefficient beam loss monitor.

Gas Ionization Chambers

We first review the properties of ionization chambers in general. At very low applied voltages, the collection of ion-electron pairs is inefficient, because of recombination before the charges reach the electrodes. As the voltage is increased, the collection efficiency usually reaches 100%, unless the density of ions and electrons is too large or the recombination rate is too high. As the voltage is raised further in cylindrical chambers with the electrons collected on the inner conductor (the preferred polarity) gas multiplication begins. There are two mechanisms for multiplication. The first is gas fluorescence near the anode producing uv light which in turn produces photoelectrons on the cathode. The second is ionization of the gas near the anode producing more ion-electron pairs. This is referred to as the proportional mode. In this mode, the multiplication is very dependent on the applied voltage, unlike the ionization chamber that has multiplication of 1.

Finally, as the voltage is raised further, the gas actually breaks down, discharging the voltage across the chamber. This is called the Geiger mode. In this case, the amplitude of the pulse is independent of the initial ionization. Because the tube voltage is discharged, the tube is "paralyzed" for 10's or 100"s of microseconds until the voltage recharges.

In cylindrical ion chambers with the inner conductor having positive polarity, more than 50% of the external signal is due to the motion of the electrons (or negative ions), and less than 50% due to motion of the positive ions. For a cylindrical ion chamber with a 6:1 diameter ratio, 75% of the total external signal is due to the motion of the electrons. This is because most of the image charges for both ions and electrons are initially on the outer electrode. The current in the external circuit is due to the motion of these image charges from one electrode to the other, as the internal charges drift to the electrodes. In the case of proportional and Geiger tubes, additional charge carriers are created near the anode, and most of the external signals are thus due to positive ions rather than to electrons.

The preference for having positive polarity on the center electrode arises from the relative drift velocities of electrons and ions. At 1 atm, electron drift velocities at 1000 V/cm are of the order of 1 cm per μs (depending on the specific gas), while for positive ions, it is of the order of 1 cm/ms. Thus when the center electrode is positive, the dominant signal is produced by the high mobility electrons, providing a dominant fast external signal, while the slow moving ions produce a relatively small external signal.

Because the number of ion pairs created per incident MIP is small (about 140 pairs per cm in argon gas at 1 atm), gas ion chambers are always used in the current-integrating (charge) mode. Typically, the calibration ranges from about 50 to 500 nanoCoulombs per rad.

The ion chamber dynamic range is limited by leakage currents at the low end, and by charge recombination at the high end[3]. Good guard-ring design will limit leakage currents to 1 pA or less. In argon ion chambers, recombination is less because the free electron does not attach to neutral ions to form negative ions. In cases where the recombination is very small, the positive ion space-charge density can inhibit ion collection, and have a similar effect[4]. The dynamic range of the FNAL chamber discussed below is limited to about 100 rads/sec (7 µA) on the high end, thus giving a dynamic range of over 10^6 to 1. Maintaining this dynamic range in the front-end electronics at these low currents is difficult.

Unlike pulse-counting beam loss monitors, current-integrating ion chambers have a very high instantaneous dose limit. Very roughly, the instantaneous dose limit is the dose rate limit (e.g., 100 rads/sec mentioned above) times the positive ion collection time (typically about 1 msec), or 0.1 rads. A pulse-type detector with a calibration of 1 Hz at 1 rad/hr would have to count at 360 MHz to measure a 0.1-rad pulse in 1 µsec.

The FNAL Argon Ionization Chamber

The FNAL argon ionization chamber[5] is an example of a conventional ion chamber developed for use around accelerators. It is a sealed-glass cylindrical ion chamber, with 10-cm long nickel electrodes, 3.81-cm outer electrode diameter and 0.635-cm inner electrode diameter. It is shown in Figure 2. The inner electrode is the anode (signal output), and the outer electrode is the cathode, biased at –2000 volts. Connections are at opposite ends of the sealed glass chamber, and a guard ring is painted on the outside of the glass to minimize end-to-end leakage currents. Its active volume is about 110 cm^3, and it is filled with argon gas at 725 mm Hg. Argon gas was chosen because the electron attachment rate to form negative ions is very small, and the electron drift velocity is about 0.5 cm/µs, thus giving a large prompt signal. Its calibration, using Eqn (5), is about 70 nC per rad. Because the chamber is sealed and there are no organic materials inside, it requires no gas replacement.

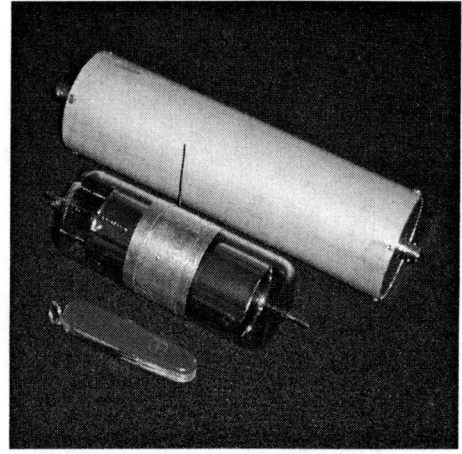

Figure 2. FNAL 110-cm^3 sealed-glass argon ionization chamber and its container.

Figure 3 shows a saturation curve for five identical ion chambers taken with a radioactive source. Note in particular that all chambers have the same output current, and that the saturation plateau ranges from about 200 volts to over 2000 volts. A beneficial characteristic of ion chambers is that the rad calibration is determined by geometry, and that the calibration is relatively independent of the applied voltage. This

Figure 3. Voltage-plateau curves for five identical modified FNAL argon ionization chambers (from Witkover and Gassner, this conference).

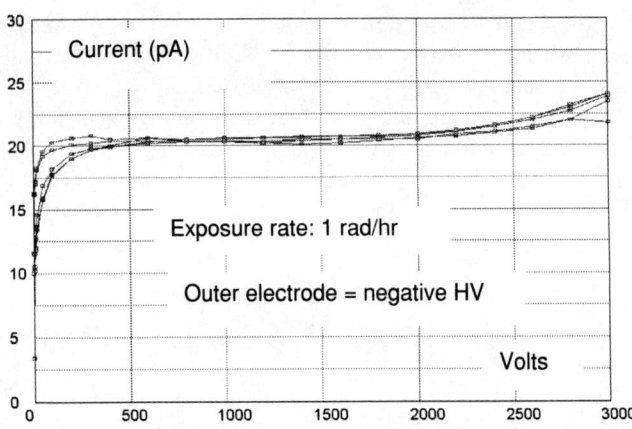

simplifies the system design in large installations, because the high voltage can be daisy-chained to many BLMs, and periodic calibrations are not required.

System readiness tests include pulsing the high voltage under computer control, and measuring the induced charge output. Because the inter-electrode capacitance is about 2 pF, a 2000-volt pulse induces about 4 nC of charge in the external circuit that can be digitized.

Figure 4 shows predicted charge-collection efficiency curves for the FNAL chamber at 1, 10, and 100 rads/sec. These curves are based on the theory of recombination in cylindrical ionization chambers[6]. This design was tested with an electron-linac pulsed radiation source up to about 1 rad instantaneous dose.

Charge collection fraction for 1,10, and 100 rads/sec dose rates vs. applied voltage, for the FNAL ionization chamber.

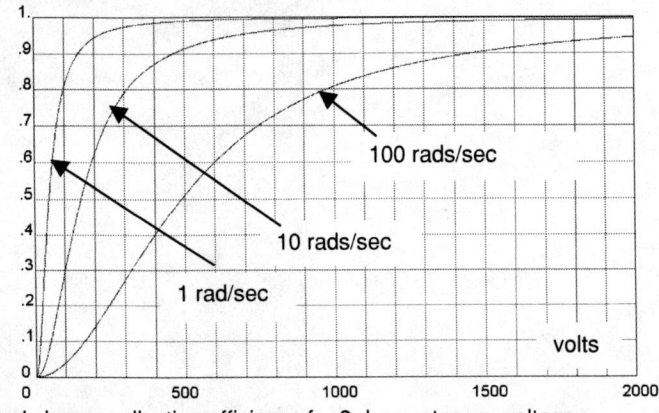

Figure 4. Predicted charge-collection efficiency for 3 dose rates vs. voltage.

The Spallation Neutron Source (SNS) beam loss at 1 GeV in the H- linac is expected to be about 1 watt/meter, which corresponds to an average dose rate of about 50 rads/hr at 30 cm. This is equivalent to about 0.25 rads/sec during the 60 Hz, 1-ms beam macropulses. Thus the FNAL argon ion chamber can monitor dose rates up to 400 times the nominal dose rate with less than about 6% recombination loss.

During commissioning of the SNS linac, an entire 600-ns, 1-GeV, millipulse could be lost at a point. The estimated rad dose at 30 cm is about 0.3 rads, which corresponds to a dose rate of about 500 krad/sec. Figure 5 shows plots of the predicted pulsed rad dose charge collection efficiency for 0.001, 0.01, and 0.1 rads vs. voltage for the FNAL ion chamber[6].

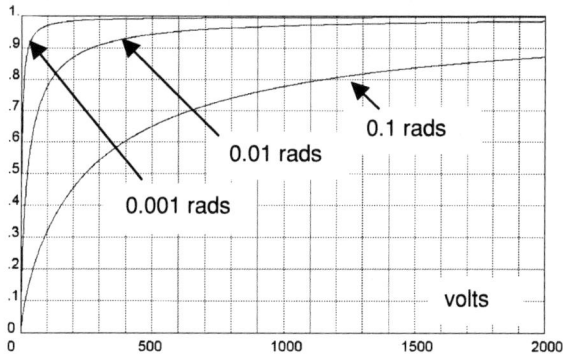

Charge collection fraction for 0.001. 0.01 and 0.1 rads instantaneous pulsed dose, for the FNAL ionization chamber.

Figure 5. Predicted charge-collection efficiency for 3 pulsed doses vs. voltage.

Long Ion Chambers (PLICs)

Panofsky long ionization chambers (PLICs) have been in use at SLAC since 1966[7]. The original PLIC was 1.5" dia. Heliax cable, 2-km long, filled with Ar/CO_2 gas. The outer conductor was grounded and the inner conductor was +HV, and the output signal was ac-coupled. Because the electron beam pulse was very short (< 2 µs), the up-beam PLIC signal (pulses traveling in the opposite direction to the beam) could be used to determine the beam-loss point to a few meters. Many variations of this original design are now in use at SLAC.

Unlike conventional ion chamber designs, variations of the original PLIC design can be very fast, and can determine loss points by time-of-flight with roughly 1-meter resolution. In Figure 6, the pulse response is shown for two PLIC designs, both using a very fast gas, Ar/CF_4, with an electron drift velocity of about 12 cm/µs[8,9].

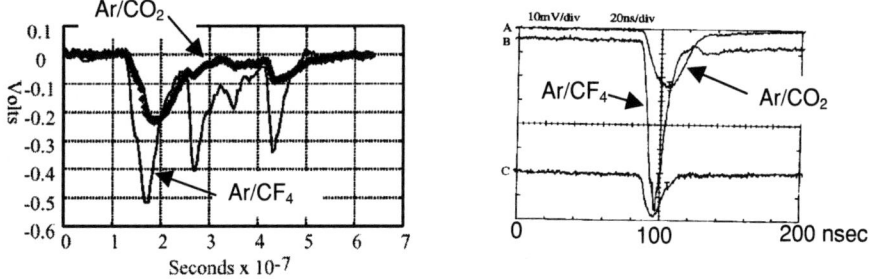

Figure 6. Fast Ar/CF_4 up-beam PLIC pulses from point losses at SLAC.

Solid State Ion Chambers (PIN Diodes)

Solid-state ionization chambers are usually reverse-biased silicon PIN diodes with frontal areas ranging from a few to 100 mm^2, and with depletion depths ranging from perhaps 100 to 300 µm. They can be used in either the current output mode or the pulsed output mode. We review the basic characteristics of two PIN diodes, the Siemens BPW 34 and the Hamamatsu S2662, used in some beam loss monitors.

Property	BPW 34	Hamamatsu S2662
Area	2.75 x 2.75 mm^2	7.5 x 20 mm^2
Depletion depth	~100 µm	~100 µm
Volume	0.75 mm^3	15 mm^3
Leakage current	~100 pA	~500 pA
Integrating mode		
Coulombs per rad	5 nC	100 nC
Rad equiv. of leakage current	70 rads/hr	20 rads/hr
Rad hardness (leakage current)	~1 Mrad	~1 Mrad
Pulse mode		
MIPs per rad	2.3E6	4.6E7
Max rads/sec (@ 10^7 counts/sec)	4 rads/sec	0.2 rads/sec
Rad hardness (spurious counts)	~100 Mrads	~100 Mrads

PIN Diode Pulse-Mode Coincidence Circuit BLM

In order to minimize the sensitivity to synchrotron radiation, two PIN diodes can be placed back-to-back, and the two pulse-output signals put into a coincidence circuit. Such a unit has been developed at DESY for use in the HERA tunnel which also has a 30-GeV electron ring[10]. The detector geometry is shown in Figure 7. Low energy photons will interact in only one PIN diode, while MIPs interact in both, producing a coincidence. The unit also has directional sensitivity. The whole unit measures about 69 mm by 34 mm by 18 mm. It is now commercially available[11].

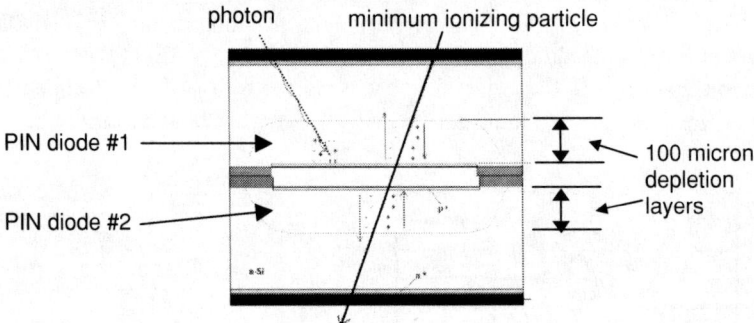

Figure 7. PIN diode coincidence circuit.

BEAM LOSS MONITORS USING LIGHT DETECTION

Detection of beam-induced light in scintillators or Cerenkov radiators represents the other most common method of monitoring beam losses. The scintillation process is also based on the Bethe-Bloch dE/dx equation. Some combinations, using the current-integrating mode, are:

Phototube	Radiator
Photomuliplier Tube	Organic scintillator (e.g., NE 102 or BC-400)
	Liquid scintillator (mineral oil based)
	Inorganic scintillators (e.g., CsI(Tl), BGO)
	Cerenkov radiator (e.g., fused silica)
	Bare PMT
Vacuum Photodiode	Scintillators as per above list
	Cerenkov radiators as per above list

Scintillation constants of some organic and organic scintillator materials are listed below[12].

Scintillator	Scintillation constant
Inorganic	
NaI(Tl)	26 eV energy loss per emitted photon
CsI(Tl)	15
BGO ($Bi_4Ge_3O_{12}$)	122
$CdWO_4$	67
CsI (unacivated)	500
Ce-activated Li glass	300
Organic	
Anthracene	60
NE-102A	100
BC-400	90
BC-517P (mineral oil)	250
Gas	
Nitrogen	1250

A useful feature of scintillators is the very fast risetime (a few to 100's of ns). Rad hardness varies from a few krads (e.g., NaI(Tl)) to about 100 Mrad (BGO, aka $Bi_4Ge_3O_{12}$). We examine two scintillator-based beam loss monitors.

The LAMPF "Paint Can" Beam Loss Monitor

The LAMPF "paint can" beam loss monitor is a 1-pint paint can filled with mineral-oil-based liquid scintillator. It uses a side-window photomultiplier (NE-4552) mounted inside the can, along with the voltage-divider resistor chain and a calibration lamp. It operates on negative HV, with a current-mode anode output. It is shown in Figure 8.

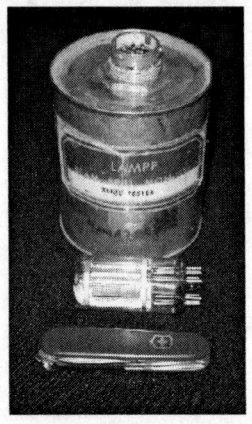

Figure 8. LAMPF "Paint Can" beam loss monitor, and NE-4552 side-window photomultiplier. . The photomultiplier, mounted inside the can, is immersed in mineral-oil liquid scintillator.

The calibration is approximately 1000 µC per rad, including factors of 250 eV of energy loss per detectable photon, 3% collection efficiency, and 20% conversion efficiency.

$$1 rad = \frac{100 \, ergs}{gram} \cdot 350 \, grams \cdot \frac{1 \, eV}{1.6 \cdot 10^{-12} \, ergs} \cdot \frac{1 \, photon}{250 \, eV} = 9 \cdot 10^{13} \, photons \, ;$$

$$9 \cdot 10^{13} \, photons \cdot 0.03 \cdot 0.2 \cdot 10,000 \, gain \cdot \frac{1.6 \cdot 10^{-19} C}{electron} = 1000 \, \mu C \quad (9)$$

Thus the units are very sensitive, relative to ionization chambers. On the downside, there is a large unit-to-unit gain variation, a large sensitivity to voltage setting, and the mineral-oil scintillator eventually turns milky and must be replaced.

The LEDA CsI(Tl) Beam Loss Monitor

The LEDA beam loss monitor is designed with a very high radiation sensitivity in order to detect beam loses from a 6,7 MeV proton beam[13]. It is a commercially-packaged 5-cm dia. By 1.25-cm high CsI(Tl) crystal (110 grams) epoxied to an end-window photomultiplier. It is shown in Figure 9. The calibration is 1 uA output for a 190 mrad/hr source, equivalent to about 19,000 µC per rad.

$$1 rad = \frac{100 \, ergs}{gram} 110 \, grams \frac{1 \, eV}{1.6 \cdot 10^{-12} \, ergs} \frac{1 \, photon}{15 \, eV} = 4.6 \cdot 10^{14} \, photons$$

$$4.6 \cdot 10^{14} \, photons \cdot 0.6 \cdot 0.2 \cdot 2,150 \, gain \cdot \frac{1.6 \cdot 10^{-19} C}{electron} = 19,000 \, \mu C \quad (10)$$

Gain curves for a few units are shown in Figure 10. Note that unlike ionization chambers, the sensitivity varies widely from unit to unit, as well as with the high voltage setting. However, the sensitivities are typically 10,000 to 500,000 times higher than ionization chambers.

Figure 9. The LEDA CsI(Tl) beam loss monitor. The calibration is about 19,000 µC per rad.

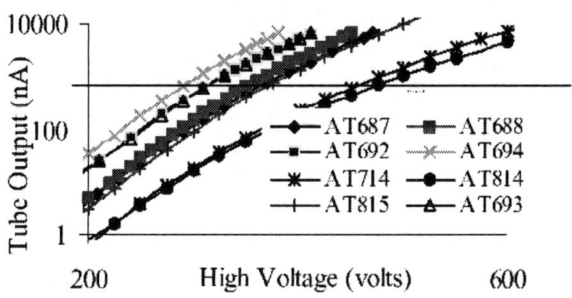

Figure 10. Gain curves for 8 LEDA beam loss monitors, at an exposure rate of 190 mrad/hr. The line at 1 µA is the calibration set point.

Scintillating Fibers and Optical Fibers

From time to time, scintillating fibers are suggested as a possible beam loss monitor. An interesting suggestion is to use Ce-activated Li glass (with Li^6) to detect neutrons. In general, the internal reflections and resultant attenuation are excessive unless a graded or stepped index fiber is used. A wavelength shifter must be used to limit self-absorption. For light produced isotropically in the fiber, only about 2% is in the cone that will be internally reflected. Lastly, the volume of the fiber is too small to produce sufficient light for most applications. Radiation darkening will probably limit the use to < 100 Mrad.

Cerenkov Radiators

Cerenkov light is the light emitted when a charged particle's velocity βc is greater then the light velocity c/n in a media with an index of refraction $n>1$. Specifically, the number of photons N in an energy range ΔE eV emitted per cm in a Cerenkov radiator is[14]:

$$N = 369.8 \left(1 - \frac{1}{n^2 \beta^2}\right) \cdot \Delta E \ photons/cm \qquad (11)$$

Cerenkov light is instantaneous, unlike scintillators, and the threshold for light output ($\beta > 1/n$) is above Compton-electron energies of several hundred keV, making Cerenkov detectors useful where there is background radiation from RF cavity X-rays or synchrotron radiation, such as high-energy electron rings and superconducting RF cavities. For a 1-GeV proton ($\beta = .875$), a fused silica radiator ($n = 1.55$), and photons between 400 and 600 nm, ($\Delta E = 1$ eV), the light output is 169 photons per cm. It is emitted in a forward cone of half-angle $\cos^{-1}(1/n\beta)$. Figure 11 shows the photon yield vs particle energy (in mass units). Compton electrons below about 150 keV will not produce any light, while 1-GeV protons or 0.5 MeV electrons produce about 169 photons/cm.

The sensitivity is much less than phototube-scintillator combinations, however. For a 5-cm diameter, 1-cm thick fused-silica radiator, a collection efficiency of 80% and a cathode efficiency of 20%, a typical response to 1-GeV protons would be

$$1 \, rad = \frac{3.1 \cdot 10^7 \, MIPs}{cm^2} \cdot 19.6 \, cm^2 \cdot \frac{169 \, photons}{MIP \, cm} \cdot 1 \, cm \, thick = 1 \cdot 10^{11} \, photons$$

$$1 \cdot 10^{11} \, photons \cdot 0.8 \cdot 0.2 \cdot 10{,}000 \, gain \cdot \frac{1.6 \cdot 10^{-19} C}{electron} = 30 \, \mu C \qquad (12)$$

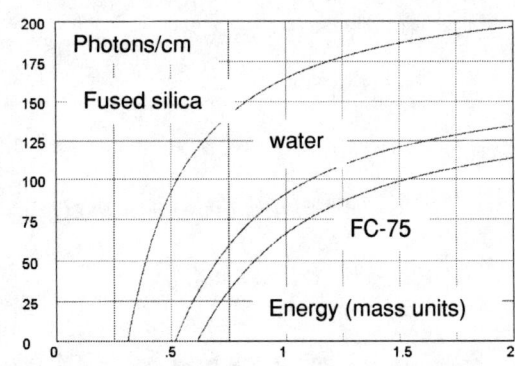

Figure 11. Plot of photons/cm vs. particle energy (in mass units) for three Cerenkov radiators:

Fused silica (n=1.55)
Water (n=1.33)
FC-75 (n=1.275)

Ionizing Radiation Backgrounds

The most common background ionizing radiation around beam loss monitors is due to RF cavity x-rays (especially from superconducting cavities). Dark-current (x-ray) radiation around copper accelerator structures can be up to 50 rads/hr. The radiation spectrum from superconducting RF cavities can extend up into the MeV region.

Another source of ionizing radiation is synchrotron radiation from electron rings. A plot of the synchrotron radiation critical photon energy is shown in Figure 12. Unlike dark-current radiation from RF cavities, synchrotron radiation is very directional, so its effects can be minimized by proper beam loss monitor placement.

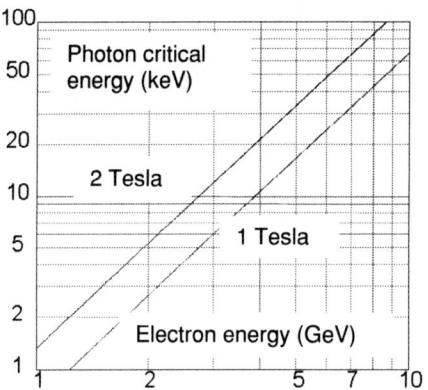

Figure 12. Synchrotron radiation critical photon energies in 1-Tesla and 2-Tesla magnetic fields, as a function of the electron energy.

In both cases, most of the radiation is sufficiently low energy so that it can be effectively shielded by high-Z material, such as lead. Tables of photon attenuation lengths are on the NIST website[1].

SUMMARY

The most widely used beam loss monitors are either of the current-integrating ionization-chamber design or the current-integrating photomultiplier-scintillator design. Both types produce a very wide dynamic range response. The phototube-scintillator combination has about a factor of 10,000 higher sensitivity to beam loss than ion chambers, but suffers from nonuniformity in unit-to-unit gains, and is very sensitive to voltage drifts.

Although ion chambers normally have poor time resolution compared to phototube-scintillator designs because of the slow electron drift velocities, special ion chamber designs using coaxial cable can achieve time resolutions of a few 10's of nsec.

In situations where there is low-energy background ionizing radiation (synchrotron radiation or RF cavity x-rays), phototube-Cerenkov radiators or PIN-diode pulse-coincidence circuits can be used. The latter are limited to a few rads/sec peak dose rates. Lead shielding can be used around any beam loss monitor to reduce the sensitivity to background x-rays.

REFERENCES

1. NIST website Physics.nist.gov. This website has tables of dE/dx and range for protons, as well as tables for photon cross sections.
2. Knoll, G.F., *Radiation Detection and Measurement* (Third Edition), Wiley,2000. See Table 5.1. This is a very good reference on radiation measurement instrumentation in general.
3. Boag, J.W., "Ionization Chambers", in *The Dosimetry of Ionizing Radiation*, Vol. **2**, Kase, K.R, Bjarngard, B.E., and Attix, F.H., Editors (Academic Press, 1987).
4. Boag J.W. and Wilson,T; *British Journal of Applied Physics* vol **3**, pages 222-229 (1952). See pages 226-227.
5. Shafer,R.E et al., The Tevatron BPM and BLM Systems, in *the Proceedings of the XII International Conference on High Energy Accelerators*, page 609 (Fermilab ,1983).
6. Reference 3, See Eqn (23) for charge collection efficiencies for continuous dose rates in cylindrical ion chambers, and Eqn (33) for charge collection efficiencies for pulsed doses in cylindrical ion

chambers, in chapter 3. This is an important reference for anyone designing an ion chamber for high dose rates.
7. Panofsky,W.K.H., SLAC Internal Tech Note TN-63-57 (1963).
8. McCormick, D., *Proceedings of the 1991 Particle Accelerator Conference*, pages 1240-42 (1991).
9. Ross, M.C. and McCormick, D. *Proceedings of the 1998 Linac Conference*, pages 192-194 (1998).
10. Wittenburg, K., in *Beam Instrumentation Workshop 2000*, pages 3-17 (AIP Conference Proceedings #546).
11. Website www.bergoz.com.
12. Reference 2, pages 226 and 235.
13. Sellyey W.C. et al., *Proceedings of the 2001 Particle Accelerator Conference*, pages 1315-1317 (2001)
14. Schiff L.I., *Quantum Mechanics* (Second Edition) McGraw Hill (1955). See Eqn(37.14) on page 271.

Optical System Design for High-Energy Particle Beam Diagnostics

Bingxin Yang

Argonne National Laboratory, 9700 South Cass Avenue, Argonne, IL 60439

Abstract. Radiation generated by high-energy particle beams is widely used to characterize the beam properties. While the wavelengths of radiation may vary from visible to x-rays, the physics underlying the engineering designs are similar. In this tutorial, we discuss the basic considerations for the optical system design in the context of beam instrumentation and the constraints applied by high-radiation environments. We cover commonly used optical diagnostics: fluorescence flags, visible and x-ray synchrotron radiation imaging. Emphases will be on achieving desired resolution, accuracy, and reproducibility.

INTRODUCTION

The unique value of direct visualization of particle beams with optics and cameras was realized very early on in accelerator engineering. With the development of quantitative imaging tools, especially the ever better availability of CCD cameras and digitizers, optical imaging has become a vital part of particle beam diagnostics today. This tutorial is designed to introduce the field to starting young engineers and builds on several tutorials presented at past Beam Instrumentation Workshops [1-3]. The readers will also find it beneficial to be informed of progress in other fields that use quantitative imaging tools extensively, such as video microscopy, astronomy, and machine vision.

Classification of Beam Imaging Techniques

The goal of modern beam diagnostics is to transfer information carried by particle beams to computer memory, faithfully and efficiently. The transfer is always performed in several steps and uses several intermediate media. Figure 1 shows a typical optical diagnostic system: A conversion device that transfers the information from particle beam to radiation beam (light, x-ray, etc.), an optical system that forms the desired radiation pattern, a read-out device (camera) that converts the radiation intensity distribution into electrical signals, and a digitizer that converts the signal into discrete intensity maps to store in computer memory.

Any hardware component or media in the system can be and has been used to categorize optical diagnostics techniques, since changes in any component or information-carrying media could dramatically alter the system's performance and design criteria. For example, designs of electron and hadron diagnostics are very different since their length scale and beam dynamics are very different. In another example, efficient conversion devices and optics need to be used to take advantage of a fast detector (streak camera). This tutorial will be biased towards electron / positron diagnostics due to the author's own limited experience.

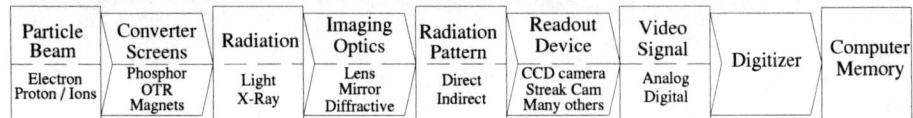

FIGURE 1. Basic components of a beam imaging system: Rectangular boxes show information-carrying media; arrow-shaped boxes show information transfer hardware.

An alternative but very important way to classify imaging techniques is by the type of point-spread function (PSF), which is the radiation intensity distribution formed with a single point source. When the PSF has a single, predominant peak, we call the optics a *direct imaging system*. The human brain readily understands images from such a system since the human eye images directly on retina. Here the transverse length scale of the PSF is characteristic of the image sharpness, and its rms length is often taken as the resolution. When the PSF is a multi-peaked pattern [4,5], often a type of interference pattern, we call the optics an *indirect imaging system*. This tutorial will deal mainly with direct imaging techniques.

The challenge facing the optics designer is best characterized by the system's *angular resolution* (viewed from the first optical element). This quantity will correctly take into account of wide variation of working distances (WD). Table 1 shows the typical dimensions of several common objects. While 1-μm bacteria are "enormous" with a short WD microscope, a 2300-km planet can be very small at an astronomical distance. As seen from the table, the angular sizes of particle beams fall in the same range of these common objects, depending on the working distance.

In the next section, we will first discuss screen-based (interceptive) beam imaging by going step-by-step through a working example, paying special attention to resolution issues and various contributing factors. In the following sections, we will discuss synchrotron-radiation-based (noninterceptive) imaging techniques, including visible light imaging and x-ray pinhole cameras.

TABLE 1. Angular Size of Several Sample Objects.

Object	Diameter	WD	Angular size	Optics	Comment
Bacteria	1 μm	0.63 mm	1.6 mrad	Microscope	Enormous
Human hair	50 μm	25 cm	200 μrad	Human eye	Fair size
Pluto	2300 km	6×10^9 km	0.4 μrad	Reflective telescope	Tiny
Electron beam	30 μm	10 cm	300 μrad	Light telescope	Fair size
		10 m	0.3 μrad	X-ray telescope	Tiny

SCREEN IMAGING SYSTEM DESIGN

The basic type of imaging system for particle beams is based on converter screens (*flags*) using scintillation/fluorescence phenomena or relativistic effects (transition radiation, Cherenkov radiation, etc.). With the appropriate choice of scintillator material, the screen can also be used for x-ray beam imaging. In this section, we will explore various aspects of design and modeling by working at an actual system [6].

The Advanced Photon Source (APS) free-electron laser system uses low-emittance electron beams. In the bunch compressor region, where the beam is strongly focused, the smallest beam is around 50 μm in rms radius, and the bunch charge is about 0.2 nC. On our wish list, we have asked for a 17-μm rms resolution, and 0.2-nC charge sensitivity. A model study also showed that 5% accuracy in beam-size measurements was needed to support the studies of the coherent synchrotron radiation effect [7]. To increase the charge sensitivity, we also asked for the highest acceptance solid angle and the largest of field of view (FOV) the vacuum enclosure allows.

Technical Specification and Consistency

Before turning the wish list into technical specifications, we perform several consistency checks. While some of them are based on basic optical principles, others merely reflect hardware limitations.

(1) FOV-to-resolution ratio

Any digitized image contains a finite number of picture elements (pixels), which imposes restrictions on the size of the FOV for a given resolution. The APS uses standard RS-170 video systems. While the horizontal pixel number can vary from 400 to 900 for different camera / digitizer combinations, the maximum number of vertical lines is 483. Using a criterion of a minimum two pixels per resolution element, we have the maximum field of view given by

$$\frac{\text{FOV}}{\text{Resolution}} \leq \frac{\text{max pixel \#}}{2} = 240. \tag{1}$$

This yields a 4-mm FOV in the vertical direction. To overcome this limitation, one could use CCD cameras with more pixels; a zoom lens to change magnification; or two cameras, one set at high resolution and one set at full field of view, sharing the light with a beam splitter or switching mirror. Once the CCD is chosen and the FOV is decided upon, the optical system magnification is then given by the ratio of pixel sizes at the image and object planes.

(2) Phase-space acceptance limit

Let us consider a simple imaging system made of two lenses, with the object and CCD chip placed at the respective focal point (Fig. 2). The magnification of the system (M) is given by the ratio of the object height (h) and image height (h'), $M = h'/h = S'/S$. Since the FOV in the image space is limited by the effective size of the CCD chip

l_{CCD}, and the F number of the lenses are also limited practically, we obtain the phase-space limitation for the entire imaging system,

$$FOV \cdot \text{Total collection angle} = \frac{2hD}{S} = \frac{2h'D}{S'} \leq \min\left\{\frac{l_{CDD}}{MF_1}, \frac{l_{CDD}}{F_2}\right\}, \qquad (2)$$

where $F_1 = D/S$ and $F_2 = D/S'$ are the F numbers of the lenses. Since the phase-space volume remains a constant through out an optical system, this limitation is fairly general in nature. It means that a large field of view and a large collection angle cannot be obtained at the same time with a single imaging system, and a compromise needs to be made. To increase total phase-space acceptance, one can use a large lens (small F number), or if the system is demagnifying ($M < 1$), use a large CCD chip.

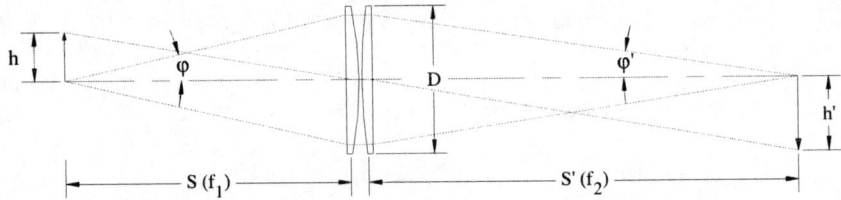

FIGURE 2. A simple imaging system using two infinity-conjugate lenses. The diameters of the lenses are D, and their focal lengths are f_1 and f_2. The object distance $S=f_1$ and the image distance $S'=f_2$.

(3) Working distance and shielding requirements

Unless the radiation generated by the particle beam is low, the CCD cameras need to be shielded to prevent radiation damage (snowy pictures). A simple but effective way is to make one or more bends in the optical path so that high-Z materials (W or Pb) can be placed to block the radiation generated at the screen from reaching the camera. For high-energy particle beams, radiation shower originating from upstream points of the accelerator should also be considered.

To perform radiation ray-tracing, all low-Z materials must be ignored. Often it is convenient to trace backwards from the camera (acceptance zones). The additional bends and shielding geometry lengthens the optical paths and reduces the F numbers of the suitable lenses, further restricting the phase-space acceptance volume.

After applying the above three constraints, we reached a set of realistic design specifications for the APS flag system (Table 2). The phase-space acceptance is based on a CCD chip 6.4 mm × 4.8 mm in size.

TABLE 2. Specifications for the APS Bunch Compressor Flag.

Camera / Optics	High Resolution	Low Resolution
Resolution	17 μm	50 μm
Working distance	125 mm	200 mm
Field of view (FOV)	6.4 × 4.8 mm^2	20 × 15 mm^2
Light collection angle	0.2 radian	0.12 radian
Phase-space acceptance	1.3 × 1 (mm-rad)^2	2.4 × 1.8 (mm-rad)^2

Selection of Converter Screen and Geometry

Converter screens are available in various types and forms, and most of them are based on ionizing-particle-induced scintillation (Fig. 3). For relativistic particles, an optical transition radiation (OTR) screen is also an option.

(1) Powdered phosphor [1,2]

A phosphor screen is formed by adhering phosphor powder to a substrate. The scintillation light reflects multiple times before escaping the powder particle, thus lighting up the entire grain. In a single layer, the resolution is limited by the average size of the grains, typically several to several tens of micrometers. These screens are economical and versatile due to their simple fabrication process. Phosphors with different spectra are available to fulfill the needs of optical system requirements. Phosphors with decay time as short as 45 ps are available for dynamic studies [1].

(2) Ceramic phosphors

Ceramic phosphor screen can be considered as sintered screen with fine grains of phosphor powder. The sizes and thicknesses of the screens are usually made to order. They are self-supporting and can be machined only with special tools. Since the grains are coupled closely optically, the spatial resolution of the screens is normally given by the size of several grains, usually in the range of 100 micrometers or more. Not many choices are available in this category. Commonly used screen materials include Chromax (Cr doped Al_2O_3) and YAG:Ce.

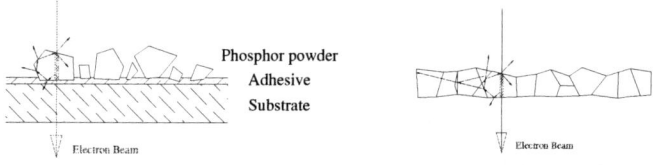

FIGURE 3. (Left) Schematics of a powdered phosphor screen. (Right) Ceramic phosphor screen.

(3) Clear plastic scintillators

Plastic scintillators have good efficiencies and short decay times, but their use in beam imaging is very rare, probably due to their low radiation damage threshold (molecular instead of ionic crystals) or poor vacuum compatibility.

(4) Translucent inorganic single-crystal scintillators

Cerium-doped yttrium aluminum garnet (YAG:Ce) was proposed as an image converter screen for electron microscopy by R. Autrata et al. in 1983 [8]. However, its actual application had been very limited due to the crystal's high cost. In the ensuing decade, large quantities of YAG crystals were produced for the laser industry and its price has dropped significantly. Interest in the translucent inorganic scintillators has been revived recently [9,10] and other crystals (YAP:Ce, LSO:Ce) have also been tested.

Compared with phosphor screens, the inorganic scintillators have significant advantages. They are economical, compatible with ultra-high vacuum, efficient in light conversion, highly resistant to beam damage, with narrow emission spectrum (reducing chromatic aberration), and most importantly, with good spatial resolution. In the case of YAG crystals, reported resolution ranges from 40 μm for high charge density beam, to 10 μm for low density ones, and even 1 μm or less for specially fabricated YAG crystals with only a several-μm-thick Ce-doped layer. A saturation-like blurring starting at a charge density of 6 – 20 nC/mm^2/bunch appears to be a significant limitation for the spatial resolution of YAG and other crystals [11,12].

(5) Optical transition radiation (OTR) screens [13]

When charged particles traverse the interface between different indices of diffraction, photons are emitted. When a metal foil is used, OTR is emitted in the specular direction from the front surface where electrons enter the metal (reflective OTR) and also in the forward direction from the back surface where electrons exit the metal (forward OTR). The radiation is radially polarized, and its angular distribution is concentrated in a cone with a radius of $1/\gamma$ and a dark center:

$$\frac{d^2N}{d\omega d\Omega} \propto \frac{\theta_x^2 + \theta_y^2}{\left(\gamma^{-2} + \theta_x^2 + \theta_y^2\right)^2}, \qquad (3)$$

where θ_x and θ_y are angles of observation from the specular / forward direction for the two types of OTR, respectively. The photon spectrum spans a wide range, and the angle-integrated photon flux in the frequency region of $[\omega_1, \omega_2]$ is given by

$$n(\omega_1, \omega_2) = \frac{2\alpha}{\pi} \ln \gamma \cdot \ln \frac{\omega_2}{\omega_1}. \qquad (4)$$

For 200-MeV electrons ($\gamma \approx 400$), about one visible photon is produced for every 60 electrons. This can be compared to $10^4 \sim 10^5$ photons generated in YAG by the same number of electrons. However, the scintillation photons are emitted in all 4π solid angles, and a practical imaging system could only capture <1% of the light, while the same system would capture a large percentage of OTR photons.

For this APS flag project, we will use one 0.1-mm-thick YAG scintillator for the converter screen at low charge intensity and one OTR screen (a 45 degree mounted metal mirror) at high intensity.

Error Management

Emittance measurements with 10% accuracy are routinely performed with basic hardware, except when the statistics are really poor. For measurements with 5% accuracy or better, all sources of error need to be considered and each one carefully managed (Table 3). Defects of screens (dopant concentration variation, etc.) and optics (scratches, oxidation, or contamination) are difficult to control or model. The tolerance budgets assigned to these sources are somewhat arbitrary, but their total effect can be measured experimentally. The statistical fluctuation of photons collected by the camera is expected to be a major source of error since each bunch contains only 0.2-nC of charge, or 1.2×10^9 electrons. For an OTR screen, this corresponds ~ 2×10^7 photons spreading out in the entire image. Assuming the accelerator is stable, successive single-bunch measurements will show the statistical fluctuation of the measured beam size.

In the next two subsections, we will discuss the modeling of the resolution and calibration and quality assurance issues.

TABLE 3. Tolerance Budget for the APS Bunch Compressor Flag.

Source of Error	Size Tolerance	Emittance Tolerance
Screen defects	1.0%	1.4%
Optics defects	1.0%	1.4%
Resolution	2.0%	2.8%
Calibration	1.4%	2.0%
Statistics	2.0%	2.8%
Total	3.5%	5%

Resolution Analysis

Resolution of optical instruments are defined differently in different fields. Microscopists like to use the Rayleigh criteria, while broadcasting video uses line pairs per screen. In beam physics, rms resolution is widely used; for Gaussian PSF, this definition results in a simple quadrature relationship between the measured source size σ, and the true source size σ_0,

$$\sigma^2 = \sigma_0^2 + \sigma_R^2, \qquad (5)$$

where σ_R is the rms width of the PSF and is called the instrument's resolution. This definition encounters two major difficulties in practice: First, a true rms calculation is prone to background noise or drift. Second, for non-Gaussian PSF, the simple quadrature relation breaks down, and a fit to the exact convolution is often used to determine the actual source size. As an example to illustrate the mathematical difficulty, the one-dimensional diffraction PSF, the sinc function, has an infinite rms width. To avoid the above technical difficulties, we propose to *define* an effective Gaussian resolution based on the quadrature relation.

Definition of Effective Gaussian Resolution

Assume that we are measuring a Gaussian source,

$$g_0(x) = \frac{A_0}{\sqrt{2\pi}\sigma_0} e^{-x^2/2\sigma_0^2}, \qquad (6)$$

with an instrument with a point spread function $h(x)$. The resultant intensity distribution is then given by

$$f(x) = \int_{-\infty}^{\infty} g_0(x-x')h(x')dx', \ a \le x \le b. \qquad (7)$$

After fitting $f(x)$ with a Gaussian function $g(x) = \dfrac{A}{\sqrt{2\pi}\sigma} e^{-x^2/2\sigma^2}$, i.e., with minimizing

$$\chi(A,\sigma) \equiv \int_a^b \left[f(x) - g(x) \right]^2 dx, \qquad (8)$$

we obtain the effective Gaussian height A and width σ. We now *define* the effective (Gaussian) resolution of the instrument as

$$\sigma_R = \sqrt{\sigma^2 - \sigma_0^2}. \qquad (9)$$

When the PSF is an exact Gaussian function, this is identical with the rms definition. However, if the PSF is not Gaussian, the resolution could be a function of source size!

Image Formation and Diffraction Point Spread Function [14]

In geometrical optics, a perfect imaging system converts a bundle of rays radiating from a single point (source) to a bundle converging to a single point (image). By the same token, a perfect imaging system in wave optics converts an outgoing single spherical wavefront centered at the source point to another converging single spherical wavefront centered at the image point O, as shown in the coordinate system in Fig. 4.

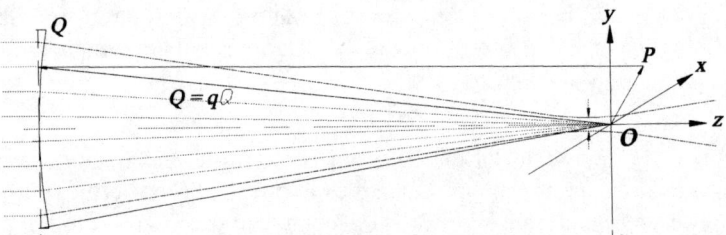

FIGURE 4. Coordinate system near the classical focus (image point) O. The z-axis is chosen to be the optical axis, y is vertical, and x is defined by the right-hand rule. Point Q is located in the exit plane of the optical system.

Using the rays passing through the classical focus as the reference path, the path difference for any line connecting a point Q on the exit surface and a point P near the focus can be obtained. Applying Huygens Principle, the light intensity distribution (PSF) near the classical focus can be expressed by a Helmholtz diffraction integral, $h_{diff}(x,y,z) = \left| A_{diff}(x,y,z) \right|^2$, and

$$A_{diff}(x,y,z) = \int A(\mathbf{q}) e^{-i\mathbf{q}\cdot\mathbf{P}/\lambda} d\Omega = \int A(\mathbf{q}) e^{-\frac{i}{\lambda}(x\cos\phi_x + y\cos\phi_y + z\cos\phi_z)} d\Omega, \qquad (10)$$

where the complex aperture function $A(\mathbf{q})$ is the light amplitude in the $\mathbf{q}(\cos\phi_x, \cos\phi_y, \cos\phi_z)$ direction far away from the focus. This integral has been well studied. Several well-known properties are listed here:

(1) The minimum beam waist is not located at the geometrical focus but between the focus and the optics. The difference is not significant in most cases unless the light cone angle falls in mrad-range or below.

(2) For uniform illumination, the PSF at the focal plane has a predominant peak with a length scale λ/θ_{max}, and many side peaks (diffraction pattern).

(3) A fairly general form of the *Uncertainty Principle* may be proved

$$\sqrt{\langle \Delta x^2 \rangle}\sqrt{\langle \Delta \theta_x^2 \rangle} \leq \frac{\lambda}{2} = \frac{\lambda}{4\pi}, \qquad (11)$$

relating the rms deviation of the transverse coordinates and momentum. The minimum size of the beam given by this expression is often referred to as the diffraction limit.

As an example, we use the *Uncertainty Principle* to estimate the resolution of OTR imaging with a cone angle of θ_{max}. Using the angular distribution of Eq. (4), it is straightforward to derive

$$\sqrt{\langle \theta_x^2 \rangle} \approx \frac{\theta_{max}}{\sqrt{(8\ln \gamma\theta_{max} - 4)/3}}. \qquad (12)$$

Other than a very slowly varying denominator, this expression is similar to the case of uniform illumination, $\sqrt{\langle \theta_x^2 \rangle} \approx \theta_{max}/\sqrt{2}$, hence we conclude that the diffraction limit of the OTR imaging is not very different from the case of uniform illumination. This estimate is consistent with the exact PSF calculation using Eq. (4) as the aperture function in Eq. (11) [15].

When the aperture is illuminated uniformly and is rectangular in shape, the PSF is a product of two sinc functions, $h_{diff}(x,y) = h_{diff}(x) \cdot h_{diff}(y)$, with each one given by

$$h_{diff}(x) = \left[\sin\left(\frac{x\theta_{max}}{\lambda}\right) / \left(\frac{x\theta_{max}}{\lambda}\right) \right]^2. \qquad (13)$$

Its effective resolution can be modeled with numerical calculation (Fig. 5). The results in the region of practical interest can be parameterized with a simple formula,

$$\sigma_{Res,diff} = 0.36 \cdot l_d \left\{ 1 + 2\left(\frac{\sigma_0}{l_d}\right)^2 \right\}^{1/4}, \quad \left(l_d = \frac{\lambda}{2\theta_{max}} \right). \qquad (14)$$

For a point source, the Gaussian fit only uses the main peak (Fig. 5). For an extended source, the convolution incorporates sidelobes into the main peak, and the effective resolution grows as source size increases.

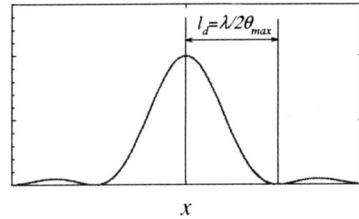

FIGURE 5. (Left) One-dimensional diffraction PSF, a sinc function. (Right) The effective Gaussian resolution of the diffraction PSF.

Charge Binning and Spilling in the CCD Pixels

All CCD camera elements have finite sizes. The charges generated over a finite area are integrated and presented as one point (binning). When the light is shining on one CCD pixel, its neighbor may also get charged due to spilled charge or scattered radiation. These effects can be modeled numerically and the results are shown in Fig. 6. The result can be roughly summarized as

$$\sigma_{R,pix} \approx 0.3 + spill\_fraction. \qquad (15)$$

In the model used here, the *spill_fraction* is the fraction of charge spilled to the nearest neighbors, evenly divided between its left and its right.

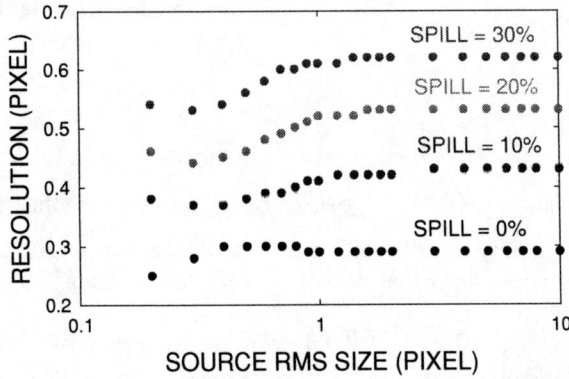

FIGURE 6. Effective Gaussian resolution due to complete binning. From bottom to top, 0%, 10%, 20%, or 30% charge of every pixel is spilled to its nearest neighbors, evenly divided to its left and right.

Effective Resolution from Defocus

If we shift the source point away from the ideal object plane by a distance z, the light cone eventually collected by the optics would illuminate a disc uniformly at the ideal object plane (Fig. 7). If the system were perfect in the sense of geometrical optics, this disc would be mapped faithfully to the camera. The PSF is thus a uniform disc with a radius $|z\theta_{max}|$. After integrating over y we obtain a one-dimensional PSF

$$h_{defocus}(x) = \frac{2\sqrt{z^2\theta_{max}^2 - x^2}}{\pi z^2 \theta_{max}^2}, \quad (x^2 \leq z^2\theta_{max}^2). \qquad (16)$$

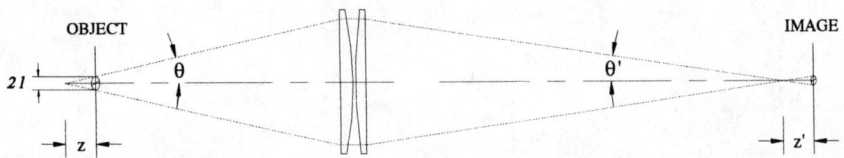

FIGURE 7. Diffusion of light due to defocus: the actual source is located away from the ideal object place by a distance z.

The effective resolution can be modeled numerically (Fig. 8). For all practical purposes when defocus is not the dominant factor, we have,

$$\sigma_{R,defocus}(x) \approx 0.5|z\theta_{max}|. \tag{17}$$

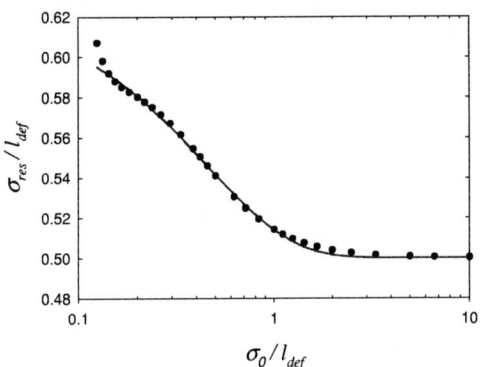

FIGURE 8. Gaussian resolution due to defocusing. Length unit $l_{def} = |z\theta_{max}|$.

Effective Resolution from Depth of Source

At normal incidence, translucent YAG screens can be modeled as uniform sources along the optical axis, extending from z_1 to z_2, with $z = 0$ at the ideal object plane. The integration of Eq. (15) along the z-axis gives a PSF,

$$h_{depth}(x;z_1,z_2) = \int_{z_1}^{z_2} \frac{2\sqrt{z^2\theta_{max}^2 - x^2}}{\pi z^2 \theta_{max}^2 |z_2 - z_1|} dz = \frac{H(x;z_2) - H(x;z_1)}{|z_2 - z_1|} \quad (x \le |z\theta_{max}|), \tag{18}$$

where

$$H(x;z) = \begin{cases} \frac{2}{\pi\theta_{max}} \left[ch^{-1}\left(\frac{z\theta_{max}}{x}\right) - \frac{\sqrt{z^2\theta_{max}^2 - x^2}}{z\theta_{max}} \right] & (|x| \le z\theta_{max}) \\ 0 & (|x| > z\theta_{max}) \end{cases}. \tag{19}$$

Its effective resolution can be numerically modeled and again parameterized approximately as

$$\sigma_{RES}^2(\rho_1,\rho_2) \approx \frac{(\rho_1+\rho_2)^2}{16} + \frac{(\rho_1-\rho_2)^2}{48} + \frac{\rho_1\rho_2(\rho_1+\rho_2)^2}{1024 \cdot \sigma_0^2}, \tag{20}$$

where $\rho_i = \frac{z_i \theta_{max}}{n}$, and n is the index of refraction of the scintillator. As expected, minimum blurring occurs when the center of the YAG is at the ideal focal plane, with the minimum resolution given by

$$\sigma_{RES,min}(\rho_1,\rho_2) = \frac{|\rho_1-\rho_2|}{4\sqrt{3}} = \frac{|z_1-z_2|\theta_{max}}{4n\sqrt{3}}. \tag{21}$$

For a source located away from the optical axis, the discs from different parts of the line source do not overlap concentrically. It results in a comet-like PSF.

Multiple reflections from front and back surfaces could increase the effective source depth. When the front surface is coated with reflective metal, it effectively doubles the source depth. Otherwise, unless the scintillator crystal or CCD camera is saturated, the reflected light adds only minor tails to the PSF since its intensity is reduced by $1/n$ by each reflection.

Contribution of Geometrical Aberrations

In geometrical optics, an ideal lens converts a bundle of rays from a source point to a bundle converging to a single image point. For a realistic optic however, the converging rays do not always pass through the image point. The deviation is the aberration, and the density distribution of these rays at the image plane forms the geometric PSF.

Detailed discussions of geometric aberrations can be found in standard optics textbooks. For beam diagnostics, two types of aberrations are important: spherical and chromatic aberrations. The former comes from the over-focusing of large-angle rays away from the optics axis, and results in focal lengths changing with pupil size. It can be reduced by using small apertures and adjusting the focus after every change of aperture. The latter comes from wavelength dependence of the refractive indices of lens materials, and results in defocusing-like blurring of the PSF. The effect can be reduced with the use of an achromatic lens, a scintillator with spectrum matching the optics, or a bandpass filter to limit the spectral width. It can be totally eliminated by the use of mirror optics.

Ray-Tracing Analysis

A ray-tracing program is not only indispensable for the analysis of the geometric aberrations, but also useful for other types of PSF analyses discussed above. Figure 9 shows a sample screen output from one ray-tracing program, ZEMAX [16]. It gives information about ray offsets from the ideal image point due to their angular and wavelength offset. Moving the source point in the transverse direction, the elongation of a spot for an off-axis source point shows the distortion of the PSF across the field of view. Moving the source point in the longitudinal direction, the change of the spot diagram shows the effect of defocus. Table 4 shows such an effort for the APS chicane flag. From the table, we make the following observations, (1) for monochromatic light, the diffraction-limited resolution dominates at the focal plane; (2) for the narrow-band optics (corresponding to the spectrum of YAG scintillation), the resolution is dominated by the chromatic aberration; (3) for the broadband optics (corresponding to the OTR spectrum), the resolution further worsens by about a factor of two; and (4) when the object is 200 μm out of focus, the defocusing starts to dominate the total resolution. We also conclude that the resolution for this design is within the budget given in Table 3.

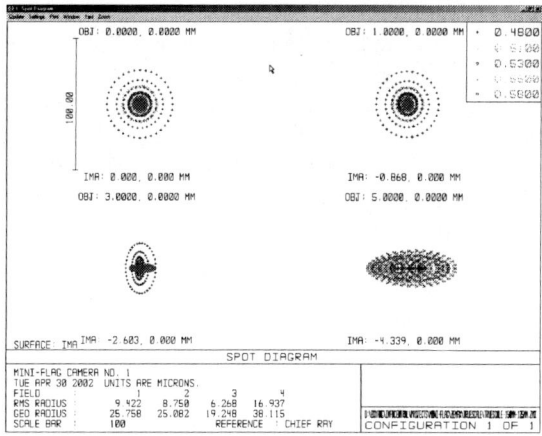

FIGURE 9. A "spot-diagram" from ZEMAX, a ray-tracing program, shows the distribution of rays at the focal plane for different wavelengths, and for on-axis and off-axis source points.

TABLE 4. Rms Radii of the PSF Calculated with ZEMAX.

Configuration	Monochromatic	Narrow band	Broadband
Wavelength (nm)	550	530±50	450-700
Diffraction peak	1.5 μm	1.5 μm	1.7 μm
Vacuum path alone	1.60 μm	5.2 μm	9.0 μm
+ 3.2 mm window	1.61 μm	5.4 μm	9.1 μm
+ Beam splitter & 90º prism	1.8 μm	6.5 μm	10.3 μm
Defocus +200 μm	10.1 μm	13.4 μm	15.1 μm
Defocus -200 μm	10.9 μm	10.4 μm	14.1 μm

We wish to make two notes at the end of this resolution analysis. First, strictly speaking, the quadrature sum rule no longer holds for more than two terms. But in practice, the system resolution is usually dominated by one or two components. Variations of other terms are not important in the overall picture. Here one can lump them with the source size before the analysis of the dominant resolution term.

Secondly, the full wave-optic treatment of light propagation through an instrument, a dream of all optics designers, has recently become available in commercial ray-tracing products. When it matures, the analytical expressions of resolution will be of only semiquantitative value, just like the geometric ray-tracing programs have already made the analytical expressions of aberration semiquantitative tools.

Calibration

Many calibration techniques are in use today, fitting a wide range of budgets and needs for accuracy: (1) known machined features in the FOV, a hole, a slot, or a set of pinholes; (2) grid patterns, especially those formed by small holes / bright spots; or (3) scanning pinholes moved by computer-controlled stages. The key to reliable calibration appears to employ features that the computer recognizes, i.e., imaging software can easily analyze dimensions of the features. For example, using matrices of small holes as calibration targets, their centroid can be used for camera scale calibration while their width can be used for resolution measurements.

The most accurate characterization technique is the use of a back-lit, scanning pinhole moved by computer controlled linear stages [6]. Scanning the pinhole in the longitudinal direction can locate the focal point and determine defocusing properties. Scanning the pinhole in the transverse direction can calibrate the camera scales and measure the resolution change across the field of view (Fig. 10). Adding variations of the wavelength, the 4-dimensional scan can be used to fully characterize the imaging system. Furthermore, the same calibration procedures can be automated and performed many times for consistency checks. Our experience shows that only the scanning pinhole technique has the reproducibility to meet the requirements in Table 3.

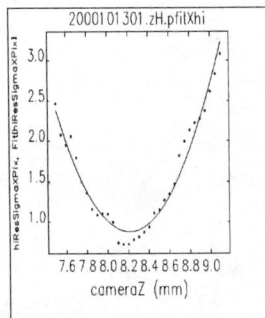 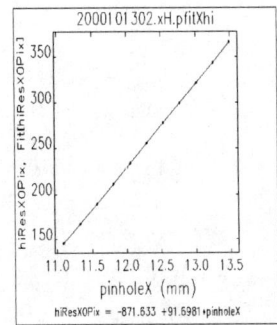

FIGURE 10. (Left) Longitudinal scans of pinhole target pinpoints the focus distance and resolution degradation due to defocusing. (Right) Transverse scans are used to calibrate the camera pixel size.

To summarize the exercise in this section, we make the following observations about typical beam flags with $\sigma_0 \geq 30\mu m$ and WD ≤ 300 mm.

(1) Requirements for the imaging optics are well within the current technological capabilities. Industrial-grade optical components are sufficient for the application.

(2) To consistently obtain 5% or better measurement accuracy, due attention should be paid to reproducible focus, reliable calibration, mechanical stability, and overall serviceability.

(3) Current crystal scintillators have high conversion efficiency, but problematic spatial resolution (blurring); while the OTR screens have good spatial resolution but low conversion efficiency. Major advances in efficient, high-resolution converter screens are highly desirable.

SYNCHROTRON RADIATION IMAGING SYSTEM

Synchrotron radiation (SR) is produced when high-energy charged particles pass a magnetic field and their trajectory is bent on a circular orbit [17,18]. The wavelength of SR spans from infrared to hard x-rays. Its use for diagnostics offer the following advantages:

(1) The diagnostic is not intrusive. Particle beams can be studied without disturbing accelerator operation.
(2) The radiation process is fast. At short wavelength, the photon pulse generated by a single electron lasts only $\sim \rho/\gamma^3 c$ (ρ = electron trajectory radius), usually lasting less than 1 fs. At the long wavelength limit, the pulse length is determined by the optics / spectrometer and could last several wavelengths. Hence the particle bench length can be measured and longitudinal dynamics can be studied in time-domain with streak cameras.

We will discuss the optical synchrotron radiation (OSR) and x-ray synchrotron radiation (XSR) separately due to the substantial differences in their instrumentation.

Optical Synchrotron Radiation Imaging

For most rings, the OSR wavelength is longer than the critical wavelength ($\lambda_c = 4\pi\rho/3\gamma^3$), and it is directed in the forward direction over a cone with the opening angle,

$$\sigma'_y = 0.73 \left(\frac{\lambda}{\rho} \right)^{1/3}, \quad (\lambda > \lambda_c). \tag{22}$$

For optical wavelength ($\lambdabar \approx 0.1$ nm), this angle is in the range of 1 to 3.4 mrad over a wide energy range, since ρ is in the range of 1 – 5 m for low-energy rings ($\gamma < 2000$) and 10 m – 40 m for high-energy rings ($\gamma > 2000$). We can also derive the diffraction-limited resolution

$$\sigma_y = \frac{\lambda}{2\sigma'_y} = 0.68 \left(\rho \lambdabar^2 \right)^{1/3}. \tag{23}$$

For $\lambda = 630$ nm, we have $\sigma_y [\mu m] = 15 \rho[m]^{1/3} \approx 15 - 50$. At high-energy rings, where the beam is smaller and the limit is higher, this constraint is very important. Much ingenuity has gone into overcoming the engineering challenges and realizing the diffraction limit.

(1) **Acceleration chamber modification:** To realize the full resolution, vertical and horizontal acceptance of the imaging system needs to be greater than $6 \cdot \sigma'_y$, often in the range of 10 – 30 mrad. Such a large opening is difficult to obtain since it conflicts with machine designers' desire of minimizing chamber discontinuities and wakefield generation.
(2) **High angular resolution:** The large particle orbit radius results in long working distance. Angular sizes of the beam often fall below 10 μrad. Mechanical supports of the optical elements need to be very sturdy. If a high flux throughput is desired, or temporal dispersion is of concern (for example, in a high-speed imaging system), an all-mirror imaging system is recommended.
(3) **Air current:** If a long transport is used, air current and density fluctuation in the air could easily deflect the light beam by microradians. An enclosed or evacuated beam path is recommended.

(4) **Cooling of the first mirror:** The synchrotron radiation power could range from several watts to kilowatts. A mirror heated in the front surface distorts significantly. A uniform heating of the surface would bend the mirror and move the focus in the y-plane (bending plane) downstream from that in the x-plane. The nonuniform heating, due to the concentrated x-ray fan on the orbit plane, creates a high-stress/strain region. It pushes the upper and lower portion of the mirror to rotate in opposite directions (Fig. 11A) and results in a vertically split image. To mitigate the problem, several approaches have been tried with partial success, all using the fact that the opening angle of the visible light is much larger than that of x-rays: half of the mirror outside of the x-ray fan [19], a grazing incidence mirror to spread out the heat load [20], an upstream blocking tube to block the x-rays, and a mirror with a slot aperture to allow the x-ray fan to pass [21].

One of the important features of the OSR imaging is its time-resolved imaging capability. Figure 11B shows sample streak camera pictures using OSR.

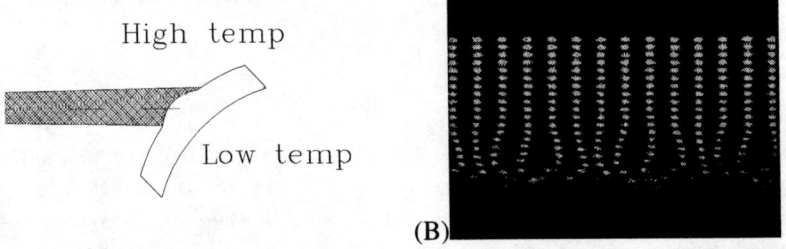

(A) (B)

FIGURE 11. (A) Deformation of mirror under synchrotron radiation heating. (B) Streak camera images of a bunch train in the APS storage ring [22].

OSR Interferometer

Before discussing the OSR interferometer, we revisit the Helmholtz diffraction integral, Eq. (11), in the one-dimensional case. This time, we divide the entrance aperture into slits of width Δ and group together the integral from pairs of symmetrically placed slits, $n\Delta \leq |\theta| \leq (n+1)\Delta$, $n = 0, 1, 2, \ldots$. The resulting amplitude from the n-th pair is

$$A_{focus}^{(n)}(x) \cdot e^{i\omega t} \approx 2\Delta \cdot A\left[\left(n+\frac{1}{2}\right)\Delta\right] \cdot \cos\left[\frac{x}{\lambda}\left(n+\frac{1}{2}\right)\Delta\right] \frac{\sin\left(\frac{x\Delta}{2\lambda}\right)}{\frac{x\Delta}{2\lambda}} \cdot e^{i\omega t}. \quad (24)$$

We can see that the amplitude is the product of three functions (Fig. 12): the light amplitude at the slits, a cosine function with (spatial) frequency determined by the angular distance of the slits $n\Delta$, and a sinc function determined by the slits' width. For the total sum to form a point-like image, all these functions need to have the same phase at the focal point, $x = 0$. When the first mirror distorts symmetrically, a phase shift is added to the time-dependent factor, $e^{i(\omega t+\phi(n\Delta))}$, and when it distorts asymmetrically, another factor is added to the spatial oscillating factor,

$\cos\left[\frac{x}{\lambda}\left(n+\frac{1}{2}\right)\Delta+\psi(n\Delta)\right]$. These additional phase shifts make it impossible to obtain a sharp, single-peaked PSF. This problem has an analogy in electronics: To faithfully amplify a short pulse, an amplifier needs to have a flat gain curve in a wide frequency band and a linear phase shift as a function of frequency, with good signal-to-noise ratio. When these conditions cannot be met, electronics engineers often use narrow band electronics to extract partial information from the pulse. Information about the shape of the pulse, or existence of side pulses is often lost in such analyses. Similarly, the OSR interferometer tries to overcome the mirror distortion by analyzing the information from one pair of slits at a time. Due to the transverse size of the source, the observed interference fringes are the convolution of the source distribution with the single particle diffraction pattern. As a result, the blur of the interference fringes are different for different spatial frequencies, hence the source size can be obtained through analyses of the fringe visibilities as a function of spatial frequency [4,23].

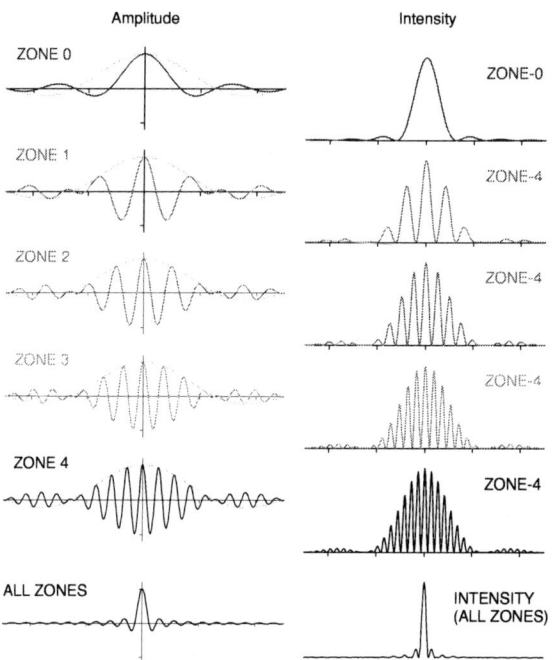

FIGURE 12. (Left) light amplitudes from symmetric slits pairs (zones) at increasing distance from the optical axis. The bottom figure shows the sum of all amplitudes, showing formation of the PSF. (Right) Light intensity at focal plane if only one zone is open. The bottom figure shows the intensity when all zones are open.

The OSR interferometer is fairly effective in overcoming the mirror distortion problem. It has rapidly developed into an important technique for beam size measurements in the past several years.

X-ray Synchrotron Radiation Imaging

For most rings, x-ray synchrotron radiation (XSR) wavelength is near or shorter than the critical wavelength. Its opening angle is smaller than that of the OSR

$$\sigma_{y'} \approx \frac{0.64}{\gamma}\sqrt{\frac{\lambda}{\lambda_c}}, \quad (\lambda < \lambda_c), \tag{25}$$

or in practical units

$$\sigma_{y'}[\mu rad] \approx 190\sqrt{\lambda[A]B[T]}, \quad (\lambda < \lambda_c), \tag{26}$$

where B is the bend magnet field. If the full wavefront is used, the diffraction-limited resolution is

$$\sigma_R[\mu m] \approx \frac{\lambda}{2\sigma_{y'}} = 0.26\sqrt{\frac{\lambda[A]}{B[T]}}, \quad (\lambda < \lambda_c). \tag{27}$$

For x-ray wavelength ($\lambda \approx 0.1$ Å), the rms opening angle is about 50 μrad and diffraction-limited resolution is less than 0.1 μm. To the author's knowledge, no one has been able to obtain such resolution for direct imaging, most likely due to the difficulty of fabricating atomic-accuracy x-ray optics. While focusing optics have been used to directly image the beam, including Kirkpatrick-Beaz mirrors and zone plates [24], it is the pinhole camera that enjoys the most popularity [25,26].

X-ray Pinhole Camera

The resolution of the pinhole camera has been discussed in a number of works. One approach is to calculate the geometric shadow and Fraunhofer diffraction separately and take the convolution of the two as the true resolution of the pinhole camera (mixed model). Another is to use a Fresnel diffraction approximation. Their results are summarized in Fig. 13. The width of the slits is in units of $\sqrt{\lambda f}$, while the resolution at the object plane is in units of $S'\sqrt{\lambda/f}$, where

$$\frac{1}{f} = \frac{1}{S} + \frac{1}{S'}, \tag{28}$$

and S and S' are the distance of the source and detector from the pinhole aperture, respectively. Note that the optimum aperture in Fresnel model is about 50% larger than that of the mixed model, and the optimum resolution is about 30% better. This can be understood in terms of the Fresnel zones: when the aperture is just slightly larger than the first Fresnel zone, all light amplitudes from the aperture are of the same sign at the image plane. Hence the central peak is the highest, and by energy conservation, the peak is also narrowest at the same time.

Figure 14 shows the pinhole camera used in the APS storage ring. The first component is a 1-mm water-cooled aperture used to restrict the radiation power load of the pinhole aperture to several watts. The pinhole apertures are located 9 m from the source. They are made of four independent tungsten blades with openings usually set at 15 μm. Temperature-regulated water flow is used to maintain the blade at a constant temperature to avoid current dependence of measurements. The vacuum window is located at 16 m, close to the detector to minimize the effect of small-angle scattering. A scintillator / optics / camera combination reads out the x-ray image.

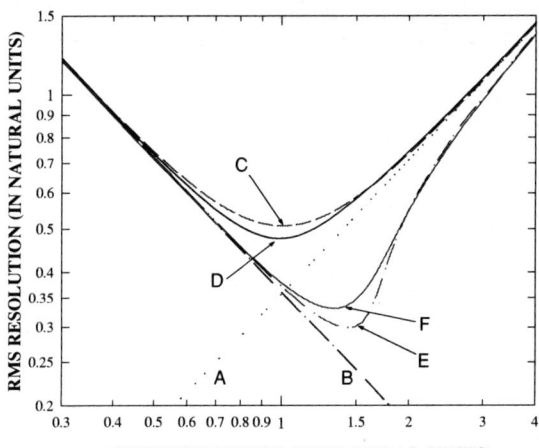

FIGURE 13. Effective resolution of pinhole cameras with different models. (A) Geometrical shadow; (B) Fraunhofer diffraction peak; (C) quadrature sum of the widths of geometric and Fraunhofer peaks; (D) convolution of geometric shadow and the Fraunhofer diffraction peak; (E) monochromatic Fresnel diffraction peak; and (F) multiwavelength Fresnel diffraction with rms spectral width equal to 30% of the center wavelength. The natural units of the aperture and the source are defined as $\sqrt{\lambda f}$ and $S'\sqrt{\lambda/f}$, respectively.

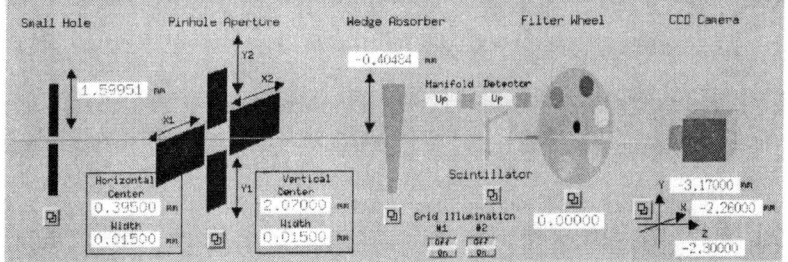

FIGURE 14. Control screen schematic of the APS storage ring pinhole camera.

ACKNOWLEDGMENTS

The author wishes to thank A. Lumpkin, G. Decker, and O. Singh for their encouragement and support. This work is supported by the U. S. Department of Energy, Office of Basic Energy Sciences, under contract no. W-31-109-ENG-38.

REFERENCES

1. Wilke, M., "Optical and X-ray Imaging of Electron Beam Using Synchrotron Emission," BIW94, AIP Proc. 333, 1994, pp. 128-147.
2. Hofmann, A., "Beam Diagnostics and Applications," BIW98, AIP Conf. Proc. 451, 1998, pp. 3-22.
3. Jung, R., "Image Sensor Technology for Beam Instrumentation," BIW98, AIP Conf. Proc. 451, 1998, pp. 74-93.
4. Mitsuhashi, T., "Spatial Coherency of The Synchrotron Radiation at the Visible Light Region and its Application for the Electron Beam Profile Measurement," Proc. Particle Accelerator Conference 1997, 1998, pp. 766-768.
5. Heimann, P. et al., "Experimental Characterization of ALS Undulator Radiation," Rev. Sci. Instrum. 66, 1885 (1995).
6. Yang, B. et al., "Design and Performance of a Compact Imaging System for the APS Linac Bunch Compressor," Proc. Particle Accelerator Conference 2001, 2001, pp. 2335-2337.
7. Borland, M et al., "A Highly Flexible Bunch Compressor for the APS LEUTL FEL," Proc. LINAC 2000, 2000, pp. 863-865.
8. Autrata, R. et al., "Single Crystal Aluminates: A New Generation of Scintillators," Scanning Electron Microscopy II, 1983, pp. 489-500.
9. Graves, W. S. and Johnson, E.D., "A High Resolution Electron Beam Profile Monitor," Proc. Particle Accelerator Conference 1997, 1998, pp. 1993-1995.
10. Rodnyi, P. A., *Physical Processes in Inorganic Scintillators*, New York, CRC, 1997.
11. Lumpkin A. H. et al., "Time Resolved Imaging for the APS Linac Beams," Proc. of LINAC 98, 1998, pp. 529-531.
12. Murokh, A. et al., "Limitations on Measuring a Transverse Profile of Ultra-Dense Electron Beams with Scintillators," Proc. Particle Accelerator Conference 2001, pp. 1333-1335.
13. Fiorito, R. B. and Rule, D. W., "Optical Transition Radiation Beam Emittance Diagnostics", AIP Conference Proceedings No. 319, 1934, pp. 21-37.
14. Born, M. and Wolf, E., Principles of Optics, Oxford, Pergamon, 1980.
15. Artru X. et al., "Resolution Power of Optical Transition Radiation: Theoretical Considerations," Nucl. Inst. Meth. B145, 160 (1998).
16. ZEMAX Optical Design Program, Focus Software, Tucson, 1992-2002.
17. Kim, K. J., "Characteristics of Synchrotron radiation," AIP Conf. Proc. 184, 1989, pp. 567-632.
18. Koch, E.-E. ed., Handbook on Synchrotron Radiation, North-Holland, Amsterdam, 1983.
19. Scheidt, K., "UV and Visible Light Diagnostics for The ESRF Storage Ring," Proc. European Particle Accelerator Conference 1996, 1996, pp. 1621-1623.
20. Fisher, A. S. et al., "Design and Initial Commissioning of Beam Diagnostics for the PEP-II B Factory," Proc. Particle Accelerator Conference 1997, 1998, pp. 2253-2255.
21. Rotela, E. R. et al., " High Precision, High Heat-Load Mirror for the APS Diagnostics Beamline," Proc. SRI95, 1995, CD-ROM.
22. Yang, B. et al., "Characterizing Transverse beam Dynamics at the APS storage Ring Using a Dual-Sweep Streak Camera," AIP Conf. Proc. 451, 1998, pp. 224-236.
23. Hiramatsu, S. et al., "Measurement of Small Beam Size by the Use of SR Interferometer," Proc. Particle Accelerator Conference 1999, 1999, pp. 492-494.
24. Keller R. et al., "Electron Beam Diagnostics Using Synchrotron Radiation at the Advanced Light Source," AIP Conf. Proc. 390, 1996, pp. 240-247.
25. Yang, B. et al., "Recent Developments in Measurement and Tracking of the APS Storage Ring Beam Emittance," AIP Conf. Proc. 546, 2000, pp. 622-630.
26. Limborg, C. et al., "A Pinhole Camera for SPEAR 2," Proceedings of EPAC 2000, 2000, pp. 1774-1776.

Fast Digitization and Digital Receiver Technology

Ralph C. Kimball

Echotek Corporation, 555 Sparkman Drive, Suite 400, Huntsville, AL 35816 USA

Abstract. The potentially lucrative wireless market has led to technological advances in mixed signal devices such as high speed, high resolution A/D and D/A converters. This same market has also driven the development of high performance multi-channel digital receiver and digital transmitter ICs. Similarly, advances in semiconductor processes, coupled with the need for reduced time-to-market, has led to the development of large, enhanced performance, in-circuit programmable logic devices. A review of the key characteristics of these mixed-signal, signal processing and programmable logic devices is presented. The application of these devices and technologies to the instrumentation of Accelerators and Storage Rings is discussed and presented by way of examples. Issues relating to the requirements associated with real-time processing, I/O throughput, reconfigurability, reliability, maintainability and packaging requirements are also addressed.

INTRODUCTION

The instrumentation requirements associated with Linear Accelerators and Storage Rings involves a broad mixture of analog and digital technologies. While the process monitoring sensors and control loop feedback driven correction elements are inherently analog, the data analysis and control loop response is increasingly dependent upon digital signal processing (DSP). This architecture leads to the need for an Analog-to-Digital (A/D) converter (or ADC) to translate the input analog sensor data into a digital data stream for processing, and for a Digital-to-Analog (D/A) converter (or DAC) to translate the digital feedback corrections to analog, thereby completing the control loop. Clearly, the characteristics of these mixed-signal devices are therefore critical to the performance assessment of the overall instrumentation subsystem.

Similarly, the computational and throughput requirements placed on instrumentation subsystem digital processing elements must be well understood. This is necessary in order to properly define DSP architectures that will provide sufficient resources and flexibility at a reasonable cost. Consideration must also be given to both centralized and distributed processing configurations in order to embrace the full breadth of instrumentation requirements, from the subsystem level to the total system level. Also, early obsolescence due to the rapid rate of advancements that are associated with today's digital technology is an ever-growing concern to the instrumentation engineer. Accordingly, product roadmaps and reliance on internationally recognized "Standards" such as ANSI, IEEE, etc. must be entered into the instrumentation design equation to mitigate the probability or impact of such events.

In view of the fact that the current trend, for both the mixed-signal and DSP elements of the above instrumentation architecture, is toward reduced geometry with increased performance, additional consideration must be given to the "packaging" of these devices. Since shrunk semiconductor process geometries result in reduced power rail voltages, most of the newer devices require multiple different DC voltages to power the analog section, the "core", and the input/output (I/O) section. Also, the reduced process geometries support smaller physical packaging with fine-pitch pins or balls, and can operate at significantly higher clock rates. These characteristics translate into the need for fine-geometry, multi-layer, stripline Printed Circuit Board (PCB) designs with special consideration given to manufacturability and testability of the end product. Furthermore, sufficient documentation must be provided for each element of the design to further ensure the utility and longevity of that product design.

A/D CONVERTER ARCHITECTURES

A number of different A/D architectures have evolved over the years, primarily driven by application requirements, semiconductor technology and cost. In general, the most popular A/Ds of today can be categorized as being a Flash, Pipelined, Successive Approximation, or Sigma-Delta design. The term "sampling A/D" is commonly used to indicate that the A/D front-end has a Sample-and-Hold (S/H) circuit to sample (or track) the input signal during one phase of the encode clock and to hold that sampled level throughout the other phase of the encode clock. Sampling A/Ds that are designed for digitizing Intermediate Frequency (IF) inputs typically have a minimum encode rate specification. This is primarily due to optimizing the S/H circuit and "droop" will result from lower frequency encode rates, owing to charge "bleed-off" in the hold circuit. Similarly, the term "under sampling" (also referred to as "super Nyquist" sampling) is used to categorize the use of an A/D to also serve as a down-conversion mixer. For a given encode clock frequency F_s, the A/D will alias the digitized signal down in bands, or "Nyquist zones" (i.e., $F_s/2$ bands), with spectral inversion occurring for the second Nyquist zone ($F_s/2$ to F_s), the fourth Nyquist zone ($3*F_s/2$ to $2*F_s$), etc. While most A/Ds are optimized for first Nyquist zone applications (i.e., DC to $F_s/2$), a number of the newer A/D converters are targeted at higher Nyquist zone applications to permit direct digitization of common communication system IFs.

Flash A/Ds utilize $2^n - 1$ comparators to produce an n-bit wide output in a single clock cycle. These devices are typically very fast and limited to about 8 to 10 bits, owing to the need for large numbers of closely matched comparators and the associated input loading. While Flash A/Ds were often used without a S/H, or the

S/H was provided externally, the current trend is to include very broad bandwidth S/H circuits integrated into the Flash A/D front-end to improve performance and to permit under-sampling high frequency inputs. Today, Flash A/Ds are typically employed in high input frequency digitization applications with clock rates from 500 MHz to >1.5 GHz, and can also be typified as being physically large (i.e., >1 square inch) and having relatively high power dissipation (i.e., 4 to 5 watts).

The Pipelined architecture takes the Flash architecture a step further by dividing the digitization process into a series of pipelined stages. Each stage employs an input S/H, an m-bit A/D, a D/A to convert the A/D digital value of that stage to an analog voltage, and a difference circuit with gain that produces a residual output. This output represents the difference of the input signal to that stage with the D/A output of that stage. This "residual" is then input to the next stage's S/H and the process is repeated. Since each stage is pipelined, new digital output values will be available at the encode clock rate once the pipeline (number of stages) is filled. The sum of the bit-widths of the "sub-ranging" stages is typically a couple of bits wider than the actual output sample bit-width "n", and this bit-width reduction is accomplished by a pipelined digital "correction" logic section. In view of the above processing, this architecture is also sometimes referred to as a "Sub-Ranging" A/D and dominates the wireless A/D market. These devices are typically small (i.e., about 0.5 square inches), relatively low power (i.e., <1 to 1.5 watts), and currently support clock rates of >100MHz at a 14-bit resolution, to >200 MHz at a 12-bit resolution.

The Successive-Approximation A/D utilizes a single comparator in conjunction with a register-driven D/A and a difference circuit. For each new conversion, the register is loaded with a ½ of full-scale value and the comparator output then represents the Most Significant Bit (MSB). Based on this MSB value, the register is then updated by a value, appropriately signed and of a ¼ of full-scale value. The process is then repeated and this is done a total of n times, utilizing smaller binary weights in progression each time, as required to produce an n-bit output. This process therefore requires a full n-clocks to produce an n-bit output, rather than the single clock per output required by the Flash A/D, or the Pipelined A/D once the pipeline is filled. This implies that the output sample data rate for a Successive-Approximation A/D is, at most, 1/n that of either of the other two architectures. Also, it should be noted that S/H droop would be significant if the circuit was not designed for long hold times relative to the A/D clock rate. Consequently, these converters are not well adapted for higher input frequency analog signal digitization applications.

The Sigma-Delta A/D architecture is significantly different in that it employs a very simple front-end consisting of a difference circuit, an integrator followed by a comparator, and a 1-bit D/A. A digital filtering logic section then follows this front-end. The D/A output is differenced with the input being digitized, and this difference is then input to the comparator. Consequently, the comparator output consists of a string of ones and zeroes, at the clock rate, that are input to both the

D/A and the digital filter logic. This bit stream of ones and zeroes is digitally filtered and decimated to produce the n-bit output sample. The combination of a very high clock rate relative to the output sample rate, with digital filtering to shape the passband, makes this architecture desirable for low frequency, narrow bandwidth, high-resolution A/D applications.

A/D SPECIFICATIONS

The theoretical Signal-to-Noise Ratio (SNR or S/N) for an n-bit A/D is the ratio of the rms full-scale, digitally reconstructed, analog input "V", to its rms quantization error (i.e., $V*LSB/\sqrt{12}$), where LSB is the Least Significant Bit. Since the rms value of the full-scale input is $0.5V/\sqrt{2}$, the theoretical maximum SNR = $2^n \sqrt{3}/\sqrt{2} = 1.225*2^n$. Since SNR is typically specified in decibels, the numerically simple and therefore commonly used expression is $SNR_{db} = 20Log(2^n \sqrt{3}/\sqrt{2}) = 20Log(2)*n + 20Log(\sqrt{3}/\sqrt{2}) = 6.021*n + 1.763$ (dB). This is also the Dynamic Range (DR) for an A/D, since DR is defined as the ideal SNR for that A/D (i.e., quantization noise limited). The "data sheet" SNR specifications for an A/D are typically extrapolated values given as a function of input signal frequency at a specific level below full-scale and for a given A/D encode rate. Similarly, the Signal-to-Noise and Distortion (SINAD) ratio specifications simply include the first N harmonics of the Total Harmonic Distortion (THD), where N is normally specified on the data sheet. The Effective Number of Bits (ENOB) is simply the reversal of the SNR calculation but is typically based on SINAD to reflect harmonic distortion (i.e., ENOB (bits) = (SINAD-1.763)/6.021).

For IF-sampling A/Ds, Aperture Uncertainty or "jitter" in the S/H circuit contributes to the noise floor owing to sample-to-sample timing variations. The theoretical SNR for an A/D limited by aperture uncertainty is SNR = $-20 Log(2\pi F_{in} T_{au})$, where T_{au} is the aperture uncertainty in seconds rms. Obviously, the effects of this jitter are more pronounced with increasing input signal frequency due to increasing dV/dt. The aperture uncertainty specification for a sampling A/D is commonly given in picoseconds rms. Typical values for examples of today's high speed, high resolution sampling A/Ds are <0.25 ps rms for the Analog Devices AD9430 (12-bits at a >200 MHz encode rate and a 700 MHz input bandwidth) and <0.1 ps rms for the Analog Devices AD6645 (14-bits at a >100 MHz encode rate and a 270 MHz input bandwidth). While these aperture uncertainties seem very small, it should be noted that, for an A/D like the AD9430 that could be used to directly digitize the typical RFs associated with Accelerators and Storage Rings, the 0.25 ps rms specification translates into a theoretical SNR of 96 dB for a 10 MHz input, to 76 dB for a 100 MHz input, and to 62 dB for a 500 MHz input. It should also be noted that the S/H associated with a sampling A/D acts essentially like a mixer in that the input signal is multiplied by the encode clock. Consequently, the above noted performance is further degraded by the presence of jitter on the encode clock. For this reason, it is imperative that the encode clock be generated by a very low phase noise source.

The Spurious-Free Dynamic Range (SFDR) of an A/D is probably the most frequently misunderstood of all A/D specifications and, as such, deserves special attention. This is sometimes due to confusing the determination of the SFDR for an A/D with that of an amplifier. An engineer at Avantek first introduced the concept of Intercept Point as an indicator of the SFDR of an amplifier in 1964. When an amplifier is operating in the linear range (i.e., below the 1 dB compression point), the levels of the spurious responses can be estimated accurately with a simple equation: SFDR = 2/3(P1-P0-10Log(BW)-NF). P1 is the input Intercept Point obtained by subtracting the amplifier gain from the output (third order) Intercept Point; P0 is the effective input noise power with no signal (i.e., -114 dBm for a 1 MHz bandwidth in a room temperature 50 ohm environment) ; BW is the system noise bandwidth in MHz; and NF is the amplifier noise figure in dB. For example, an amplifier with a third order Intercept Point of +38 dBm, a noise figure of 4 dB and a gain of 16 dB would have a 1 MHz bandwidth SFDR of: 2/3(+38-16-110) = -88 dBm, or SFDR = 88 dB. It should be noted that increasing the bandwidth will reduce the SFDR and decreasing the bandwidth will increase the SFDR.

The SFDR of an A/D is typically specified as the ratio of the rms signal amplitude to the rms value of the peak spurious spectral component, and the peak spurious component may or may not be a harmonic. SFDR may be specified as dBc (i.e., dB below the carrier), in which case it will degrade as the input signal level is lowered, or it may be specified as dBFS (i.e., always referred back to the converter full-scale input level). The important point here is that there is no mention of bandwidth in the A/D SFDR definition, unlike that for the amplifier SFDR definition. Consequently, the A/D SFDR specification, as given on an A/D's data sheet, is simply based on the highest spur level observed on a DFFT associated with a given data set or sets, for the specified input and operating conditions (i.e., for the specified Nyquist zone).

It should also be noted that the SFDR specification for an A/D could be significantly greater than the A/D's ideal SNR (i.e., DR). For example, the Analog Devices AD6644 is a 14-bit, 65 MSPS (Million Samples Per Second) A/D with a DR = 6.021 * 14 + 1.763 = 86.06 dB, but has single-tone and multi-tone SFDR specifications of >90 dB and >100 dB, respectively. This "phenomenon" is most easily explained by the concept of "processing gain". For today's highly linear A/Ds, like the AD6644, input signal variations that are less than the resolution (an LSB step) can be accurately extracted from the noise by digitally processing multiple samples. The signal of interest is essentially buried in the noise that is randomizing the A/D states, such that the periodicity of the signal (i.e., correlated signal) can be differentiated from the (uncorrelated) "white" noise.

It is for this reason that "dither" noise is sometimes added to the A/D input signal to improve the A/D's linearity, and therefore improve its SFDR. This is typically accomplished by adding white noise, band-limited to a few hundred kHz, to the input signal to be digitized. This practice is primarily limited to digital receiver applications or other applications where sharp digital filtering of the digitized sample stream is accomplished by post-processing. This is due to the fact that this noise must be placed in a portion of the spectrum such that it can be subsequently removed by digital filtering. The minimum input power level for the noise being added is normally dependent on the specific A/D and its architecture. Dither is most frequently used with sub-ranging A/Ds and the minimum power level is generally just enough to cause the input signal to cross one or more sub-ranges, thereby mitigating localized nonlinearities. Dithering by adding white noise is most advantageous when dealing with relatively pure single-tone inputs, since multi-tone and noisy single-tone inputs are essentially self-dithering.

Processing gain is realized in the Frequency domain by the DFFT imparting a gain of 10Log(number of points) (dB) for complex samples, thereby dropping the noise floor in each frequency bin and exposing the low level correlated signals. Increasing the A/D sampling rate also lowers the noise floor, because the noise spreads out over more frequencies, while the total integrated noise remains constant. In the time domain, decimating digital filters can therefore impart a processing gain of up to 3 dB for each halving of the input noise bandwidth. This is accomplished by digitally lowpass filtering the input data sample stream at rate F_{in} and then reducing the output data sample rate to F_{out} (note: F_{out} must be consistent with the new output bandwidth sample rate requirement). The digital lowpass filter and data rate decimation process is generalized as yielding a S/N improvement equal to 10Log(decimation), but it should be noted that this assumes that the input noise bandwidth is being reduced proportional to the decimation factor. For example, the digital lowpass filter must have a (passband + transition band) ±0.125 F_{in} to permit a decimated output sample rate $F_{out} = F_{in} / 4$, thereby reducing the noise bandwidth by 4 and realizing a processing gain (S/N improvement) of up to 6 dB.

While there are numerous other important A/D specifications, they will not be addressed here in the interest of brevity. Instead, the focus is on today's high speed, high-resolution A/Ds, since these devices are most applicable to direct digitization of the input RF and IF signals associated with Accelerators and Storage Rings. There is, however, one observation with regard to these devices that relates to the above discussion of noise floor and SFDR. These A/Ds can, in the absence of an active input signal, output digitized noise floor sample streams that contain just two or three distinct values. Accordingly, some correlation can exist within these digital sequences, and are manifested as "spurs" appearing on an otherwise white noise floor, when viewed on the associated DFFT. These spurs are easily identified since, inputting a signal that often may even be at a level below one LSB, will cause that signal to appear and the noise floor to then be white.

D/A CONVERTERS

It is apparent from the previous discussion of A/D converter architectures that D/As are an integral part of A/Ds, and critical to the performance characteristics of those devices. Additionally, the use of stand-alone D/As is required to complete the digital processing based control loop topology. Again, it is tacitly assumed that high-resolution devices utilized in relatively low frequency correction applications are well understood and that the emphasis is to be placed on high frequency Accelerator and Storage Ring analog control requirements.

Most of today's high speed, high resolution, D/As are current-based designs, whereby the digital input code is translated into a balanced differential output current-pair that is linearly related to that input code. The resolution of an n-bit D/A is then a function of the full-scale current setting (typically about 20 ma) divided by $2^n - 1$. A number of the newer high-resolution D/A designs employ multiple switched current segments, somewhat analogous to the sub-ranging A/D architecture, to improve linearity and increase dynamic performance. The differential current outputs can be converted to voltage outputs by resistive terminations if signal reconstruction down to DC is required, or more commonly for the IF applications being addressed, by transformers configured to match a given load impedance (normally 50 ohms) and to perform a differential to single-ended voltage conversion.

The "glitch" energy (i.e., narrow unwanted spikes on the output occurring at the update clock rate) used to be a D/A specification of major concern and designers often placed a S/H after the D/A to sample the output, after the glitch interval, and to hold that level during the next sample update. The more recent D/A designs can be typified as having very low glitch energy (i.e., <5 pV-s) and it is increasingly common for this specification to be omitted from D/A data sheets. This is due in part to the fact that these devices are typically being used in band-limited IF output applications whereby this relatively low glitch energy is well filtered. Accordingly, these device data sheets opt for dynamic performance characterization and emphasize SFDR, SNR, etc., much the same as for the high speed, high resolution A/Ds. However, it should be noted that, unlike for an A/D, the SFDR for a D/A is specified as the difference, in dB, between the rms amplitude of the output signal and the peak spurious signal over the specified bandwidth.

Since the D/A-output reconstructed waveforms consist of a series of discrete samples that are updated at the clock rate, F_{out}, the output analog response exhibits a sinx/x roll-off that, for the first Nyquist zone, will vary from 0 dB at DC, to -4 dB at $F_{out}/2$. Additionally, the output waveforms will be replicated in each successive Nyquist zone, with spectral inversion occurring for the second Nyquist zone, the fourth Nyquist zone, etc., as was the case for A/Ds. Unlike the A/D's, where the level of the replicated images was determined by the A/D's front-end analog input response, the sinx/x roll-off continues for the replicated D/A-produced images. Hence, the response will be -13 dB at $3F_{out}/2$, -18 dB at $5F_{out}/2$, etc. which explains why D/As are normally associated with relatively low IF output frequencies. In fact, the maximum output frequency to be reconstructed by a D/A is commonly designed to be <$0.25F_{out}$ in order to keep the sinx/x roll-off to <0.22 dB.

In view of the fact that the digitally reconstructed output waveform sinx/x response is a function of the D/A output sample update rate, the D/A manufacturers have placed major emphasis on increasing the supported update rate, while maintaining good dynamic device performance characteristics. Low power, high resolution (12-bit to 16-bit) devices, such as the Analog Devices AD97xx family of D/As, are currently available with output sample update rates to 400 MSPS. However, these devices would impose serious I/O bandwidth requirements were it not for their inclusion, within the devices, of interpolating digital Finite Impulse Response (FIR) filters. This digital processing consists of increasing the sample clock rate and inserting samples of zero value between the valid input samples, lowpass filtering the resultant sample stream, and applying gain to correct for the drop in signal level owing to the inserted zero values.

For example, interpolation by 4 would increase the sample stream data rate "F" by 4 to F_{new} and insert 3 zero values between each original F sample rate input value. The new sample stream would then be lowpass filtered with a FIR filter design that provides a passband of DC to <$\pm 1/8F_{new}$ and a stopband from $\pm 1/8F_{new}$ to $\pm(F_{new}/2)$. Since the insertion of 3 zero values along with each original input value reduces the resultant output level to one fourth that of the original level, a gain of 4 would be applied to the output sample data stream. This is accomplished by shifting each binary sample value left two bits toward the MSB (i.e., multiplication by 4). These D/A devices are referred to as "interpolating" D/As, and this architecture is dominating the more recent D/A designs because it mitigates the I/O bandwidth requirements associated with achieving a flat output response. In fact, these designs, particularly when employing higher interpolation rates, even target applications at the common communications IF (70 MHz) market. This is possible because reconstructed output waveforms, with frequency content as high as 100 MHz, still fall within the 0.22 dB flat output response at D/A output update sample rates of <400 MHz.

DIGITAL RECEIVERS

The classic analog heterodyne (or super-heterodyne) receiver typically employs a low-noise amplifier followed by a mixer that is driven by a Local Oscillator (LO) to downconvert the input RF to a lower IF. The mixer is then typically followed by one or more analog filters to band-limit the IF output (i.e., eliminate the unwanted mixing product terms or "images"). To obtain a high SFDR, these mixers typically must have a high IP3 and therefore require high LO drive levels. Also, depending upon the conversion loss of the mixers and insertion loss of the filters, additional IF amplification stages may be required. The IF signal A(t) is then converted to Inphase (I) and Quadrature (Q) signals via a 90-degree (quadrature) hybrid (i.e., produces a 0 degree and a 90 degree phase-shifted output signal pair) and downconverted to baseband. The quadrature components are: I (t) = A (t) cosine (2πF (t)) and Q (t) = A (t) sine (2πF (t)), where F is the downconversion frequency. These quadrature signals are then Square-law (diode) detected and lowpass filtered. For single-sideband applications, the 90-degree hybrid and mixers are more commonly replaced by an I&Q demodulator. This device combines two balanced mixers and a 90-degree hybrid in a single device that permits down conversion with quadrature outputs that are selectable as either the upper-sideband, or the lower-sideband, with rejection of the carrier (LO) and the unselected sideband.

This conversion to quadrature signals is necessary because a Square-law detector is simply a device whose output voltage is proportional to the square of its input voltage and, therefore, contains no phase information. Consequently, two signals in quadrature are required, each Square-law detected, to produce a Baseband complex signal (i.e., "video output pair") that preserves both the amplitude and the phase information, as required for subsequent processing. When that subsequent processing involves digital demodulation or information processing, these quadrature Baseband signals are digitized by A/D-pairs to produce quadrature (I and Q) digital data streams, as required.

A digital receiver (DR) performs the exact same function, but does it digitally, and potentially, much better. In an ideal world, a DR would utilize a single A/D to directly digitize the input RF signal. In the real world, the A/D is utilized at the earliest possible point in the classic heterodyne receiver chain, typically at an IF stage. However, it should be noted that advancements in A/Ds, as discussed above, make the delineation of what is IF and what is RF fuzzy, at best. Consequently, let's assume that the RF associated with Accelerators and Storage Rings can be directly digitized by an appropriately selected high speed, high-resolution A/D.

Digitization Requirements

It is first necessary to realize that the A/D must be digitizing a band-limited signal so that out-of-band signals will not alias into the desired information bandwidth. Consequently, some means of active or passive bandpass filtering must exist within the input path prior to the A/D. Next, it must be assured that the input signal will not exceed the full-scale input to the A/D, or clipping and distortion will result. Also, the peak anticipated input signal level should approach the A/D's full-scale input level with acceptable "headroom", in order to utilize the full dynamic range of the A/D. This may or may not require gain. These A/Ds typically have differential inputs, high differential input impedance (i.e., >1 kohm), and an input DC offset requirement. In view of this, wideband transformers having turns-ratios of 1-to-1 or higher are commonly used to drive the A/D differential inputs and, therefore, some gain may be realized via the transformer turns-ratio selection. Any remaining gain requirements would then have to be met with external amplification.

Digital Mixing

The digital "real" data stream out of the A/D at a sample rate F_s is directly converted to a Baseband "complex" quadrature data stream via a digital downconversion process (i.e., quadrature hybrid and analog mixers, or I&Q demodulator, equivalent). In the past, this downconversion process often used a digital processing "trick", owing to logic resource and speed limitations. Note that if the desired downconversion frequency is $-F_s/4$ (i.e., centering the desired frequency span at DC), the mixing process only requires four digital samples at that downconversion frequency. If these digital samples corresponded to the phase points (0, -90, -180, -270) degrees (i.e., negative frequency rotation sequence), the resultant cosine sequence would be (1, 0, -1, 0), and the associated sine sequence would be (0, -1, 0, 1). Therefore, only the sign of the input data must be changed periodically to generate the desired I and Q output data streams and, since the sine and cosine waveforms used for the mixing process are "perfectly" reconstructed by the alternating 1's and 0's sequences, the SFDR is beyond question. Similarly, using the sine and cosine sequences for counterclockwise rotation will result in a $+F_s/4$ upconversion. This is useful when the lower image is to be centered at DC, since that image will now alias to DC. The resultant quadrature data are then lowpass FIR filtered and decimated. It is common for the $\pm F_s/4$ up/downconversion process to be combined with the FIR filtering as a single process.

The $F_s/4$ up or downconversion requirement that the sample frequency be exactly four times the downconvert frequency is very restrictive and, therefore, not very desirable. Nevertheless, these "tricks" eliminate the need for true digital multiplication, are logic-efficient and support higher clock rates than would otherwise be possible, and are therefore still commonly included in today's Commercial digital receiver devices. Also, one other digital processing "trick" is worth mentioning at this time, because it is frequently used to invert a real data stream spectrum. Simply stated, mixing the input real data stream with a frequency at $F_s/2$ inverts the real spectrum and, for a digital mixer, this can simply consist of a (1,-1) multiplication sequence. This sequence then reduces to simply changing the sign of every other input data sample. This spectral inversion is normally used to reverse a previous spectral inversion that occurred during the digitization process (i.e., second Nyquist zone digitization, etc.). However, it should be noted that any DC component of the input data stream will mix with $F_s/2$, resulting in replicated terms at $\pm F_s/2$, and all multiples thereof.

A more flexible digital downconversion process can be realized by the use of a Numerically Controlled Oscillator (NCO). An NCO performs the LO downconversion to quadrature without a fixed sample frequency to downconversion frequency requirement. The NCO consists of a digital sine and cosine generator, with the computations being done by a Coordinate Rotation Digital Computer (CORDIC) algorithm that does the sine and cosine computations necessary to generate 0 and 90 degree phase-shifted LO-equivalent digital waveforms. The computations are based on an iterative half-angle process, are pipelined, and take a few more clocks than the number of bits of precision specified. However, once the pipeline is filled, new digital sine and cosine outputs then result for each and every clock.

The input parameter to the CORDIC is a two's complement (i.e., signed) binary value between zero and $\pm 2^{(n-1)}$, where n is the number of bits of precision, that represents the desired phase for which the sine and cosine values are to be computed. Thus, an input stream of digital phase values, at a given clock rate, translates into digitally synthesized sine and cosine waveforms that are the desired upconversion or downconversion LO frequency. The input two's complement digital number to the CORDIC algorithm is computed, for each clock cycle, as a "Phase Value" (PV) that is the running n-bit sum of a "Center Frequency Step" (CFS) and a "Phase Offset" (PO). The CFS, which is actually a phase increment to be added each clock, is computed as CFS = $(F_s/F_c) \cdot 2^n$, where F_s is the desired LO frequency to be synthesized and F_c is the digital clock update rate. Also, remember that the 2^n value is signed, so the range is $\pm 2^{(n-1)}$, and that values ranging from 0 to $(2^n -1)$ represent a complete rotation around a unit circle. The sign of this phase increment determines the direction of phase rotation around the unit circle, clockwise or counterclockwise, corresponding to a negative or a positive LO frequency, respectively.

The NCO Phase Offset value is typically zero, but can be used to impart a fixed (starting) phase offset on the synthesized output. It is commonly used to "initialize" the PV accumulator via a "sync pulse", as might be required at the start of a coherent pulse or sweep. Also, the PO feature can be used to time-interleave one or more NCOs, thereby permitting the synthesis of even higher LO frequencies. Since the PV is the sum of the CFS value and the PO value, with 2^n bit precision, this accumulator performs modulo math, and the highest frequency that can be synthesized by a single NCO is $\pm F_c/2$. However, this is not a problem since the A/D clock rate and NCO clock rate are normally equal, and so the A/D data will alias at $F_c/2$. If the required LO frequency is higher than one half the NCO clock rate (i.e., A/D clock rate is a multiple of the NCO clock rate), then two NCOs can be time-interleaved to synthesize LO frequencies up to the NCO clock rate, three NCOs can be time-interleaved to synthesize LO frequencies up to one and one half the NCO clock rate, etc.

The digitally synthesized sine and cosine outputs from the CORDIC algorithm are then used to multiply the input A/D real data stream, A_{in} to generate the I and Q quadrature output data streams I (t) = A_{in} cosine (PV (t)) and Q (t) = A_{in} sine (PV (t)). The SFDR associated with the digitally synthesized sine and cosine waveforms is a function of how "pure" these waveforms are. The purity is a function of the numerical precision employed and, for an n-bit binary word size, ranges from perfect (i.e., as would be the case for $F_s/4$ phase increments), to the limit of the CORDIC precision. These variations from correct sine and cosine values, to values that are numerically limited approximations to the correct values, gives rise to spurious spectral components. Therefore, there is a direct relationship between the SFDR and the numerical precision associated with an NCO design. Some NCO designs within Commercial digital receiver devices include provisions for amplitude and/or phase "dither" to mitigate the effects of repetitive accuracy errors, but for NCOs implemented in FPGAs, it is usually easier to simply increase the numerical accuracy employed to increase the SFDR.

The CORDIC output precision for the sine and cosine values is typically set to about one half that of the PV value used to generate them. For example, CORDICs utilizing a 32-bit PV value, with 16-bit sine and cosine output values, can typically provide a worst-case SFDR of >96 dB. Similarly, utilization of a 36-bit PV value, with 18-bit sine and cosine output values, can typically provide a worst-case SFDR of >108 dB, or better. These examples are given because 16-bit by 16-bit and 18-bit by 18-bit "hardware" multipliers, optimized for speed and resource utilization, are common in today's programmable devices. Also, it should be noted that these are "signed" multiplies and that the A_{in} data should therefore be two's complement values, MSB-justified. If the A_{in} data precision is less than the sine or cosine precision (i.e., multiplier input precision), then the unused multiplier bits associated with the A_{in} input should be the LSBs and should be tied low (input zeroes).

Digital Filtering

The NCO output I and Q Baseband quadrature data streams contain both sum and difference frequencies and must therefore each be lowpass filtered to remove the unwanted sum frequencies. Since it is very desirable to highly over-sample the wanted information bandwidth to achieve substantial processing gain, the overall (Nyquist) bandwidth may be substantially greater than the information bandwidth of interest. For multi-channel wireless applications, this translates into the need for lowpass filters with high decimations and very sharp transition bands. These needs can be met through the use of one, and typically more than one, decimating digital filter. Commercially available single-channel, and more recently, quad-channel digital receiver chips from Vendors such as the Analog Devices AD6640, Graychip (now TI) GC4016, and Intersil (formerly Harris) HSP50216, typically employ one or more cascaded integrator-comb (CIC) filters, followed by one or more FIR filters, to meet these requirements.

CIC filters do not require multipliers and have very limited logic requirements. Their structure consists of N digital integrator stages operating at the input data rate f_s, followed by N digital comb stages that have a differential input sample delay M, that is normally 1 or 2 clocks, operating at the decimated (output) data rate $f_o = f_s/R$, where R is the decimation factor. The CIC filters associated with Commercial digital receiver devices typically have a minimum decimation of from 2 to 8, and can provide very high decimations (i.e., up to 65536 for the HSP50216). However,

CIC filters also typically have a large transition band, owing to the (unity gain) Power (magnitude squared) frequency response: $P(f) = (1/(M*R)^N)^2 (\sin(\pi M f R/f_s) / \sin(\pi f/f_s))^{2N}$, where f is the input signal frequency.

This expression is commonly rewritten as: $P(f) = (\sin(\pi M f) / \sin(\pi f/R))^{2N}$ and is usually plotted as: $P(f) dB = 10 \log(P(f))$, for frequencies from DC to one half the output sample rate (f_o). However, some Vendors prefer to display this form of the frequency response as a plot, from DC to f_o that then mirrors about $f_o/2$, or from $-f_o/2$ to $+f_o/2$ that is then symmetrical about DC. Also, these frequency responses are commonly plotted vs. f/f_o, since the actual frequency response for digital filters is essentially a normalized function of the output sample rate and so it doesn't matter what the actual output sample rate is. This is an important point since, for any given digital filter design (i.e., not only CICs, but also FIRs, etc.), the frequency response of the filter scales with the sample rate.

Commercial digital receiver devices have primarily been designed for wireless communications applications, where a relatively narrow bandwidth single channel is to be extracted from a large, multi-carrier frequency band. For such applications, the high decimations that can be provided by CIC filter utilization translates into a significant reduction in logic resource requirements in conjunction with potentially high processing gains. The one or more FIR filters that follow the CIC filter section are then utilized to provide the sharp transition bands that are required in order to extract the narrowband channel information, while rejecting adjacent channel information that would otherwise alias into the passband. These FIR filters also typically provide additional decimation and, therefore, additional processing gains.

A FIR filter is the digital equivalent to an analog transversal filter and, as such, outputs the sum-of-products for a number of weighted taps, spaced across a delay line. The tap weights are the filter coefficients, and the tap spacing (delay) is determined by the input data clock rate (note: symmetrical coefficient FIR filters are phase linear). The decimation by "m" can consist of simply throwing away (m-1) out of every m output samples (i.e., the resultant sum-of-products generated for each clock cycle). Alternatively, the additional (m-1) available clock cycles can be used to facilitate either the processing of additional taps, or to permit a reduction in the logic resources that are required to process the given number of taps. FIR filter hardware design tools are available from numerous sources such as the programmable logic Vendors (i.e., Altera, Xilinx, etc.), as well as hardware-independent FIR filter design (coefficient computation) tools from Vendors like MATLAB and organizations such as IEEE. Basically, the "flatness" of the passband, the width of the transition band, and the attenuation provided in the stopband, are functions of the filter design methodology, in conjunction with the number of taps and the numerical precision employed. A common FIR filter design technique uses window functions (i.e., Blackman, Hamming, etc.) to reduce sidelobes. Another popular method, Remez (or Parks-McClellan) algorithm, is based on optimum approximations for passband and stopband cutoff frequencies, that allows tradeoffs between transition band and passband ripple.

FIR filter logic design can be typified as being either a "polyphase", or a "fully parallel" (transversal) implementation. Polyphase simply means that different coefficients and different data are used at different times. These filters typically employ a fixed number of taps with fixed coefficients, fixed decimation, and are normally designed with the FPGA Vendor's tools. Polyphase FIR filter designs are popular because their implementation requires a minimum of logic resources, but at the price of being rather inflexible. A fully parallel FIR filter design can actually utilize any one of a number of different logic implementations in order to support either fixed or User-programmable coefficients, allow for selection of a coefficient set from multiple such coefficient sets, permit a variable number of taps, support variable decimations, etc. In general, these designs are User-created and then compiled for the target FPGA, require significantly more logic resources than polyphase implementations, but offer as much application flexibility as required.

It is important to note that large numbers of very fast "built-in" hardware multiplier/accumulators have been included as an integral part of the more recent FPGA device designs (i.e., the "Stratix" family from Altera and the "Virtex-II" family from Xilinx). This trend is due to the level of hardware processing parallelism that is possible within FPGA designs, and that cannot be attained by existing "traditional" DSP and CPU devices with fixed-architectures and instruction-sets. Also, these FPGA-implemented hardware multiplier/accumulator logic elements are based on highly efficient core logic designs that combine to form 18-bit by 18-bit (or higher) precision units with little overall performance degradation (i.e., speed reduction or additional peripheral logic requirements). This, in combination with the increased logic, routing, and memory resources offered by these newer devices, makes the utilization of fully parallel FIR filter designs more practical to implement. This is very appealing from an applications point-of-view, owing to the above noted increase in flexibility that can be realized by such FIR filter designs. FPGA filter implementations can result in increased performance at significantly reduced cost, even if the FPGA is operated at clock rates well below that of today's GHz DSPs and CPUs, not to mention the additional potential system-wide benefits such as simplified real-time processing requirements, reduced I/O requirements, etc.

DIGITAL TRANSMITTERS

The classic single-sideband analog transmitter is the reverse process of that for the classic heterodyne receiver. Therefore, it consists of band limited I and Q Baseband signals that are to be upconverted to a final output RF, via one or more intermediate stages (i.e., IFs). In its simplest form, where the RF output is also the first IF, a single I&Q modulator is driven by both quadrature inputs, configured for selection of either the upper-sideband or the lower-sideband, and an LO that is at the appropriate output RF frequency. When the input I and Q data is digital, two

D/As are required to convert these input data streams to analog quadrature signals. Consequently, analog lowpass filtering of both quadrature signals is then required to remove the replicated images associated with a sampled process.

For a digital transmitter, the I&Q modulator is replaced by an NCO configured to perform an "imageless" upconversion. The upconversion is imageless because the multiplication is accomplished entirely in the complex frequency domain (i.e., I and Q complex digital input data streams multiplied by NCO complex LO waveforms). Thus, the entire Baseband is simply translated up in frequency by the programmed LO frequency. At this point, additional signal processing requirements become dependent upon the final RF output requirements. If higher RF output frequencies are required than can be reasonably provided directly by a D/A, then two D/As are usually employed to convert the IF I&Q data streams for utilization by a subsequent analog (I&Q modulator) upconversion stage. In this event, both D/A outputs must be lowpass filtered to remove the sampling-related spectral replications.

If the RF output frequency band is not beyond that which can be directly supported by a D/A, then either of the digitally (imageless) upconverted quadrature data streams can provide the desired information bandwidth. The selected (I or Q) output sample stream will therefore be centered at the desired RF (LO) frequency, and will be output as a real data stream for reconstruction by that D/A. However, it should be noted that the complex multiplications are no longer necessary since only the real outputs need be computed (i.e., the required functionality is now equivalent to that of the analog I&Q modulator). Having to only compute the real output: IF (t) = I (t) cosine (PV (t)) - Q (t) sine (PV (t)), where PV is the NCO Phase Value (i.e., LO), reduces the number of digital multipliers required for the output data sample computation by one half (i.e., from 4 to 2). This essentially reduces the transmitter NCO logic requirements to that of the digital receiver, but with an LO and quadrature pair (I and Q) as the inputs producing a real IF output, as opposed to the receiver's LO and real IF inputs producing quadrature pair (I and Q) outputs.

The digital filtering requirements associated with digital transmitters is significantly different from those associated with digital receivers. As previously noted, the reconstructed analog output from a D/A exhibits a sinx/x roll-off that is output sample rate dependent. Therefore, the highest possible output sample rate should be employed to mitigate the roll-off over the desired output information bandwidth. Since the output spectrum is replicated at every multiple of the sample rate and must therefore be band-limited, increasing the output sample rate also simplifies (analog) filtering requirements.

The input digital sample rate F_s can be increased by a factor R (i.e., output sample rate $F_{out} = R\, F_s$) via interpolation. Commercially available digital transmitter devices, that are the counterpart to the digital receiver devices, typically utilize a combination of one or more interpolating FIR filters to increase the data rate. The FIR filters provide sharp transition bands to maximize the stopband and provide good out-of-band rejection. One or more interpolating CIC filters, with relaxed stopband requirements, then follow the FIR filters in order to efficiently support high interpolation factor requirements. However, it should be noted that, while the decimating CIC filters are unconditionally stable, interpolating CIC filters, if disturbed, can go unstable, resulting in broadband, high level white noise. This problem is avoided if there is no rounding employed within the integrator stages, since it is the unbounded growth of small errors in the integrator stages that cause the instability. However, the restriction on rounding translates into a word size growth requirement that, in turn, increases logic resource requirements. Also, it should be noted that noise (i.e., bit errors) within the integrator stages can still cause instability. One Commercial four-channel digital transmitter device (the GC4116 from Graychip) has a patented feature ("auto flush") that detects CIC instability and automatically re-initializes the filter.

The sample rate increase must also be sufficient to support the desired output IF sample rate requirements. In view of the limitations on today's D/A clock rates, the IFs that can be directly supported by digital transmitters are typically <100 MHz (i.e., <$0.25F_{out}$, and F_{out} <400 MHz), and the lower the IF frequency, the better the performance. In some cases, the designs take advantage of the spectral replications associated with a sampled process. In these cases, the NCO usually is programmed to place the desired output frequency in a higher Nyquist zone, with consideration given for the sinx/x D/A output response (i.e., the peak of the response will be 13 dB down at 1.5 F_{out} and 18 dB down at 2.5 F_{out}).

Another common digital transmitter D/A feature is "zero-stuffing", whereby the D/A output section is clocked at twice the input sample rate with "zero" (mid-scale) value outputs being inserted between each input sample. This stretches the sinx/x response at the expense of 6 dB in output power level at DC. Similarly, these D/As may also include an inverse sinx/x filter that is usually an 11-tap, symmetric, FIR filter that employs binary-weighted coefficients, so it is logic efficient and fast. This filter flattens the D/A output sinx/x response to < 0.1 dB) from DC to 0.45 F_{out} at the expense of a 3.815 dB insertion loss. Also, it should be noted that changing the sign of the interpolated data stream, commensurate with the $F_s/2$ digital mixing sequence previously discussed, results in spectral inversion. This sequence is sometimes combined with an interpolation filter to create a highpass response as a D/A device feature (i.e., AD9772A, etc.) that is useful in direct IF output applications.

PRODUCT DESIGN CONSIDERATIONS

The transformation of a design concept into a viable Accelerator or Storage Ring instrumentation product, using today's technology, is worthy of discussion. Such products must be reasonable to manufacture, reliable, and maintainable, in order to be cost effective. Additionally, consideration should be given to multi-use product designs to afford commonality, not only within a given facility, but also across multiple facilities. This is a far more reasonable goal today than it was in the past, owing to the increased bandwidth of today's mixed-signal devices, combined with the dynamic in-circuit programmability and burgeoning resources of current and emerging FPGAs.

Such product designs must start with schematic capture, via one of the numerous Commercial products such as ORCAD, etc., utilizing standardized and verified parts libraries and associated mechanical footprints. Notes should be present on the schematic to reflect operational limits, special component placement requirements, critical signal routing constraints, controlled impedance requirements, decoupling, thermal considerations, etc. Since these designs will employ programmable logic devices (i.e., EPLDs and FPGAs), and may also include programmable ASICs (i.e., digital transmitter or receiver chips, DSPs, etc.), register maps and supporting documentation requirements then arise in support of the software development associated with hardware debug and test, applications program interface (API) development, etc.

The schematic capture package also generates a Bill of Materials (BOM) that lists part designators, vendor part numbers, quantities, etc., as well as a "netlist" file that provides the component routing (interconnect) information and identifies the mechanical footprint for each component. The netlist file is imported to a PC Board Layout package, such as PCAD, etc., that creates the component footprints for subsequent placement by the PC Board layout person, and "rubber bands" the required interconnects (routes), based on the netlist interconnect data. The layout person first creates the mechanical board outline, reads in the netlist data, assigns the layers (multi-layer boards are required for the designs being addressed), and then proceeds with initial component placement.

For purely digital boards, it is fairly common practice for the layout person to assign routing "rules" to be used by an "auto-router" that actually routes the board design. This practice is not generally acceptable for densely routed layouts because these routers typically drop too many vias (i.e., small plated-thru holes that provide between-layer routing paths) to thereby simplify the routing from point A to point B. However, the vias actually reduce the available PC board routing area by virtue of being holes that block routing paths, and so the routers keep asking for more layers in order to create more routing paths. There are limits on the number of layers that are "practical" for PC board designs, based on controlled impedance requirements, board thickness (i.e., to fit a card guide or connector), cost, etc. In general, PC board design is one of the remaining areas where it is hard for an auto-router software package to beat a skilled PC board layout person. This is especially true when it comes to dense, high speed, mixed-signal designs.

However, these board layout packages do typically offer features that greatly enhance the efficiency with which such designs can be "hand routed", making hand routing a viable, though more expensive alternative, but one that can yield a significantly better product. Additionally, the design verification features (tools), that are an integral part of such packages, are essential to the development of a quality, well controlled, and well documented product. The resultant output

(Gerber) files, in conjunction with board lay-up drawings, are then available for electronic transfer to a "board house" (i.e., PC board manufacturing facility).

The selection of a board house for a specific design is not simply cost-based, because different board houses have different capabilities. The small physical sized, lead pitch, ball spacing, etc. of today's components commonly translate into dense boards with fine line-spacing, very small (or micro) vias, buried and blind vias, very flat surface features, etc. These requirements may well narrow the field of board houses that can meet the requirements without, or with acceptable up-charges. For high speed, mixed signal designs, one must add controlled impedance to this list of requirements. Similarly, the potentially high component cost and typical low manufacturing volume associated with the type of products being addressed here, is such that a "net list" board test (i.e., Gerber file based) capability like that of a "Probot" becomes an essential board house requirement in order to ensure receipt of "good" boards.

Next, the requirements associated with "board-stuffing" must be addressed. Again, the small geometries encountered in dealing with today's components mandates the use of "pick-and-place" equipment, re-flow ovens (and re-flow profiles), and fine-tipped or hot air soldering equipment for touch-up (re-work) under suitable magnification. Board inspection is also somewhat more involved. For example, Ball Grid Arrays (BGAs) must either be x-rayed, optically inspected (ERSA Scope, for example), or tested via a JTAG boundary scan. Board stuffing houses (i.e., contract manufacturers) capable of meeting these requirements often refuse work orders below a minimum quantity that may well be in the hundreds to thousands. However, there are fully qualified houses that will service "prototype" or "limited production" quantity orders, albeit at an up-charge.

EXAMPLE BEAM INSTRUMENTATION PRODUCT DESIGN

So, how does one put this all together? Perhaps the best approach would be to do an example Storage Ring Beam Instrumentation product design. Let us assume that the product is to support two requirements, the first being bunch-by-bunch current monitoring, and the second being beam steering. For the purpose of generality (i.e., to be in about the middle of common Storage Ring RFs, we will assume that the RF is 360 MHz, the Storage Ring diameter is about 80 meters, and we therefore can have a maximum of 300 current bunches. Thus, the revolution frequency F_{rev} for a bunch will be 1.2 MHz. For sensors, we will assume that we have four BPMs, spaced evenly about a cross section of the ring, and the outputs have been amplified, if required, and band-limited via filters. A "trigger" will also be required to identify the "bunch number 1" time. We will first address the bunch-by-bunch current monitoring requirement.

To provide bunch-by-bunch current monitoring, a sample must be taken at the appropriate time, relative to the start of a RF cycle, for each and every bunch, with these bunches being spaced at the RF interval. For this example, that requirement translates into the need for a 360 MHz A/D encode rate, sampling a 360 MHz analog (RF) BPM output, and thereby creating sets of 300 digital samples, corresponding to the maximum of 300 discrete bunches that can be distributed about the Storage Ring. These sets of 300 samples will then repeat at the revolution frequency (i.e., 1.2 MHz x 300 samples = 360 MSPS). It is assumed that provision for co-adding up to "n" bunch samples for each bunch is required (i.e., adding bunch 1 sample for revolution 1, with bunch 1 sample for revolution 2, etc., and similarly for bunch 2 samples, etc.). It is also assumed that all four BPMs must be processed such that the data from the four BPMs can be used in support of beam steering.

First, we must select an appropriate A/D, one that provides the best resolution and can directly digitize a 360 MHz input. This is beyond the capability of existing 14-bit A/Ds, and the best 12-bit A/D choice is the Analog Devices AD9430. This device supports a maximum encode rate of >210 MHz (rated performance is to 210 MSPS, but it works well beyond this), and has a full-power analog input bandwidth of 700 MHz. Two of these devices must be time-interleaved to support the 360 MSPS requirement. This is accomplished by inputting the 360 MHz reference clock (sinusoidal waveform at +4 dBm) to a differential PECL receiver followed by a PECL flip flop with a preset control (required to support initial time alignment). The virtually perfect 50% duty cycle differential outputs then drive the A/D encode inputs 180 degrees out of phase with respect to each other (i.e., both A/Ds encode at 180 MHz, but on opposite edges of that clock). The estimated SINAD at a 180 MHz sample rate for a 360 MHz input is >58 dB. The Aperture Uncertainty (jitter) is 0.25 ps rms, so the jitter-limited S/N is 65 dB.

The first (in time) AD9430 will output an LVDS sample stream at 180 MSPS that is translated by the FPGA into two LVTTL data streams, corresponding to samples for bunches 1 and 3, then 5 and 7, etc. Similarly, the second AD9430 outputs samples for two data streams with bunches 2 and 4, then 6 and 8, etc. These four 90 MSPS data streams will be processed by an FPGA that is configured to have four dual-port memories, corresponding to the four input bunch sample data streams. Each of the four dual-port memories will be organized as 128 by 36 bits and support the co-adding of samples for 75 bunches (i.e., memory 1: address 1 is bunch 1, address 2 is bunch 3, etc.) so four memories support sample data for the 300 bunches. At the start of each "n" co-add sequence, the samples are simply written to the appropriate memory addresses. For each of the remaining (n-1) sample data sets, the new sample is added to that read out of the memory, and the new sum is then written back to that same memory address. These four concurrent processes therefore execute at a 90 MHz rate. Upon completion of the "n" co-add sequences, the sample sum data is read out of memory and the new sample data is written into memory, and the process repeats. The output sum data are written to a FIFO buffer within the FPGA for subsequent output via the selected Bus Interface (i.e., VME, PCI, RACEway, etc.).

One or more dual-port memories can also be configured to accept individual (selected) bunch sample data streams in support of a true "decimating digital filter" feature, as opposed to a simple co-add. Thus, as the sample data from a given bunch is written cyclically through the memory, it can be concurrently processed by a decimating FIR filter configured to have as many taps, and as much precision, as required for the desired passband, transition band, and stopband. This can be taken one step further. A "digital receiver" feature can also be added via NCO downconversion of an individual bunch sample data stream. This frequency translation (up or down) can be by any frequency up to one half the effective sample rate (i.e., half the revolution frequency), and the quadrature bunch sample data can then be FIR-filtered and decimated, as appropriate.

This A/D and FPGA architecture can be replicated four times, on a single PC board to process the data from all four BPMs. This permits data processing, within an FPGA, that utilizes all four sources of bunch sample data in support of beam steering. Intellectual Property (IP) FPGA core designs are readily available and simplify the implementation of complex logic functions (i.e., Logarithms, Complex DFFTs, Rectangular-to-Polar, etc.). Thus, the "standard" beam position algorithms can be implemented with relative ease and both time-domain and frequency-domain processing can be supported.

Since the 360 MSPS time-interleaved A/Ds alias at multiples of 180 MHz, a signal at 361 MHz will alias to 1 MHz, as will a signal at 359 MHz. If this is a problem, then a bandpass filter with an asymmetric response can be used to band-limit the BPM output such that a band can be passed that is from a known "clean" portion of the spectrum. A digital receiver can be included in the design to extract the desired high frequency "span of interest" from the A/D-pair output sample stream. Time-interleaved NCOs perform the programmed downconversion and the output quadrature Baseband data is then digitally filtered and decimated, as required. At this input data rate, the NCO, CIC, and FIR filter functionality must be implemented within the FPGA, because this data rate exceeds that of currently available commercial digital receiver devices. However, these real-time parallel processing requirements are easily implemented within an FPGA.

Since the FPGAs are in-circuit programmable and can be reconfigured from memory (i.e., normally "Flash devices"), a given hardware design can be tailored to multiple applications. Flash memory devices can provide sufficient storage capacity for multiple designs and can also be rewritten via the Bus Interface. Consequently, all features of the above example design could be implemented individually, or in combination, providing an application-flexible common design. The processing afforded by such high speed, mixed signal, FPGA-based designs, can significantly reduce the external processing and I/O requirements that are associated with Accelerator and Storage Ring instrumentation. Additionally, this approach can minimize the instrumentation system complexity, yielding increased reliability, all while providing enhanced performance, flexibility, and commonality, at a potential cost savings.

Invited Talks

10th Beam Instrumentation Workshop
May 6-9, 2002

FNAL Instrumentation: Lessons Learned

James L Crisp

*Fermilab[1]
PO Box 500
Batavia, IL 60540*

Abstract. Experience gained during the recent commissioning of the Main Injector accelerator and the Recycler storage ring at Fermilab will be discussed. Some of the more interesting problems involve; ground differences, cabling for bpm and multiwire systems, electromagnetic noise, and magnetic shielding

LOCAL CURRENTS

"Ground"

According to Mr. Webster, definitions of "ground" include:
- The position or portion of an electric circuit at zero potential with respect to the earth.
- A conducting connection to such a position or to the earth.
- A large conducting body, as the earth, used as a return for electric currents and as an arbitrary zero of potential.

However, ohm's law requires "ground" to have different voltage potentials that depend on how currents flow from one part to another.

Main Injector Bending Magnet System

The Fermilab Main Injector uses conventional magnets to contain the beam as it accelerates from 8GeV to 150GeV. Twelve 1KV power supplies in series with 344 bending magnets are ramped to 9375 amps in about ½ second. To build the desired waveform, each supply ramps up separately in about 30 milliseconds. Bending magnets have a series impedance of $0.8m\Omega$ and 2mH with a coil to ground capacitance of 30nf.

---

[1] Operated by Universities Research Association, Inc.

TABLE 1. Main Injector bending magnet system.

344 Bending magnets	0.8 mΩ and 2 mH each
capacitance to ground	30 nf each magnet
12 power supplies	1KV each supply
Maximum bend current	9375 Amps
1/2 second ramp overall	30 msec each supply

As the supplies ramp up, the changing voltage to ground induces "local currents" through the magnet capacitance to ground. These currents flow along the path of least resistance such as beam pipe, signal cables, and cable tray. Local currents make the bend buss current different for each supply and magnet. Generally, the bend magnet current at only one point is monitored and controlled. At Fermilab, orbit distortions lead to the realization that magnet current was not the same around the ring and depended on ramp waveform, power supply turn on order, and geometry of the bend magnet system. Local currents also produce significant potential differences in the ground system that result in problems with some instrumentation systems.

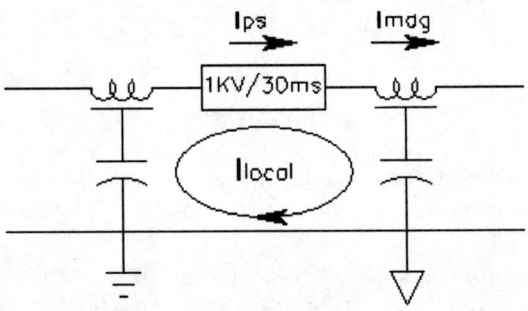

FIGURE 1. Power supply with two magnets in series. Local currents are driven by the changing voltage to ground. Even though all of the magnets and supplies are in series, they don't necessarily all have the same current.

Vacuum Chamber Impedance

If the vacuum chamber of the Main Injector were cut in one place, the impedance across the cut could be estimated as shown below.

$$R_{tot} = \frac{C_M}{C_P t}\rho = 5.6 \, \Omega$$

$C_M = \textit{machine circumference} = 3320 \textit{ meters}$

$C_P = \textit{pipe circumference} = (0.0254)*13.9"$ (1)

$t = \textit{wall thickness} = (0.0254)*0.0595"$

$\rho = \textit{resistivity of stainless} = 9e-7 \, \Omega-m$

The inductance of the 2 mile diameter ring is estimated below:

$$L = \mu_o \frac{\pi a}{2} N^2 = 1 mH.$$

$$\mu_o = 4\pi \times 10^{-7} \quad (2)$$

$$a = radius \approx 528 \; meters$$

$$N = 1 = number \; of \; turns$$

Magnet laminations surround the beam pipe and increase this inductance. With a typical "H" magnet lamination geometry, the reluctance for the path surrounding the pipe will be about 4 times that for the normal bending field. The beam pipe has only one turn compared to about 10 for the normal magnet coils. The increase in inductance attributed to the laminations of all 344 magnets is estimated below.

$$L_{la\min ations} = 344 \, (2 \, mH) \frac{1}{4} \left(\frac{1}{10} \right)^2 = 1.7 \, mH \quad (3)$$

The corner frequency defined by the ring resistance and inductance is $R/2\pi L = 11 Hz$ (5.6Ω, 2.7mH). The ring is resistive below this frequency and inductive above. The inductance from the magnet laminations will tend to steer beam image currents through the beam pipe and local currents through other pathways such as instrumentation cables.

The stainless steel beam pipe wall thickness (.0595") is one skin depth at 100 KHz. At frequencies below this, the magnetic fields from the beam current will leak through the wall and begin to produce measurable potential differences that follow the beam around the ring. For 1e11 particles distributed within 1/7 of the circumference (1.6usec) the voltage to ground would be:

$$V_{ground} = \pm \frac{1}{2} 1e11 \frac{1.6e-19}{1.6e-6} \frac{5.6\Omega}{7} = 4 \, mV \quad (4)$$

Estimate of Local Current

For one power supply turning on, the voltage to ground on the magnet buss has a simple distribution. Symmetry suggests the average voltage to ground is about +¼ of the supply voltage on ½ of the ring and -¼ on the other half. This voltage will induce currents through ¼ the capacitance and ¼ of the resistance. The resistance is assumed to be only that of the stainless beam pipe. The Fermilab Main Injector also has a ground conductor around the circumference. The capacitive coupling to ground through the magnets will make the ground voltage proportional to dV/dt as estimated below.

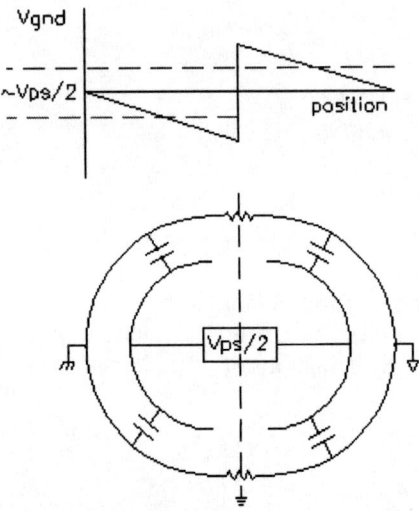

FIGURE 2. Voltage to ground of the magnet buss with one power supply. With a single power supply turning on, symmetry suggests the average voltage to ground is about +¼ of the supply voltage on half of the ring and –¼ on the other half.

$$V_{ground} = \frac{1}{4}\frac{V_{ps}}{\Delta t}\frac{C_{tot}}{4}\frac{R_{tot}}{4} = 30\ mVolts$$

$$V_{ps} = 1\ KV$$

$$\Delta t = 30\ m\sec \quad (5)$$

$$C_{tot} = 344 \times 30\ nf$$

$$R_{tot} = 5.6\ \Omega$$

Measurement of Local Current Induced Voltages

The measured voltage between the tunnel ground and the service building ground is shown below. The measurement was made by connecting both the center conductor and the shield of a coaxial cable to the beam. The shield of the service building end was connected to ground through the chassis of the oscilloscope. The impedance of the test cable shield is large compared to all of the other ground connections and does not disturb the measurement.

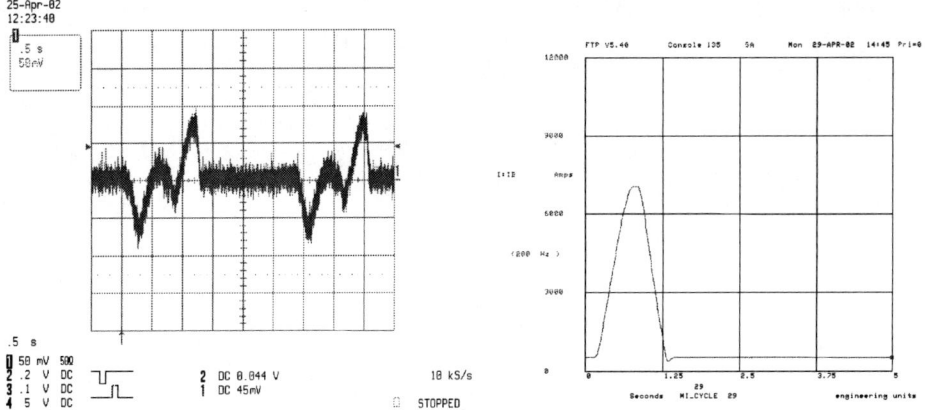

FIGURE 3. Measurement of the voltage between the tunnel and service building grounds during the magnet ramp. The voltage measured on the tunnel ground induced by two consecutive ramps is shown on the left (50mV, .5sec per division). One magnet ramp is shown on the right (3000Amps, 1.25 sec per division). The voltage is proportional to dV/dt.

Effects of Voltage Differences Between Grounds

The voltage between tunnel ground and building ground can induce errors into the signal path of instrumentation cables. The voltage will appear across the shield impedance. For equal source and receiving impedance, 1/2 of the noise voltage will appear across the load resistor as shown below.

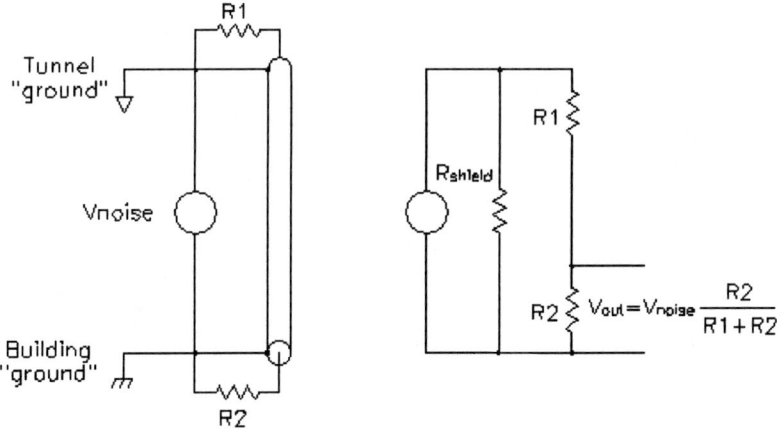

FIGURE 4. Schematic of a coaxial cable between the tunnel ground (or beam pipe) and the building ground.

The noise voltage across the load resistor can be reduced by inserting a transformer as shown below. Typically, these are made by wrapping as many turns as possible of coaxial cable on a torroidal tape wound core.

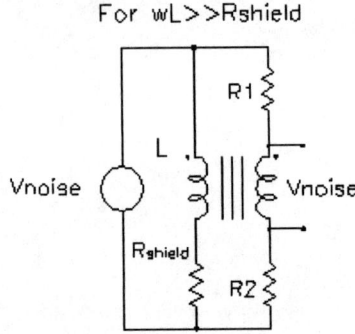

FIGURE 5. Schematic demonstrating the benefit of torroidal cores. For wL>>Rshield the noise well be removed from the load impedance.

For an inductance L of the core, and total cable shield resistance Rs, the amount of noise reduction is given below.

$$\frac{V_{out}}{V_{noise}} = \frac{R_S}{R_S + j\omega L} = \frac{1}{1 + \frac{j\omega L}{R_S}}$$

$$\text{for } f \gg f_o = \frac{R_S}{2\pi L} \tag{6}$$

$$\frac{V_{out}}{V_{noise}} \approx \frac{f_o}{f}$$

The core has no effect below the corner frequency fo=Rs/2πL. Above this frequency, the noise is reduced by fo/f. Such a core was not effective above 50 KHz because of turn to turn capacitance and associated resonance's in the windings themselves. Common mode filters using such an approach can be effective but only for a limited frequency range. The core must be inserted into the signal cable between the tunnel and building grounds. Well intentioned, but less informed, people have inserted a transformer inside the relay rack with the cable shield grounded were it entered the rack. If the shield of the cable is grounded before the torroid, it does no good.

TABLE 2. Typical common mode filter using a torroidal core

Arnold T8027 tape wound Supermalloy core	3" OD, 2" ID, 1" Ht	104 uH/N^2
500 ft RG58 cable	4 Ω/1000ft	2 Ω
30 turns on core		94 mH
fo	$R/2\pi L$	3.4 Hz
60 Hz noise reduction	60/3.4	17.6 (or 25db)
Maximum frequency		50 KHz

Fast Kicker Magnets

Fast kicker magnets are another source of voltage differences between grounds. At Fermilab, fast kicker magnets have rise times of 20 to 100 nsec and pulse widths of 1.6 to 20 usec. Several RG220 cables are used in parallel with up to 600 amps in each cable to carry the pulse to the magnet in the tunnel. The impedance of the RG220 cable shield is 0.05Ω for a modest cable length of 200ft. The 600 amp return current will make 30 volts on it's way back to the power supply! The shield of the RG220 is tied to the building ground upstairs and to the magnet case in the tunnel. If the magnet case is connected to the tunnel ground or beam pipe, this 30 volt signal will drive currents through any instrumentation cables connected between the tunnel and the building.

If the magnet case is left floating, beam image currents will produce electromagnetic waves as they cross from one side of the magnet to the other. The kicker magnet becomes an antenna broadcasting the beam frequency with substantial power. This rf interference can induce substantial signals into nearby signal cables.

This is a difficult problem because the fast rising kicker currents have frequency components as high as the beam. The decision to ground kicker magnet cases has not been resolved at Fermilab.

Multiwires

Multiwire Signals and Ground

Multiwires are secondary emission profile monitors made from wire grids. The small charges induced on the wires by the passing beam are integrated and the relative amplitudes on the wires in the grid are measured to produce a transverse beam profile. A 50 um wire centered on the beam collects 1 charge for each 2000 protons. This depends on many things but is typical for the beam density at Fermilab. The 40 picocoulombs collected from 5e11 protons will produce a 0.4 volt signal in a 100pf integrator. The signal to noise ratio is often several hundred and results in good profiles.

FIGURE 6. Multiwire (or secondary emission) profile monitor. Horizontal wires are placed on one side with vertical wires on the other. Each plane has 48 wires. This one has 2mm spacing.

The collected charge is stored in the cable capacitance and is later bleed into the integrating capacitor through a resistor. If the cable shield were connected to ground both at the tunnel and the service building the difference between grounds would swamp the signal. Just 1 mV difference between grounds would integrate to 5 volts with the 10KΩ 100pf integrator used.

The solution to this problem lies in exploiting the essentially infinite beam source impedance. The beam will induce the same secondary emission for any wire voltage within reason. Figure 7 indicates that with the shield grounded in the tunnel, AC noise can be induced in the load impedance above the corner frequency given by the cable capacitance and the load impedance. For a 500 foot 20pf/ft cable into 10KΩ the corner frequency would be 1.6KHz. If the shield is not connected to ground in the tunnel then noise voltage becomes Rload/Rbeam, essentially zero. The signal return currents will flow through what ever ground connections it can find. Because the beam source impedance is so large the return current path can have substantial impedance without affecting the profile measurement.

Unfortunately, cables are run through cable trays with many other conductors that can have substantial "signals" on their shields. Capacitive coupling between the shield of the multiwire cable and it's environment can still induce noise.

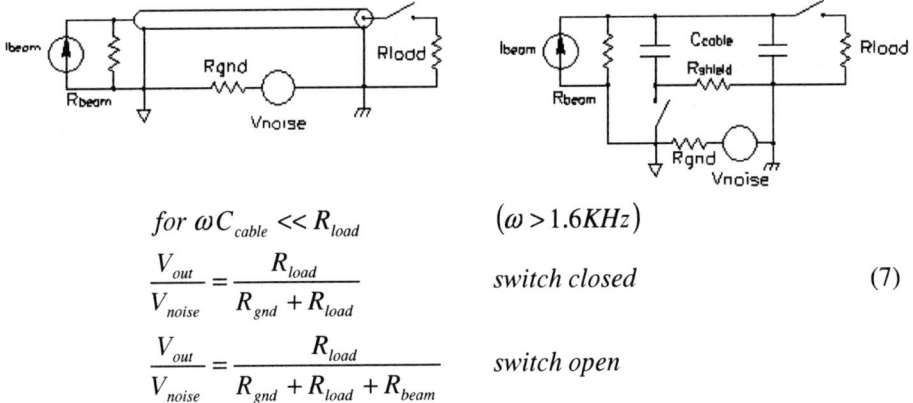

$$\text{for } \omega C_{cable} \ll R_{load} \qquad (\omega > 1.6 KHz)$$

$$\frac{V_{out}}{V_{noise}} = \frac{R_{load}}{R_{gnd} + R_{load}} \qquad \text{switch closed} \qquad (7)$$

$$\frac{V_{out}}{V_{noise}} = \frac{R_{load}}{R_{gnd} + R_{load} + R_{beam}} \qquad \text{switch open}$$

FIGURE 7. Multiwire noise analysis.

Multiwire Cable

Originally Fermilab used custom bundles of RG174 coaxial cable to instrument multiwires. The original multiwires measured only the horizontal or vertical profiles using 24 wires. The latest design measures both horizontal and vertical profiles with 48 wires in each plane. The cost and burden of terminating these bundles precluded their use and it was decided to switch to a shielded ribbon cable. Figure 8 indicates how the ribbon is folded into a round shape and shows the braided shield. The "flat to round" cable is significantly cheaper and is much easier to terminate using mass connectors. The geometry constraints leads to a significant reduction in wiring errors. However, we still manage to get the polarity wrong on a regular basis.

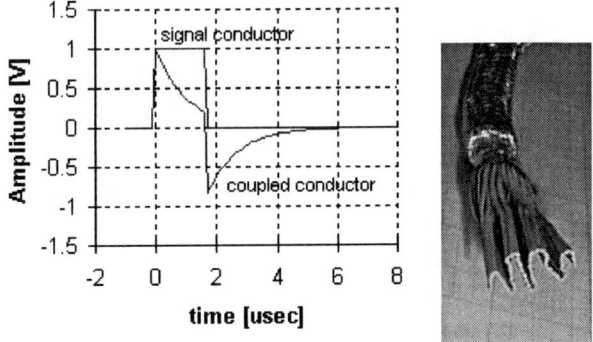

FIGURE 8. Effect of capacitive coupling between signal conductors in Multiwire cable.

The multiwire scanner simultaneously integrates all 96 channels. One feature of the firmware running the scanner is that if any one channel exceeds a limit, integration is stopped for all channels. Unfortunately, capacitive coupling between conductors in the "flat to round" cable can cause ghost profiles or satellites. Figure 8 suggests what a capacitively induced signal might look like on a single conductor. If the integration window is closed before the capacitively coupled signal integrates to zero, then a net signal will be measured and a ghost profile will appear as shown in Figure 9. The normal integrate window is sufficiently long to make capacitive coupling insignificant. The advantages of "flat to round" cable outweigh this undesirable feature.

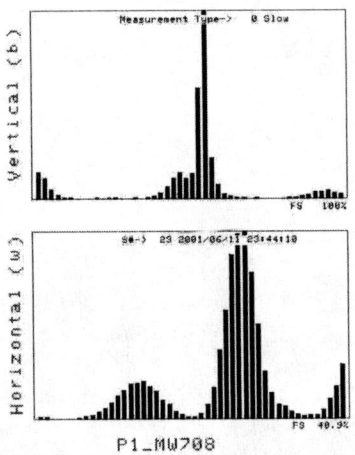

FIGURE 9. If the integration window is closed before the capacitivly coupled signal averages out, satellite profiles are left in the profile.

BPM's

Resonant BPM's

The Fermilab Recycler uses a split tube BPM. An 11 inch section of elliptical beam tube is sliced at an angle to form the two BPM electrodes. The electrodes are held inside of a concentric vacuum chamber. The system is configured to measure the 3rd harmonic of the bunch spacing at 7.5MHz. To improve the signal to noise ratio, a resonator is formed with the plate to ground capacitance and an inductor. A preamplifier is located near the BPM and connected to it through a two foot cable. The plate impedance at resonance is 1.5 KΩ and is determined by the preamp input impedance as shown in Figures 10 and 11.

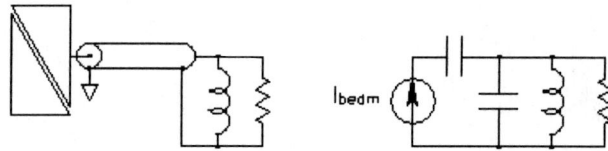

FIGURE 10. Beam Position Monitors used in the Fermilab Recycler are made to resonate at 7.5MHz by connecting the capacitive plate in parallel to an inductor through a short cable.

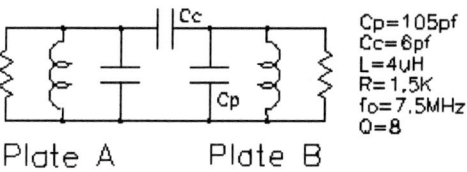

Cp=105pf
Cc=6pf
L=4uH
R=1.5K
fo=7.5MHz
Q=8

FIGURE 11. The beam position monitor plates are coupled through the small capacitance between them. Coupling will be significant if the plate to ground impedance is comparable to the coupling impedance.

Plate to plate capacitance causes problems. An impulse or swept sine wave applied to one plate is coupled to the other plate, changing measured position. At 7.5 MHz, the coupling impedance (1/jwC = 3.5KΩ) is comparable to the 3KΩ preamplifier input impedance (1.5KΩ each side). The coupling impedance can be reduced by shunting it with an inductor between the two electrodes, Equation 8. Coupling will be reduced by a factor of Q for the resonant circuit. Figure 12 shows the coupling with and without the inductor for an impulse and a swept sine wave applied to one electrode.

$$Q = 2\pi \frac{pk\ Energy\ Stored}{Energy\ Lost/Cycle} = 2\pi \frac{\frac{1}{2}CV^2}{\frac{1}{2}\frac{V^2}{R}\frac{1}{f}} = \omega_o RC \qquad (8)$$

$$R = Q\left(\frac{1}{\omega_o C}\right)$$

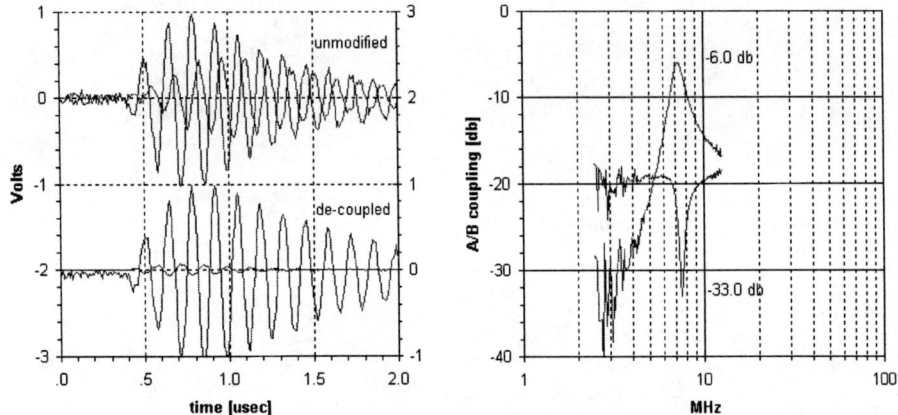

FIGURE 12. Measurements of coupling before and after adding the de-coupling inductor in time and frequency. On the left, the signal on both the driven and coupled electrodes are shown for both the coupled and de-coupled condition. On the right, only the coupled plate is shown for both the coupled and de-coupled condition. Coupling was reduced by 27db at 7.5MHz.

BPM Cables

The Main Injector at Fermilab has 208 bpm's that use 262000 feet of RG8/U cable (49.6 miles)! The longest cable length is 1400 feet and the shortest is 120 feet. The cable was pulled as pairs to obtain similar length and facilitate phase matching. The spread in propagation velocity required pulling an extra 2% or 30 feet to each bpm. A number of bpm's were outside of the specification and required adding cable to one side to obtain matched delay through the pair. Figure 13 indicates the percent difference between cables in a pair as they were pulled in. The cable did not meet the manufacturers specification ±2% variation in velocity.

Figure 13 also shows the measured characteristic impedance for one pair of cables. Only a half dozen were measured with similar results. Again, the cables did not meet the manufacturer's specification of ±2Ω.

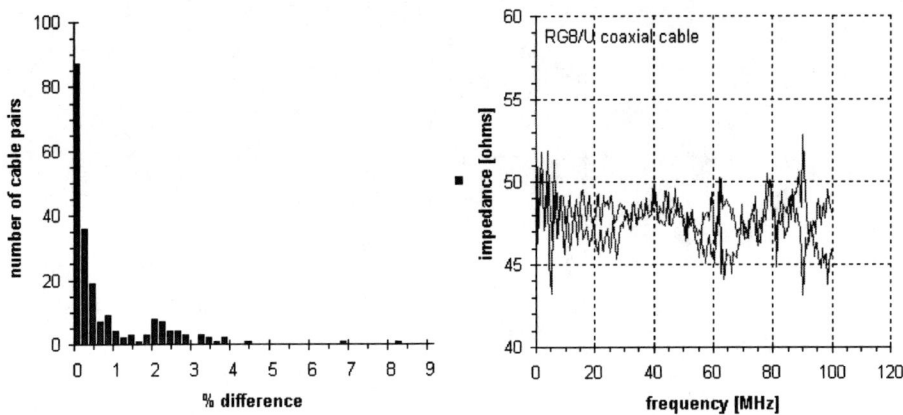

FIGURE 13. Distribution of the percent difference in cable length for the 208 pairs of cables used for the Fermilab Main Injector BPM's. Measured characteristic impedance of one pair of RG8/U BPM cables.

After installing the bpm's in the Booster to Main Injector transfer line we were plagued with complaints of poor accuracy from one particular bpm. Eventually what was found is that one cable of the cable pair had a resonance very near the frequency used by the bpm system. Figure 14 shows the measured characteristic impedance as a function of frequency. The marker is placed at the beam frequency component used by the bpm system, 52.813MHz. The problem was corrected by replacing the pair of cables. The offending cable was closely inspected but no obvious problem could be found. The cable was cut into sections with all sections displaying the same resonance. It is speculated that perhaps the cable reel sat on one side and placed periodic deformations at just the right spacing to make a resonance at exactly the worst possible frequency. A diameter of 2.3 feet corresponds to ½ wavelength at 52.813 MHz. This is consistent with the diameter of the reels used. Cables from other nearby bpm's had similar resonance's at higher and lower frequencies as if they were to come from different depths of the reel. A problem in the manufacturing process is also possible.

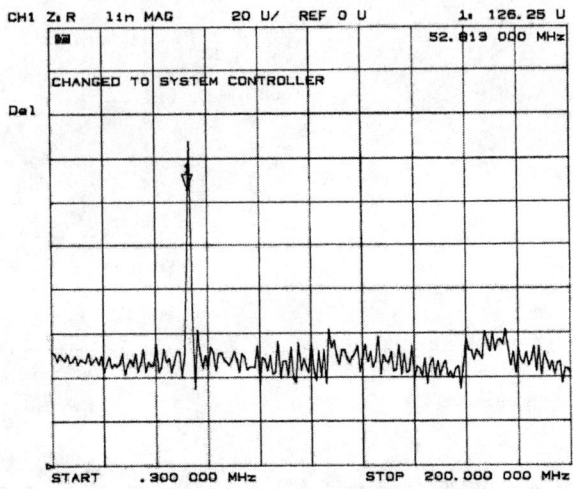

FIGURE 14. Characteristic impedance of a problem BPM cable. The beam bunch frequency being measured (at the marker) was nearly at the peak, making position measurement very sensitive.

Magnetic Interference

DC Current Transformer

A Bergoz Parametric Current Transformer was purchased and installed in the Fermilab Recycler to measure average beam charge. The device worked satisfactory until it was moved. The magnetic field from the now nearby Main Injector bend buss induced a substantial error in the measured beam charge, Figure 15. According to the manual, the transformer has a 100 uAmp/gauss radial sensitivity to magnetic fields. Typical magnetic fields from the buss are a few gauss. The error in Figure 15 is about 1e10 and corresponds to 1 mAmp or 10 gauss. A magnetic shield was constructed with 1/16 inch thick high permeability mu-metal. The shield was 9 inches in diameter and 18 inches long. The shield reduced the magnetic field by 22db, Figure 16.

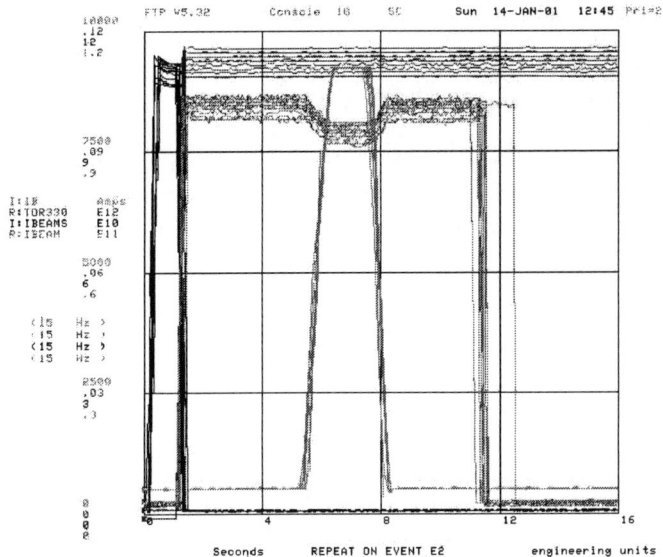

FIGURE 15. Beam charge in Main Injector is transferred into the Recycler at about 1.2 seconds. The Main Injector ramps between 5 and 8 seconds. The measured Recycler beam charge is affected by the current in the Main Injector bend buss.

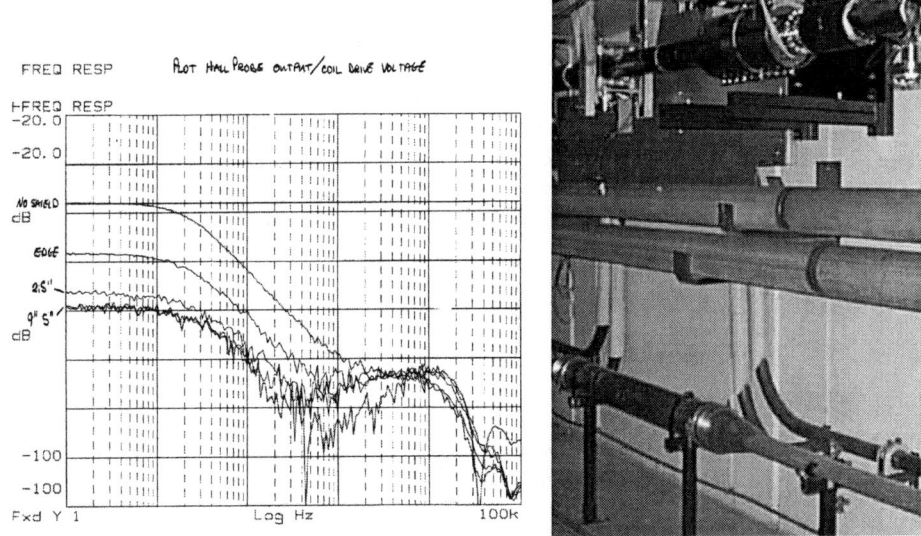

FIGURE 16. A high permeability mu-metal shield was placed around the Bergoz Parametric Current transformer. The Main Injector bend buss is shown in the background attached to the enclosure wall. The plot on the left indicated the measured magnetic field as a function of frequency without the shield, at the edge of the shield, and at 2.5, 5, and 9" from the edge. The shield is 9" in diameter.

Diagnostics For Recirculating And Energy Recovered Linacs[1]

Geoffrey A. Krafft and Jean-Claude Denard[2]

Thomas Jefferson National Accelerator Facility
12000 Jefferson Ave., Newport News, VA 23606

Abstract. In this paper, the electron beam diagnostics developed for recirculating electron accelerators will be reviewed. The main novelties in dealing with such accelerators are: to have sufficient information and control possibilities for the longitudinal phase space, to have means to accurately set the recirculation path length, and to have a means to distinguish the beam passes on measurements of position in the linac proper. The solutions to these problems obtained at Jefferson Laboratory and elsewhere will be discussed. In addition, more standard instrumentation (profiling and emittance measurements) will be reviewed in the context of recirculating linacs. Finally, and looking forward, electron beam diagnostics for applications to high current energy recovered linacs will be discussed.

INTRODUCTION

Small recirculating electron accelerators, operating at beam energies below 1 GeV, have existed for many years starting with the development of the microtron. In the past decade, several GeV scale recirculating accelerators have been built, including the Bates recirculating linac at MIT, the Mainz cascaded racetrack microtrons, and the 5-pass superconducting recirculating linacs at Jefferson Lab. These accelerators have now operated long enough that many standard beam diagnostic techniques have been used and developed, and it is worthwhile to review the progress that has been made on new, and not so new, beam diagnostic elements.

A schematic diagram of the Continuous Electron Beam Accelerator Facility (CEBAF) is shown in Figure 1, and a summary of the main accelerator parameters is given in Table 1. CW beam, originating in an injector, is recirculated up to five times through each superconducting radiofrequency (SRF) linac. The beam may be directed into up to three experimental halls simultaneously, the beam current in each hall being at the third subharmonic of the fundamental RF frequency of 1497 MHz. The lowest operating energy is about 0.8 GeV, the present full energy is nearly 6 GeV, and a cost-effective upgrade to 12 GeV is possible and planned. The combination of five-pass recirculation, a three-laser photocathode source, and subharmonic-rf-separator-based extraction enables simultaneous delivery of three beams at different but correlated energies.

---

[1] Work supported by United States Department of Energy under contract DE-AC05-84ER40150.
[2] Present Address: Synchrotron Soleil - Bat. 209H – Centre Universitaire Paris Sud – BP34 – F 91898 Orsay Cedex

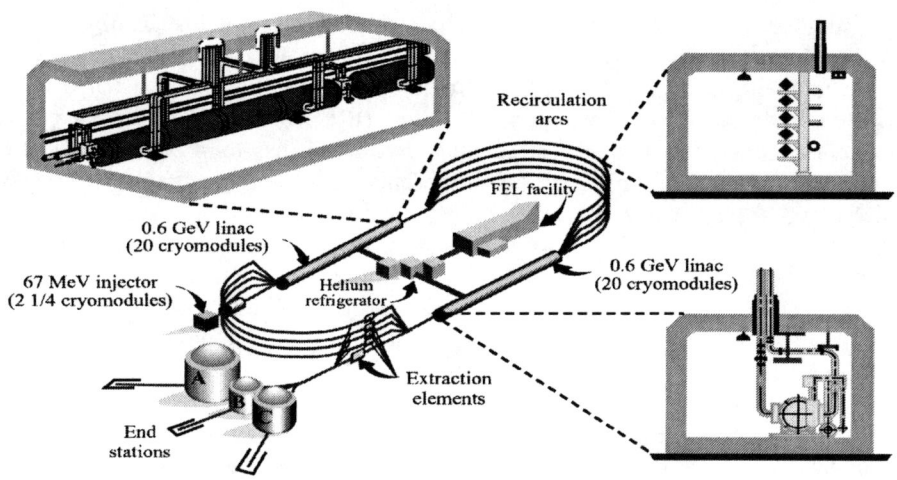

FIGURE 1. Jefferson Laboratory's CEBAF accelerator.

The most radical innovations in CEBAF are the choice of SRF technology and use of multipass beam recirculation. Neither had been previously applied on the scale of CEBAF. In fact, until LEP II came into operation, CEBAF was the world's largest implementation of SRF technology. Beam recirculation minimizes the cost of the SRF linacs to achieve a given energy, and has been executed with large enough bend radii to keep open the possibility of future energy upgrades.

TABLE 1. CEBAF Accelerator Parameters.

Item	Value	Unit
Beam Energy	0.8-6	GeV
Beam Average Current	<180	microAmps/Hall
Normalized *rms* Emittance	<1	mm mrad
Repetition Rate	499	MHz/Hall
Charge per Bunch	<0.4	PC
Extracted *rms* Energy Spread	$<10^{-4}$	
Transverse *rms* Beam Size	<100	microns
Longitudinal *rms* Beam Size	60(200)	microns(fsec)
Beam *rms* Angular Spread	$<1/\gamma$	

As shown in Table 1, recirculated linacs like CEBAF can be designed for short bunch operation naturally, because the beam orbit does not actually close on itself as in a ring. Many techniques for measuring, monitoring and adjusting bunch length have been developed as part of the project, allowing routine operation of this accelerator at bunch lengths of order 100 fsec. When it comes to actually operating a recirculated linac, some of its properties must be regulated in ways somewhat differently than in typical linacs or rings. Therefore, it is worth sharing some of the techniques found useful for stabilizing and maximizing beam performance in such accelerators.

In addition to such high energy accelerators, increasing interest in high current recirculating accelerators, particularly those that are energy recovered, is developing. At the moment, the highest average current recirculated linac in existence is the Jefferson Lab Infrared Free Electron Laser (IRFEL). An energy recovered arrangement was chosen for this accelerator because it has long been realized that energy recovered operation may provide the only path to high average power (MW scale) in an FEL, and the idea needed to be explored quantitatively on a relatively small-scale recirculated linac. Two attendant benefits of choosing to energy recover this particular linac were: (1) the RF power required to continuously run the FEL is greatly reduced, and (2) the disposal problems associated with the spent electron beam after the FEL interaction is considerably alleviated. The device proved to be a useful test stand for developing beam diagnostics, as will be mentioned in the body of this work.

In this paper we will discuss longitudinal phase space measurement, manipulation, and control; recirculation path length adjustment; multipass beam position monitoring; and operational control ideas, especially data obtained from rapid, time synchronized toggling measurements. Next, options for beam profiling in such devices are given. The paper concludes with an introduction to the types of advanced diagnostics that may be required for future high current recirculated linacs.

LONGITUDINAL DYNAMICS AND PHASE SPACE MEASUREMENTS

Unlike the Bates recirculator, where the beam energy spread requirement does not dictate the need for short bunches, or the Mainz racetrack microtrons, where the primary determinants of the extracted longitudinal phase space are the initial matching and the longitudinal stability of the phase oscillations in the accelerator, the CEBAF accelerator from the beginning was designed to have an extremely small relative energy spread [1]. The approach taken in the design was to: (1) design a continuous beam electron injector that produced very short bunch lengths, (2) design recirculation optics that were isochronous (R_{56}=0) and achromatic, and (3) choose to operate the linacs on the crest of the accelerating wave. There is no longitudinal focusing in the accelerator proper, and small single bunch energy spread arises from the small electron pulse length interacting with the curvature of the crested RF wave, assuming each of the beam passes is made at the same RF phase. In this section we discuss how to make and observe the short bunches, in the next we discuss how to make the different beam passes have simultaneous time-of-arrival.

In a recent review, the requirements on bunch length for producing a small energy spread are given [2]. In order to have less than 2.5×10^{-5} *rms* energy spread from the RF curvature, and to have some control margin to allow fast fluctuations in the RF controls and the energy spread that they generate, a bunch length of 1.8 RF degrees (3 psec *rms*) was specified for CEBAF. In the course of detailed longitudinal matching designs of the injector it was found that bunch lengths as short as 0.1 degree (200 fsec *rms*) were possible, and of course desirable as they reduce the overall single bunch energy spread.

Three somewhat different types of measurements were done in order to verify such short bunches are possible and have been obtained at CEBAF. The first, both chronologically and with greatest utility operationally, was the phase transfer monitor [3]. Its principle and operational use are discussed at length and quite thoroughly elsewhere [4]. In brief, a longitudinal beam pick-up is used to obtain a beam derived RF signal that is mixed against the RF master oscillator to detect time-of-arrival. By modulating the beam chopping phase and simultaneously detecting and recording time-of-arrival, the phase transfer technique is used to produce a graph of output bunch phase against input bunch phase, including the non-linearities in the transfer response. By properly correcting nonlinear behaviors in the transfer function plots, one is in reality correcting and adjusting non-linear distortions in the longitudinal phase space. When properly done, all the different longitudinal slices of the longitudinal phase space at the chopping point are adjusted to have the same time-of-arrival at the location of the pick-up. The pick-up cavity is strategically placed just prior to the first acceleration to 45 MeV, after which little further longitudinal motion arises due to relativistic effects. The resolution of the technique is below 200 fsec, which corresponds to 60 microns in bunch length.

Chronologically the next method developed was a monitor based on coherent synchrotron radiation (CSR) [5-7]. From a continuous operation and monitoring point of view, non-invasive beam diagnostics are preferred, and those based on measuring synchrotron light are preferred to those based on transition radiation for this reason. Because the bunch length is of order 100 microns fully compressed, and because the linac beam must be bent through an injection magnetic chicane anyway, narrowband coherent synchrotron radiation detection was used in order to provide a continuous bunch length (change) monitor. CSR was detected through a bandpass filter with center wavelength of 500 microns and transmission width of 20%. Such a monitor, based on sensitive Schottky diode detection, was able to detect signals at down to 1 microA beam current easily [8]. Some care must be taken when applying techniques based on spectrum detection of bunch shapes, because it is actually highly unlikely that the bunch shapes in linacs, recirculating or not, are Guassian, or that calculations of the bunch shape based on Guassian distributions will lead to meaningful results [5,9]. So even at the earliest stages of the CEBAF work on coherent detection techniques, it was understood that some other method would be needed to obtain the bunch shape accurately.

The third method used to analyze the bunch length and distribution shape was the so-called zero phasing method. As indicated schematically in Figure 2, in this time domain method, one (or several) RF accelerating cavities are phased close to the zero-crossings of the accelerating fields, and the beam longitudinal distribution is analyzed by transverse measurement by a beam profile monitor located at a location of non-zero dispersion [10]. Data at both crossings and with the RF in the zero-phased cavities off is gathered, and differences in the observed distributions between positive going and negative going crossings can be used to infer information about the slope of the longitudinal phase space distribution. At CEBAF, zero-phasing measurements are considered the standard baseline technique, from which the CSR monitor could be calibrated for an actual bunch length measurement. Such cross calibrations apply to the extent that distribution shape changes are small during the period of monitoring, a circumstance that can only be verified a posteriori, by zero-phasing measurements before and after the period of beam monitoring.

At Jefferson Lab, the smallest bunch length, determined by a Guassian fit to zero-phasing longitudinal distribution data, was 85 fsec *rms*. It was verified that CSR emission was highest when the bunch length was shortest in this machine setup [11]. Many years after this work, it was verified that the standard operations mode for the CEBAF accelerator has *rms* bunch length from 150-200 fsec, over the entire current operating range in the accelerator, well below the requirements [12]. The main disadvantage of the zero-phasing method is that it is highly invasive to normal beam operations because special RF setups have to be introduced to make the measurements. On the other hand, the information that is derived from the method is quite detailed as to the actual longitudinal distribution. Recently the method has been used to determine time dependent beam parameters (slice parameters) from the beam emerging from an RF gun at Brookhaven National Laboratory. The time resolution reported was 8 fsec in a bunch whose duration is several psec [13].

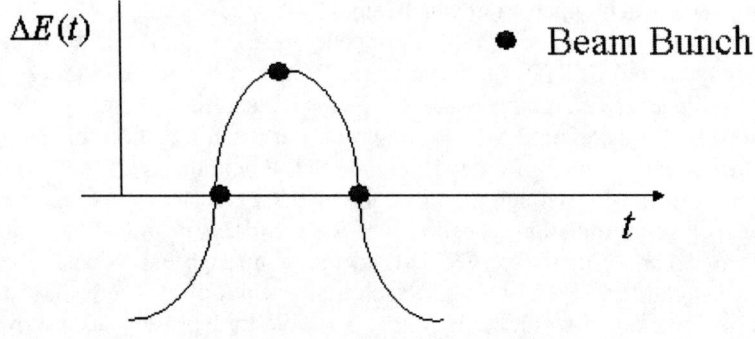

FIGURE 2. RF settings for zero phasing measurements of bunch length.

RECIRCULATION BEAM PATH

The next quantity that must be properly set is the recirculation beam path. The requirements are again given in Reference [2]. In order to have a small contribution to the energy spread of the beam, each pass of the beam must have the time of arrival within a few degrees of the energy crest phase for all the passes simultaneously. One way to achieve this condition is to: (1) measure and crest the bunches passing through each of the accelerators in CEBAF on the first pass, followed by (2) making sure that each bunch of each subsequent pass arrives at the same time as the first pass beam within the requirement. This approach is taken in the chronologically first method of path length control discussed next. Presently, a newer, non-invasive and continuous method of path length control is being investigated and implemented and will be described in the following subsection. Those readers used to storage ring practice of adjusting the accelerator frequency to adjust for path length shifts should recall that such an idea is probably straightforward to implement on a two pass recirculated linac, but is not of obvious utility for more passes as the differing path lengths of the passes are likely to be unequal, and not adjusted properly by a single frequency change. It also might be operationally challenging to be adjusting the operating frequency of many superconducting cavities on a continuous or semi-continuous basis.

Direct Measurements By Time Of Arrival

The first method for path length control involved periodic cresting of the bunches on the accelerating fields of the linacs in the first passes, followed by precision relative time-of-arrival monitoring of the higher pass beam bunches. Cresting, which presently happens no more frequently than once a day during quiet operations, now proceeds automatically [14]. No additional beam instrumentation is need beyond a BPM or viewer located in a dispersed location for observation of the crest condition.

Special instrumentation was developed for the higher pass relative time-of-arrival monitoring that is based on the phase transfer measurement electronics. The outputs from longitudinal pickup cavities installed inside the recirculation loop are amplified and fed into one side of several fundamental frequency mixers, the other sides being fed by a fundamental frequency signal derived from the master oscillator. The first signal can be phase shifted (electronically) so that on average, the mixer acts as a phase detector.

A special beam condition is established to perform the measurement. The once around time for beam in CEBAF is 4.2 microseconds. If a 4 microsecond beam pulse goes through the 5 passes of the accelerator, the signal out of the longitudinal pickup will come in successive 4 microsecond chunks spaced by 4.2 microseconds, one for each pass of the machine [15]. Ideally, when all the arrival times are identical, the output from the phase detector is constant and zero if the phase shifter is properly set. If there are level changes in the output of the phase detector, one can easily infer the beam phase shift on the pass of interest. In theory, it merely remains to digitize the signal out of the phase detector at the correct time corresponding to each pass. In current practice, the device is a little more accurate because the full waveform out of the phase shifter is digitized at 10 MHz, and the phase shift levels are determined by software analysis of the complete record. Accuracy is increased because many data points are averaged to determine the phase shift of each pass [16].

The performance of this device is very good. Even at the standard measurement mode operating current of 3 μA, the precision is 0.1 degrees or 200 fsec in the individual readings. Averaging within a single pass waveform increases the precision by about 5. These measurements correspond to changes in the machine circumference of about 10 microns.

Beam Based Phasing By Phase Modulation

The main weakness of the direct method of phase error determination is that it requires the machine to be placed in a pulsed beam mode and is therefore invasive to normal beam delivery. It would be far better to have a non-invasive method of phase monitoring that can operate with the standard CW beam delivered to users. Because of the development of feedback systems that operate at high enough sample rate [17], it is possible to consider placing various modulations on the beam, that will be removed before delivery of the beam to users, but which also have beam diagnostic purposes. In fact, the best way to check that one is on crest of a linac is to phase-modulate the linac and make sure that the energy change is zero for each pass of the beam through each linac. As shown schematically in Figure 3, this principle is used for the continuous phase monitoring capability that has been recently developed at CEBAF [18]. If the bunch is crested, as shown in the Figure, there is no energy change at the modulation frequency, and a small signal at twice the modulation frequency due to the curvature of the RF wave on crest (the modulation amplitude is expanded for clarity in Figure 3). On the other hand if there is in fact a phase offset between the beam and the linac RF, there is an energy modulation produced by the phase modulation at the modulation frequency, that is to first order linear in the phase displacement.

In order to obtain information that discriminates errors that arise in the separate CEBAF linacs, the phases of the two linacs are modulated at two different frequencies: 383 Hz and 397 Hz. In practice, the amplitude of modulation is small enough (0.05 degrees at 1497 MHz) that the feedback system can adjust away the energy centroid shift, and small enough that there are no other deleterious effects generated by the modulation. Utilizing lockin techniques to generate the phase modulations, BPMs at dispersed locations in each pass beamline at the end of the second CEBAF linac are observed for beam motion at the lockin frequencies. After calibration, observed lockin outputs correspond to the summed phase shifts (from all previous passes) from each linac. As there should be no phase shift anywhere, a simple matrix calculation gives the individual phase shifts to be adjusted on each pass [19].

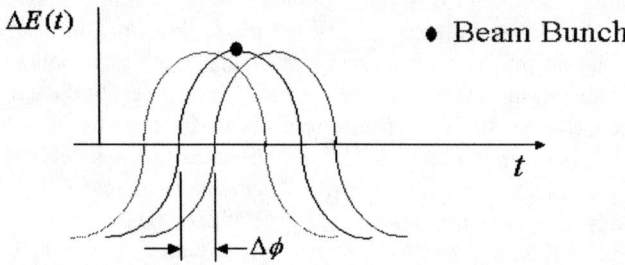

Bunch "Crested" when $d\Delta E / dt = 0$

FIGURE 3. Beam based phasing by linac phase modulation. The modulation amplitude $\Delta\phi$ is 0.05 degrees at 1497 MHz. The beam phase offset is proportional to the same-frequency energy modulation.

MULTIPASS POSITION MEASUREMENTS

As the beam propagates along the linac, each beam pass present at a beam position monitor excites the device. Ideally, the beam position monitor should be able to give the user information on the beam position on individual passes. How is the individual pass information to be obtained? Four basic methods have been suggested to solve this problem: (1) measure properly timed beam pulses (or gaps!), (2) distinguish individual passes by time-of-arrival of the beam pulses originating on the monitor electrodes, (3) utilize pseudorandom sequences with orthogonal time delay properties combined with beam modulation to separate out signals from the different beam passes, and (4) measure positions with monitors of two different frequencies.

The first method involves providing a beam pulse and reading the beam position monitor electronics at a time when only one beam pass is in the monitor. Thus, the measurement is much like the direct time-of-arrival measurement discussed above. As long as the detection electronics have response times short compared to the once-around time, which is 4.2 microseconds for CEBAF, measurement and interpretation of the results is straightforward. A variant, that is better for application to a high duty factor accelerator, but is slightly more complicated to analyze and calculate and has reduced dynamic range, is to blank off the beam for times short compared to the once around time, and synchronize the digitization of the monitor electronics with the arrival of the "gap" on the different passes [20]. It should be noted that in order for this method to succeed, the BPM trigger or reading system must have a means to delay the read time to account for the fact that individual monitors should be read at different times at different longitudinal locations in the accelerator.

The second method, at this stage mainly theoretical, applies only in cases where the charge-per-bunch is high enough that measurable beam pulses are produced on the BPM electrodes, and in situations, as in the IRFEL, where there is a substantial time difference between the accelerating beam pass and decelerating beam pass because only a small fraction of the accelerating buckets are filled. Because the accelerating and decelerating beam pulses are separated in time, a sufficiently precise timing method is able to discriminate the information from the separate beam passes. Such a method is, from a timing point of view, much more sensitive to time jitter between digitization trigger and beam arrival than in the first method; now the relevant time scale to distinguish the bunches is the time of bunch separation, not the once around time. Measurements on the Jefferson Lab IRFEL have shown that accelerating and decelerating beam pulses are indeed distinctly observable on an oscilloscope, but the next-level design of a suitable detection and digitization system has not been completed at this time. For accelerators designed with more than two passes, it will be essential to have the various accelerating and decelerating passes in different RF buckets in order to apply this method.

The third method is probably the best method to apply to diagnosing CW beam, but is the most difficult conceptually and the most difficult to implement. It has long been known that various bit sequences may be produced that are pseudorandom (equal numbers of zeros and ones) with various beneficial properties in addition. In particular, it is possible to produce sequences that when digitally mixed with themselves yield non-zero output, but when mixed with time-delayed versions of themselves yield zero output. As the once-around time from the accelerator produces a fixed time delay between the beams on different passes for a machine like CEBAF, an obvious application of such sequences is to modulate the beam current at a small level in a pseudorandom way where the "states" change once in a once-around time, and detect electrode readings once in a once-around time. The data stream is analyzed by mixing (both analogue and digital mixing have been proposed) the data stream with the pseudorandom code synchronized with the first pass beam to get the first path beam position information, by mixing the data stream with the pseudorandom code shifted by the once around time to get the second pass beam position information, by mixing the data stream with the pseudorandom code shifted by twice the once around time to obtain third pass beam information, and so on. Of course the code must become longer the larger the number of passes and the time response of the monitor will become correspondingly longer as the code length increases.

A final possibility, that is particularly clean when applied to an energy recovered linac (ERL) with one accelerating pass and one decelerating pass and every bucket filled, is to have two sets of BPMs at different frequencies (or a single BPM that can detect both fundamental and second harmonic): the fundamental and twice the fundamental. It is clear that if the position of the two beam passes is identical, then the fundamental is not excited at all and twice the fundamental is excited with an amplitude twice as large. On the other hand, if the two beams were displaced oppositely, the fundamental would be excited and the second harmonic would be zero. So by taking a slightly more complicated difference over sum formula, both positions of both beams may be determined. Note that in general the sensitivity of the BPMs at the different frequencies might be quite different and must be adjusted for.

Up to now, only method 1 above has been used at CEBAF. It clearly is of no help for continuous monitoring of the CW beam delivered during normal beam operations.

FAST DIFFERENCE MEASUREMENTS

From a beam diagnostics point of view, one of the advantages of a recirculated linac is that the beam orbit does not close on itself. Therefore, the process of verifying the beam optics of the focusing lattice can be somewhat simpler than in rings. Most of the optics are measured utilizing some sort of differencing scheme. For example, difference orbit measurements throughout the accelerator are performed by activating sets of steerers in the accelerator and observing the beam orbit changes on beam position monitors (BPMs). Likewise, the dispersion characteristics of the bend regions and arc may be evaluated by taking differences in the orbits between two separate energy states. Another example in the longitudinal direction, the isochronicity of the bending arcs may be evaluated by measuring the time-of-arrival phase at two different beam energies. In the earliest days of the CEBAF accelerator, these types of measurements were either done by hand changes in the perturbing elements, or by gathering data using simple computer scripts to make the changes and record the data for off-line analysis [21]. The idea, similar to the old batch processing methods on computers, was that only after analysis would one make adjustments to the accelerator. Such a scheme, because of the time it takes and because of the difficulty in maintaining beam conditions between measurements, becomes totally impractical for any adjustments except the simplest, and can severely limit ones ability to do more sophisticated types of correction algorithms.

Following the computer analogy, more modern interactive computer programs based on real time systems were developed to allow users to do work where one could easily change parameters and observed changes in the results calculated by the computer. Such an approach is effective to the extent that all calculations are completed on a time-scale short compared to a human response time because one reduces the dead time of the human user. The fast optics difference measurements were developed to be rapid enough that an operator (user) can perform adjustments and observe results in real-time so that he/she can quickly converge to an accelerator condition that is known to give reliable beam operations.

The earliest attempts at fast differencing occurred during the commissioning of the first CEBAF arc. In order to determine the R_{56} of the first arc, which should nominally be zero, a 70 Hz square-wave modulation was placed on the amplitude of a linac cavity and the time-of-arrival was observed with an RF phase detector [21]. As the pulsed tuneup beam was line synched, the output from the phase detector consisted of two levels corresponding to the phases of the two energy beams. Setting R_{56} to zero could be done in real time by adjusting magnets in the arc beam transport system that were at positions of non-zero dispersion so that the phase difference was zero. In addition, it was checked that two spots appeared on viewers at positions with non-zero dispersion corresponding to the two energy states when the modulation was present. This means an easy way to set the dispersion to zero, a step needed as part of machine setup, is to modulate and adjust the quadrupoles that control dispersion so that two spots become one. An operator can perform this chore rapidly, because of the real-time nature of the response of the accelerator.

A more quantitative system was developed based on 30 Hz toggling of the 60 Hz tune-up beam [22]. For difference orbits, the beam is toggled between two states with two sets (one for each transverse direction) of air-core magnets, and for dispersion the beam is toggled in two energy states. The data acquisition is synchronized to the toggling, so the sign of the dispersion or matrix element is unambiguously determined at the location of every beam position monitor in the machine. The dispersion adjustment is easy conceptually: the operator must zero out dispersion outside any bend region of the accelerator. Setting up a confined difference orbit (and by implication a confined beam orbit) is done by calculating the difference between the Courant-Snyder invariant measured by the toggling system, and the design invariant, and doing adjustments so that the growth in amplitude is minimized. Through the use of this system, all nine passes of the accelerator optics can be adjusted so that there is no growth in Courant-Snyder invariant in a period of about an hour. After performing this procedure, it is almost always the case that the measured machine aperture (as measured by increasing the amplitude of the toggling steerers) is quite large compared to the beam size, and the beam optics is properly contained.

To complete this section, it should be mentioned that in addition to the benefit to the operator, having fast differencing capability allows optics experts to make measurements far more complete, thorough, and sophisticated with the toggling system than without it. This type of work, which tends to be "batch mode" due to limited machine development time, proceeds more rapidly if greater quantities of relevant data may be gathered within a given time period.

BEAM PROFILING TECHNIQUES

In this section references are given to some of the beam profiling techniques used on the CEBAF accelerator. Almost all of the beam viewers in CEBAF are made from phosphorescent materials, and all such beam viewers in the linac regions where multiple beam passes are on a common beam line have a transverse hole near the center of the viewer. Having such a hole is important to allow lower pass beams through the viewer in an unobstructed and unperturbed way, while retaining the ability to observe the higher pass beams. The lower pass beams are merely cleanly steered through the viewer holes. The advantages of phosphorescent viewers are that a common material and viewer design work throughout the machine energy range in its most useful application: observing whether beam halos or tails have formed on the beam, and which are usually corrected through the machine optics checks mentioned above, or by measurement and correction of RF phases in the injector. The disadvantages of the viewers are two-fold. The main disadvantage is that such a viewer cannot remain inside the CW electron beam without the viewer being destroyed. Consequently, only a pulsed low duty factor "tune-up" beam can be analyzed with these devices. Even if we would like to analyze the tune-up beam in detail, it is well known that phosphorescent devices are easily saturated, leading to ballooning of the beam distribution on measurement.

Both of these problems are "solved" by implementing optical transition radiation (OTR) viewers. Because the OTR radiator emits radiation prompt with the transition and the light propagates only in vacuum, and because the optical power emitted is much lower than for phosphorescent devices to begin with, the saturation characteristics and the linearity of the emission with beam current of the transition radiators are much superior, making the development of a profiling instrument somewhat easier and more accurate. It has also been possible to develop foils thin enough, the foils at CEBAF are ¼ micron in width, so that the electron beam is practically unperturbed by its interaction with the radiator, and so that the radiator easily survives the maximum CW beam current in the accelerator [23].

OTR viewers were first installed at CEBAF to perform an experiment to verify that the resolution limits of such viewers did not go as $1/\gamma$, as had been postulated. In fact, a 4 GeV beam was analyzed with an OTR viewer with a resolution of 60 microns, where the camera was sensitive from 300-600 nm [23]. After this experiment was successfully concluded, most subsequent viewer installation, which was usually motivated by the desire to perform a detailed profile measurement, has been of OTR devices. That the OTR viewers could monitor high current CW beam [24] is demonstrated by the data in Figure 4. The energy spread and energy centroid position, as measured by forward OTR profiling, is plotted against time for a period of two weeks, during which time there was a user requirement to keep the energy spread small in the Hall C beam. The relative energy spread was held at 5×10^{-5} during the whole period of running, and the viewers, in combination with a rapid digitization and analysis capability, provided the experimenters with a continuous record of the beam size. The only real problems were that for some beam parameters (low beam energy and high beam current) it is possible to trip off the machine protection system by the beam loss, and simultaneous use of the monitor and a downstream Compton beam polarimeter is sometimes not possible. All of the viewers on the Jefferson Lab IRFEL use optical transition radiation, and the viewers on the multipass beam line all have holes.

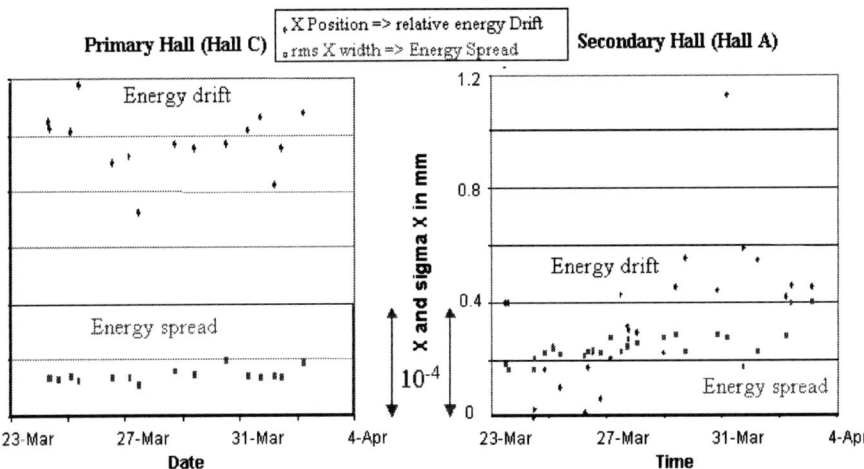

FIGURE 4. Energy Centroid and Energy Spread Measured by Optical Transition Radiation at CEBAF.

Although an experiment has been performed to check high energy diffraction transition emission by a hole on the CEBAF beam-line, no detailed experiments have yet been performed to check whether some of the theoretical details of the diffraction radiation patterns are observed in practice [25]. Diffraction radiation, as an alternative to OTR, has the nice feature that higher average beam currents may be monitored in a more non-invasive way. A technique recently developed at storage rings, involving interferometry of synchrotron radiation, is being investigated as a non-invasive continuous, and relatively fast profile monitor.

To conclude this section, it is worth mentioning that there are many wire scanners in the CEBAF accelerator. The advantages such devices have by comparison to OTR viewers are that they have somewhat better resolution than the viewers and it is thought, because electronic measurements are made, that they are more linear with the beam current. The primary disadvantages of this technique are that a single measurement takes about 45 sec, limited by the scan speed, and that one is not in a position to measure fully CW beam with wire scanners, limited by blinding of the machine protection systems due to loss of electrons scattered in the wire. These devices have found to be very useful for emittance measurements, establishing beam optical functions at various machine locations, and for measuring, checking, and correcting details in the beam profile. Due to slow scan speed and the extra time it takes to perform a full "quad-scan" emittance measurement, measurements with multiple scanners (4 or 5) at different longitudinal locations is the preferred method of doing such measurements in a machine where beam-time is a precious commodity.

FUTURE WORK

Within the past few years, work in the field of recirculated linacs has grown due to widespread interest in utilizing such accelerators in a variety of applications. Many of these applications are summarized in the review previously referenced [1] and in a recent International Committee on Future Accelerators Beam Dynamics Newsletter [26]. A substantial number of the new applications rely on applying the notion of energy recovery: the beam recirculation is arranged to have first an accelerating pass (or several accelerating passes), followed by a decelerating RF phase pass of the beam (or several passes) in order to reuse the energy resident in the beam. Such an arrangement allows substantial average beam currents to be accelerated to high energy for the application. Also, such schemes allow the total RF power cost to run the facility to be reduced far below what it would be had the linear accelerator not been energy recovered. Such statements may be quantified by defining a power multiplication factor

$$\kappa = P_{beam}^{ave} / P_{rf} \qquad (1)$$

where P_{beam}^{ave} is the average beam power at highest energy and P_{rf} is the continuous RF power needed to accelerate the beam. It is easy to show that for a storage ring, κ approaches the ratio of the synchrotron radiation damping time to the revolution time, a number of order or greater than a thousand for a typical ring. The most recent superconducting energy recovered linac designs have κ from 200 to 500; the overall accelerator efficiency is getting to values approaching those in storage rings.

The other advantage of the recirculating linac arrangement has already been discussed. Just as is true for conventional linacs, the opportunity exists for manipulating the longitudinal phase space in the accelerator to produce very short electron bunches. The combination of high average beam power, produced in high frequency streams of relatively low charge electron bunches, with ultrashort bunches (of order 100 fsec) will probably be uniquely obtained in an energy recovered linac.

It is expected that high average current energy recovered linacs will begin to look like storage rings as regards to their electron beam diagnostics. Beam diagnosis based on detection and analysis of synchrotron light should be readily achieved with results and performance similar to present day storage rings. As the overall beam size in present ERL designs is only somewhat smaller than in storage rings, beam position monitoring and profiling will undoubtedly build on storage ring experience, and feedback and stabilization systems will undoubtedly follow storage ring developments.

New developments are anticipated in several areas: (1) beam profiling and emittance measurement at low beam energies (i.e., beam energies small enough that synchrotron light is not a possibility), (2) bunch length monitoring and control, especially with highly compressed bunches, and (3) measurement of electron beam halo. All of these measurements are particularly difficult if a non-invasive measurement of the high average power beam is desired.

At present it seems that the best way to diagnose an electron beam profile at low energies is to use a residual gas ionization profile monitor. Such a device can operate in a non-invasive way and preliminary calculations indicate reasonable profile integration times at the high average CW currents anticipated. It seems that the best way to monitor the bunch length will be through spectrum analysis of the rather copious coherent synchrotron radiation that is emitted by the high average power electron beam. An alternate idea, gaining increasing numbers of adherents recently, is to detect and spectrum analyze coherent diffraction radiation from a diffracting aperture that the high current beam can pass through without causing damage. There is interest in utilizing such diffracting holes to obtain direct measurements of beam emittance by analyzing the modulation depth of the far-field diffraction radiation pattern.

As the average circulating beam power in energy recovered linacs is huge, it will be imperative to make sure that only the smallest portion of the beam is lost in transit through the accelerator. For example, a relative beam loss less than 10^{-5} is necessary for a recently proposed recirculated linac light source at Cornell [27,28] to operate properly. Beam halos at this level are not routinely measured on other electron accelerators. Initial measurements will be performed on a prototype injector for this machine, one of whose purposes is to develop a strong case that high average power electron beams with small haloes can indeed be produced. Measurements will proceed by placing radiators with different hole sizes in the halo regions, and by measuring the scattered electrons by sensitive radiation detectors. It will be most important to understand which parts of the injector (where most of the halo must originate) are most important for determining halo, and finding ways to minimize the amount of halo generated.

ACKNOWLEDGMENTS

This work was accomplished through the dedicated efforts of a large range of Jefferson Lab staff. The authors would like to thank all those who helped build and perform bunch length measurements, path length measurements and adjustments, feedback systems and beam stability work, and who developed the toggling measurements into their final form. The large number of individuals involved makes it impossible to list them all here. Just the same, Karel Capek should be singled out for special thanks, as he has been involved with practically every beam diagnostics project at Jefferson Laboratory.

REFERENCES

1. Leemann, C.W., Douglas, D. R., and Krafft, G. A., "The Continuous Electron Beam Accelerator Facility: CEBAF at the Jefferson Laboratory," in *Annual Reviews of Nuclear and Particle Science*, edited by C. Quigg, Palo Alto, CA: Annual Reviews, Volume 51 2001, pp. 413-450.
2. Krafft, G. A., Denard, J.-C., Dickson, R. W., Kazimi, R., Lebedev, V. A., and Tiefenback, M. G., "Measuring and Controlling Energy Spread in CEBAF" in *Proceedings of the XXth International Linac Conference*, edited by A. W. Chao, SLAC-R-561, Palo Alto, CA: Stanford Linear Accelerator Center, 2000, pp. 721-725.
3. Krafft, G. A., "Status of the Continuous Electron Beam Accelerator Facility" in *Proceedings of the 1994 International Linac Conference*, edited by K. Takata et al., Tsukuba, Japan: National Laboratory for High Energy Physics, 1994, pp. 9-13.
4. Krafft, G. A., "Correcting M_{56} and T_{566} to obtain very short bunches at CEBAF" in *Proceedings of the Microbunches Workshop*, edited by E. B. Blum, M. Dienes, and J. B. Murphy, AIP Conference Proceedings 367, New York: American Institute of Physics, 1996, pp. 46-55.
5. Krafft, G. A., Wang, D., Price, E., Feldl, E., Porterfield, D., Wood, P., and Crowe, T., "Coherent Synchrotron Radiation Detector for Non-Invasive Subpicosecond Bunch Length Monitor" in *Proceedings of the 1995 Particle Accelerator Conference and International Conference on High Energy Accelerators*, IEEE 95CH35843, Piscataway, NJ: Institute of Electrical and Electronic Engineers, 1995, pp. 2601-2603.
6. Wang, D. X., Krafft, G. A., Price, E., Wood, P., Porterfield, D. and Crowe, T., "A Fast Coherent Synchrotron Radiation Monitor for the Bunch Length of the Short CEBAF Bunches" in *Proceedings of the Microbunches Workshop*, edited by E. B. Blum, M. Dienes, and J. B. Murphy, AIP Conference Proceedings 367, New York: American Institute of Physics, 1996, pp. 502-511.
7. Wang, D. X., "Electron Beam Instrumentation Techniques Using Coherent Radiation" in *Proceedings of the 1997 Particle Accelerator Conference*, edited by M. Comyn et al., IEEE 97CB36167, Piscataway, NJ: Institute of Electrical and Electronics Engineers, 1998, pp. 1976-1980.
8. Wang, D. X., Krafft, G. A., Price, E., Wood, P. A. D., Porterfield, D. W., and Crowe, T. W., *Appl. Phys. Letters* **70**, 529-531 (1997).
9. Krafft, G. A., "Diagnostics for Ultrashort Bunches" in *Beam Diagnostics and Instrumentation for Particle Accelerators DIPAC 97*, edited by A. Ghigo, M. Giabbai, and G. Possanza, LNF-97/048(IR), Frascati: Laboratori Nazionali di Frascati, 1997, pp. 48-52.
10. Wang, D. X., and Krafft, G. A., "Measuring Longitudinal Distribution and Bunch Length of Femtosecond Bunches with RF Zero-Phasing Method" in *Proceedings of the 1997 Particle Accelerator Conference*, edited by M. Comyn et al., IEEE 97CB36167, Piscataway, NJ: Institute of Electrical and Electronics Engineers, 1998, pp. 2020-2022.
11. Wang, D. X., Krafft, G. A., and Sinclair, C. K., *Phys. Rev. E*, **57**, 2283-2286 (1998).
12. Kazimi, R., Sinclair, C. K., and Krafft, G. A., "Setting and Measuring the Longitudinal Optics in CEBAF Injector" in *Proceedings of the XXth International Linac Conference*, edited by A. W. Chao, SLAC-R-561, Palo Alto, CA: Stanford Linear Accelerator Center, 2000, pp. 125-127.
13. Graves, W. S., private communication. See also his various contributions to *Proceedings of the 2001 Particle Accelerator Conference*, edited by P. Lucas and S. Webber, IEEE 01CH37268, Piscataway, NJ: Institute of Electrical and Electronics Engineers, 2001
14. Tiefenback, M. G., and Brown, K., "Beam-Based Phase Monitoring and Gradient Calibration of Jefferson Laboratory RF Systems" in *Proceedings of the 1997 Particle Accelerator Conference*, edited by M. Comyn et al., IEEE 97CB36167, Piscataway, NJ: Institute of Electrical and Electronics Engineers, 1998, pp. 2271-2273.
15. Krafft, G. A., Crofford, M., Douglas, D. R., Harwood, S. L., Kazimi, R., Legg, R., Oren, W., Tremblay, K., and Wang, D., "Measuring and Adjusting the Path Length at CEBAF" in *Proceedings of the 1995 Particle Accelerator Conference and International Conference on High Energy Accelerators*, IEEE 95CH35843, Piscataway, NJ: Institute of Electrical and Electronic Engineers, 1995, pp. 2429-2431.
16. Hardy, D., Tang, J., Legg, R., Tiefenback, M., Crofford, M., and Krafft, G. A., "Automated Path Length and M_{56} Measurements at Jefferson Lab" in *Proceedings of the 1997 Particle Accelerator*

Conference, edited by M. Comyn et al., IEEE 97CB36167, Piscataway, NJ: Institute of Electrical and Electronics Engineers, 1998, pp. 2265-2267.
17. Dickson, R., and Lebedev, V. A., "Fast Digital Feedback System for Energy and Beam Position Stabilization" in *Proceedings of the 1999 Particle Accelerator Conference*, edited by A. Luccio and W. MacKay, IEEE 99CH36366, Piscataway, NJ: Institute of Electrical and Electronics Engineers, 1999, pp. 646-648.
18. Lebedev, V. A., Musson, J., and Tiefenback, M. G., "High-Precision Beam-Based RF Phase Stabilization at Jefferson Lab" in *Proceedings of the 1999 Particle Accelerator Conference*, edited by A. Luccio and W. MacKay, IEEE 99CH36366, Piscataway, NJ: Institute of Electrical and Electronics Engineers, 1999, pp. 1183-1185.
19. Tiefenback, M. G., "On-line Measurement and Tuning of Multi-pass Recirculation Time in the CEBAF Linacs" in *Proceedings of the 2001 Particle Accelerator Conference*, edited by P. Lucas and S. Webber, IEEE 01CH37268, Piscataway, NJ: Institute of Electrical and Electronics Engineers, 2001, pp. 553-555.
20. Powers, T., Doolittle, L., Ursic, R., and Wagner, J., "Design, Commissioning, and Operational Results of Wide Dynamic Range BPM Switched Electrode Electronics" in *Proceedings of the 1996 Beam Instrumentation Workshop*, edited by A. Lumpkin and C. Eyberger, AIP Conference Proceedings 390, New York: American Institute of Physics, 1998, pp. 257-265.
21. Chao, Y., Crofford, M., Dobeck, N., Douglas, D., Hofler, A., Hovater, C., Krafft, G. A., Legg, R., Perry, J., Price, E., Suhring, S., Tiefenback, M., van Zeijts, J., "Commissioning and Operation Experience with the CEBAF Recirculation Arc Beam transport System" in *Proceedings of the 1993 Particle Accelerator Conference*, edited by S. T. Corneliussen, IEEE 93CH3279-7, Piscataway, NJ: Institute of Electrical and Electronics Engineers, 1993, pp. 587-589.
22. Lebedev, V. A., Bickley, M., Schaffner, S., van Zeijts, J., Krafft, G. A., and Watson, W. A., *Nucl. Inst. and Methods* **A408**, 373-379 (1998).
23. Piot, P., Denard, J.-C., Aderley, P., Capek, K., and Feldl, E., "High Current CW Beam Profile Monitors Using Transition Radiation at CEBAF" in *Proceedings of the 1996 Beam Instrumentation Workshop*, edited by A. Lumpkin and C. Eyberger, AIP Conference Proceedings 390, New York: American Institute of Physics, 1998, pp. 298-305.
24. Denard, J.-C., Piot, P., Capek, K., and Feldl, E., "High Power Beam Profile Monitor with Optical Transition Radiation" in *Proceedings of the 1997 Particle Accelerator Conference*, edited by M. Comyn et al., IEEE 97CB36167, Piscataway, NJ: Institute of Electrical and Electronics Engineers, 1998, pp. 2198-2200.
25. Rule, D. and Fiorito, R., "Noninterceptive Beam Diagnostics Based on Diffraction Radiation" in *Proceedings of the 1996 Beam Instrumentation Workshop*, Edited by A. Lumpkin and C. Eyberger, AIP conference Proceedings 390, New York: American Institute of Physics, 1998, pp. 510-517.
26. Krafft, G. A., and Zhang, Y., "Advances on Recirculated Linac Light Sources," in *ICFA Beam Dynamics Newsletter No. 26*, edited by K. Hirata and J. M. Jowett, International Committee for Future Accelerators, 2001, pp. 7-8, and other references throughout this newsletter.
27. Gruner, S. M., Bilderback, D., Bazarov, I., Finkelstein, K., Krafft, G., Merminga, L., Padamsee, H., Shen, Q., Sinclair, C., Tigner, M., *Rev. Sci. Inst.* **73**, 1402-1406 (2002).
28. Bazarov, I., Belomestnykh, S., Bilderback, D., Finkelstein, K., Fontes, E., Gray, S., Gruner, S., Krafft, G., Merminga, L., Padamsee, H., Helmke, R., Shen, Q., Rogers, J., Sinclair, C., Talman, R., Tigner, M., "Study for a Proposed Phase I Energy Recovery Linac (ERL) Synchrotron Light Source at Cornell University", CHESS Technical Memo 01-003, JLAB-ACT-01-04, 2001 (unpublished)

Tune Measurement in RHIC[1]

M. Brennan, P. Cameron, P. Cerniglia, R. Connolly, J. Cupolo, W. Dawson, C. Degen,
A. DellaPenna, J. DeLong, A. Drees, D. Gassner, M. Kesselman, R. Lee, A. Marusic,
J. Mead, R. Michnoff, C. Schultheiss, R. Sikora, and J.Van Zeijts

Brookhaven National Laboratory, Upton, NY 11973, USA

Abstract. Three basic tune measurement methods are employed in RHIC; kicked beam, Schottky, and phase-locked loop. The kicked beam and 2GHz Schottky systems have been in operation since the first commissioning of circulating beam in RHIC in 1999. Preliminary PLL measurements utilizing a commercial off-the-shelf lockin amplifier were completed during that run, and the resonant BPM used in that system also delivered 230MHz Schottky spectra. With encouraging preliminary results and the thought of tune feedback in mind, a PLL tune system was implemented in the FPGA/DSP environment of the RHIC BPM system for the RHIC 2001 run. During that run this system functioned at the level of the present state-of-the-art in tune measurement accuracy and resolution, and was successfully incorporated into a tune feedback system for use during acceleration. Each of the tune measurement systems has particular strengths and weaknesses. We present specific and comparative details of systems design and operation. In addition, we present detailed tune measurements and their utilization in the measurement of chromaticity and the implementation of tune feedback. Finally, we discuss planned upgrades for the RHIC 2003 run.

Introduction

RHIC is a superconducting two-ring synchrotron, with the capability to accelerate and store particle species ranging from protons to Gold. All species heavier than protons must cross transition during acceleration. The effect of intra-beam scattering (IBS) on emittance grows with the square of charge, so that for heavier species such as Gold the luminosity lifetime at store is limited by longitudinal beam loss out of the bucket due to IBS. Ramp rate requirements and hysteresis effects in the superconducting magnets limit the machine cycle time from injection to store and back to injection to a minimum of about 30 minutes under optimum conditions. Without beam cooling, fast and efficient machine cycles are essential to maximize the integrated luminosity. The primary requirement for tune measurement in RHIC is defined by this need for fast and efficient machine cycles.

A variety of problems are encountered in the development of an acceleration ramp. The first and most fundamental is that the machine model does not permit to accurately set the tunes to a predetermined value, but rather that the tunes during a ramp attempt must be measured and corrected, either by feedforward during the next ramp attempt, or by tune feedback during acceleration. The situation is complicated by the fact that lattice optics change up the ramp; due to aperture limitations beam is injected with $\beta^*=10$, and beta squeeze in the IRs is accomplished during ramping. As a result, when running without tune feedback a good many unsuccessful attempts,

---
[1] Work performed under the auspices of the U.S. Department of Energy

often over the span of many weeks, have been required in the development of a successful ramp, and a ramp remained successful only so long as machine conditions didn't change.

Beyond the problem with model accuracy, which does not appear to be surmountable in the foreseeable future, there are many additional problems to overcome in ramp development. Early in the ramp there are fast changes in tune and chromaticity due to snapback [1]. Through the ramp good chromaticity control is essential, but not always present during ramp development. The head-tail instability requires negative chromaticity below transition and positive above. Even with the correct sign, large chromaticities are harmful, first because the fast decoherence makes tune measurement difficult or impossible, and second because the large linewidth results in resonance overlap and beam loss. Beam loss drives currents in pickup electrodes, which often obscure beam signals at the time when measurements are most needed. Transition crossing presents its own special set of problems for tune measurement. The need for precise tune measurement to confirm the chromaticity sign change is made more difficult by transverse size oscillations and beam loss in dispersive regions (where the tune pickups are located) as a result of longitudinal quadrupole oscillations following the phase jump. The dynamic range problem always present in tune measurement (observation of the small difference signal at the betatron frequency in the presence of the much larger signal due to beam offset and sum mode at the revolution frequency) is sometimes aggravated by orbit changes at transition. Coupling often complicates the measurement and interpretation of tune data. And finally, there remain all the usual problems of data acquisition and processing, integration with the control system, and operator friendliness. Each of the tune measurements systems has its' own specific strengths and weaknesses in dealing with the problems outlined above.

	tune accuracy	small chrom req'd?	timing req'd?	coh/ incoh?	chrom on ramp?	coupling?	comments
BPMs	10^-3	yes		coh		need big kick	from inj oscillations
ARTUS	10^-3	yes	yes	coh	decoh (sign?)	close approach	emittance growth
HF Schottky	10^-3	no	no	incoh	sideband width	from line shape	continuous, non-pert
LF Schottky	10^-4	no	no	incoh	sideband width	line presence	continuous, non-pert
PLL	10^-5	yes	no	incoh?	1 Hz radial mod	line presence	cont, non-pert, sensitive
QMM	?	?	no	incoh			inj matching, sensitivity?
Head-Tail	10^-3	yes	yes	coh	data analysis?		parasitic to ARTUS
AC Dipole	?	?	no?	coh?			DX BPM electronics?

TABLE 1. Characteristics of various tune measurement methods.

The table above outlines characteristics of various tune measurement methods. There has been experience with all these methods (except measurement of the quadrupole moment) in the first two RHIC beam runs. The first five methods will be discussed in greater detail in the following sections

The ARTUS System

The acronym ARTUS [2] is derived from "A Rhic Tune measurement System". Betatron oscillations are excited with a fast transverse kicker magnet and beam positions are recorded from a BPM. The fractional tunes are extracted by performing a FFT analysis. The BPM assigned to the tune meter resides at a location with high horizontal betatron function and moderately high vertical betatron function (at the Q3 magnets at 2 o'clock). The capability of multiple turn-by-turn kicks is included to ensure the needed signal amplitude at all beam energy settings. The readout electronics and controls are installed in a VME crate in the 1002 service building.

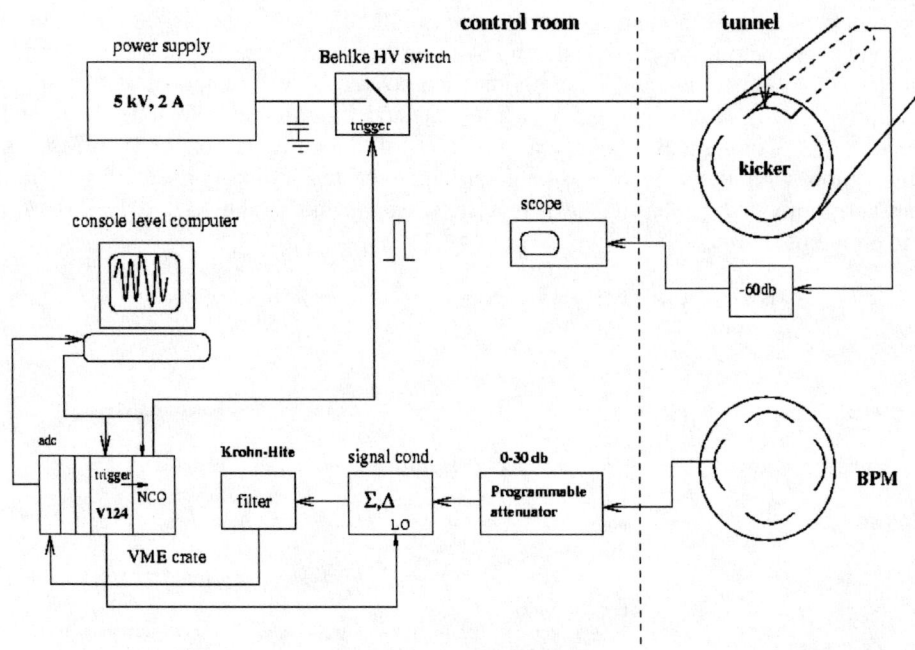

FIGURE 1. ARTUS block diagram

Each ring has two kicker modules with four 2-m stainless steel striplines, allowing both horizontal and vertical kicks. The two kickers are connected in series to provide 4 m of stripline. Each stripline subtends an angle of 70 degrees with an aperture of 7 cm. The assembly is designed to give 50Ω impedance when opposing lines are driven in the difference mode. Single pulses can power each of the four planes independently. The kick pulses are generated by fast FET switches, producing pulses that are approximately 140 ns long. Single bunch excitation is possible with even up to 120 bunches per ring. All switches for all striplines in both rings are charged by one 5kV/2A power supply. The kick angle after one pulse with 3 kV received by an ion going through the kickers is approximately 10 µrad at injection energy ($\gamma \cong 10$).

Figure 1 shows BPM signal processing and kicker triggering. The FET switches are triggered by a TTL pulse of 200 ns width from a numerically-controlled oscillator (NCO) board. The NCO outputs pulses with the required phase and a remotely settable frequency of up to 20 MHz. The phase and frequency resolutions are 0.09 degrees and 11.6 mHz respectively. By selecting a NCO frequency close to the betatron frequency the beam is kicked resonantly, enhancing the effect on the beam significantly compared with a single kick. A set point equal or very close to the betatron frequency was shown to kick the beam out of the ring if the number of turns was too high. In order to control the total number of kicks, the NCO is triggered synchronously with the beam using an in-house beam sync trigger V124 board. Other channels of the same V124 board trigger the data acquisition from the BPM. This board allows the tune measurement to be triggered by any event broadcasted on the beam synchronous link or on demand. Thus the tunes can be easily correlated any time with any other instrumentation.

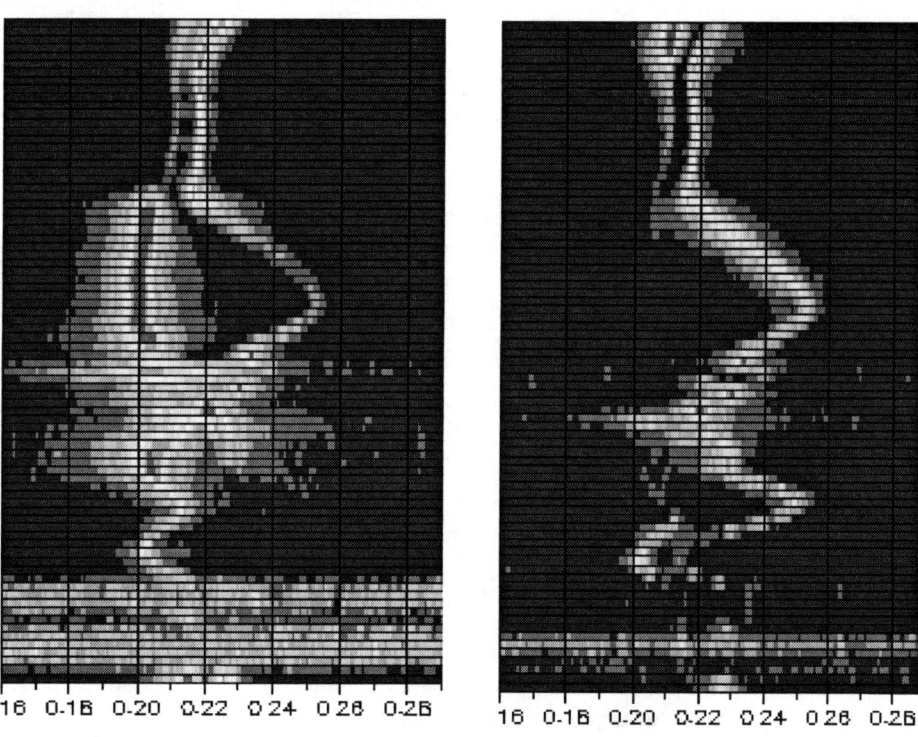

FIGURE 2. ARTUS tune measurement during an acceleration ramp

Figure 2 above shows ARTUS tune measurement results during an acceleration ramp in September of 2001. Horizontal tune is in the left panel, and vertical in the right. The beginning of the ramp is at the bottom of the figure, transition is about 1/4 of the way up, and flat-top is at the top. Several interesting observations can be made about this data. For the first 20% of the ramp the horizontal signal is obscured by broadband noise due to beam loss, and the vertical signal is absent, perhaps due to large chromaticity. When tune information does appear the horizontal tune is brushing the 1/5 resonance, and the horizontal signal is stronger than the vertical in the vertical spectrum. Horizontal tune again brushes 1/5 about 30% into the ramp, and then the spectrum becomes broad and somewhat confusing until midpoint. Shortly after midpoint horizontal tune sits on 1/5 for about 30 seconds, while vertical briefly walks onto the 1/4 resonance. For most of the last quarter of the ramp horizontal and vertical signals are virtually identical. The effect of coupling is somewhat confusing here. Finally, it appears that the tunes cross shortly before the end of the ramp. This ramp illustrates some of the difficulties of interpreting ARTUS spectra, as well as the fact that tunes and chromaticities were not under control in September, despite the fact that the run had been in progress for several months. Improvements planned for the RHIC 2003 run include moving the kicker to a region of higher beta, utilizing separate PUEs for horizontal and vertical to maximize pickup beta in both planes, and modifying the analog front end to simplify timing.

The High Frequency Schottky System

Two high-frequency cavities from Lawrence Berkeley National Laboratory [3] are used to detect Schottky signals from both beams. The transverse modes are TM120 and TM210 at 2.069±0.002GHz. They have measured Q of 4700, and are separated by 4 MHz. A longitudinal mode is at 2.741GHz. The signals are down-converted to 2MHz and amplified in the tunnel, then transported to an external 10MHz bandwidth FFT analyzer. Data is provided to the control system through LabVIEW communicating with the FFT analyzer via TCP, as well as through a remote Xterm scope application.

The usefulness of the 2GHz Schottky system during acceleration of Gold beams is limited by the large width and resulting overlap of the revolution and betatron lines at and near injection energies, where the relativistic slip factor is large. In addition, the 0.4% increase in RHIC revolution frequency during ramps results in line movement of 8MHz at 2GHZ during the ramp, causing rapid sweeping of the spectral lines across the 400KHz wide cavity resonance. The possibility of using a beam-synchronous frequency for down conversion was investigated and discarded because of bandwidth problems in the available frequency multiplier, and more significantly because of the timing system interface required to implement the line-hopping needed to track the cavity resonance as it would then sweep under the stationary spectrum. A consequence was that averaging could not be used to decrease noise during ramps. Solutions to the problems arising out of the non-stationary spectrum were also hampered by limitations in the interface in the FFT analyzer, which permitted transfer of spectra at a maximum rate of about 1Hz. Within the limitations of available memory, some of these deficiencies could be overcome by utilizing the FFT analyzer in time capture mode. In this mode it digitized as fast as possible for a given resolution bandwidth, then replayed the capture buffer to post-process the FFT off-line. In a typical ramping setup about 10 seconds of data could be saved in capture mode. Figure 3 shows a time capture of the first successful transition crossing in RHIC.

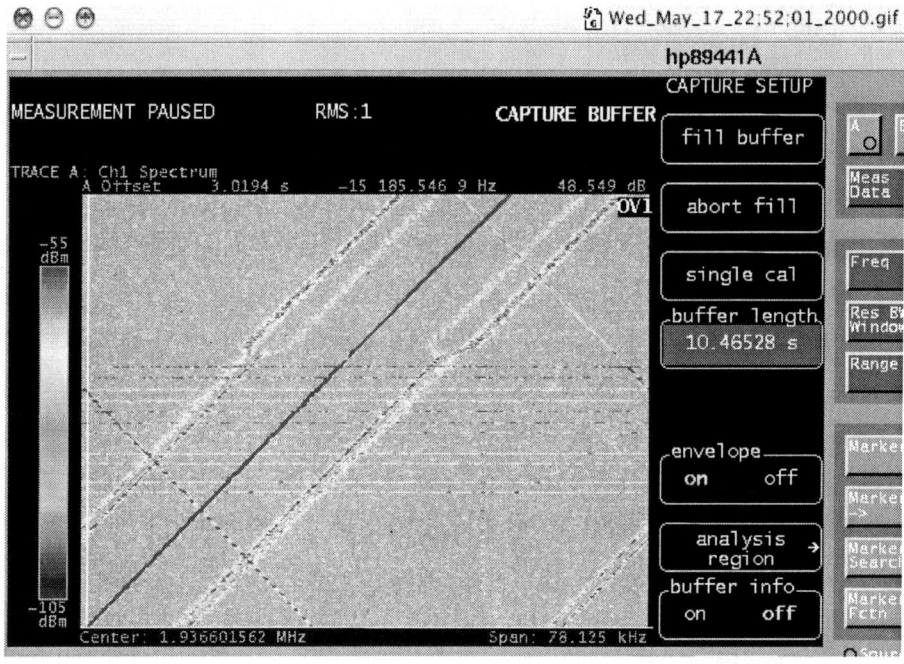

FIGURE 3. HF Schottky measurement of first successful transition crossing

The horizontal axis spans a spectrum width of 78.125KHz. The vertical axis is time, the top of the figure is about 1s before transition, and the bottom is about 2s after transition. The revolution line and betatron sidebands are sweeping from upper right to lower left. The frequency sweeping results from the fact, as mentioned above, that the local oscillator used for down-conversion was not beam synchronous. Two sets of betatron sidebands of unequal intensity appear, caused by weak coupling of the tunes. At transition the sidebands cross due to nonzero chromaticity. It is clear that the signs of vertical and horizontal chromaticity were opposite at transition. The broadband noise in the Schottky spectrum after transition has a period of 0.08 seconds, which corresponds to the period of bunch length oscillations observed by the wall current monitor at the same time. These oscillations were just sufficient to drive the tails of the transverse distribution into the beampipe walls in high dispersion regions (including the location of the Schottky cavities) at times of maximum intrabunch momentum spread, causing currents that excited broadband noise in the cavity. The phenomenon of broadband noise in the Schottky spectrum during beam loss is frequently observed at RHIC, for instance at transition in Figure 4 below.

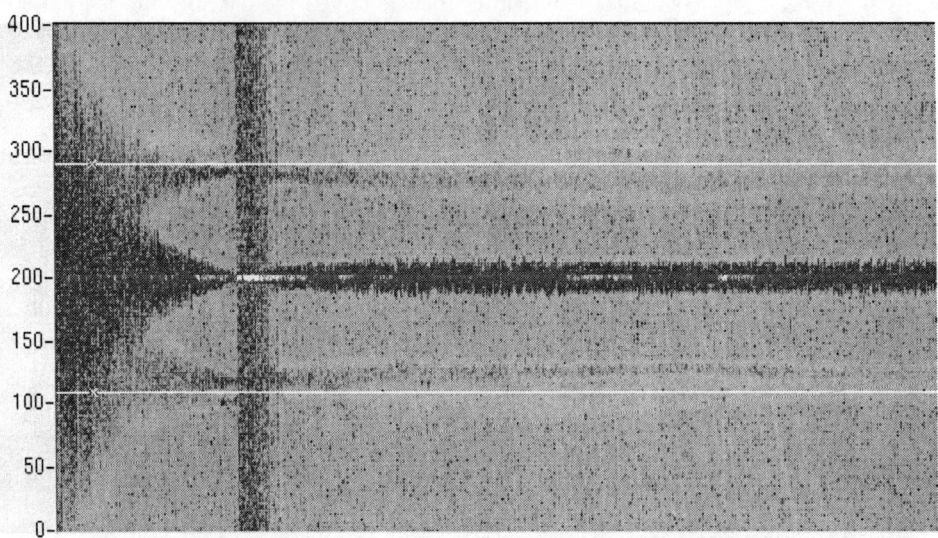

FIGURE 4. HF Schottky measurement of tune up the ramp

During RHIC 2000 a LabVIEW application that centered the revolution line in each raw spectrum and permitted visual averaging of the resulting spectrogram was created to partially circumvent the FFT box difficulties discussed above. Figure 4 is a spectrogram acquired during a complete ramp from $\gamma=10.3$ to $\gamma=70$. The vertical axis covers a frequency span of 78KHz, the RHIC revolution frequency, so that only a single revolution line and set of betatron sidebands are seen. The white horizontal lines are markers generated to indicate a fractional tune of 0.225. The beginning of the ramp is at the left of the figure. Betatron sidebands are not resolved until the relativistic slip factor becomes small about 30 seconds into the ramp. All lines become narrow as transition is approached, where by definition the revolution frequency is the same for particles of differing momentum. Broadband noise is observed immediately following transition. Sidebands remain clearly resolved until the end of the ramp, allowing easy measurement of tune. Note the asymmetry between sidebands in the latter part of the ramp, indicating large chromaticity, and the presence of lines from both planes due to coupling. The betatron linewidth is quantitatively expressed [4] as:

$$\Delta f = f_0 \ \Delta p/p \ [(n \pm \nu) \eta \pm \xi]$$

where the revolution frequency $f_0 \sim 78$KHz, momentum spread $\Delta p/p \sim .001$, harmonic number $n \sim 26500$ at 2.07GHz, tune $\nu \sim .23$, slip factor η varies from $-.008$ at injection to $.002$ at store, and chromaticity ξ is typically a few units either positive or negative. The chromatic contribution to linewidth adds to the upper sideband and subtracts from the lower.

In addition to the spectrogram displays of the previous figures, it has proven useful [5] to construct stripchart displays of tune, chromaticity, momentum spread, and transverse emittance as measured from the Schottky spectra.

Several improvements are planned for the RHIC 2003 beam run. Amplifier saturation due to beam steering offsets and the effect of the 200MHz storage RF is frequently encountered, and greatly diminishes the reliability of the data. Beam offsets are present because of an aperture restriction that forces unconventional steering to get good collisions at the adjacent IP. Efforts are underway to locate and remove the obstruction. Sensitivity to beam loss might also be helped by this. An improved VME/DSP based data acquisition system is planned to overcome the limitations of the HP89410 DSAs. These instruments are good studies tools, but suffer from high cost, poor data accessibility, and poor integration with the control and (perhaps more significantly) timing systems. In addition, they seem to sometimes provoke networks data storms. The improved controls and timing interface gained in the VME/DSP based data acquisition system will permit tracking of the cavity resonance with a beam-synchronous local oscillator, so that S/N can be straightforward improved with signal averaging. Finally, improvements to chromaticity and emittance calculations are planned.

The Low Frequency Schottky System

Initial motivation for the development of the LF Schottky came from dissatisfaction with the comparatively poor frequency resolution of the HF Schottky system. Implementing a resonant pickup at a frequency that is an order of magnitude lower will result in momentum-dependent linewidths that are an order of magnitude lower. The quarter-wave 50 ohm shorted striplines of a standard RHIC BPM were resonated [6] in the lowest-order difference mode at ~240MHz by coupling them with a half wavelength section of 3/8" heliax. Achieving optimal coupling (defined as Qloaded = Qunloaded/2) to the quarter wave points was accomplished with a quarter wave transmission line impedance transformation to 50 ohms. Fine tuning was accomplished with capacitors at the end of additional quarter wave stubs. The difference signal from a hybrid was filtered, amplified, and brought out of the tunnel.

Figure 5 shows data taken during a ramp when beam was lost shortly after transition. The horizontal axis spans a spectrum width of 78.125KHz. The vertical axis is time, with the top of the figure at the start of the ramp, and the bottom shortly after transition. The revolution line is at the center. Unlike the HF Schottky, all lines are clearly resolved at injection energy. Due to coupling, signals from both planes are visible, and it appears that the tunes crossed a third of the way to transition. At transition both tunes shift down, and chromaticities are the same sign and approximately the same magnitude in horizontal and vertical. Sidebands around the revolution line at transition are probably due to oscillation of the RF loops as beam intensity drops below that required for stability. Sharp lines are also prominent at the quarter and at .375. In general sharp resonance lines are most often observed at frequencies that oddly enough correspond to the traditional British fractions, and the amount of spectral power present correlates with beam loss. An accelerator physics explanation of these lines is not yet available.

Improvements to the LF Schottky for the RHIC 2003 beam run are similar to those planned for the HF Schottky. Position information from adjacent BPMs will be used to center the moveable pickup on the beam, reducing dynamic range and saturation problems. Sensitivity will be improved by moving the pickup to a region of higher beta.

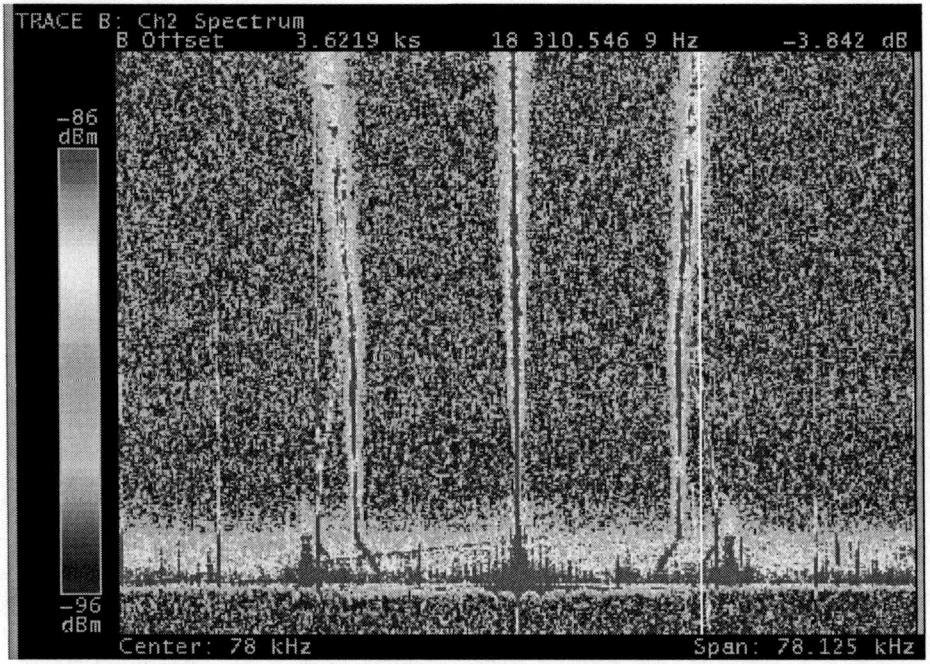

FIGURE 5. LF Schottky measurement of tune up the ramp

An improved VME/DSP based data acquisition system is planned to overcome the constraints of the HP FFT boxes. Because the signal from the LF Schottky pickup is also used for tune feedback and there was only a single local oscillator, LF Schottky spectra were not available simultaneously with PLL tune measurements. The signal will be split and separate beam synchronous LOs will be available next run. Finally, improvements to tune, chromaticity, coupling, and emittance calculations are planned.

The Phase Locked Loop Tune Measurement System

The PLL utilizes signals from the LF Schottky pickup. The primary difficulty in constructing a high sensitivity transverse pickup is the dynamic range problem that results from trying to see signals at the Schottky level in the presence of the coherent beam spectrum, which is typically at least 100dB stronger. In designing the PLL tune measurement system for RHIC 2001 we dealt with this problem in several ways. We placed the pickup resonance well above the coherent spectrum, at 8.5 times the 28MHz acceleration RF. We resonated only a difference mode so that the sum mode coherent signal remaining at the pickup frequency would not enjoy enhancement of its power by the Q of the difference mode. We utilized a moveable BPM so that the remaining difference mode coherent signal at the revolution harmonic due to beam offset could be minimized. We bandpass filtered the output of the BPM with a high-Q cavity filter before the first stage of amplification to avoid saturation. And finally, we employed a 1 KHz bandwidth high-Q filter at the baseband 78KHz input to the digitizer to get rid of the revolution line ~15KHz away.

At the core of the PLL tune system is a custom numerically controlled oscillator [7] sitting in VME and clocked from the 28MHz low level RF system. All frequencies in the tune system are thus synchronous with the beam. To simplify the discussion that follows, it is accurate only within the fractional portion of the betatron tune at harmonic 3061. To this level of approximation, the output of the NCO is at harmonic 96 of the 78 KHz revolution frequency. When the loop is locked and after x32 frequency multiplication, the output of NCO C is at ~238MHz (i.e RFx8.5, or harmonic 3060) plus the betatron frequency. These frequencies (harmonic 3060 and harmonic 1) are mixed in a suppressed carrier single sideband modulator. The output is at the betatron line above harmonic 3061, and is highpass filtered before entering a 10W class A amplifier. The output of the amplifier drives the 1m long 50 ohm kicker striplines through a difference hybrid and about 100m of heliax into the tunnel. The kicker excitation travels with the beam through the betatron-tune-dependent phase shift between the kicker and the resonant pickup. Pickup output at 238MHz is bandpass filtered, boosted by 30dB, and again transported via 100m of heliax to the mixer, whose output is again at 78KHz. The signal is delivered to the high impedance input of a Dynamic Signal Analyzer for FFT analysis and display, as well as to the 50 ohm input of the analog front end for amplification and filtering. By including the betatron frequency in the local oscillator for up and down conversion, the tune signal is always nominally at the same frequency (78KHz), and the need for a tracking filter at the input to the digitizer is eliminated.

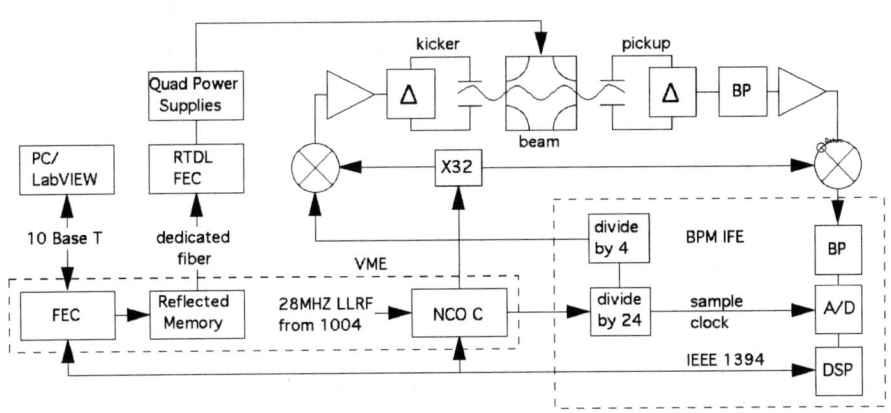

FIGURE 6. PLL/Tune Feedback Block Diagram

The digitizer clock is generated by a divide-by-24 in the gate array of a modified RHIC BPM module [8]. The 78KHz signal which is up-converted, phase shifted by the beam tune and down-converted, is generated by an additional divide-by-4 to permit a simple I/Q demodulation [9] of the signal. The data is processed in the DSP of the BPM module. The functions performed by the DSP include I/Q demodulation, phase compensation during the frequency swing of acceleration, loop gain/linewidth compensation during the relativistic slip factor swing of acceleration, signal averaging/lowpass filtering, and NCO control. The processing is performed on blocks of data, whose length is typically 8KB. Update of the NCO is at around 30Hz. The DSP communicates with VME via IEEE1394. High-level control of the PLL system is accomplished with a MacIntosh running LabVIEW, communicating with VME via ethernet. The functions performed by the LabVIEW program include writing setup parameters, calculating and writing the loop lock indicator, and beam transfer function (BTF) measurement. BTF measurements were used to determine the amount of phase shift compensation required during ramping. A typical BTF is shown in figure 7.

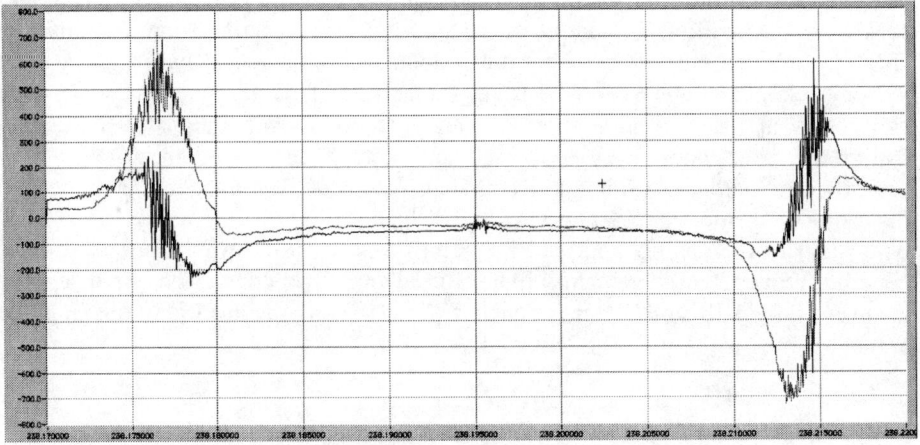

FIGURE 7. BTF at Injection Energy

The revolution line is the slight disturbance at the center of the image, and its smallness indicates that beam was well centered in the pickup. The difference in linewidth between the upper and lower sidebands indicates chromaticity was small and negative. What appears to be fuzz or noise in the signal is the structure of the synchrotron satellites.

Figure 8 shows horizontal and vertical tune as measured by both the PLL and Artus during a ramp. The lower black continuous trace is the horizontal PLL, and the blue dots that overlay it are Artus measurements. The PLL appears unperturbed while Artus is delivering instantaneous kicker power ~80dB above the PLL excitation at random phase every two seconds. The upper red trace is vertical PLL, and the green dots are vertical Artus. Agreement between PLL and Artus is generally quite good. The left vertical scale is fractional tune. The right vertical scale is beam intensity, and refers to the blue line that starts at the upper left of the image. It shows significant beam loss early in the ramp, which is probably a result of the tail of the horizontal tune distribution crossing the 1/5 resonance. In an effort to measure chromaticity, during this ramp the radial beam position was modulated by 200μ at 1Hz. The modulation pattern was on for 3 seconds, then off for 3 seconds. The resulting tune modulation is clearly visible near the end of the ramp in the horizontal. If one looks closely this pattern can also be discerned in both horizontal and vertical at other times in the ramp. A detailed analysis is presented elsewhere [11]. The large variance in the PLL data in the second half of the ramp is probably due to a combination of high loop gain and beam-beam tune shift.

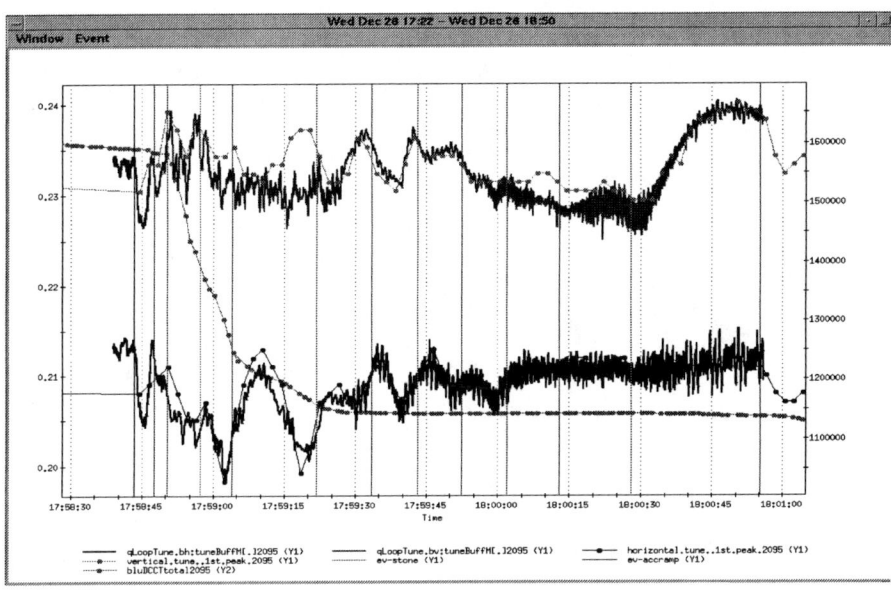

FIGURE 8. Artus and PLL tunes up the ramp

In addition to beam-beam tune shift, several other interesting features are illustrated in the ramp of figure 9. At the time of this ramp a hardware problem (since corrected) existed which caused loop phase on a given ramp to be arbitrary modulo $\pi/2$. Autolock was initiated by setting a window around the injection tune as measured by Artus, then sampling the phase within that window with loop gain very high. The average sampled phase was then taken to be the beam phase, the loop gain was turned down, the window was opened, and the loop would lock. On this particular ramp the initial vertical tune from Artus was incorrect, and the loop locked lower when the window was opened. Early in the ramp, as the lower horizontal tune rose and approached the vertical, the vertical lock jumped to the opposite side of the horizontal line, then jumped back as the horizontal tune moved away. This is typical of behavior often seen in the PLL, namely that it is not stable as the tunes approach each other. Around 16:07 the effect of beam-beam tune shift appears dramatically in the vertical as the collision point of the unclogged beams sweeps alternately into and out of the intersection region. At the same time the horizontal signal appears to be smoothed. This probably is the result of very large chromaticity, which causes the beam transfer function portion of the loop gain to be small and lowers the loop bandwidth, giving the effect of low-pass filtering. The ramp ends at about 16:07:30. The vertical tune is then chirped as the rings are first cogged, then un-cogged and slipped to properly align the abort gaps, then re-cogged. Beam-beam tune shifts throughout are about .002.

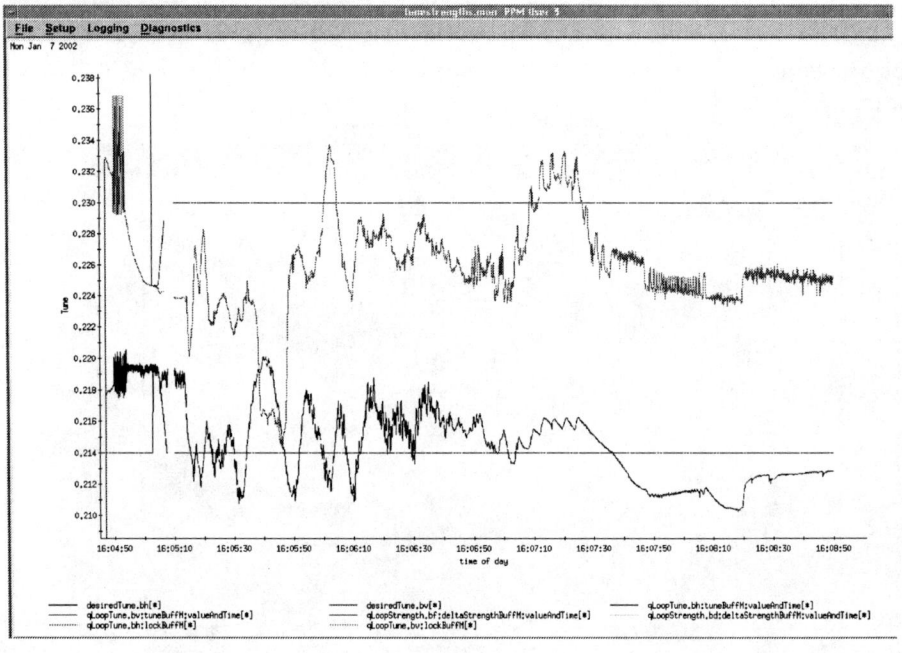

FIGURE 9. A ramp illustrating several features of PLL measurement

Tune Feedback

The tune feedback control loop [11,12] is implemented as a digital control loop running in a power PC Front End Computer in VME. The horizontal and vertical tunes are converted to horizontal and vertical strengths through a matrix that relates the desired tune change to strengths. The horizontal and vertical strengths are not independent since this matrix contains cross terms. These strengths are then used to calculate the required magnet currents. As shown in the block diagram of figure 4, magnet coefficients are calculated in the Front End Computer local to the PLL tune measurement, then transported via reflected memory and a dedicated fiber optic line to the power supply building 600m away, where quadrupole currents are written to the power supplies via the Real Time Data Link.

Data from the first successful ramp with tune feedback is shown in figure 10. As in figure 7, the vertical scale at the right and the blue trace correspond to beam current. Losses early in the ramp do not differ much from those in figure 8, and are probably due again to large horizontal tune spread overlapping the 1/5 resonance. The large excursion in vertical about one third the way up the ramp was due to a power supply problem, and resulted in about 15% beam loss. Without tune feedback the ramp would

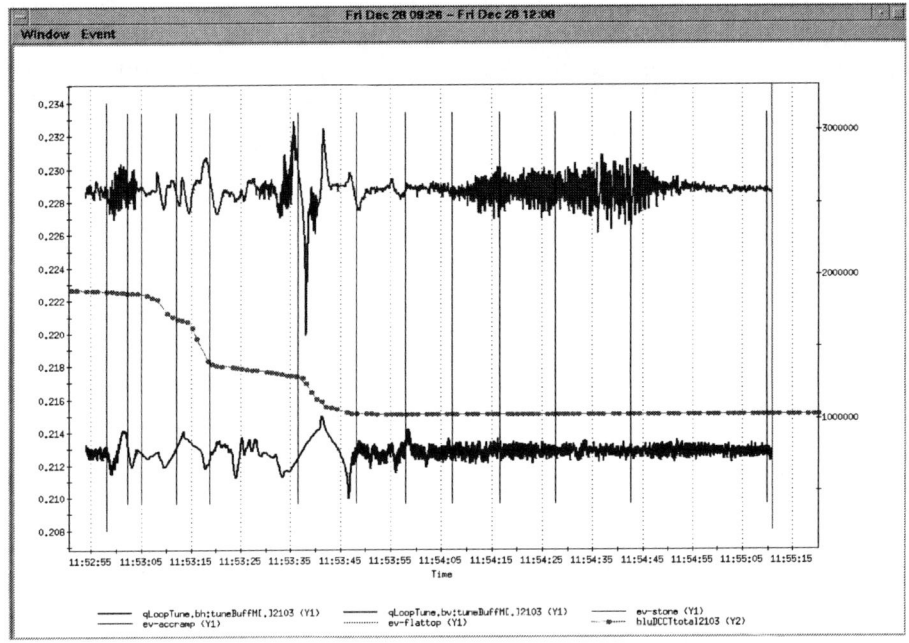

FIGURE 10. First successful ramp with tune feedback

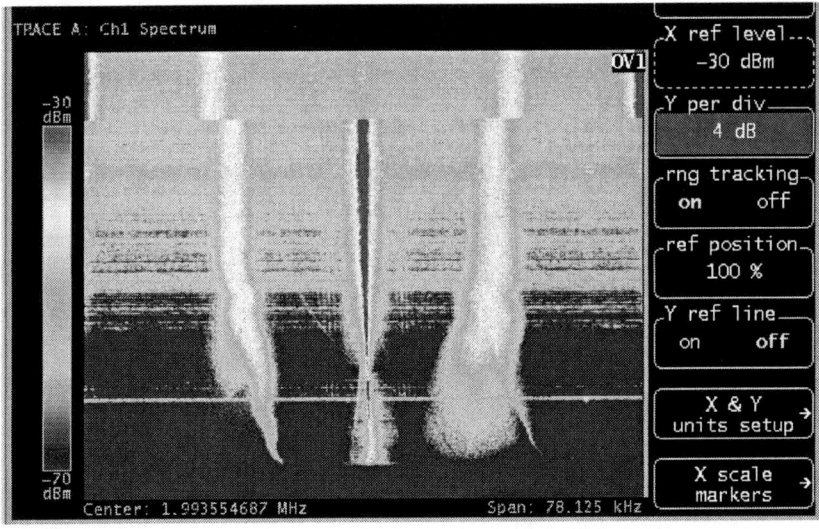

FIGURE 11. Down-ramp tune and chromaticity from HF Schottky

have aborted at this point. Tune feedback was attempted on four additional up-ramps before the end of the run, with one failure in the final second of the ramp due to excessively large horizontal chromaticity.

Near the end of the polarized proton run the greater portion of a shift was devoted to down-ramps. The motivation was to decelerate polarized beam and re-measure polarization at injection energy, where the analyzing power of the p-Carbon polarimeter is known, to get a lower limit on polarization at 100GeV, where the analyzing power is not known. Figure 11 shows data from the horizontal HF Schottky during an attempted down-ramp. The ramp begins at the discontinuity near the top of the image. Tune regulation is good for the first half of the ramp. As chromaticity gets large and coupling appears the combination of reduced BTF/ loop gain and phase information from the opposite plane causes tune control to suffer. The huge chromaticity becomes evident from the striking difference in upper and lower sideband widths. Perhaps the most interesting aspect of the down-ramp effort was that it revealed a divergence in the view of just what kind of tool one possesses in the PLL/tune feedback, and how this tool might best be used. The controversy arose over whether tune feedback is a ramp development tool, or a control that is engaged after the ramp development effort is complete. The bulk of the effort was done without tune feedback, and down-ramps were not successful. The benefits of this failure were that the importance of chromaticity in tune feedback became evident to all, and clear thought was stimulated on the nature of tune feedback as a ramp development tool.

Several improvements in PLL/tune feedback are planned for RHIC 2003. The system will benefit from the pickup improvements mentioned in the LF Schottky section. In ramp development emittance growth is a secondary concern, so rather than kicker excitations of less than 1W that were typical of RHIC 2001, the full 10W of amplifier power can be applied. Kickers are also being moved to a region of larger beta to better utilize the available kicker power. To coherently kick all bunches in a multi-bunch fill, the excitation frequency must be Q+v (sum of integer and fractional tunes) plus an integer multiple of the bunching frequency [13]. This condition was not observed during the last run, will be observed during the next, and will result in more efficient excitation (and may remove some ambiguities in phase). Moving beyond pickup and kicker improvements, the BPM module-based data acquisition system will be replaced with a VME-based FPGA/DSP system. Baseband frequency will be shifted from 78KHz to 455KHz, permitting the use of easily available very sharp ceramic filters to remove the adjacent revolution line, as well as resulting in an additional 6dB of processing gain. Improved digital filtering will be implemented. The VME-based system will permit operation from the control room rather that the diagnostics building, improving communications and accelerating the development of operator familiarity. Finally, a considerable effort is underway to model system behavior, including PLL behavior in Matlab and beam behavior via UAL [14].

Summary

A variety of sophisticated tune measurement systems exist in RHIC. This is the result of significant effort by a great many individuals over the span of several years, and that effort continues in the form significant improvements for the coming beam run. The plan for RHIC 2003 is to commission the machine from day one with tune feedback. This will require the best possible operation of all tune, chromaticity, and coupling measurement systems.

Acknowledgements

The authors would like to express their gratitude to the many individuals who supported the design, development, and operation of the systems described in this paper. We are particularly grateful to Mike Harrison for his initiative in the collaboration with Berkeley to create the HF Schottky system, and to Tom Shea for his excellent accomplishment in building the foundation for RHIC Beam Instrumentation.

References

1. W. Fischer et al, "Beam-Based Measurement of Persistent Current Decay in RHIC", BNL/RHIC/C-A/AP 32, Nov 2000.
2. P. Cameron et al, "ARTUS: A Rhic Tune monitor System", BNL/RHIC/AP/156, July 1998. http://www.agsrhichome.bnl.gov/AP/ap_notes/rap_index.html
3. W. Barry, J. Corlett, D. Goldberg, D. Li, "Design of a Schottky Signal Detector for Use at RHIC", EPAC98, Stockholm. http://accelconf.web.cern.ch/AccelConf/e98/PAPERS/WEP16G.PDF
4. D. Boussard, "Schottky Noise and Beam Transfer Function Diagnostics", CERN 95-06, p749.
5. P. Cameron et al, "Schottky Measurements During RHIC 2000", PAC2001, NY.
6. M. Kesselman et al, "Resonant BPM for Continuous Tune Measurement in RHIC", PAC2001, NY.
7. J. DeLong et al, "Synthesizer-Controlled Beam Transfer from the AGS to RHIC", PAC2001, NY.
8. T.J. Shea et al, "DSP Based Data Acquisition for RHIC", PAC95, Dallas.
9. C. Ziomek and P. Corredoura, "Digital I/Q Demodulator", PAC95, Dallas.
10. S. Tepikian et al, "Measuring Chromaticity along the Ramp Using the PLL Tune-meter in RHIC", EPAC2002, Paris.
11. P. Cameron et al, "Tune Feedback at RHIC", PAC2001, NY.
12. C. Schultheiss et al, Real-Time Betatron Tune Control in RHIC, EPAC2002, Paris.
13. K. Lohman et al, "Q-Monitoring in LEP", CERN/LEP-BI/88-45, p7.
14. N. Malitsky and R. Talman, AIP 391 (1996).

Laser Beam-Profile Monitor Development at BNL for SNS[1]

R. Connolly, P. Cameron, J. Cupolo, D. Gassner, M. Grau,
M. Kesselman, S. Peng and R. Sikora

Brookhaven National Lab
Upton, NY, USA

Abstract. A beam profile monitor for H⁻ beams based on laser photoneutralization is being developed at Brookhaven National Laboratory (BNL) for use on the Spallation Neutron Source (SNS) [1]. An H⁻ ion has a first ionization potential of 0.75eV and can be neutralized by light from a Nd:YAG laser (λ=1064nm). To measure beam profiles, a narrow laser beam is passed through the ion beam neutralizing a portion of the H⁻ beam struck by the laser. The laser trajectory is stepped across the ion beam. At each laser position, the reduction of the beam current caused by the laser is measured. A proof-of-principle experiment was done earlier at 750keV. This paper reports on measurements made on 200MeV beam at BNL and with a compact scanner prototype at Lawrence Berkeley National Lab on beam from the SNS RFQ.

INTRODUCTION

Photoneutralization of H- beams [2,3,4] has been used for measuring beam parameters and for beam manipulation. The first ionization potential of an H- ion is 0.75 eV which is the energy of a 1.67 µm photon. As shown in fig. 1, any photon with λ<1.6 µm can neutralize an H- ion. In these applications light from a laser is used to mark a portion of the beam. Downstream from the laser interaction point the beam has three components: H- ions, neutral atoms and unbound electrons. A magnetic field is used to separate one or two of these from the rest of the beam and measurements are made on the remaining beam.

Laser marking of the beam has been done in three ways. The first is to use very short light pulses to neutralize a small phase slice of the entire cross section of the beam. This technique was developed at Los Alamos National Lab to measure longitudinal emittance [5]. Light from a Q-switched Nd:YAG laser was passed through a pulse slicer and frequency doubler to produce 23ps-long pulses. These short light pulses passed through the H- beam. The charged beam was deflected into a beam stop and a time-of-flight measurement was made on the neutralized beam component to measure momentum spread. A clever modification on this idea using a mode-locked laser and spectrometer was proposed but never built [6].

A second marking technique is to neutralize the entire cross section of the beam with a laser pulse several rf periods long. At Los Alamos this was done to measure the transverse emittance of beams at the exit of an rf cavity [7]. The beam power was too

---

[1] SNS is managed by UT-Battelle, LLC, under contract DE-AC05-00OR22725 for the U.S. Department of Energy. SNS is a partnership of six national laboratories: Argonne, Brookhaven, Jefferson, Lawrence Berkeley, Los Alamos, and Oak Ridge.

great to intercept the full beam with a slit but allowing the beam to drift would introduce space-charge emittance growth. A laser neutralized the full cross section of the beam at the exit of the cavity and then a magnet removed the charged beam. A slit and parallel-channel collector was placed after the clearing magnet. Since the measured beam was neutralized at the cavity exit the actual phase space there could be determined by transforming the measured phase space through a simple drift with no space-charge corrections.

At Fermilab a laser has been used to place a notch in the beam when it sweeps over a Lamberston magnet to reduce activation of the magnet [8]. In this case the neutralized beam hits a beam stop and the charged beam is bent 90° down the transport line. A 5ns, 99% notch was produced.

The third marking technique is to focus the laser light into a narrow ribbon and neutralize a small transverse slice of the beam. The transverse profile can be measured by translating the laser 'wire' across the beam and, at each position, measuring the size of the effect it makes on the beam. A measurement which would collect the removed electrons was proposed by D.R. Swenson et al [9]. This paper reports on efforts to develop a laser profile monitor (LPM) at Brookhaven National Lab which measures the beam current notch created by the laser pulse [10].

Profiles of the SNS H⁻ beam will be measured in the medium energy transport line (MEBT) between the radio frequency quadrupole (rfq) and the linac entrance, along the linac, and in the linac-ring transport line. Stepped carbon-wire scanners are the primary profile diagnostic. However beam heating will limit wire scanners to tuning and matching applications with either the beam pulses shortened or the current reduced. Also there are concerns about placing wires near the superconducting cavities where wire failure can cause cavity damage. These concerns have motivated the effort to develop a laser profile monitor (LPM) which is noninvasive.

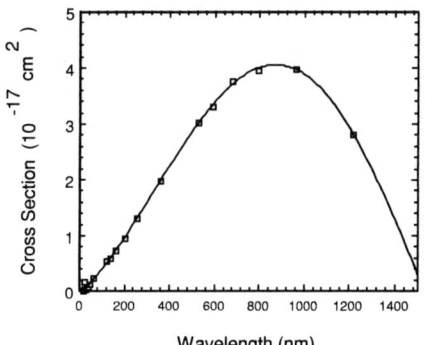

FIGURE 1. Calculated cross section for H⁻ photoneutralization as a function of photon wavelength. Data are from a table in ref. [2].

PHOTONEUTRALIZATION

Figure 1 shows the photoneutralization cross section as a function of photon wavelength in the center-of-mass frame. If the laser beam crosses the H- beam at a lab angle of θ_L the photon energy in the moving frame is Lorentz shifted by the amount [11,12],

$$E_{CM}=\gamma E_L[1-\beta\cos(\theta_L)] \quad (1)$$

For the SNS laser installation θ_L will be 90° so at the full energy of 1GeV the center-of-mass photon energy will be double the lab energy. For these measurements and probably for the final SNS installation the laser will be a Q-switched Nd:YAG laser operating at its fundamental of λ=1064 nm so at the full energy of 1GeV the neutralization cross section will be about 70% of the low energy cross section.

The fraction of beam ions which get neutralized passing through the laser beam is,

$$f_{neut} = 1 - e^{-\sigma(E)Ft} \quad (2)$$

Here $\sigma(E)$ is the energy-dependent cross section, F is the photon flux, and t is the time the ion is in the light beam. The photon flux in the moving reference is also transformed the same as the photon energy [13],

$$F_{CM}=\gamma F_L[1-\beta\cos(\theta_L)] \quad (3)$$

The neutralization fraction from a given laser initially drops with increasing beam energy then becomes almost flat from 400MeV to 1GeV. In this range the decreased reaction cross section from the Lorentz boosted photon energy is approximately offset by the Lorentz boost in photon flux.

For example, the laser on the SNS MEBT experiment produces a 20ns-long pulse with an output energy of 50mJ. It is focused to a rectangular spot 1mm wide by 3mm along the beam. The approximate variation of neutralization fraction with beam energy this laser will produce is shown in fig. 2. This calculation does not include the Lorentz shift of neutralization cross section. Higher power lasers are required for higher beam energies to achieve the same signal level in the detector.

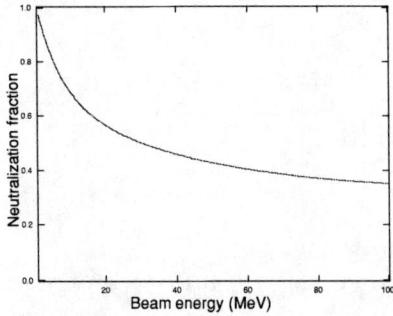

FIGURE 2. Calculated neutralization fraction vs. beam energy for a 20ns-long, 50mJ laser pulse focused to a spot size of 1mm x 3mm.

THE 750 KeV EXPERIMENT

Our first profile measurement was made on the BNL linac between the rfq and the first drift tube linac tank, fig. 3. A light pulse from a Q-switched Nd:YAG passed through the 750keV H⁻ beam from the linac rfq neutralizing most of the beam the light passed through. A downstream current transformer measured a dip in the beam current which was proportional to the fraction of the beam hit with the light, fig. 4. The laser beam was stepped across the ion beam and the profile constructed by plotting the depth of the current notch vs. laser beam position.

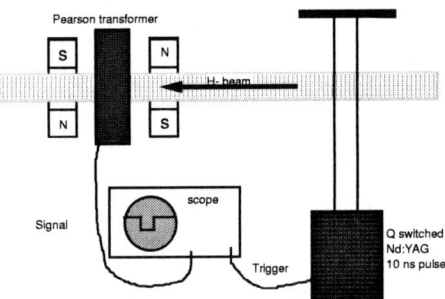

FIGURE 3. Laser scanner experiment on BNL linac. The first of two 10 Gm dipole magnets removes the free electrons from the beam and the second straightens the beam.

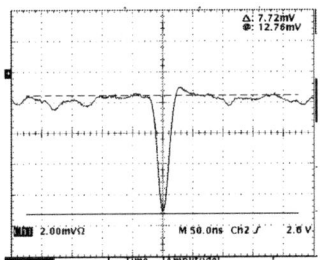

FIGURE 4. Scope trace of the current transformer signal showing notch created by the laser pulse.

The arrangement of the laser and optics on the linac beamline is shown in fig. 5. A CFR200 laser from Big Sky Laser [14] was mounted on a shelf at the top left. Three 45° mirrors were mounted inside the vacuum on linear motion feedthroughs. The top-left mirror was used to switch between vertical and horizontal scans and the other two did the scanning. The top-right mirror scanned horizontally and the bottom-left mirror scanned vertically. Both scanning mirrors are shown with arms to hold lenses. In this experiment the lenses were not installed.

The CFR200 puts out 200 mJ pulses that are about 20 ns long. Without lenses the beam diameter is about 0.6 cm giving a photon flux of $1.9 \times 10^{26}/cm^2 s$. About 97% of the ions passing through the center of the laser beam were neutralized. The laser is triggered 400µs before the measurement is to made. The laser then returns a timing pulse synchronous with the Qswitch firing which was used to trigger a scope.

When an ion is neutralized the free electron continues to move along with the beam. These electrons have to be removed from the beam to measure a current drop. In an accelerator installation this is accomplished by either rf cavities or quadrupoles but in the experiment the current transformer had to be placed in the same vacuum chamber as the laser optics. For this reason we placed two weak permanent-magnet dipoles on either side of the transformer. The pole tips are 2.5cm square and 5cm apart and the field is about 400 G. The first magnet deflects the electrons from the beam and the second one straightens out the beam.

The data were taken by moving the mirrors manually and measuring the notch depth on an oscilloscope set to average 15 shots. We measured a maximum notch depth of about 40% on the horizontal scan. If the laser beam power was uniformly distributed over the spot the maximum notch depth should have been closer to 60%. Based on this we conclude the laser power was not uniform over the spot.

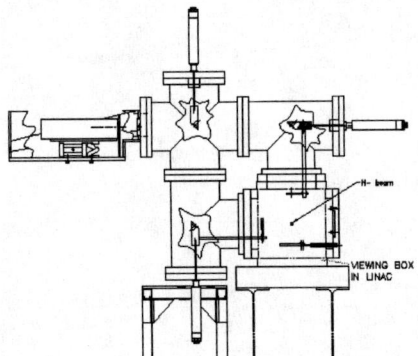

FIGURE 5. Laser scanning assembly installed on linac beamline. View is looking up beamline.

Measured Profiles

Figure 6 shows the measured horizontal and vertical profiles. In each plot the measured points are indicated by markers and the curve is a gaussian fit to the data.

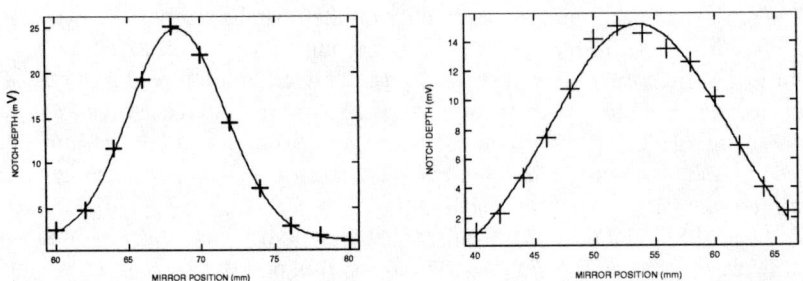

FIGURE 6. Measured horizontal (left) and vertical beam profiles.

The rms widths of the two fitted curves are $\sigma_x=3.32\pm0.05$ mm and $\sigma_y=7.3\pm0.6$ mm. These values agree with expectations from previous measurements at this location, however for this experiment there was no profile measurement by another method. We had a carbon wire installed in the beam box but we were operating parasitically during a production run and we were never able to get the short beam pulses necessary to prevent damage to the wire.

THE MEBT EXPERIMENT, 2.5MeV

Based on the promise of the BNL linac experiment we designed a laser platform which would attach to one of the wire-scanner chambers on the SNS MEBT at Lawrence Berkeley National Lab. Mounted on the platform, fig. 7, are a 50mJ/pulse laser head, a lens holder, and three linear actuators which move 45° mirrors in a more compact but otherwise identical arrangement to the linac experiment of fig. 5. There are four profile measurement stations. Each has a wire scanner and four windows for laser-beam access. We installed the platform on the most upstream chamber after the MEBT was under vacuum and ready for first beam.

A 300mm-focal-length cylindrical lens is mounted directly in front of the laser head and the two optical path lengths from the lens to the beam center are the same. The lens produces a 1mm wide by 3mm long light ribbon across the beam producing a measurement window of rms width 0.3mm. Since the measured rms width of the horizontal profile is 1.48mm the width of the laser beam caused about 2% broadening of the profile.

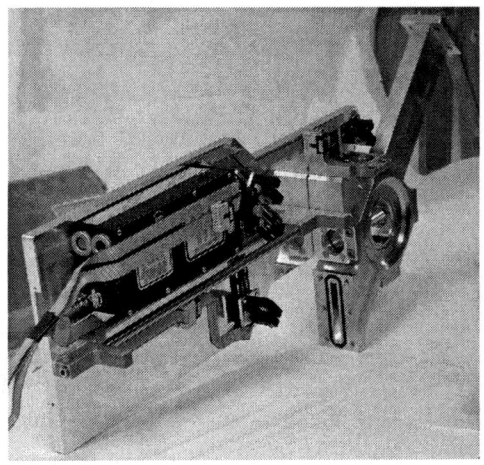

FIGURE 7. Laser scanning platform mounted on the SNS MEBT wire scanner chamber.

The signal was detected with the existing beam transformer at the end of the MEBT. The BCM signal was split and our half was fed into a LeCroy LT374L scope with ethernet connection [15]. A math channel on the scope was used to average for several pulses to reduce noise and rf pickup. The rf pickup could have been greatly reduced with the use of a 50MHz low-pass filter as was done on the 750keV experiment but none was available. In the profiles shown in fig. 8, the signals from 25 beam pulses were averaged giving about 40dB signal/background in the beam center.

The experiment was controlled in Labview. The program switched the laser, moved the mirrors, initialized the scope for each new position, and read the data. Cursors on the scope were set manually around the pulse. For each set of averaged data the program summed the channels between the cursors, summed an equal number of channels before the pulse, and subtracted these two 'integrals' to give one data point in the profile.

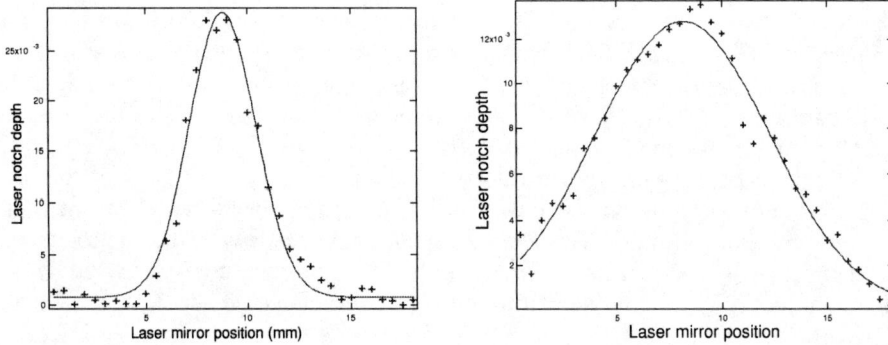

FIGURE 8. Beam profiles measured on the SNS MEBT with the laser profile monitor. The horizontal profile (left) has a measured width of σ=1.60±0.04mm and the vertical profile has σ=4.16±0.16mm.

MEASUREMENTS AT 200 MeV

After the measurements on the BNL linac at 750keV the entire apparatus shown schematically in fig. 5 was moved to the high energy end of the linac to measure 200MeV beam. It was installed in the linac-AGS transfer line which is no longer used for beam transport, fig. 9. The 200mJ laser hear is mounted on the covered shelf at the top right. The beam line chamber also had the carbon wire installed.

In this installation there were two cylindrical lenses which moved with each mirror. A 300mm-focal-length lens produced a waist perpendicular to the ion beam and a 50mm lens spread the light beam longitudinally to reduce power density on the laser beam stop. The light beam which crossed the ion beam was 1mm wide by 20mm long. The calculated neutralization of beam passing through the laser light is 72%.

For this experiment the goal was to use stripline beam position monitors (BPMs) to measure the laser notch. A single-plane RHIC BPM was installed before and after the beam box. In the superconducting linac of SNS there are BPMs between rf tanks with a current transformer at the exit of the linac. Using the striplines as detectors gives us access to a upstream and downstream detector spaced by a single rf structure.

Using the transformer for signal detection will require good transmission through the full linac before profiles can be made at any point.

FIGURE 9. LPM measurement station at 200 MeV in the BNL linac.

To measure the notch we adjust the phase and attenuation of two stripline signals and combine them with a hybrid to produce a nulled signal in the absence of laser neutralization. The laser pulse causes a signal imbalance which appears either as a wide or narrow spot in the rf envelope depending on how the signals are combined. Different from the transformer measurements, now the signals are bipolar with no dc component. To integrate the laser notch, the scope data are passed through a Labview VI which takes the absolute value of each data point.

Also different from the first two experiments, we have had extremely noisy beam at 200MeV. However with care we have been able to match the beam pattern from the upstream and downstream BPMs and produce a null signal with over 20dB common mode rejection. Figure 10 shows a scope trace of the laser notch in the nulled signal.

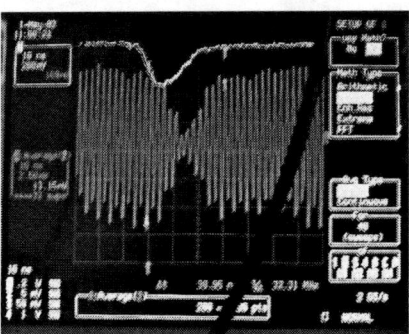

FIGURE 10. Laser notch in the beam rf envelope picked up on the BPM striplines. The top trace is the signal from a photodiode near the laser head.

Experimental Difficulties

The first two experiments were conducted at low energy, where we had beamline access with the beam on, and inside clean vacuum systems. At 200 MeV we have had to operate remotely from outside the tunnel. Also we have been in a section of beam line which normally is not used to transport beam so we have had limited beam time.

There was a period of two months between installation of the experiment and available beam during which time the laser controller was in the tunnel. Less than a month after we started getting beam time the laser controller failed from radiation and had to be replaced. The beamline is thirty years old and is slightly contaminated with pump oils. All the optics are in the vacuum and, during the wait for beam, they became contaminated with oil. When we started taking data the laser light burned the oil on the optics forming milky patches which scattered the light. During the early measurements the signal continued to get worse over time. We discovered the damaged optics when replacing the damaged laser.

Profile Measurements

For several reasons including those mentioned above the measurements at 200MeV have not been as clean as the low-energy ones. The first profile data at 200 MeV, fig. 11, made with 100µA of beam from the polarized source, was taken by hand by measuring the notch depth with the scope cursors.

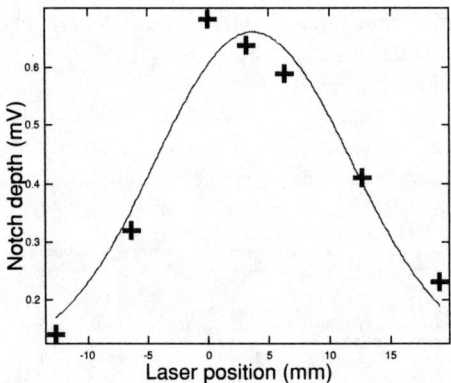

FIGURE 11. Profile of 100µA polarized beam. The width of the fitted gaussian is 8.1±2.2 mm.

After the polarized run we had several weeks of 400µs-long, 10mA beam pulses before the laser failed from radiation. During the early portion of this time we were able to get clean laser signals but as time progressed the signal progressively got worse. In hindsight we realize this was from the oil contamination problem.

Figure 12 shows a LPM-measured profile and a wire-scanner profile taken the day before the LPM profile. We were unable to get a LPM profile and wire profile at the same time because the wire scanner failed after one use. The linac had been returned to the set points of the previous day although between the two times we had beam the linac had been switched back to normal production operation. The measured widths are: $\sigma_{LPM}=5.3\pm1.3$mm and $\sigma_{wire}=3.51\pm0.34$mm.

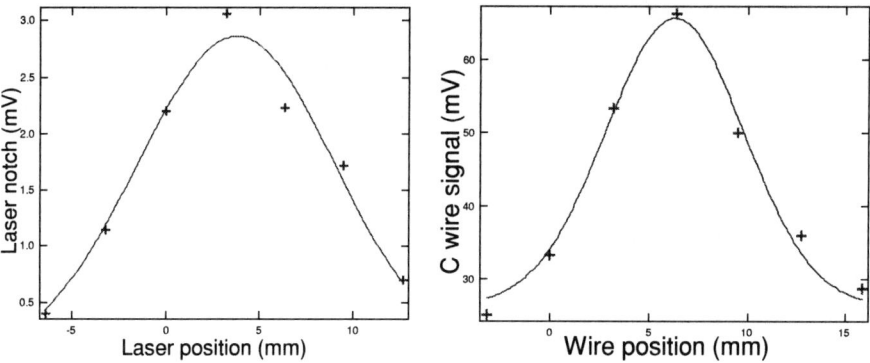

FIGURE 12. Beam profile taken with a wire scanner (right) and a beam profile taken the next day with the laser scanner (left). Profiles were taken on different days but the linac had been restored to the set points of the previous day.

Our last beam run before this paper produced a fully-automatic horizontal profile scan. The Labview software was the same as that used at LBNL with the beam current transformer. Since the stripline data are bipolar with no dc component we added an absolute value VI to rectify the scope data then summed the data between cursors to produce a notch integral. Figure 13 shows the rectified scope trace of the laser notch. Figure 14 shows the laser profile and a profile taken with the carbon wire at the same time. With the laser the width of the fitted gaussian is $\sigma=2.82\pm0.76$mm and the wire measured $\sigma=3.61\pm0.22$mm.

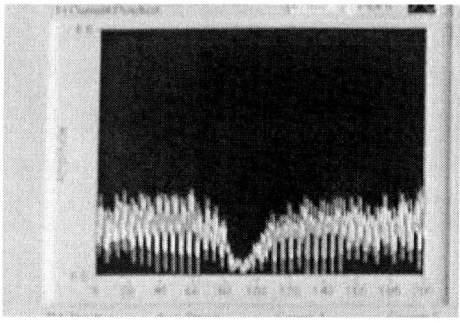

FIGURE 13. At each laser position the scope averaged forty traces. The scope was read into Labview and passed through an absolute value VI to rectify the bipolar signal. Cursors on the scope marked the notch channels and all of the data between the cursors were added to produce an integral.

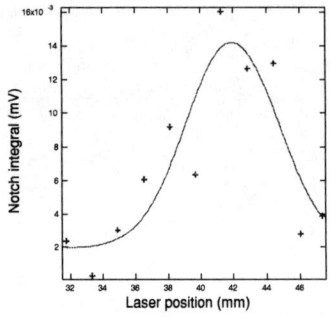 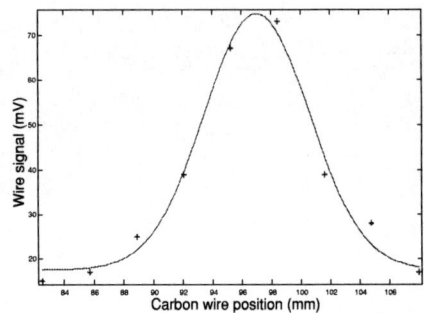

FIGURE 14. The fit to the laser date on the left gives a width of $\sigma=2.82\pm0.76$mm. The wire profile on the right measured $\sigma=3.61\pm0.22$mm.

DISCUSSION

Transverse profiles of H- beams can be measured by scanning a laser beam across the ion beam and detecting the notch in the beam current downstream. This technique is attractive because no components are in the vacuum, profile measurements can be made without disrupting machine operation, and measurements can be made on high power beams. As we demonstrated on the SNS MEBT, profile measurement capability can be added to an operating accelerator if a suitable window exists and a downstream current transducer is available.

Q-switched Nd:YAG lasers are perfect for these measurements on beams with energies up to about 1GeV. These lasers are readily available with a wide range of output energies. Lasers with pulse energies of close to a Joule are available with compact laser heads attached to power units by umbilicals which make them suitable for mounting on compact platforms on beamlines.

Our experiments have placed the laser controller and cooling unit next to the beamline for convenience. Two laser controllers have failed from radiation. Any installation of a LPM in a radiation area has to have the controller in a nonradiation area. We do not know the radiation doses the laser heads can take. The plan for SNS is to place the entire laser outside of the tunnel and transport the light by mirrors or fiber optics.

The two LPM experiments which used beam current transformers for beam-current detection produced extremely clean signals with very little set up time. The measurement at 750keV gave a signal/noise ratio at beam center of 25dB and the MEBT experiment gave 40dB. Measurements on 200MeV beam with BPM striplines have been made but we need to make improvements in the data processing.

ACKNOWLEDGEMENTS

The authors thank Brian Briscoe and Vinnie LoDestro at the BNL linac. Also we thank Alex Ratti, Larry Doolittle and John Staples at LBNL for letting us try the MEBT platform during commissioning. SNS personnel who have become involved include Saheed Asadi, Warren Grice, Sasha Alexander and Tom Shea.

REFERENCES

[1] http://www.sns.gov/
[2] J.T. Broad and W.P. Reinhardt, Phys. Rev. A14 (6) (1976) 2159.
[3] M.J.Ajmera and K.T. Chung, Phys. Rev. A12 (2) (1975) 475.
[4] M. Daskhan and A.S. Ghosh, Phys. Rev. A28 (5) (1983) 2767.
[5] W.B. Cottingame, G.P. Boicourt, J.H. Cortez, W.W. Higgins, O.R. Sander and D.P. Sandoval, Proc. 1985 Particle Accelerator Conf., IEEE Trans. Nucl. Sci. NS-32 (1985) 1871.
[6] V.W. Yuan, R.C. Connolly, R.C. Garcia, K.F. Johnson, K. Saadatmand, O.R. Sander, D.P. Sandoval and M.A. Shinas, "Measurement of longitudinal phase space in an accelerated H⁻ beam using a laser-induced neutralized method", Nucl. Instr. and Meth. A329 (1993) 381-392.
[7] R.C. Connolly, K.F. Johnson, D.P. Sandoval and V. Yuan, "A transverse phase-space measurement technique for high-brightness H⁻ beams", Nucl. Instr. and Meth. A132 (1992) 415-419.
[8] R. Tomlin, "Ion Beam Notcher Using a Laser", FERMILAB-Conf-01/117-E (2001).
[9] D.R. Swenson, E.P. MacKerrow and H.C. Bryant, "Non-Invasive Diagnostics for H- Ion Beams using Photodetachment by Focused Laser Beams", Proc. 1993 Beam Instrumentation Workshop (Santa Fe). A.I.P. Conf. Proceedings, (319), 343.
[10] R. Connolly, P. Cameron, J. Cupolo, M. Grau, M. Kesselman, C-J Liaw and R. Sikora, "Laser Profile Measurements of an H- Beam", Proc. 5th European Workshop on Beam Daignostics and Instrumentation, (Grenoble), http://www.esrf.fr/conferences/DIPAC/DIPAC2001Proceedings.html
[11] R.E. Shafer, "Laser Diagnostic for High Current H- Beams", Proc. 1998 Beam Instrumentation Workshop (Stanford). A.I.P. Conf. Proceedings, (451), 191.
[12] R. Tomlin, "Laser Stipping of Relativistic H- Ions with Practical Considerations", FERMILAB-TM-1957, (1995).
[13] S.M. Shafroth and J.C. Austin, Eds., *Accelerator-Based Atomic Physics Techniques and Applications*, A.I.P. Press (1997) Chap. 6.
[14] Big Sky Laser Technologies, Inc., Bozeman, MT 59715.
[15] LeCroy Corp., Chestnut Ridge, NY 10977.

Short Bunch Beam Diagnostics

Patrick Krejcik

Stanford Linear Accelerator Center
2575 Sand Hill Rd, Menlo Park CA 94025

Abstract. With the emergence of 4th generation FEL based light sources there is now considerable interest in both producing and characterizing ultra-short (<100 fs) electron bunches. Knowledge of the extremely high peak current in a short bunch is required to diagnose the SASE (self amplified stimulated emission) process. Measuring the femtosecond duration of the pulse is inherently interesting, particularly for experimenters using the beam to measure fast phenomena (e.g. femto-chemistry). Diagnostic techniques that have the necessary femtosecond resolution will be reviewed: These include high-power RF transverse deflecting structures that "streak" the beam in the accelerator allowing the bunch length to be recorded on a profile monitor. Electro optic crystal diagnostics use the electric field of the electron bunch to modulate light thereby exploiting the femtosecond technology of high bandwidth visible lasers. Coherent synchrotron radiation (CSR) from dipole magnets and optical diffraction radiation (ODR) both result in radiation with wavelengths of the order of the bunch length and hence in the terahertz band which can be detected by a variety of techniques. The role of each of these techniques is discussed in terms of its application at the Linac Coherent Light Source (LCLS) and the Short Pulse Photon Source (SPPS) currently under construction at SLAC.

INTRODUCTION

The Linac Coherent Light Source (LCLS) [1] to be built at SLAC utilizes electron bunches as short as 80 femtoseconds rms to generate self-amplified stimulated emission (SASE) X-ray radiation in a FEL. The production and tuning of these ultra-short bunches is critical to the performance of the LCLS but the measurement of such short bunch lengths is a considerable challenge that cannot be met, for example, with conventional streak camera technology. The technologies and limitations to various bunch length measurement techniques are reviewed in this paper together with plans for developing and testing these techniques at SLAC.

The new Short Pulse Particle Source (SPPS)[2] at SLAC offers a nearer term opportunity to test and compare these different diagnostic techniques with bunches as short as 30 fs rms, far shorter than anything so far produced in a high energy electron accelerator. The locations of the SPPS and LCLS installations are shown in figure 1 in relation to the other accelerators at SLAC. SPPS will only produce spontaneous x-ray radiation from its undulator but the peak spectral brightness from this source, at 10^{24} photons s^{-1} mm^{-2} mr^{-2} (0.1% bandwidth), still exceeds that of any existing x-ray source.

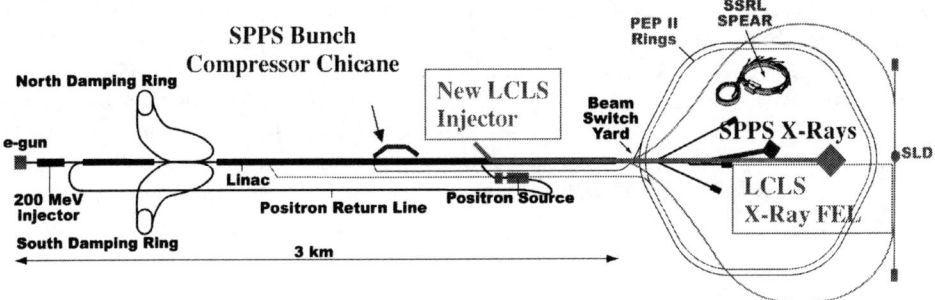

FIGURE 1. The locations of the SPPS linac bunch compressor chicane is shown superimposed on the SLAC site together with the LCLS facility which utilizes the last one third of the linac to generate coherent x-rays from a FEL located downstream of the beam switch yard.

To illustrate the application of the various short bunch diagnostic techniques the accelerator layout of the LCLS is shown in figure 2 with indicators for the locations of the various devices. Bunches of approximately 3 ps rms length are produced in the RF photoinjector and accelerated to 150 MeV where a magnetic chicane compresses the bunches to an rms length of 630 fs. A second bunch compressor is located at the 4.5GeV point where the bunch attains its final bunch length of 75 fs.

At the injector, confirmation of the critical initial longitudinal distribution of the bunch is independently diagnosed with both electro-optic profiling of the bunch and with a short, 0.55 m long S-band transverse deflecting cavity. The electro-optic system can make use of the already available high-bandwidth Ti:Sapphire laser system for the injector to easily achieve the necessary sub-picosecond resolution to characterize the gun pulse. The relatively low 150 MeV energy of the injector means that only a short section of transverse deflecting structure is required to streak the beam on a downstream profile monitor using about 1 MW of power split off from one of the injector klystrons.

A second transverse deflecting cavity is also located in the high-energy linac with a length of 2.44 m and is powered with 25 MW from a separate klystron.

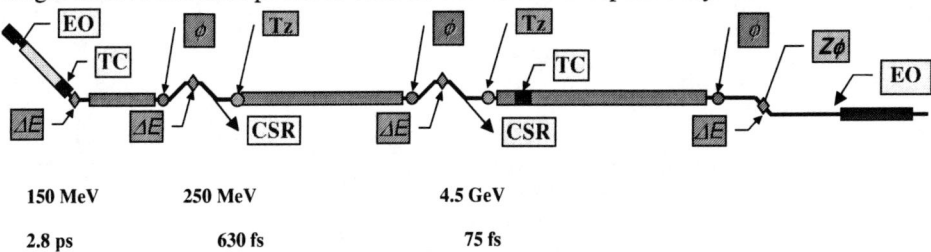

FIGURE 2. A schematic layout of the LCLS accelerator and bunch compressor system showing the types and locations of the various diagnostics to measure bunch length and characterize the longitudinal phase space of the beam: Electo-Optics (EO), Transverse Cavity (TC), Terahertz power monitors (Tz), Coherent Synchrotron Radiation monitors (CSR), Energy spread monitors (ΔE), Beam Phase monitors (φ), and Zero-phase measurement locations (Z φ).

The energy and energy spread of the beam are measured on beam position monitors and profile monitors at all the dispersive locations where the beam is bent. The very short bunches generate a coherent component to the synchrotron light in the bunch compressor chicane bends so the light will also be diagnosed to monitor the bunch length at these locations. Tuning and feedback control of the chicane bunch compressors is facilitated by rapid, pulse-to-pulse measurement of the CSR power by narrow band terahertz detectors. Beam phase monitors are located at each accelerating sector. The accelerating section downstream of the second bunch compressor will be used for zero-phase crossing measurement of bunch length in conjunction with the profile monitor located in the final bend before the undulator.

The SPPS linac bunch compressor chicane [3] is similarly equipped with short bunch diagnostics to tune and measure its performance.

Each of these techniques will be described in the following sections and their potential for reaching the necessary resolution in the LCLS and SPPS will be discussed. The full temporal characterization of the electron bunch also involves an assessment of the timing jitter and how it can be measured on a pulse-to-pulse basis.

TRANSVERSE DEFLECTING CAVITIES

The principal of the transverse cavities is shown in figure 3 where it can be seen that a strong longitudinal-to-transverse correlation, or crabbing, can be imposed on the bunch when the cavity is operated close to zero phase crossing of the applied RF voltage. This technique of measuring bunch length was in fact used during the early commissioning years of the injector at SLAC[4] and in more recent times has been proposed as a method for measuring very short bunches[5]. In order to test this technique we have just completed recommissioning one of the original SLAC deflecting structures[6] by installing it in the SLAC linac where it can be used with a variety of beams during normal accelerator operations[7].

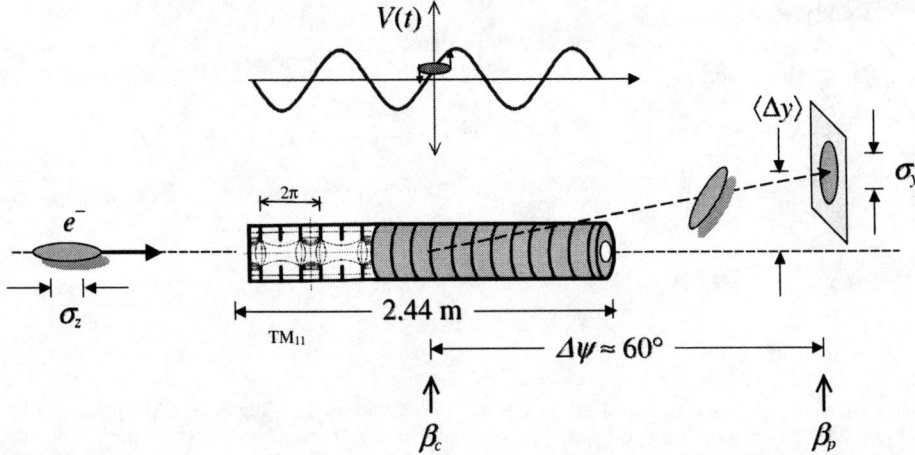

FIGURE 3. The transverse deflecting cavity can be used to crab the bunch so that its projected transverse beam size measured on a downstream profile monitor is proportional to bunch length.

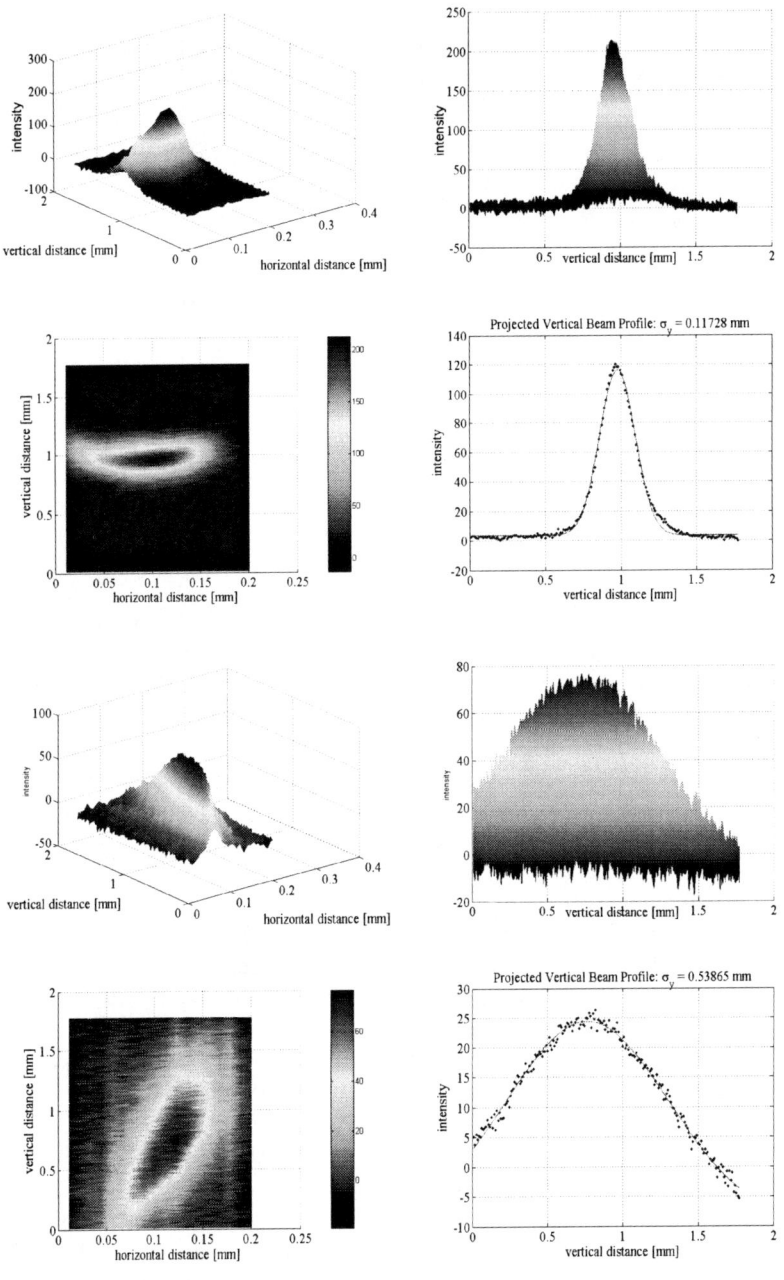

FIGURE 4. Profile monitor images and projected vertical image profile with the RF transverse cavity off (left) and with the RF on (right) showing how the beam is streaked.

A profile monitor screen (p) is positioned at a suitable transverse betatron phase advance from the deflecting cavity (c). The beam size measured on the screen is the quadrature sum of the initial vertical beam size, $\sigma_{y0}^2 = \beta_p \varepsilon_{y0}$ and the RF induced crabbing.

$$\sigma_y = \sqrt{\beta_p \varepsilon_{y0} + \sigma_z^2 \beta_c \beta_p \left(\frac{2\pi e V_0}{\lambda E_0} \sin \Delta \psi \cos \varphi \right)^2} \tag{1}$$

The advantage of this technique is that a large deflecting voltage, V_0 can be achieved at short RF wavelength, λ by applying high peak RF power supplied by the same type of klystron as is used to power the normal accelerating sections.

A recent test of this technique with a 2.44 m long 2856 MHz device at SLAC[7] showed that 10 μm bunch length resolution could be achieved with a RF deflecting amplitude of 17.5 MV on a bunch with a normalized vertical emittance of 3*10-5 m rad. Images of the streaked beam are shown in figure 4.

In addition to measuring the rms bunch length, information can also be gained on the bunch length distribution. The measured profile is a convolution of the actual longitudinal distribution in the bunch with the vertical beam size at the screen as determined by the transverse emittance. When the longitudinal distribution is non-Gaussian, due to the bunch being over compressed for example, the measured profile is also non-Gaussian. Particle tracking[8] of the LCLS beam show that quite complex longitudinal bunch distributions are to be expected as a result of the compression dynamics together with the strong wakefields experienced in the linac that are excited by the short bunch. Simulation[9] of the transverse beam profile from the RF deflecting cavity shows that adequate resolution can be achieved to reveal most of the details of the complex longitudinal distribution of the LCLS bunch.

The RF deflecting cavities offer such a powerful resolution of the beam properties along the longitudinal z-coordinate of the bunch that it is reasonable to consider measuring other properties as a function of the "longitudinal slice" position along the bunch. Consider, for example, that the horizontal width of the beam can be measured at different vertical positions on the screen and hence at different positions along the bunch, as shown in figure 5a. If this is combined with a standard emittance measuring technique such as varying the strength of an appropriate upstream quadrupole then the horizontal emittance of each of the slices along the bunch can be determined.

Another technique is to use a profile monitor screen in a region of large horizontal dispersion, after a bending magnet, so that the horizontal beam size becomes a measure of energy spread in the beam. Recording the horizontal beam size along the vertical axis of the profile monitor gives a measurement of what is referred to as the slice energy spread of the bunch. The slice emittance and slice energy spread together with the bunch length and charge are key parameters in the performance of an FEL.

Although this bunch length measurement technique disrupts the beam it is possible to power the device on and off on a pulse-by-pulse basis so that it can operate with a 1% or less duty cycle and minimize the impact to the downstream users.

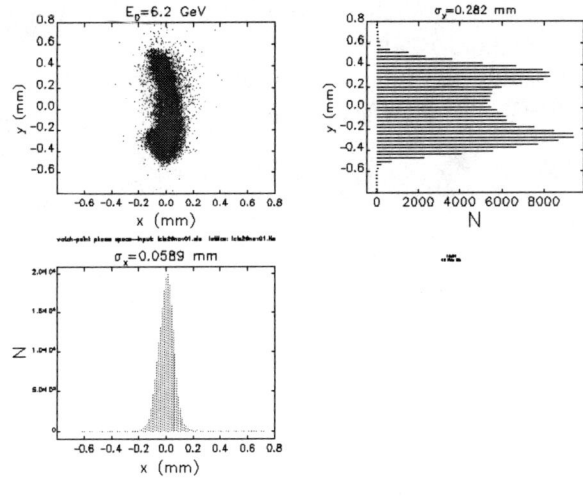

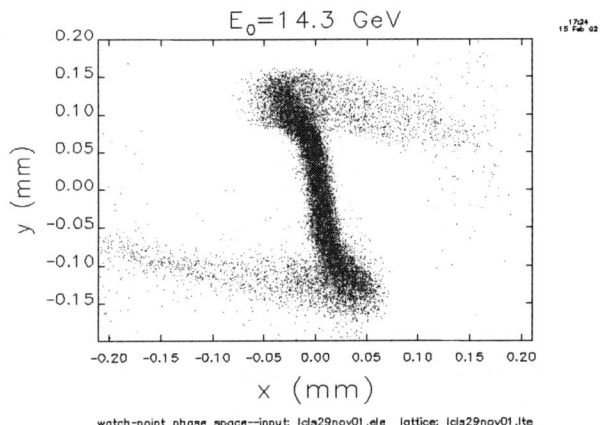

FIGURE 5. a. Expected longitudinal slice profile for the LCLS beam (from P. Emma, ref. 1 and 9) from which the horizontal slice emittance could be reconstructed, and b. the slice energy spread obtained by viewing the streaked beam at a high dispersion profile monitor location (ibid).

ZERO-PHASE CROSSING MEASUREMENTS

A more invasive measurement of the bunch length distribution uses the RF accelerating cavities that are part of the linac. Rather than accelerate the beam on crest the bunch is moved to the RF zero-phase crossing where the accelerating field seen by each particle changes rapidly along the bunch length thereby inducing a correlated energy spread, or chirp, along the bunch, as shown in figure 6. The bunch length which was first rotated onto the energy axis by the RF is now transformed into the spatial coordinate by observing the transverse beam profile on a down-stream high-dispersion profile monitor.

The measured horizontal beam size σ_x at a profile monitor location where the dispersion is η_x is the quadrature sum of the beam size σ_{x0}, with the RF off, and the contribution from the energy spread:

$$\sigma_x = \sqrt{\sigma_{x0}^2 + \sigma_z^2 \eta_x^2 \left(\left(\frac{dE}{dz} \right)_0 + \frac{2\pi e V_{rf}}{\lambda_{rf} E_0} \cos \varphi_{rf} \right)^2} \qquad (2)$$

where $(dE/dz)_0$ is the initial energy spread in the beam.

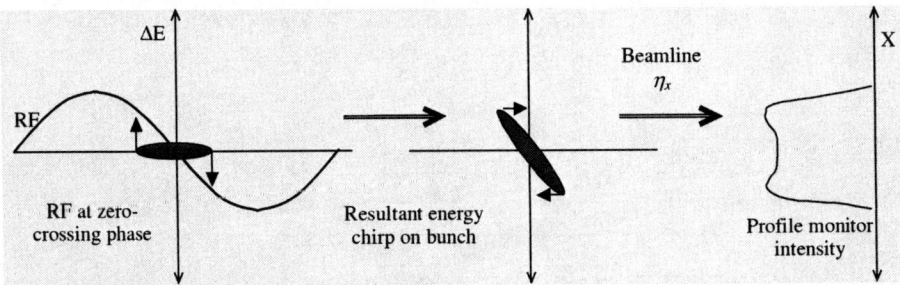

FIGURE 6. A rotation in longitudinal phase space is induced by moving the bunch off crest to the zero-phase crossing where the resultant correlated energy spread is measured on a high-dispersion location profile monitor.

Good bunch length resolution can be obtained if the induced correlated energy spread is much greater than the incoherent energy spread in the bunch. For this reason it is usually necessary to use a large portion of the linac in this off-crest mode.

This technique has been used extensively in short bunch applications such as FEL injectors to measure the bunch length[10,11]. The same technique described above for measuring the slice emittance with the transverse cavity can also be used in conjunction with the zero-phase crossing measurement.

This technique relies on a rotation in longitudinal phase space of the bunch, induced by the RF. In order to distinguish this from wakefield-induced energy correlations in the bunch it is possible to rotate the bunch in opposite directions by performing the measurements on the two zero-phase crossings of the RF in turn. The wakefield contribution will be seen to either aid or oppose the RF induced energy spread and can thus subtracted out of the bunch length reconstruction.

The more invasive nature of zero-phase crossing measurements means they are usually only performed during initial tune up of the accelerator or to cross check other bunch length measuring techniques.

ELECTRO OPTIC TECHNIQUES

At extremely short bunch lengths where the previous techniques begin to give out through lack of RF power we can move to other technologies. It is possible to take advantage of the advances in laser science where femtosecond technology has been developed. Very short electron bunches have very high electric fields associated with them which can be used to modulate a laser pulse. The problem of measuring the electron bunch length then moves to one of measuring the laser pulse modulation, for which several techniques exist.

As an illustration, we consider the setup shown schematically in figure 7, where an electro-optic (EO) crystal responds to the electric field from the electron bunch.

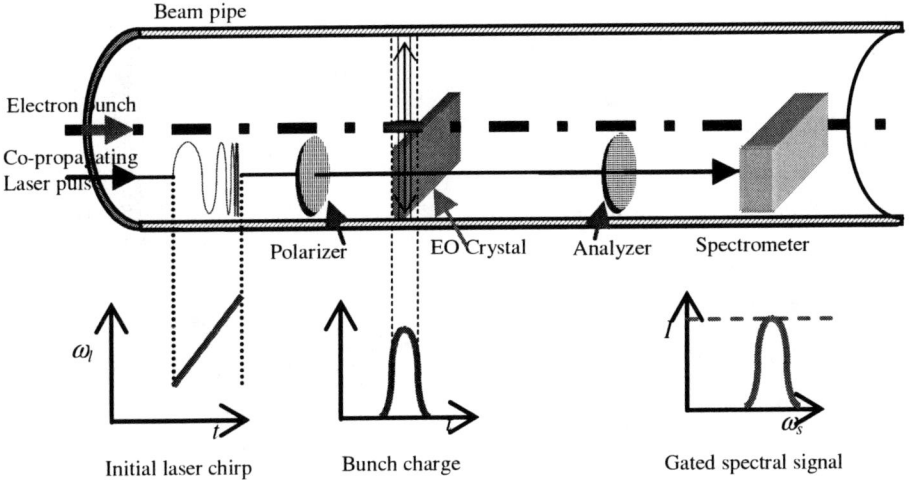

FIGURE 7. The electric field from a bunch modulates the polarization in an electro optic crystal and gates the transmission of a chirped laser pulse.

The electric field of the bunch alters the optical retardation along one axis of the birefringent crystal according to

$$P = \varepsilon_0 \left[X_1 E + X_2 E^2 + X_3 E^3 + \ldots \right] \qquad (3)$$

The optical transmission through a polarizer-analyzer pair is modulated as the polarization vector in the crystal rotates under the influence of the electric field of the bunch. By superimposing a chirp on the incoming laser pulse the time structure of the electron beam induced modulation can be measured in the frequency domain with a spectrometer.

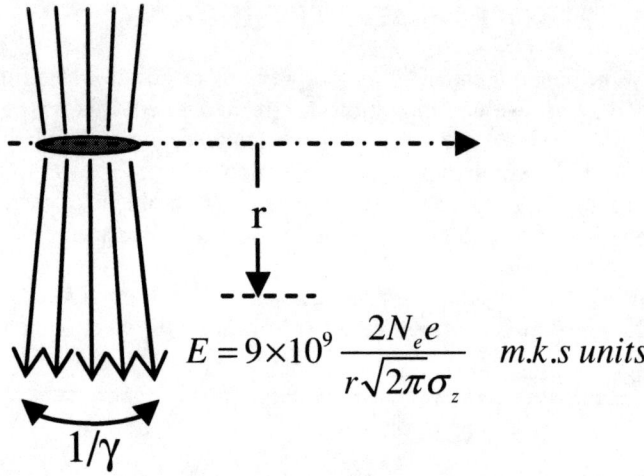

FIGURE 8. Point charge approximation for the electric field at a distance r from a short, relativistic bunch.

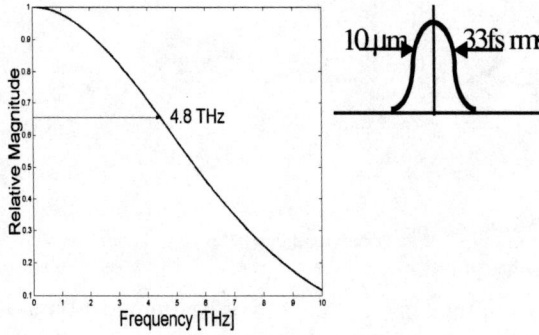

FIGURE 9. The bandwidth of a Gaussian bunch extends into the THz band for short (10μm) bunches.

The electric field of an ultra-relativistic bunch of N_e electrons at a distance r, from the crystal is shown in figure 8 for a Gaussian bunch of length σ_z[12].

At high energies the opening angle, $1/\gamma$, is very small so the crystal can be placed some distance from the beam without loss of bunch length resolution. The field is also quite strong, for example a 10 μm long bunch with 1 nC charge will generate a 72 MV m^{-1} field at a distance of $r = 1$ mm. At low energies around the gun and injector the crystal needs to be brought into closer proximity with the beam to maintain good bunch length resolution.

In a beam pipe the high frequency component of the wakefields contain the bunch length distribution information. The bandwidth of radiation produced from a Gaussian bunch, given by its Fourier transform,

$$F(\omega) = e^{-\frac{\omega^2 \sigma^2}{2}} \quad (4)$$

extends into the terahertz region for the ultra-short bunches considered here, as shown in figure 9.

COHERENT RADIATION FROM A SHORT BUNCH

Terahertz radiation is generated by the bunch through a variety of means. Special cases of wakefields for short bunches are coherent transition radiation and coherent diffraction radiation. In addition, the synchrotron radiation from an ultra-short bunch in a bending field has a pronounced coherent component at wavelengths comparable to the bunch length.

Coherent synchrotron radiation

The power spectrum of the incoherent synchrotron radiation is proportional to the number of electrons, N, in the bunch whereas the coherent term is enhanced by N^2, as indicated by equation (5) from reference[13]

$$P(\lambda) = P_0 N \{1 + NF(\lambda)\}$$
$$F(\lambda) = \left| \int_{-\infty}^{+\infty} S(z) e^{2\pi i / \lambda} dz \right|^2 \quad (5)$$

The power spectrum, taken from Wang [13], shown in figure 10, peaks in the terahertz band. It is strongly dependant on bunch length so that a power meter tuned on a narrow wavelength band, for example 0.2 mm in the case shown in figure 10, would produce a clear signal inversely proportional to bunch length. A number of detectors are available for THz radiation, including Schottky devices and germanium photodetectors whose band can be extended into this range. Bandpass filters can also be fabricated at these wavelengths, which should be centered at a wavelength corresponding to the nominal bunch length for an application.

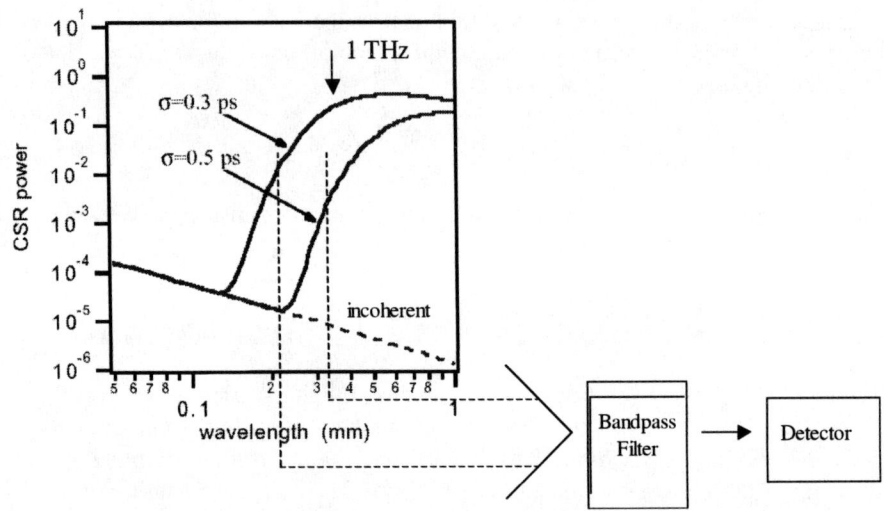

FIGURE 10. Power spectrum of coherent synchrotron radiation calculated for two different bunch lengths (after Wang [13]).

Measurements of the power in the THz band with detectors such as these, or bolometers, do not yield any phase, or bunch shape, information. The measurement still performs the useful function of providing an rms measurement of the bunch length, which is useful in tuning applications along the length of the accelerator.

The phase and arrival time of the bunch can, however, be measured by a fast gating technique using femtosecond laser technology. The output from a Ti:Sapphire laser, for example, is directed onto a semiconductor substrate of GaAs or radiation damaged Si-on-sapphire, generating photo electrons. Metallized electrodes on the surface forming a Hertzian dipole at THz wavelengths allow current to flow when the laser pulse and THz radiation are coincident in time. Such devices, shown in figure 11, are referred to as Auston switches [14] and are used in THz pump probe applications.

Coherent diffraction radiation

The THz detection techniques described above for CSR can also be applied to the detection of coherent diffraction radiation in the beam pipe. It is anticipated that a number of these relatively simple devices would be used along an FEL accelerator to aid in bunch length tuning and feedback control. CSR power would be monitored at the bunch compressor bends and CDR at the diagnostic station preceding the undulator. Pump probe timing measurements with THz radiation could be conveniently made where high bandwidth laser light is already available, at the injector for example, or at the undulator entrance where FEL experiments also employ femtosecond lasers for pump probe characterization of the radiation.

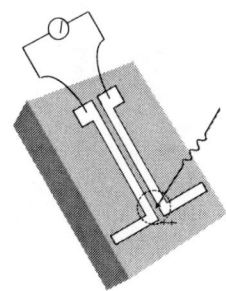

FIGURE 11. An Auston switch comprises of metallized electrodes forming a dipole on a semi-conductor substrate.

ACKNOWLEDGMENTS

The contributions of Paul Emma to the beam dynamics of short bunch production, tuning and diagnostics are gratefully acknowledged.

This work is supported under the United States of America Department of Energy contract DE-AC03-76SF00515.

REFERENCES

1. "LCLS CDR", SLAC-R-593, April 2002.
2. M. Cornacchia et al., "A Subpicosecond Photon Pulse Facility for SLAC", SLAC-PUB-8950, LCLS-TN-01-7, Aug 2001. 28pp.
3. L. Bentson, P. Emma, P. Krejcik, "A New Bunch Compressor Chicane for the SLAC Linac to Produce 30-fsec, 30-kA, 30-GeV Electron Bunches", EPAC-2002, Paris, France, June 3-7, 2002. p.682-685
4. R.H. Miller, R.F. Koontz, D.D. Tsang, "The SLAC Injector", IEEE Trans. Nucl. Sci., June 1965, p. 804-8.
5. X.-J. Wang, "Producing and Measuring Small Electron Bunches", Proc. of the 1999 Part. Acc. Conf., New York, NY, March 1999.
6. O. H. Altenmueller, et. al., "Investigations of Traveling-Wave Separators for the Stanford Two-Mile Linear Accelerator", The Review of Scientific Instruments, Vol. 35, Number 4, April 1964.
7. R. Akre, L. Bentson, P. Emma, P. Krejcik, "Bunch Length Measurements Using A Transverse RF Deflecting Structure In The Slac Linac", EPAC-2002, Paris, France, June 3-7, 2002. p. 1882-1884.
8. M. Borland et al, "Start-To-End Jitter Simulations of the Linac Coherent Light Source", Proc. of the 2001 Part. Acc. Conf., Chicago, Il, June 2001, p. 2707-2709.
9. P. Emma, J. Frisch, P. Krejcik, LCLS-TN-00-12, Aug. 2000.
10. D. X. Wang, and G. A. Krafft, "Measuring Longitudinal Distribution And Bunch Length Of Femtosecond Bunches With RF Zero-Phasing Method", Proc. of the 1997 Part. Acc. Conf., Vancouver, May 1997, p. 2020.
11. W.S. Graves et al.," Ultrashort Electron Bunch Length Measurements at DUVFEL", Proc. of the 2001 Part. Acc. Conf., Chicago, Il., June 2001, p. 2224-2226.
12. J.D. Jackson, "Classical Electrodynamics", 2nd Ed., John Wiley & Sons, New York, 1975.
13. D. X. Wang, "Electron Beam Instrumentation Techniques Using Coherent Radiation", Proc. of the 1997 Part. Acc. Conf., Vancouver, May 1997, p. 1976-1980.
14. D. H. Auston, K. P. Cheung, and P. R. Smith, "Picosecond photoconducting Hertzian dipoles", Appl. Phys. Lett. **45** (3), p. 284-286 (1984).

Contributed Talks

10th Beam Instrumentation Workshop
May 6-9, 2002

Reconfigurable Instrumentation Technologies, Architectures and Trends

Rok Uršič

President and Founder
Instrumentation Technologies
Srebrničev trg 4a, SI-5250 Solkan, Slovenia
rok@I-tech.si, www.I-tech.si

Abstract. Reconfigurability is liberating radio-based instrumentation devices from chronic dependency on hard-wired characteristics of the radio front end. Today the evolution toward practical "reconfigurable" instrumentation is accelerating through a combination of approaches. This evolution is challenging analog, digital and software designers and the associated product development process. The goal of this paper is to give a short overview of the latest technologies, architectures and trends in this field.

INTRODUCTION

The focus of this paper is on instrumentation systems that process radio frequency (RF) signals. The family includes beam position monitors, beam current monitors, tune systems, low-level RF systems, transverse and longitudinal feedbacks and similar. This is not a comprehensive overview or a tutorial on reconfigurability. It is a vision that builds on my experience.

Motivation

The idea is extremely simple but ambitious: to build a system that can "grow" with an accelerator and support new requirements or applications without changing hardware. To realize this idea, we need reconfigurable systems.

The idea is not new and has been around in the telecommunication sector for a while under the guise of software defined radio. Through the 1970s and 1980s, radio systems migrated from analog to digital in almost every respect from system control to source and channel coding to hardware technology. In the early 1990s, the software radio revolution began to extend these horizons by liberating the radio-based services from chronic dependency on hard-wired characteristic of the radio [1]. These kind of systems are standardized in manufacturing, which translates into lower cost-per-unit, but customized in application, which translates into flexibility and future proof solution.

Space for Creativity and Innovation

System architecture is the creative ground for beam instrumentation designers. Building blocks are, on the other hand, sophisticated integrated circuits. The beam instrumentation designer has no influence on design, cost and performance trends in this field. Developments are governed by fierce competition in the economy of scale markets like telecommunications, radar, ultrasound and similar. Even though the instrumentation designer cannot influence developments in this field, it can certainly benefit from them.

TECHNOLOGIES

In 1965, Gordon Moore, one of the founders of Intel Corporation, predicted that the number of transistors that could be constructed on a unit area of silicon would double every 18 months [2]. Known as "Moore's Law", this rule-of-thumb has formed the basis for predictions on such diverse electronic phenomena as the capacity of memory devices, the capabilities of 3D graphics accelerators, and the performance of microprocessors and DSPs. A second aspect of Moore's law is his prediction that the doubling of transistors would be achieved for the same price. This means that if we continue to use a constant number of transistors, then the price-per-transistor will halve every 18 moths as new device technologies become available, which translates into lower product costs to the end user.

Furthermore, a lesser-known section of the famous Moore's paper deals with the linear electronics. He stated: "Integration would not change linear systems as radically as digital systems". In other words, the cost/performance gap between analog and digital world will widen with time motivating design engineers to implement more and more functions in digital domain.

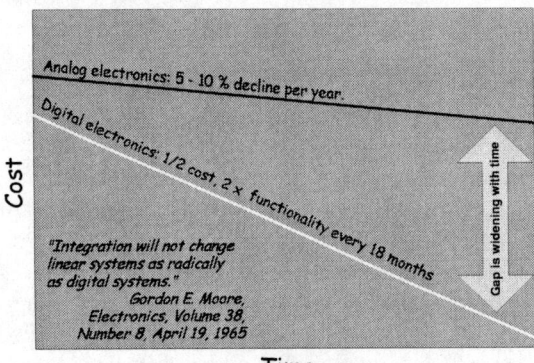

FIGURE 1. One of the consequences of Moore's Law is that cost/performance gap between analog and digital electronics is widening with time.

Analog to Digital Converter Integrated Circuits

The wideband ADC is one of the fundamental components of the reconfigurable instrumentation. It is the key building block that connects analog and digital domains.

The ADC is a hybrid, mixed signal device with analog and digital sections. The cost trend for ADC is not following either of the two curves shown in the graph in figure 1. Analysis of this technology is outside the scope of this paper. Walden [3] studied the relationship between ADC performance and technology parameters.

TABLE 1. State-of-the-art commercially available ADC (2002).

Model No.	Manufacturer	Number of bits	Sampling Frequency [MHz]	3 dB Bandwidth [MHz]
MAX1420	Maxim	12	60	400
CLC5958	National Semiconductor	14	65	210
AD6645	Analog Devices	14	105	250
AD9226	Analog Devices	12	65	750
AD9244	Analog Devices	14	65	750
AD9433	Analog Devices	12	125	750
ADS2807	Texas Instruments	12	50	270

Digital Processing Integrated Circuits

There are three established families of digital signal processing integrated circuits: application specific integrated circuits (ASIC), field programmable gate arrays (FPGA) and general-purpose digital signal processors (DSP). A new breed of devices/technology, re-configurable computers, is gaining momentum in the wireless market. Despite the fact that this technology will have to prove its performance and market sustainability, we included them in table 2 due to their lucrative potential.

The cost/performance trend for digital signal processing devices approximately follow the Moore's law for digital integrated electronics. Figure 2 compares these technologies from hardware adaptability, programmability and performance point of view. Before choosing a particular technology or combination of technologies for a specific design, however, system designer must also consider other characteristics like interconnect capacity, size-power tradeoffs, degree of required parallelism, etc. Discussion on those topics is outside the scope of this paper and can be found in [1].

TABLE 2. A coarse comparison of different signal processing technologies.

	ASIC	DSP	FPGA	Re-configurable computers
Hardware adaptability	-	-	+	+
Programmability	-	+	-	+
Performance	+	o	o	+

ARCHITECTURES

Conventional instrumentation designs implement algorithms in a hard-wired analog circuitry. They are optimized to perform specific functions. Reconfigurable designs, on the other hand, take advantage of system architectures that combine analog, digital and software domains. As one can see from figure 2, which shows the evolution of super-heterodyne receiver, more and more functions are implemented in digital. This requires new radio system architecture. It also requires a different approach to radio system design. Design and development team must be able to integrate broad range of engineering and management disciplines in order to design develop and lunch a successful reconfigurable instrumentation device.

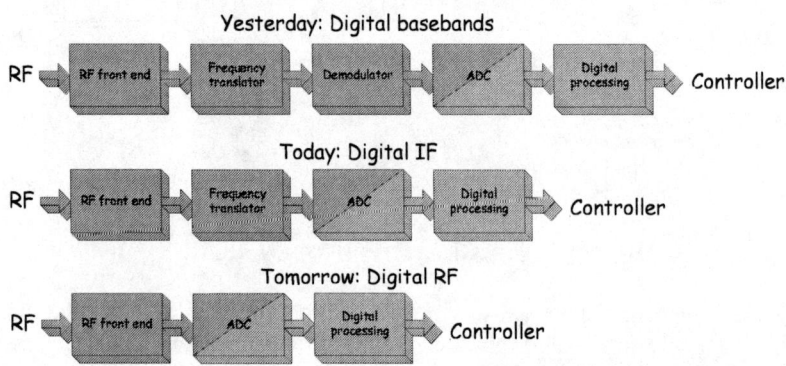

FIGURE 2. The evolution of superheterodyne receiver. The boundary between analog and digital is moving towards the "antenna".

The Interface Question

An important aspect of any architecture is how the system interfaces with external world. In case of beam instrumentation devices this is relatively straightforward as long as the interfaces are analog. This design practice, which was the norm for years in this field, has a significant advantage; it offers a clean interface between an instrumentation device and a control system. However, it has a significant disadvantage; it hinders development of more sophisticated re-configurable systems. Analog interface does not allow straightforward integration of digital signal processing hardware. If we accept the idea that instrumentation people are responsible for the performance of their systems, then the cleanest interface is on the responsibility boundary.

The Performance Question

As soon as we accept the idea that digital signal processing is an integral part of an instrumentation system, we need a new set of performance metrics. Conventional systems were specified with such parameters as accuracy, resolution, bandwidth, and similar. For reconfigurable systems we need to add parameters such as throughput, latency time, real-time capability, batch processing depth, processing power and similar.

Another important aspect regarding reconfigurable systems is that performance must always be associated with firmware revision. New features or even new applications can be downloaded to a system without changing hardware.

Evolution Towards a Clean Digital Interface

The following three figures show a possible interface evolution scenario for reconfigurable instrumentation devices. The aim is to bring the level of "cleanliness" of the analog interface to the digital domain and in this way facilitate proliferation and smooth integration of cutting edge digital signal processing technologies into instrumentation devices.

As noted before, analog interface was a norm for years. It has the disadvantage that it does not allow integration of digital signal processing into instrumentation devices.

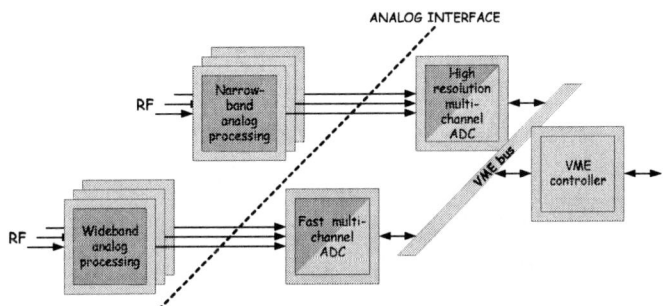

FIGURE 3. Yesterday: analog interface.

The next step toward the evolution of a clean digital interface is interface at the driver/backplane level. It has a significant advantage that it opens the possibility for integrating digital signal processing into instrumentation devices. It is, however, a challenge from the system integration point of view. The gray area of responsibility between instrumentation and controls is significant and requires good collaboration between the two groups.

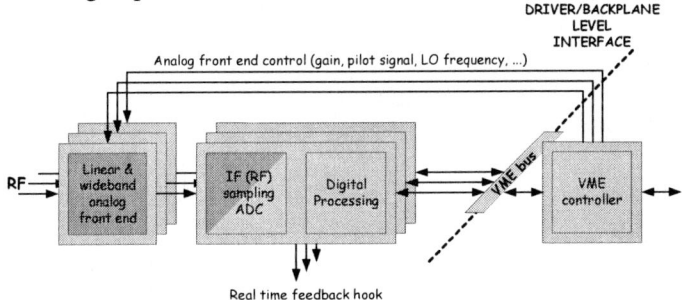

FIGURE 4. Today: driver/backplane interface.

A possible scenario, which brings a brand new perspective on the process of developing reconfigurable instrumentation devices, is shown in figure 5. The enabling technology is Ethernet. System integration in such configuration is simplified, because individual cards can have their own processor, operating system and memory and can communicate independently with other cards. Because nodes can be operating-system agnostic, integration is no longer required at the driver/backplane level but ascends to the network and transport layers, which means significant time savings and simpler design models.

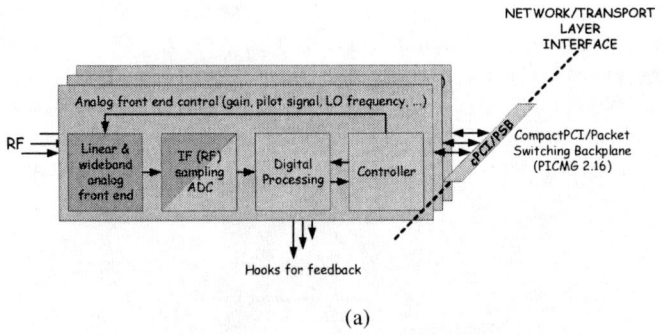

(a)

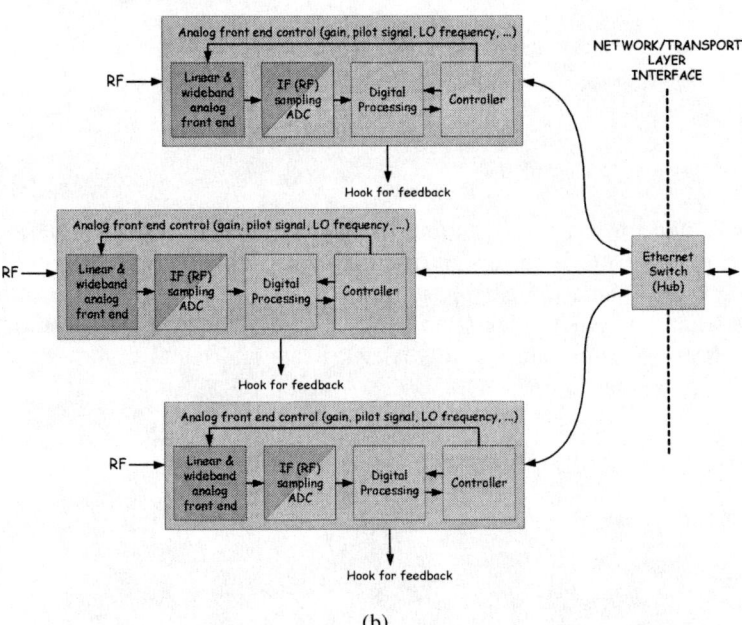

(b)

FIGURE 5. Tomorrow: (a) Ethernet within the chassis based on PICMG 2.16 standard for cPCI or (b) stand alone solutions provide a brand new perspective on the process of developing reconfigurable instrumentation systems.

TRENDS

Life Cycle Customization

Accelerators are complex machines. Their performance depends strongly on the quality of instrumentation support. New requirements are generated throughout the life cycle of any accelerator, which in turn requires better performance or even additional functionality from instrumentation systems. Historically this resulted in difficult, expensive and time consuming upgrades. Reconfigurable instrumentation offer future proof solution that facilitates simple and low cost software customization.

Fixed-functionality product

Life cycle	
Development cycle	Difficult and expensive hardware customization cycle
	Operational cycle

Reconfigurable product

Life cycle	
Development cycle	Simple and low cost software customization cycle
	Operational cycle

⟶ Time

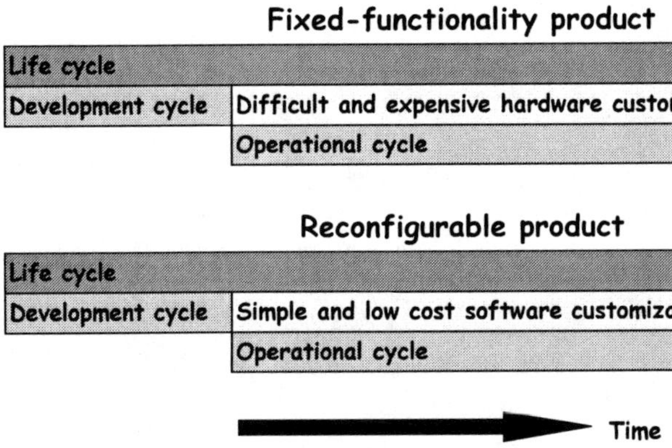

FIGURE 6. Reconfigurable products offer simple and low cost customization throughout the product life cycle.

Commercial Of The Shelf (COTS)

Specialized knowledge, skills and tools in different engineering areas are needed to develop, manufacture, supply and provide technical support for the state of the art reconfigurable products. The resource allocation is, in most of the cases, beyond the capabilities of a single instrumentation group at particle accelerator facilities. In addition, the development process for reconfigurable devices is complex and more expensive that in the case of developing simple analog modules. Standardization in manufacturing allows suppliers of COTS modules to achieve better quality and price per module by distributing development cost over a larger volume.

On the other hand, customization is better done by users. They know the issues, they know the machine and they can tailor the solutions to their specific needs. In order to do that efficiently they need good tools and support.

Quality Tools and Support

Two key prerequisites for a successful future of reconfigurable products are quality tool(s) and good technical support. The quality tools should allow modeling and simple, abstraction level customization of the product in laboratory. They should also facilitate verification of developed models on a real hardware. Good technical support helps new users to get acquainted with a system. It should also provide system life cycle support regarding repairs, spare parts, firmware upgrades, etc. The COTS model will be successful only if those two prerequisites are met.

CONCLUSIONS

Widening gap in the cost/performance trend between analog and digital integrated electronics provide the foundation for reconfigurable instrumentation systems. These systems offer in a single integrated solution benefits that were not achievable before with conventional hard-wired analog modules. New system architectures that will allow simple system integration, support life cycle customization by means of user-friendly tools will drive the reconfigurable beam instrumentation revolution.

REFERENCES

1. Joseph Mitola III, *Software Radio Architecture, Object-Oriented Approach to Wireless Systems Engineering*, New York: John Willey & Sons, Inc., 2000, pp. 1-31.
2. Gordon E. Moore, *"Cramming more components onto integrated circuits"*, Electronics, Volume 38, Number 8, April 19, 1965.
3. Walden, R., *"Analog to digital converter survey and analysis"*, JSAC, New York: IEEE Press, February 1999.
4. Ursic Rok and Raffaele De Monte, *"Digital Receivers Offer New Solutions for Beam Instrumentation"*, Proceedings of the 1999 Particle Accelerator Conference, New York, 1999, pp. 2253-2255.
5. John Peters., "PICMG 2.16 CompactPCI/Packet Switching Backplane Specification" TechFocus, June 2001, pp. 92-93.

Optical method for mapping the transverse phase space of a charged particle beam

R.B. Fiorito[1,2], A.G. Shkvarunets[2] and P.G. O'Shea[2]

[1]TR Research Inc., 3 Lauer Terrace, Silver Spring, MD 20901
[2]Institute for Research in Electronics and Applied Physics, University of Maryland, College Park, MD 20742

Abstract. We are developing an all optical method to map the transverse phase space map of a charged particle beam. Our technique employs OTR interferometry (OTRI) in combination with a scanning pinhole to make local orthogonal (x,y) divergence and trajectory angle measurements as function of position within the transverse profile of the beam. The localized data allows a reconstruction of the horizontal and vertical phase spaces of the beam. We have also demonstrated how single and multiple pinholes can in principle be used to make such measurements simultaneously.

INTRODUCTION

The "pepper pot" technique commonly used to map out the transverse phase space of a charged particle beam employs collimators to physically segment the beam into beamlets, which are then allowed to freely drift downstream onto an imaging device. The mean trajectory angle and divergence of each beamlet are obtained from a knowledge of the aperture size, drift distance, size and position of the beamlet image. From this information a map of the beam's transverse phase space can be constructed.

The effectiveness of the technique relies on the ability of the collimator to stop part of the beam. However, as the energy of the beam increases, thicker and/or higher atomic number materials must be used in the collimator, resulting in larger and more costly devices. Eventually permeability of the collimators to high energy particles renders the method ineffective. In addition, the technique is especially difficult to implement for small beam sizes produced by high brightness accelerators. Therefore, an alternative to the standard pepper pot technique is clearly needed.

We are developing an all optical approach to phase space mapping, which does not rely on the collimation of the beam itself, but rather on masking the optical radiation produced by the beam. The results are equivalent to those obtained with the conventional pepper pot device. We refer to the general concept of using beam based optical radiation for this purpose as *optical phase space mapping* (OPSM) or the *optical pepper pot*. The important advantages of this method are: 1) minimally or non invasive devices such as thin foils or magnetic fields can be used to generate the optical radiation; 2) optical radiation can be easily transported away from the beam environment; 3) well developed optical processing techniques can be utilized to extract beam parameters, 4) many types of optical radiation can be employed, e.g. synchrotron, edge, diffraction and transition radiation. In this paper we show how optical transition radiation (OTR) can be utilized for OPSM.

OTR has proven to be a versatile and effective diagnostic for measuring the spatial distribution, rms divergence, and rms emittances of relativistic electron and proton beams [1]. In particular, a two foil OTR interferometer [2] produces an interference pattern (OTRI), which is highly sensitive to the beam's energy, divergence and trajectory angle, and can be used to diagnose these quantities. Simultaneous imaging of the transverse profile when the beam is focussed to an x or y waist condition, combined with measurement of rms x or y divergences, allows a determination of the orthogonal (x,y) *rms emittances* of the beam [3]. Both time integrated and time resolved rms emittance measurements have been made.

The concept of applying OTR diagnostic techniques to map the beam transverse phase space, i.e. measuring the divergence and trajectory angle as a function of position within the beam profile for any beam envelope condition, was first proposed and patented by us in 1992 [4]. Recently this concept has been demonstrated experimentally in a proof of principle experiment, which utilized a 100 MeV linac located at the Lawrence Livermore National Laboratory [5]. In the LLNL experiment the divergence and trajectory angle of a *single* optically masked portion of the beam were measured and compared to those of the entire beam. These data allowed an estimation of the Courant-Snyder parameters and phase-space ellipse tilt angle and thus allowed a more accurate determination of the beam emittance than was previously possible with standard rms measurements taken at a beam waist.

The results of this experiment indicated that it should be possible to produce a complete map of the transverse phase space of charged particle beam using OTRI and optical masking. However, the experiment did not attempt to produce a phase space map. Furthermore, the large amount of scattering produced in the front foil of the LLNL OTR interferometer dominated the measured divergence and made the extraction of the beam divergence from the data difficult. The work described here produces an actual transverse phase space map and makes use of a improved interferometer design, which allows a more accurate measurement of the beam divergence.

THEORETICAL BACKGROUND

Transition radiation is produced by a charged particle as it passes between media with different dielectric constants, for example a metallic foil in vacuum. The angular distribution of this radiation reveals details about the particle's energy, position and direction with respect to the vacuum-foil boundary. The radiation is radially polarized - a fact which can be used to separate and determine the horizontal and vertical beam divergences and trajectory angles.

OTR interferences are generated from two parallel foils inclined at $\psi = 45^0$ and separated by a distance $L \gtrsim L_V(\theta,\lambda) \equiv (\lambda/\pi)(\gamma^{-2}+\theta^2)^{-1}$, the vacuum coherence length, which is defined as the distance over which the particle's field and the OTR photon differ in phase by one radian. Interference occurs between the forward OTR from the first foil, which is reflected by the second foil, and backward (reflected) OTR from the second foil. These sources are coherent because the forward directed photon is co-moving with the relativistic particle and remains in phase with it for a distance of the order L_V. The total spectral-angular intensity of two foil OTRI is given by

$$\frac{d^2I}{d\omega d\Omega} = 4 \frac{F(\theta,\psi)\alpha\gamma^2}{\pi^2 \omega} \frac{\zeta^2}{(1+\zeta^2)^2} \sin^2[\frac{L}{2L_V(\theta,\lambda)}], \qquad (1)$$

where $\omega = 2\pi f$, f is the frequency of the observed TR photon, θ is the angle of observation in a plane perpendicular to the velocity vector of the particle $\vec{v}$ for forward TR and in the direction of specular reflection for backward TR, $F(\theta,\psi)$ is the Fresnel reflection coefficient, which for a highly conductive foil is approximately equal to unity for all polarization components, $\zeta = \gamma\theta$ is the scaled observation angle and $\alpha \simeq \frac{1}{137}$ is the fine structure constant.

An ensemble of particles produces angular and spectra distributions which are modified by the superimposed contributions of individual particle trajectories. The effect of rms beam divergence can be calculated by convolving the spectral-angular intensity Eq. (1) with a distribution of particle trajectory angles, e.g. a Gaussian distribution. Then by fitting the measured angular distribution to the theoretical profile produced from the convolution, one can determine the rms beam divergence. This method can, in principle, be used at any beam energy and there is no absolute lower bound to the divergence that can be measured. This is due to the fact that the normalized rms emittance of a beam $\epsilon_{rms}^n = \gamma\theta_{rms}r_{rms}$ is an invariant quantity. As the beam energy increases, the rms divergence θ_{rms} decreases as γ^{-1}, which is also the angle of peak intensity of OTR. Therefore, the percent change in the OTR angular pattern due to the beam divergence is independent of the beam energy. However, a practical limit is imposed by the resolution and sensitivity of the optics and imaging device used to display the interference pattern. Using conventional electronic imaging cameras (SIT, CCD and CID's), we have been successful in measuring normalized rms divergences $\sigma_{rms}^n = \gamma\theta_{rms} \approx 0.05$.

In addition, the trajectory angle of the beam being observed can be measured by observing the centroid of the angular distribution of OTRI, which is centered about the direction of specular reflection ($\theta = 0$). If the observed beam is moving at some angle with respect to this direction, the centroid of the pattern will be shifted away from $\theta = 0$ and can be accurately determined [5].

The far field OTR interference pattern can be observed with a lens-camera system, by placing the camera at the focal plane of a lens whose focal length is chosen to match the field of view desired, which is usually several times γ^{-1}. The interferences must be observed through a narrow band filter in order to be distinct and usable for beam diagnostic measurements.

EXPERIMENTAL METHOD

OTR phase space mapping experiments were performed at the Naval Postgraduate School's S band rf linac. The beam energy was 95 MeV, the average current was 0.1-0.3 microamps and the macropulse repetition rate was 60 pps. Two pairs of magnetic quadrupoles were used to produce the desired beam focussing condition at the site of the interferometer mirror.

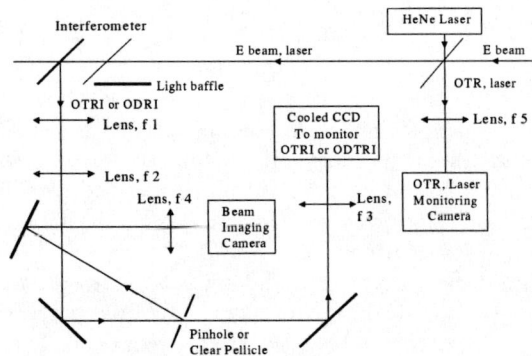

FIGURE 1. Experimental setup.

A schematic of the experimental setup is shown in Figure 1. The optical system was designed to collect an angular field of view $\Delta\theta \approx 8/\gamma$ ($\sim$40 mrad) for all points in the beam profile. A HeNe laser and a 50/50 aluminized Kapton beam splitter placed about 1 meter upstream from the interferometer were used to align the optics and the electron beam with the optical axis.

The OTR interferometer consisted of an aluminum front foil frame, which housed two circular aluminum foil mounting rings, and a silicon mirror coated with 1000 Angstroms of aluminum. A 0.7 micron thick aluminum foil was mounted on the bottom ring of the frame and a 5 micron thick, rectangular aperture copper micromesh (750 lines per inch, 33 micron period) was mounted on the top ring. The foil/mesh-mirror spacing was 25.4 mm. The solid foil thickness and composition were chosen to minimize the contribution of electron beam scattering. The 0.7 micron Al foil produces a calculated rms scattering angle of about 0.1 mrad at 95 MeV, which is much less than the rms beam divergence, $\theta_{rms} \simeq 1$ mrad, which had been previously measured by us using OTRI. Thus, the beam divergence dominates the total measured divergence, the reverse of the situation in the LLNL experiment.

Figure 2. shows a side view of the interferometer seen here through a 6 inch diameter fused silica window on the observation side of a six way cross. The foil

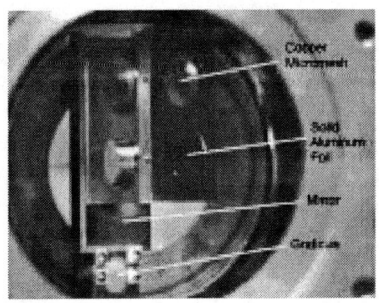

FIGURE 2. Side view of the OTR interferometer.

holder and mirror were parallel and oriented at 45^0 with respect to the beam direction. The circular mesh and foil holders are seen on the right side and in reflection from the mirror on the left side. Note the clear position of the mirror visible just below the reflection of the foil holder and the circular graticule directly below the mirror.

A linear actuator vertically positioned the assembly in one of four positions: 1) the graticule position, which allowed a measurement of the magnification of the optics, 2) the clear mirror position, which allowed the laser to be reflected and also as a target position for the production of OTR at the mirror position, 3) the aluminum foil position, used to generate OTRI and 4) the mesh position, which, when used with the laser, produced a far field diffraction pattern; this was used to calibrate the angular field of view of the cooled CCD camera.

The OTRI were optically transported by a pair of achromatic lenses, $f_{1,2}$, = 240 and 480 mm respectively, and focussed to obtain a two fold magnified image of the beam's spatial distribution at the site of the mask. The unmagnified rms beam radius r_b is about 2 mm. To preserve linearity in both position and angle the spacing between the lenses $f_{1,2}$ was set equal to the sum of their focal lengths.

The optical mask, a 1 inch diameter aluminized pellicle with a 1 mm circular pinhole at its center, was positioned at the image plane of $f_{1,2}$. The mask was mounted on a remote controlled x, y translator whose positioning accuracy was less than 10 μm. The OTRI pattern emerging from the pinhole was observed in the focal plane of lens $f_3 = 480mm$ by an Apogee Instruments, Peltier cooled, CCD array, which was programmed to integrate from 30 sec., in the case of whole beam OTRI, to 6 minutes for pinhole OTRI. To register the pinhole position with respect to the beam image, the rear of the mask is imaged by lens f_4 =90 mm and a standard CCD camera. To obtain the OTRI pattern from the entire beam, the mask was replaced by a clear pellicle beamsplitter whose reflectivity was about 10%. A 650 nm x 70 nm optical interference filter was used to obtain the OTR interferograms.

In a separate experiment, OTR interferences from two 1 mm diameter pinholes holes, placed 2 mm apart, were observed. A cylindrical lens, whose focal length = 250 mm, was placed at this same distance downstream from the mask. The cylindrical lens and pinholes were arranged so that light passing through the pinholes was dispersed only in the vertical direction, i.e. perpendicular to the horizontal axis of the lens and parallel to a line through the center of the two pinholes. Two other lenses, which replaced the lens f_3 shown in Figure 2, whose focal lengths were180 and 200 mm, respectively, were used to image the focal plane of the cylindrical lens onto the cooled CCD. With this system, two horizontally separated OTRI patterns could be simultaneously imaged and used to obtain divergences from each of the two pinhole positions.

RESULTS AND DISCUSSION

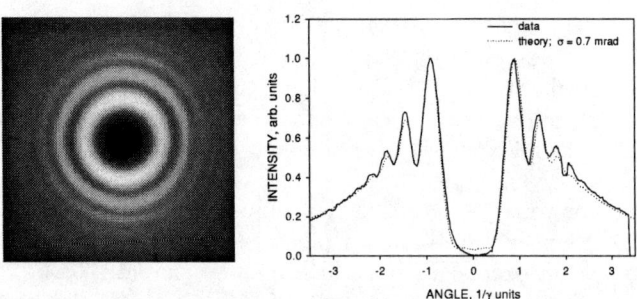

FIGURE 3. Left: OTRI at a y (vertical) beam waist; right: a vertical scan of the OTR interferogram shown on the left.

Figure 3. (left) shows an OTR interferogram taken at a y waist condition and (right) shows a vertical scan of the y waist OTRI, together with the theoretical fit. The fitted value of the rms y divergence of the beam under this condition, $\sigma_y = 0.7 \pm 0.05$ mrad. Similar analysis of an OTR interferogram of the whole beam taken at an x waist produced an rms x divergence $\sigma_x = 1.0 \pm 0.1$ mrad.

In order to perform phase space mapping, the beam was focussed to produce an arbitrary elliptical distribution at the mask position. Then a 1 mm diameter pinhole aperture was scanned across the beam image and the divergence and trajectory angle was measured at each position of the pinhole using OTRI. The number of data points here was limited by the signal to noise ratio, which was about two for the 1 mm pinhole images. The signal was limited by the amount of light which could be collected by the 1 mm pinhole aperture and the total OTR emitted by the $0.1\mu A$ beam; the background was primarily due to x rays, gamma rays and neutrons passing through the two inch thick lead and two inch thick polyethylene shielding surrounding the cooled CCD camera.

Figure 4. shows the beam image and pinhole superimposed for five different locations within the beam profile.

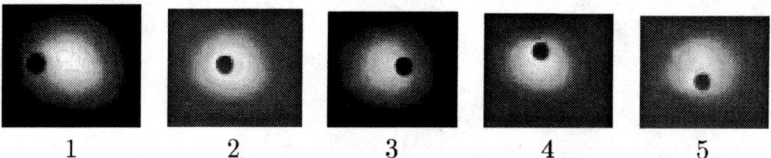

FIGURE 4. Five pinhole positions each registered with respect to the image of an arbitrarily focussed beam.

We obtained the divergence and trajectory shifts with respect to the second position (#2 above) for each of the five positions shown above. Figure 5. shows the data presented as two phase space maps: left, vertical phase space and right, horizontal phase space. The vertical width of the rectangles in the maps gives the local divergence and the centroid gives the local trajectory angle.

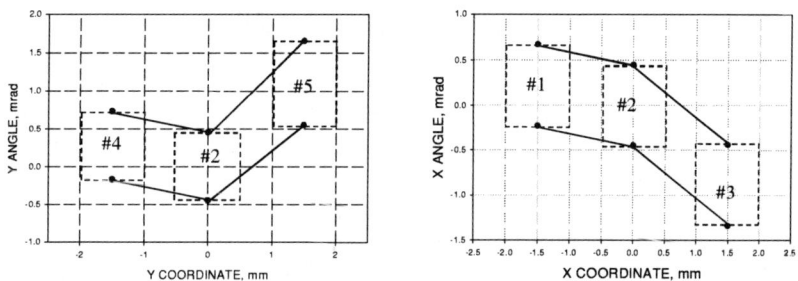

FIGURE 5. Vertical (left) and horizontal (right) phase space maps.

Unfortunately, the high background level and the instability of the beam over the long exposure times (6 minutes) used for the pinhole measurements severely limited the accuracy of the measurements. The local x and y values of the divergence presented in Figure 5. are nearly equal $\sigma_{x,y} = 0.9 \pm 0.05$ mrad. This value is larger than the measured y divergence of the entire beam $\sigma_y = 0.7$ mrad, and approximately equal, within the experimental uncertainty, to the measured value of the x divergence $\sigma_x = 1.0$ mrad. The measured pinhole divergences, therefore, should only be considered to be upper bounds. The angular trajectory data have smaller uncertainties and both show well defined trends; we interpret these data to represent the actual local beamlet trajectory angles.

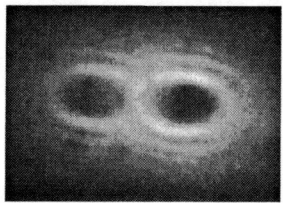

FIGURE 6. Double pinhole interferograms.

Figure 6. shows an image of the OTR interferences simultaneously observed from a double pinhole mask. The image was acquired by integrating over a time interval of 6 minutes. The inclination of the axis joining the center points of the two images is believed to be due to slight misalignment of the cylindrical optics. Line scans taken through each of the centers of each image along a line perpendicular to the line joining the centers of the images show that the local vertical divergences are approximately equal to within experimental uncertainty ($\sigma_{y1,2} = 0.7 \pm 0.05$ mrad), but that the trajectory angles from the two sampled portions of the beam are not the same. The trajectory angle shift of the beamlet observed by the right pinhole in Figure 6. with respect to that observed by the left pinhole is 1.8 mrad. Again, although close to the measured divergence of the entire beam, the vertical divergence values can only be considered to be upper limits.

CONCLUSIONS

We have shown that maps of the transverse phase space of a 95 MeV relativistic electron beam can be constructed using an all optical method. To demonstrate the technique we employed optical transition radiation interferences and an optical mask consisting of a single scanned pinhole or two fixed pinholes. From an analysis of OTRI patterns emerging from the single or double pinhole optical mask, referenced to position within the beam image, localized divergence and trajectory angles were made and used to construct horizontal and vertical phase space plots. Our results demonstrate an optical analogue to the standard pepper pot method which is commonly used to map the transverse phase space of charged particle beams. In future studies this method will be refined to produce more precise time integrated and possibly time resolved optical phase space maps.

ACKNOWLEDGMENT

This work was supported by US DOE STTR Grant No. DEFG02-01ER86 130.

REFERENCES

1. D.W. Rule, Nucl. Instrum. Meth. B24/25, 901 (1987).
2. L. Wartski, S. Roland, J. Lasalle, M. Bolore, and G. Filippi, J. App. Phys. 46, 3644 (1975).
3. R. B. Fiorito and D. W. Rule, "Optical transition radiation beam emittance diagnostics," in: *Beam Instrumentation Workshop*, ed. R.E. Shafer, AIP Conf. Proc. No. 319, (AIP Press,1994).
4. R. B. Fiorito and D. W. Rule, US Patent 5120968, "Emittance measuring device for charged particle beams", issued June 9, 1992
5. G. P. Le Sage, T.E. Cowan, R.B. Fiorito and D.W. Rule, Phys. Rev. ST Accel. and Beams, Vol 2, 12802 (1999).

Beam Diagnostics in the SNS Linac

M.A. Plum, L. Day, S. Ellis, R. Hardekopf, R. Meyer Sr., J. O'Hara,
J. Power, C. Rose, R. Shafer, M. Stettler, J. Stovall

Los Alamos National Laboratory, Los Alamos, NM 87544, USA

Abstract. Most of the design work for the Spallation Neutron Source (SNS) linac beam diagnostics instrumentation is complete, and we are now entering the construction phase. Some instrumentation has already been delivered and tested at the Lawrence Berkeley Laboratory front-end systems tests. In this paper we will discuss the SNS linac beam diagnostics instrumentation designed by Los Alamos National Laboratory, and some of the early performance results. We will briefly mention the general layout of diagnostics in the SNS linac, then focus on two systems of special interest: the beam position monitors in the drift tube linac, and the wire scanner actuators in the superconducting linac.

INTRODUCTION

In the Spallation Neutron Source (SNS) facility, H⁻ beams are accelerated to 2.5 MeV in an RFQ, to 87 MeV in a drift tube linac (DTL), to 186 MeV in a coupled cavity linac (CCL), and finally to 1000 MeV in a superconducting linac (SCL). The 60 Hz, 1-ms, 36 mA peak current beam pulses are chopped into 690-ns long segments with a 1 μs period to give an average current of 1.4 mA and an average beam power of 1.4 MW.

The intense beam and the superconducting environment place special requirements on the beam diagnostics instrumentation [1,2]. Intercepting diagnostics cannot survive the entire 1-ms long beam pulse, especially at the lower beam energies. Diagnostics located nearby the superconducting RF cavities must also be ultra clean and highly reliable.

In the SNS linac there are beam position monitors (BPM), wire scanners (WS), beam current monitors (BCM), beam loss monitors (BLM), and energy degrader / Faraday cups (ED/FC), with quantities shown in Table 1. Also shown in Table 1 is the division of effort between Lawrence Berkeley National Laboratory (LBNL), Brookhaven National Laboratory (BNL), and Los Alamos National Laboratory (LANL). Laser-based profile monitors [3], in-line slit and collector emittance stations, and bunch shape monitors are also being developed.

The beam current monitor system is based on fast current transformers (FCT) from Bergoz Instrumentation, and custom built electronics. The electronics package uses the same PCI motherboard as the BPM system (see discussion below), but with a different analog front end, designed and fabricated by Brookhaven National Laboratory.

TABLE 1. Quantities of beam diagnostics devices in the SNS linac.

Device	MEBT	DTL	CCL	SCL	Resp. Lab
BPM	6[1]	10	12	32	LANL
WS	5[2]	5	10	32	LANL
BCM[3]	2	6	2	1	BNL
BLM		6	24	58	BNL
ED/FC		5	1		LANL

[1]Pickups designed and fabricated by LBNL.
[2]Pickups designed and fabricated by BNL.
[3]DTL and CCL BCM transformers supplied by LANL.

The beam loss monitor system [4] is also designed and fabricated by BNL. It is based on argon-filled ion chambers and custom built electronics packaged in VME modules.

The Energy Degrader / Faraday cup system, designed and fabricated by LANL, will be used to set the phase and amplitude of the DTL tanks. As shown in Fig. 1, each ED/FC unit has a graphite or copper plate of unique thickness in front of a Faraday cup. The plate, or "energy degrader", stops all the beam particles below a pre-determined cutoff energy suitable for a particular DTL tank. This ensures that a good phase and amplitude scan can be made without interference from beam particles that are not properly accelerated. To prevent contamination of the Faraday cup signal from secondary electrons a bias ring is located between the energy degrader and the Faraday cup. About –200 V will be sufficient to repel them. Graphite inserts are used in the Faraday cups to increase the maximum allowable beam power.

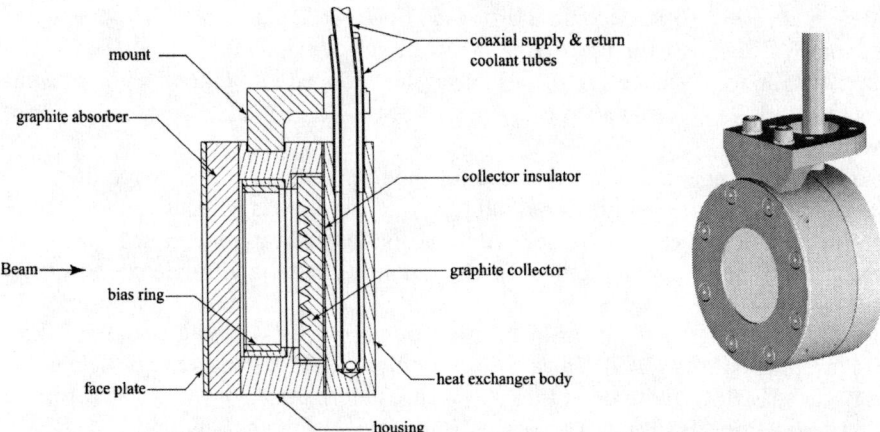

FIGURE 1. Images of an ED/FC unit. Left: cross section line drawing. Right: isometric image.

All the BPMs in the linac are of the dual-plane shorted-stripline design, with apertures varying between 25 and 70 mm diameter. The BPMs in the MEBT have 22° lobe angles to allow them to fit inside the quadrupole magnets. The CCL and SCL lobe angles span 45°, with the former located inside quadrupole magnets and the latter located toward the middle of each warm inter-segment region between the cryomodules. The DTL BPM pickups will be discussed in detail in the next section.

All the BPMs use the same electronics, which will also be discussed in detail in the next section.

The wire scanner actuators in the MEBT, DTL, CCL, and D-plate are based on linear actuators designed by LANL and purchased from Huntington Mechanical Laboratories, Inc. Each actuator fork has three 32-micron diameter carbon wires that can be biased to about ±100 V. The wires are offset by at least enough to ensure that there is only one wire at a time within ±2 rms of the beam center. The SCL wire scanner actuators will be discussed in detail in a later section.

In addition to all the above production diagnostics, a temporary Diagnostics Plate [1] is also being fabricated to commission the linac up to the end of DTL tank 1. The D-Plate diagnostics comprise three BPMs, one wire scanner, one BCM, one ED/FC, one view screen, an eight-segment halo scraper, a slit and collector emittance station, and a full power beam stop. Once the D-plate has fulfilled its function, which is expected to last about 90 days, it will be removed. Then DTL tank 2 will be installed in its place, and DTL commissioning will continue using only the production diagnostics permanently mounted in the DTL.

A diagnostic system of special interest is the DTL BPM system. The SNS is the first linac to incorporate BPMs inside DTL drift tubes. The SCL WS system is also of special interest because it employs actuators with formed bellows and a fork mounted on the end of a pivoting arm. These two systems will be discussed in detail in the following sections.

THE DTL BPM SYSTEM

The SNS DTL comprises six 402.5-MHz DTL tanks with FFODDO lattices. With this lattice there are many drift tubes without quadrupole magnets inside, thus permitting other uses for the empty drift tubes, such as steering magnets and BPMs. The inside bore of a drift tube is normally a smooth tube, with a length that gradually increases with beam energy. The DTL tank 1 drift tubes are too short to place BPMs inside them, but once in DTL tank 2, they are sufficiently long. There are two BPMs each in DTL tanks 2 through 6. Each BPM has equal-length 32 mm-long electrodes that subtend 60-deg. angles and with apertures equal to the normal 25-mm drift tube bore aperture. The electrodes and the housing are made of copper to match the normal drift tube material.

Of course the DTL tank is flooded with high power rf, and there is a concern that the rf will swamp any beam position signals. Simulations [5] with MAFIA [6] show that the rf interference is minimized to acceptable levels by placing the feedthrough ends of the shorted stripline electrodes in the centers of the drift tubes, as shown in Fig. 2. We further reduce the rf interference by processing the BPM signals at 805 MHz (the DTL tank frequency is 402.5 MHz). Right-angle Kaman SMA feedthroughs and Kapton-insulated coaxial cables are used to carry the signals up to the tops of the drift tube stems. To ensure good performance at high radiation levels, we replaced the PTFE (Teflon™) dielectric inside the SMA cable connectors at the drift tube end with polyimide (Vespel™), and we replaced the dielectric inside the SMA cable connectors at the top of the drift tube stem with polyethylene.

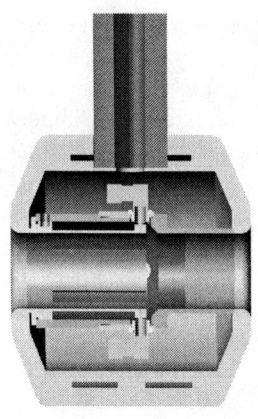 

FIGURE 2. Left: cross section of a DTL BPM inside a drift tube. Right: photograph of a DTL BPM.

The signals are then transported to the equipment aisle on ¼-inch Heliax cables to a signal processor, which has three basic components – the analog front end (AFE), the digital front end (DFE), and the PCI motherboard. The motherboard, shown in Fig. 3, fits into a standard PCI slot in a rack mounted personal computer. The AFE downconverts the BPM signals to 50 MHz, then the DFE samples the resultant signals at 40 MHz to generate I/Q pairs. The digitized data is then uploaded into the PC memory and analyzed with a LabVIEW program to generate beam position and beam phase information. This data is then transferred to the EPICS operating system over an Ethernet link. In other locations in the SNS linac, such as the CCL and SCL, the processing occurs at 402.5 MHz to avoid interference from the 805 MHz rf used in these portions of the linac. Some specifications of the electronics are shown in Table 2. An example of some bench-test data for the electronics is shown in Fig. 4.

TABLE 2. Some specifications of the SNS BPM system.

Position accuracy	±1% of aperture radius
Position resolution	±0.1% of aperture radius
Phase accuracy	±2° of processing frequency
Phase resolution	±0.2° of processing frequency
Dynamic range	65 dB
Bandwidth	5 MHz

The first real-life use of the electronics occurred in April 2002 during the Front End Systems tests at Lawrence Berkeley National Laboratory (LBNL), where the SNS ion source, RFQ, and MEBT were designed and built. The first DTL BPM pickups will not see any beam until late 2003. However, tests with full power rf in DTL tank 3 are planned for November 2002 to test the rf interference.

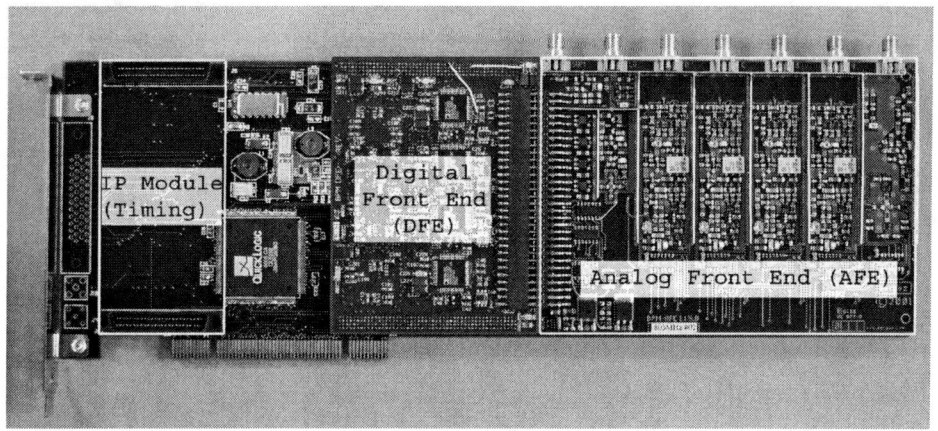

FIGURE 3. Photograph of a BPM PCI card, with AFE and DFE daughter cards.

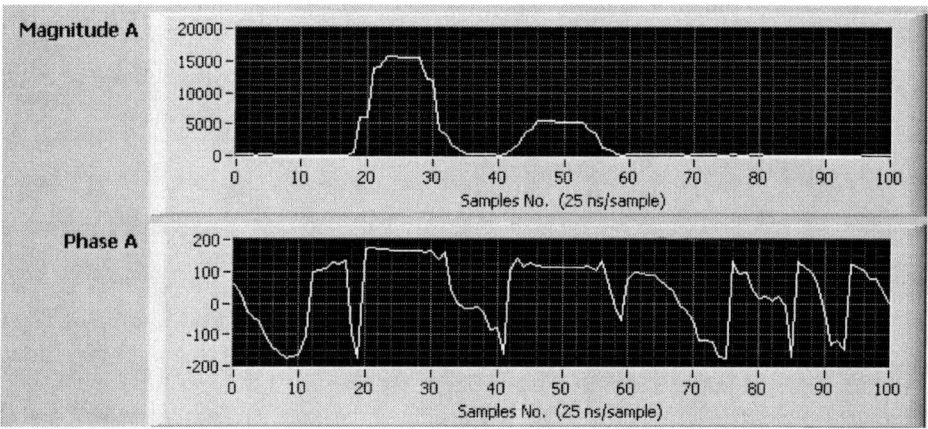

FIGURE 4. Example of some bench test data for the BPM electronics. Top trace: the self calibration pulse is injected first into the electronics, then into the cable. The amplitude difference of the two peaks is a measure of the cable loss and the time difference is a measure of the cable length. Bottom trace: phase measurement (data are valid only when the top trace is non zero).

THE SCL WS SYSTEM

The superconducting portion of the SNS linac has 23 cryomodules each containing three or four 6-cell rf cavities (there is room in the beam tunnel for 9 additional cryomodules for future upgrades). Between each cryomodule is a 1.6-m long warm inter-segment region that contains two quadrupole magnets, a vacuum pump, a BPM, and a wire scanner mounted on a vacuum box. Due to the close proximity to the superconducting rf cavities, any component in the inter-segment region must be compatible with the 10^{-9} Torr vacuum system, must have a very low particulate count, and must be very reliable. Vacuum breaches and particulate migrations into the rf cavities could potentially cause serious damage to them. Maintenance of any beam diagnostics is also very difficult, since any operation that requires opening the vacuum system may involve removing the entire inter-segment beam line to a clean room. All components must also be suitable for bake out at 250 °C for four hours. Other wire scanner system requirements are shown in Table 3.

TABLE 3. Some specifications of the SNS WS system.

Width measurement accuracy	±10% when fitted with a shape
Min beam pulse width (MEBT, DTL, CCL)	50 µs
Min beam pulse width (SCL)	100 µs

To satisfy these requirements we chose a design [7] that employs a formed bellows. The usual welded bellows were deemed to be inherently difficult to clean and more likely to form vacuum leaks. The drawback to using formed bellows is their limited stroke, leading to bellows that would be impractically long in the usual linear actuator design. For this reason we chose a pivoting actuator, so that the wire scanner fork sweeps through a gentle 20-deg. arc as it passes through the beam, as shown in Fig. 5. The arc is small enough that it does not seriously degrade our ability to individually measure horizontal, vertical, and diagonal beam profiles. The bellows design was extensively modeled with the COSMOS computer program to optimize the material, the wall thickness, the convolution dimensions, and the length. Our model results were also verified by the bellows vendor. The final choice was a bellows 21-cm long, 0.1-mm thick, and made of 304L stainless steel.

As mentioned above, the H⁻ beam is too intense to allow use of the wire scanners during full power operation. The duty factor must be cut back enough to prevent damage to the signal wires. We evaluated [8] 32-micron diameter carbon (the largest diameter commercially available), 20-micron diameter tungsten, and 125-micron diameter tungsten wires, and found that only the carbon wires will survive 100 µs beam pulses (the minimum requirement), and even then only for 10 Hz operation. Peak temperatures for 1 and 10 Hz operation are shown in Table 4 for critical points along the linac (a critical point is defined to be where the beam size and energy deposition per particle conspire to create the worst case wire heating for a particular region of the linac). These temperatures were computed with a computer model that includes the effects of both proton and electron energy deposition and a temperature-dependent heat capacity. Only radiative cooling is assumed. When carbon wires are used and the signal is taken off the wire (as opposed to measuring prompt radiation from wire-induced beam loss), the maximum temperature [9] is limited to 1,600 K due to thermionic emission currents.

We chose to measure beam profiles by measuring the current on the signal wire, so the wires are biased to about 100 V (either positive or negative) to control the secondary electrons. In most cases the wire will be negatively biased to drive off the secondary electrons, but in some cases it may be more favorable to positively bias the wire. An example of such a case is where the signal caused by secondary electron emission is almost exactly cancelled by H⁻ particles stopping in the wire, which, for 32-micron diameter carbon wire, occurs at a beam energy around 2.2 MeV. The signal processor has been designed to accommodate either positive or negative biasing, and bi-polar wire signals.

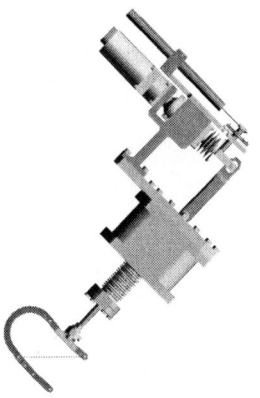
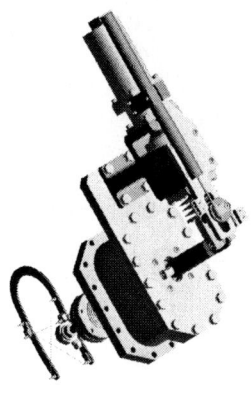

FIGURE 5. Left: side view of the SCL wire scanner actuator. Right: isometric view.

Three 32-micron diameter carbon wires are mounted on the wire scanner fork, offset from one another so that only one wire at a time will be within ±2 rms of the beam center. The third wire will be used to measure x-y correlations. The wires are mounted with a collet design developed for the LEDA experiment at LANL. They allow for a simple mounting system that keeps the wires under tension.

TABLE 4. Some wire scanner signal wire temperatures at various critical points along the linac.

Location	Energy (MeV)	horiz. beam size (rms, cm)	vert. beam size (rms, cm)	Temp (°C)	Temp (°C)
MEBT[0]	2.5	0.09	0.16	1370[1]	
DTL 1	7.5	0.092	0.17	730[1]	1180[2]
DTL 2	22.9	0.086	0.15	530[1]	950[2]
DTL 3	39.8	0.10	0.11	530[1]	940[2]
DTL 4	56.5	0.073	0.13	580[1]	1010[2]
DTL 5	72.5	0.11	0.11	530[1]	940[2]
DTL 6	86.8	0.14	0.071	620[1]	1055[2]
CCL[3]	88 – 186	0.11	0.11	592[1]	1019[2]
SCL[4]	186 – 1000	0.16	0.16	320[5]	665[6]

[0]WS #5. [1]1 Hz, 50 us, 26 mA beam. [2]10 Hz, 50 us, 26 mA beam. [3]106 MeV. [4]200 MeV. [5]1 Hz, 100 us, 26 mA. [6]10 Hz, 100 us, 26 mA.

The accuracy of a beam size measurement depends in part on the positioning accuracy of the signal wire. We tested the actuator over a limited range of 4 mm (the range of the theodolite alignment system at our disposal), moving the actuator in both directions to include the backlash effects. The data showed less than ±0.13 mm of position error, which, together with the electronics error[*] is adequate to meet the requirement to measure the rms beam size with an accuracy of ±10%. We also cycled the actuator about 10,000 times to simulate 30 years of normal operation, and we did not observe any problems.

---

[*] The error on each signal level measurement due to the electronics is less than 1% of the value of that measurement plus 2% of the signal level measured at the center of the beam.

As in the case of the BPM system, the wire scanner electronics is based on a custom-designed signal processor sampled by a PC running LabVIEW. The PC transmits the beam profiles to the EPICS control system over an Ethernet connection. The same PC also controls the motion of the actuator.

SUMMARY

The beam diagnostic challenges presented by the SNS linac primarily involve the high beam power and the constraints imposed by the superconducting rf cavities. Due to the limited space available in these workshop proceedings, we chose to focus on just two diagnostics systems of particular interest – the DTL BPM system and the SCL WS system. Both system designs are now complete. The DTL BPM fabrication is just finishing up, and the SCL WS actuator fabrication is just getting underway. Prototype electronics for both of these systems were tested at the LBL FES tests earlier this year and found to perform well. First beam through the DTL BPMs is expected around summer 2003, and first beam through the SCL WSs is expected around fall 2004.

REFERENCES

1. Shea, T.J. et. al., "SNS Diagnostics" in *Beam Instrumentation Workshop 2000*, edited by K. Jacobs and C. Sibley, AIP Conference Proceedings 546, New York, 2000, pp. 132-146. Also Hardekopf, R.A. et. al., "Beam Diagnostic Suite for the SNS Linac," ibid., pp. 410-418. Also Gassner, D. et. al., "Spallation Neutron Source Beam Los Monitor System," ibid., pp. 392-400. Also Kesselman. M. et. al., "SNS Project-Wide Beam Current Monitors," ibid., pp. 464-472.
2. J.F. Power et. al., "Beam Position Monitors for the SNS Linac," in *Proceedings of the 2001 Particle Accelerator Conference*, Chicago, IL, 2001, IEEE cat. no. 01CH37268. Also Liaw, C.J. and Cameron, P. "Carbon Wire Heating due to Scattering in the SNS," ibid. Also Plum, M.A. et. al, "Diagnostics Plate for SNS Linac Commissioning," ibid. Also Shea, T.J. et. al., "SNS Accelerator Diagnostics: Progress and Challenges," ibid. Also Hardekopf, R.A. et. al., "Wire-Scanner Design for the SNS Superconducting-RF Linac," ibid.
3. Connolly, R. et. al., "Laser Beam-Profile Monitor Development at BNL for SNS", these proceedings.
4. Gassner, D. et. al., "Preliminary Design of the Beam Loss Monitoring System for the SNS," these proceedings.
5. Kurrenoy, S., "Beam position monitors for DTL in SNS linac," SNS memo SNS-104050000-TD0003 - R00.
6. MAFIA Release 4.24, CST GmbH, Darmstadt, 2000.
7. Hardekopf, R.A. et. al., "Wire-Scanner Design for the SNS Superconducting-RF Linac," in *Proceedings of the 2001 Particle Accelerator Conference*, Chicago, IL, 2001, IEEE cat. no. 01CH37268.
8. Plum, M.A., "SCL wire scanner signal wire choice," Jan. 16, 2002, SNS Tech Note SNS-104050200-TR0010 - R00.
9. Gilpatrick, J.D., "LEDA Beam Diagnostics Instrumentation: Measurement Comparisons and Operational Experience," in *Beam Instrumentation Workshop 2000*, edited by K. Jacobs and C. Sibley, AIP Conference Proceedings 546, New York, 2000, pp. 401-409.

Streak Camera Characterization Using a Femtosecond Ti:Sapphire Laser.

Mario Ferianis and Miltcho Danailov

Sincrotrone Trieste, S.S. 14 km. 163.5, 34012 Trieste, ITALY

Abstract. At ELETTRA a Streak Camera system is in operation since 1999. Several measurements have been performed so far on the diagnostic bending magnet beam line of the Storage Ring. This instrument has been also widely used during the Storage Ring FEL commissioning to fully characterize both the electron beam and the FEL radiation down to 190nm and pulse length of $7.7ps_{FWHM}$. As the FEL pulse duration is approaching few picoseconds, it becomes important to study the performance of the streak camera in this regime which is very close to its temporal resolution limits. We therefore use the newly available femtosecond Ti:Sapphire laser delivering sub-50fs pulses to fully characterize the Streak Camera in short pulse operation. Using both infrared laser light and its second harmonic in the blue we study the effects of incident light wavelength and bandwidth on the resolution, as well as the linearity of the sweeps (linear single sweep and double sweep with synchroscan) on full-scale extension. The results are presented in this paper, together with preliminary measurements of the laser locking to the external Radio Frequency.

INTRODUCTION

Streak camera (SC) is a widely used diagnostic tool in the accelerator community [1]. At ELETTRA a SC has been installed in 1999 [2], mainly intended as a diagnostic tool for the storage ring (SR) longitudinal dynamics. The SC has been a fundamental diagnostic tool for the commissioning of the European UV/VUV Storage Ring FEL project at ELETTRA [3] and it is currently used for the further development of this project. In some conditions, the length of the FEL pulses is approaching the temporal limit of the camera; therefore it was important to find the operating conditions giving the best temporal resolution.

Recently, the laser laboratory at ELETTRA has been equipped with a Ti:S femtosecond laser and autocorrelator for laser pulse measurement, one of its main applications being the development of a photo-cathode electron gun [4]. The successful operation of such an electron gun will require highly stable synchronization to the Radio Frequency signal driving the accelerating structures. A set of measurements with the streak camera was devised with the purpose of characterizing both the streak camera performances as well as to study the properties and parameters of the frequency-locking unit. We present hereafter a set of measurements demonstrating a sub-2ps resolution of the camera and sub-ps jitter in the timing of the laser.

STREAK CAMERA MEASUREMENTS AT ELETTRA

The streak camera at ELETTRA is installed on a bending magnet light port, which is specifically used for diagnostic purposes. This instrument has been designed and manufactured by Photonetics (presently Optronis [5]) according to ELETTRA specifications (e. g. Synchroscan Frequency of 250MHz).

Description of the Streak Camera

The SC is equipped with two orthogonal deflection axis and with UV-graded input optics. The S20 photo-cathode frequency response goes down to 200nm. Thanks to a synchroscan frequency of 250MHz, consecutive bunches of a multibunch beam can be acquired in single shot. The fastest deflection speeds are 25ps/mm in synchroscan mode and 10ps/mm in single sweep mode. The ligth intensity at the input is reduced by a set of neutral density (N.D.) attenuators. A fast opto-electronic shutter [2], based on a Pockels Cell, reduces the background light due to the undeflected photons (no photo-cathode gating is used), which appear with the fastest secondary sweeps.

Measurements on the Storage Ring: Synchrotron and FEL Radiation

The typical bunch length vs. bunch current and bunch length vs. RF cavity voltage dependancies are periodically measured. Longitudinal multi bunch instabilities corresponding to different machine operating conditions have been effectively characterized using the SC. Recently, a new measurement activity has started with the synchrotron light pulses directly focused onto the photo-cathode (2x8mm) to acquire also the transverse bunch motion (horizontal or vertical).

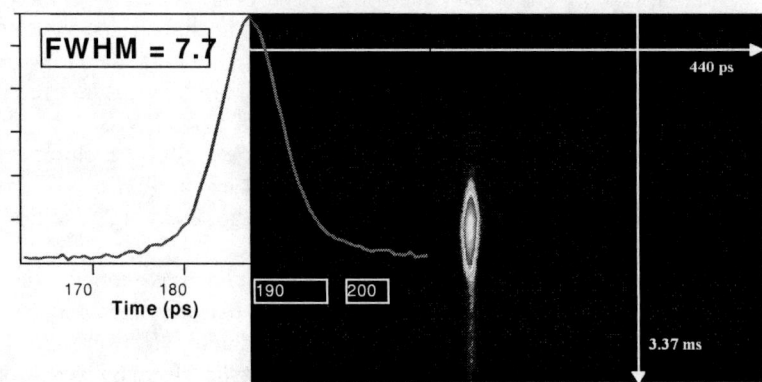

FIGURE 1. Example of synchroscan single shot acquisition: FEL radiation at 196nm, with a FWHM=7.7ps, vertical axis: 3.37ms full screen; horizontal axis: 441ps full screen.

Being the SR-FEL located close to the diagnostic bending magnet, the FEL radiation (fig. 1) has been easily driven to the SC input for direct measurement [6].

FEMTOSECOND MEASUREMENT SET-UP

The streak camera has been located on an optical table adjacent to the one of the laser (see figure 2, left). Both tables are not isolated from ground vibrations.

The Femtosecond Laser

The source of ultrashort pulses is a Kerr-lens Mode Locked (KLM) Ti:S laser pumped by a 5W continuous wave intracavity frequency-doubled Nd:YVO$_4$ laser (Millenia, Spectra Physics). The 4 mm long Ti:S crystal is placed in a standard X-fold cavity allowing for soft aperture KLM (details on the cavity design can be found e.g. in [7]). It was normally used at 4.3W of pump power giving about 400mW average power in mode-locking regime. The cavity length was adjusted so as to provide a repetition rate of 100MHz. Dispersion control was achieved with two intracavity fused silica prisms, allowing for pulse duration in the 40fs range and bandwidth around 40nm FWHM, centered at 810nm. A typical intensity autocorrelation trace (obtained by a FR-103 autocorrelator, Femtochrome Research, Inc) is shown on fig. 2 RIGHT (scale 2μs/div).

FIGURE 2. LEFT: Picture of the measurement laboratory: in foreground, the streak camera is visible with some of its optical components; the femto-second laser (black box) with its peripheral units is located on the second table, in background. **RIGHT:** Plot of the autocorrelation function corresponding to a 42fs$_{FWHM}$ pulse.

Given the calibration factor of the autocorrelator (31fs/μs) and assuming a sech^2 pulse shape one obtains 42fs for the FWHM duration averaged over few hundred pulses. A small fraction of the beam was reflected by a glass plate and directed to the Pockels Cell system by two metallic mirrors. The second-harmonic was generated by a 1mm thick lithium triborate (LBO) crystal. Due to the narrow spectral acceptance and group-velocity mismatch of this crystal the SH pulses are estimated to be in the 150fs range, which can still be considered a δ-function time signal input for the streak camera.

Even if the mode-locking mechanism of the laser is a passive one, it allows the synchronization of the laser to an external RF signal by the use of a phase-locking loop technique [8]. This technique is based on active control of the cavity length driven by an error signal provided by a phase detector: at this stage we have used a Timing Stabilizer (CLX-1100, Time-Bandwidth Products). It implements the frequency locking by acting on the back mirror position using fast piezo for fine movement around a zero position adjusted by a picomotor drive. During normal operation, the phase-noise measured at the output of the phase detector of the CLX-1100 was in the range 0.6÷1ps.

Timing and Synchronization Set-up

The main issues of the timing and synchronization system are:
- to provide the trigger and synchroscan signal to the streak camera
- to provide a stable reference signal to the laser timing synchronization system
- to assure the synchronization and the time stability of these signals at the picosecond level

The allowed frequencies for the Synchroscan deflection unit range from 249.2MHz to 251.0MHz; the frequency of the reference signal for the laser timing synchronization system is 100MHz. Therefore, we decided to adopt a Master Oscillator (Rhode&Schwartz SML02) frequency of 500.0MHz and to divide it by 5 to obtain the 100MHz reference for the laser. The same Master Oscillator drives the Synchroscan deflection Unit, after a "divide by 2" and a band-pass filtering for increased spectral purity, as well as the long division chain (÷(8640xM), M programmable) generating the gate signals ($\cong$10Hz), used to form the trigger pulse and for the Pockels Cell HV driver. The complex architecture of this timing system [2] stems from the necessity of providing very low frequency, low jitter trigger signal to the SC adopting, where possible, timing units already developed [9] at ELETTRA. Only the Auxiliary Board has been specifically designed for the SC timing system.

DESCRIPTION OF THE MEASUREMENTS

The purpose of the measurement is the complete characterization of the streak camera taking advantage of the femtosecond laser, which requires the synchronization of the femtosecond laser to the Master Generator signal.

General Considerations

The Optronis streak camera is specified to have a resolution of $2ps_{FWHM}$, in Single Sweep mode. The femtosecond Ti:Sapphire laser delivers sub-50fs pulses at a wavelength of 810nm, with a rate of 100MHz. It is easy to estimate that even after propagation through the optical components (Pockels cell, camera optics, etc) the laser pulse width remains below 300fs for both the fundamental and second harmonic wavelengths. Therefore, the following statements hold:

i) the laser pulse appears to the streak camera as a δ function. As a consequence, the laser pulse width cannot be measured, though its time stability can be observed and measured

ii) the operation of the streak camera with a δ function will reveal its true time resolution and how it is affected by different parameters
iii) the time stability (jitter) of the streak camera sweeping can be measured provided the laser pulse stability has been previously measured in the same configuration by other means

The effects of the proper setting of the input optics according to the incident light wavelengths have been observed, operating the SC in Focus Mode (deflection stopped). The two different modes of operation of the streak camera have been tested: synchroscan and single sweep. Special emphasis has been given to the latter as it is the most accurate one (most suited to the femto-second laser) and the most demanding for the triggering. The measurements are summarized in Table 1.

Table 1: summary of the streak camera measurements with the femto-second laser

Measured quantity	Streak Camera Mode / Single shot	Varied parameter	Notes
1) Resolution	Synchroscan	Pinhole diameter	D=200, 100 and 50μm
2) Laser locking process	Synchroscan sequence	Laser control loop open/close	
3) Resolution	Single sweep	Pinhole diameter	D=200, 50 and 30μm
4) Resolution	Single sweep	Light intensity	N. D. =0, 0.5 and 0.99
4) Resolution	Single sweep	Input wavelength	λ=810, Δλ=40nm λ=405nm, Δλ=10nm
5) Jitter of Sweep ramp	Single sweep Accumulation=10	Start of fast ramp	
6) Accuracy and Linearity	Single sweep	Delayed Pulses	Optical Delay ΔL= 0, 1.5 and 3.0mm

Data analysis

Depending on the operation mode, the output of a fast streak camera measurement of a periodic short optical pulse is an image with one or several spots, for single sweep (see fig. 3 right) and synchroscan (see fig.3 left) respectively.

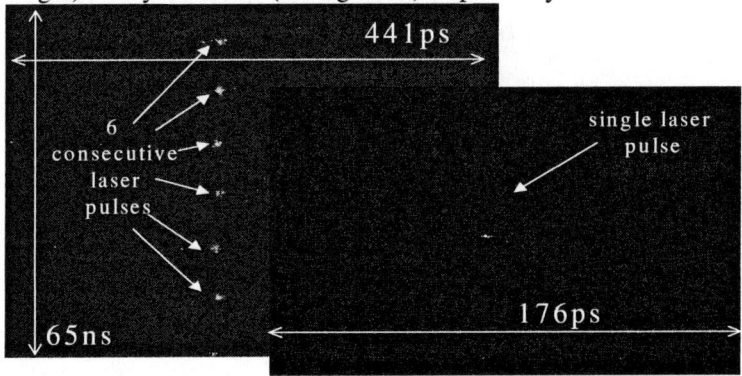

FIGURE 3. Streak Camera acquisitions of the femto-second laser. **LEFT**: synchroscan single shot of 6 consecutive laser pulses; vertical axis: 65ns full screen; horizontal axis: 441ps full screen. **RIGHT**: single sweep shot of one laser pulse; horizontal axis: 176ps full screen.

Therefore, for each synchroscan acquisition, the average FWHM of the laser spots has been computed. Then, the average FWHM values of different synchroscan acquisitions, relative to the same measurement set-up, have been averaged again. For single sweep acquisitions the FWHM values from different acquisitions under the same conditions have been averaged and the RMS computed. In fig. 4 the dispersion for seven single-sweep single-shot acquisitions is shown: the average FWHM is equal to 2.94ps, while the RMS of this distribution is 0.47ps.

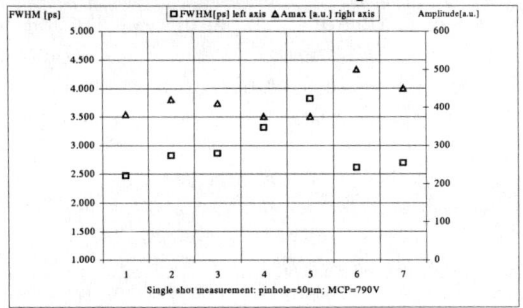

FIGURE 4. Plot of data dispersion for Single-Sweep single-shot acquisitions of the fs laser.

In fig. 5, a typical SC acquisition of laser pulse and its gaussian fit ($\sigma=1.3$ps).

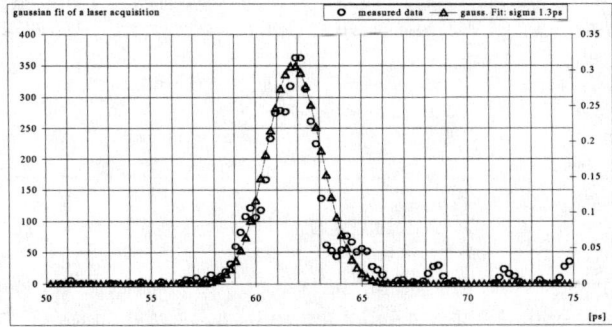

FIGURE 5. Streak Camera acquisition of laser pulse (circles) and its gaussian fit (triangles): $\sigma=1.3$ps.

Results and discussion

The obtained results are in good agreement with the specifications provided by the manufacturer: a comparison is reported in Table 2. In single sweep mode, as the photoelectrons inside the streak tube experience only a single deflection, there is less distorsion and the minimum resolution is achieved (resolution<2ps).

Table 2: summary of the measurement results and comparison with specification data

Measured quantity	Streak Camera Mode	Value	Specification data
Resolution	Synchroscan	3.5ps FWHM	<3.5ps FWHM
Resolution	Single sweep	1.4ps FWHM	<2ps FWHM
Linearity	Single sweep	±1.25%	±5%
Jitter	Single sweep	<5ps RMS	<5ps

The beneficial reduction of pinhole diameter has been observed in both operating modes (fig. 6 and 7): the reduction of the photocathode emitting area reduces the spot size of the e-beam on the phosphor screen, and in addition minimizes the geometrical aberrations inside the streak tube.

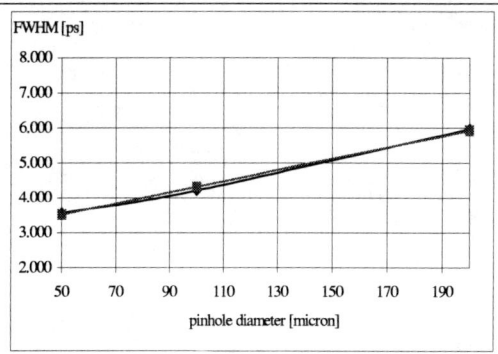

Pinhole Diameter [μm]	No. of Synchroscan Acquisitions (see Data Analysis)	AVG [ps]	RMS [ps]
50	7 (each 8 spots)	**3.54**	0.30
100	4 (each 8 spots)	**4.20**	0.44
200	4 (each 8 spots)	**5.96**	0.33

FIGURE 6. Plot of the resolution vs. pinhole diameter in synchroscan mode. The linear fit is: M0+M1*DIA$_{pinhole}$ with: M0=2.7; M1=16*10^{-3}.

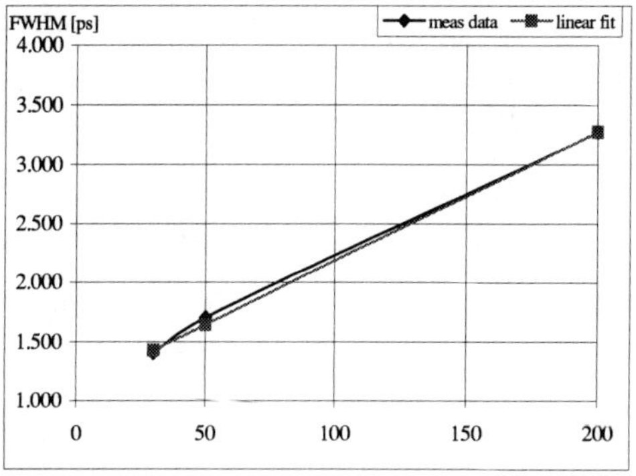

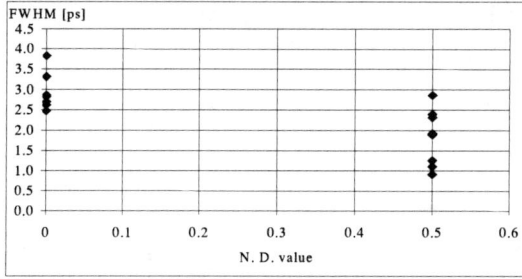

Pinhole Diameter [μm]	No. of Single Sweep Acquisitions (see Data Analysis)	AVG [ps]	RMS [ps]
200	17 (each 1 spot)	**3.54**	0.30
50	9 (each 1 spot)	**1.702**	0.75
30 (slit, 0.8mm high)	4 (each 1 spot)	**1.4**	0.19

FIGURE 7. Plot of the resolution vs. pinhole diameter in single sweep mode. The linear fit is: M0+M1*DIA$_{pinhole}$ with: M0=1.1; M1=10.9*10^{-3}.

The influence of the light intensity impinging on the photocathode was studied using different N.D. attenuators in front of the input optics, while keeping the same pinhole and MCP voltage. As can be seen in fig.8, there is an improvment at low input light levels, which can be attributed to reduced space charge effects.

	No. of Single Sweep Acquisitions	AVG [ps]	RMS [ps]
	ND=0.5; 9 acq.	**1.702**	0.75
	ND=0; 7 acq.	**2.94**	0.47

FIGURE 8. Resolution vs. N.D. attenuation in single sweep mode.

Next figure illustrates the effect of wavelength on temporal resolution. From these preliminary (λ_0=810nm and 405nm) measurements the effect of the larger energy spread induced by higher energy photons can be observed. This test will be completed as soon as the third harmonic of the fs laser will be available.

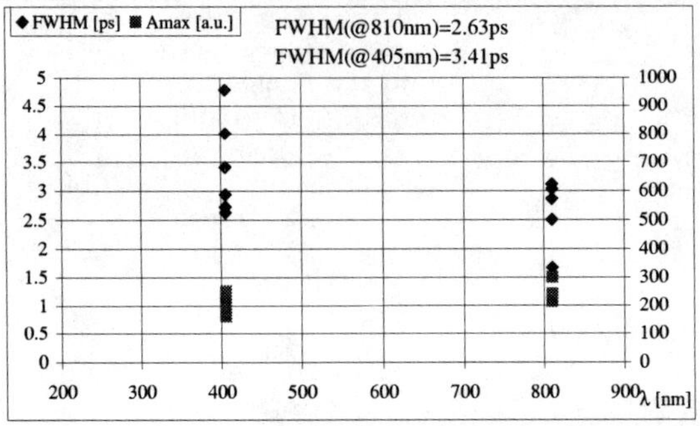

No. of Single Sweep Acquisitions	AVG [ps]	RMS [ps]
810nm: 5acq $\Delta\lambda$=40nm	2.63	0.58
405nm: 7acq $\Delta\lambda$=10nm	3.41	0.84
Diff	+29%	

FIGURE 9. Resolution vs. input wavelengths. N.D.=0 (relative comparison). Pinhole=50µm.

The accuracy and the linearity of the fastest single sweep have been checked using an optical delay line. The optical path length variation ΔL (translation stage/10µm resolution: 33fs) was: 0mm, 1.5mm and 3.0mm yielding a delay (2ΔL) between pulses of 0, 10ps and 20ps. The maximum error is 2.5% (0.7ps over 27.8ps). Due to different optical delays most part of the sweep has been used during this measurement so that we checked the accuracy and the linearity as well.

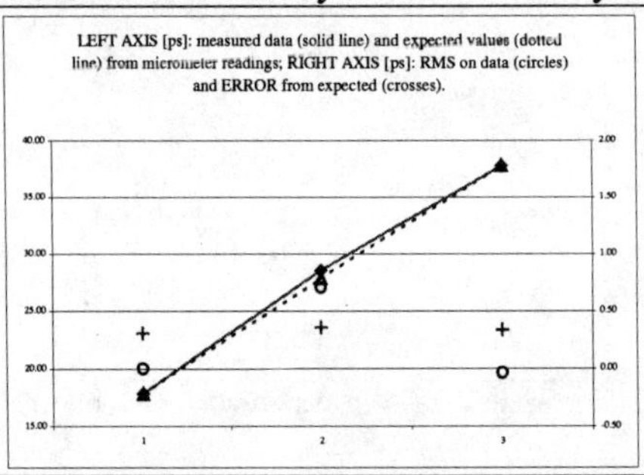

Δ Path Length [mm]	Average Pulse Length [ps]	Avg. Delay [ps]	RMS On Delay [ps]
0	2.53	17.8	0.31
1.5	2.72	28.50	0.36
3.0	3.06	37.76	0.34

FIGURE 10. Plot of the measured ΔT [ps] for three different optical lengths. Pinhole=200µm

Electrical Measurements on Timing and Laser Stability

For the electrical measurements, listed in Table 3, the 50GHz TEK sampling scope CSA803A has been used with the 25GHz photodiode (1411 by New Focus).

Table 3: summary of the stability measurements			
Scope Input	Scope Trigger	Measured Jitter	Notes
RF Master 500MHz	TCK, 57.78kHz	910fs, RMS	Accumulation time for statistics = 30 seconds
Laser 100MHz reference	TCK, 57.78kHz	1.62ps, RMS	
25GHz Photodiode	TCK, 57.78kHz	660fs, RMS	
25GHz Photodiode	Streak Camera Trigger, 10Hz	~4ps, PK-PK (estimated from scope plot)	The frequency of the trigger to the scope is too low for an RMS measurements

The measured jitter between the trigger TCK and the 100MHz laser reference signal is 1.62psRMS, whereas the jitter between the same trigger TCK and the laser light pulse, acquired by the photodiode, is 0.66ps (fig. 11 left). This data are consistent with the low pass filter of the stabilization loop of the CLX1100 unit, which filters the "high" frequency jitter that is eventually present on the 100MHz signal.

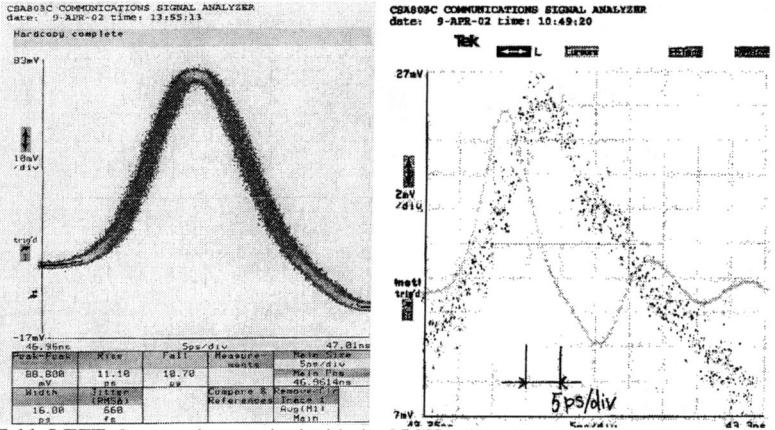

FIGURE 11. LEFT: Laser pulse acquired with the 25GHz photodiode and the 50GHz sampling scope. Jitter$_{RMS}$=660fs. **RIGHT**: same photodiode signal but with 10Hz trigger.

In spite of its ultra low frequency (10Hz), the same trigger signal of the SC has been used to reconstruct the photodiode laser pulse (fig.11 right) with a jitter=4ps$_{pk-pk}$. This data together with the streak camera accumulation mode acquisitions showing a total jitter of 18.4ps$_{pk-pk}$, give (subtracting in quadrature) a preliminary value of 4.4ps$_{RMS}$ jitter of the streak sweep unit (<5ps$_{RMS}$ on the specs). As this is a crucial point for future applications of the SC this parameter will be fully characterized vs. the trigger signal parameters (t$_{rise}$ and amplitude). The most viable option for reducing it to the picosecond level appears the adoption of an optical switch enabled sweep ramp.

ACKNOWLEDGMENTS

The Authors are grateful to D. Bulfone, for careful manuscript reading, and to the Instrumentation Group people who helped in setting up this experiment.

REFERENCES

1. Scheidt, K., Review of Streak Cameras for Accelarators: Features, Applications and Results, Proceedings of EPAC 2000, Vienna Austria, 2000, pp. 182-186.
2. Ferianis, M., The ELETTRA Streak Camera: Set-up and First Results, Proceedings of DIPAC 1999, Chester UK, 1999, pp. .
3. Walker, R.P. et al., The European UV/VUV Storage Ring FEL PROJECT at ELETTRA, Proceedings of EPAC 2000, Vienna Austria, 2000, pp. 93-97.
4. G.D'Auria et al., The FABRE Project: Design and Construction of an Integrated Photo-Injector for Bright Electron Beam Production, Proceedings of EPAC 2000, Vienna Austria, 2000, pp. 1681-1683
5. Optronis (former Photonetics), 77694 Kehl Germany, http://www.optronis.com
6. M. Trovo` et al., Operation of the European storage ring FEL at ELETTRA down to 190 nm, Proceedings of 23rd International FEL Conference, Darmstadt, Germany, Aug. 2001;
7. M.T.Asaki et al, Opt.Lett. **18** (1993), 977.
8. M.J.W.Rodwell et al, IEEE J.Quantum Electron. **25** (1989), 817
9. Ferianis, M. et al., The New Timing System for the ELETTRA Linac, Proceedings of EPAC 1996, Sitges, Spain, 1996.
10. CLX-1100 User Manual May 2001, Time-Bandwidth Products, www.tbwp.com.

Advanced Intraundulator Electron Beam Diagnostics Using COTR Techniques[*]

A. H. Lumpkin, W. J. Berg, S. Biedron, M. Borland, Y. C. Chae,
R. Dejus, M. Erdmann, Z. Huang, K.-J. Kim, J. Lewellen, Y. Li,
S. V. Milton, E. Moog, D. W. Rule[†], V. Sajaev, and B. X. Yang

Advanced Photon Source, Argonne National Laboratory, Argonne, IL 60439

Abstract. A significant advance in intraundulator electron-beam diagnostics has recently been demonstrated based on coherent optical transition radiation (COTR) imaging. We find signal strengths from a microbunched beam in a UV-visible free-electron laser to be several orders of magnitude higher than that of incoherent optical transition radiation. In addition we report that the far-field images of COTR interferograms carry information about beam size and asymmetry, divergence, and pointing.

INTRODUCTION

One of the standard means of imaging electron beams is the use of optical transition radiation (OTR) as the conversion mechanism [1-4]. Its inherently good spatial resolution and ultrafast time response must be balanced with its lower conversion efficiency in some applications with bright beams. However, in the case of microbunched electron beams, we report a significant advance in intraundulator diagnostics based on the signal enhancements and structure observed in coherent optical transition radiation (COTR). Longitudinal microbunching of the electron beam occurs as it copropagates with the emitted synchrotron radiation within the undulator. The density modulation develops at the fundamental wavelength of the light, which leads to a growing fraction of electrons emitting in phase, or coherently. A favorable instability results in an exponential growth of the light generation in a self-amplified spontaneous emission (SASE) free-electron laser (FEL) [5-8]. The signal strength now goes as the square of the number of particles involved $(b_n N)^2$, where b_n is the amplitude of the Fourier component of the electron distribution with spatial frequency k_n, and N is the total number of particles. We have routinely had to use neutral density (ND) filters providing attenuation factors of 10^4 to 10^5 (!) for COTR from 200-pC electron beams that have been microbunched. This would be an unheard of scenario with OTR imaging, and the light yield exceeds most scintillators.

In the case of the Advanced Photon Source (APS) FEL operating in the UV-visible regime, we have used standard CCD imaging cameras to obtain near-field, far-field, and spectral information on the COTR [9-12]. In addition to the unprecedented signal

---

[*] Work supported by U.S. Department of Energy, Office of Basic Energy Sciences under Contract No. W-31-109-ENG-38.
[†] Carderock Division, NSWC, West Bethesda, MD 20817.

strength in the near-field images (used to measure beam size), we have found noticeable structure and θ_x-θ_y asymmetry in the far-field images plus narrow-band spectral emission. In our two-foil geometry, we see clear COTR interferometer images that are explained by the product of a bunch form factor (developed from the beam size) and the single electron interference pattern [13]. The spacings of the interference fringe peaks act like an internal "measurement grid" for beam size (as well as carrying information via fringe visibilities about beam divergence). For our optics we actually infer better beam size sensitivity (σ~30 µm) from the COTR fringe visibility than from the beam size image directly! A brief description of our experiments and some results to date will be presented.

EXPERIMENTAL BACKGROUND

These experiments were performed at the APS using beam accelerated to 217 MeV by the S-band linac normally used as part of the injector system for the 7-GeV storage ring. We have obtained data with both the thermionic rf gun, with an 8-ns-long macropulse, and the photocathode (PC) rf gun, which generates a single micropulse at 6 Hz. The guns [14,15] and linac are described elsewhere. The beam is transported to the set of undulators (maximum of nine) in the low-energy undulator test line (LEUTL) tunnel [16]. A schematic of the experiment is given in Figure 1. It is not to scale, and there is an approximately 40-m transport line between the three-screen emittance station and the entrance of the undulators.

When all nine undulators are installed, the magnetic structure length is 21.6 m. The properties of the undulators and diagnostic stations have been previously described [17]. Very briefly, the undulator cells have a period of 3.3 cm and length L=2.4 m.

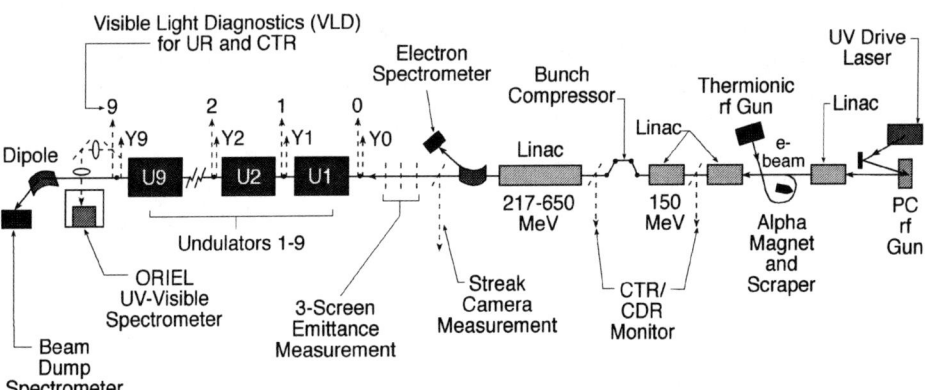

FIGURE 1. A schematic of the APS SASE FEL and the ten intraundulator diagnostic locations.

They have a fixed gap with a field parameter K=3.1. There is approximately a 0.38-m space after each undulator. This space is used for diagnostics and focusing and steering elements before the first undulator and after each of the installed undulator sections. A schematic of these stations is shown in Figure 2. The screens on the first actuator include positions for a YAG:Ce/mirror, a mirror at 45°, and a thin (6 µm) Al foil mounted with its surface normal to the beam direction. This thin foil serves two functions: (1) to block the stronger, visible undulator radiation (UR) and (2) to generate OTR or COTR as the e-beam transits the foil/vacuum interfaces. A digital camera views the e-beam images from the YAG and the reflected undulator radiation from the mirror. The second actuator, located 63 mm downstream, involves a retractable mirror at 45° to the beam direction. Another digital camera and moveable lens provide both near-field and far-field (focus at infinity) imaging. This visible light detector (VLD) system thus provides both beam size and angular distribution data. Both neutral density (ND) filters and bandpass (BP) filters are selectable by up to three filter wheels in front of the cameras.

As indicated in Figure 2, a remotely controlled pick-off mirror and lens system can be used to redirect the UR or COTR to an Oriel UV-visible spectrometer. Spectral effects were observed including the onset of sideband production after FEL saturation. For the purposes here, it was important to verify the COTR narrow-band spectrum centered around the fundamental SASE wavelength. Chromatic effects in spatial focus or the time-domain are avoided with COTR.

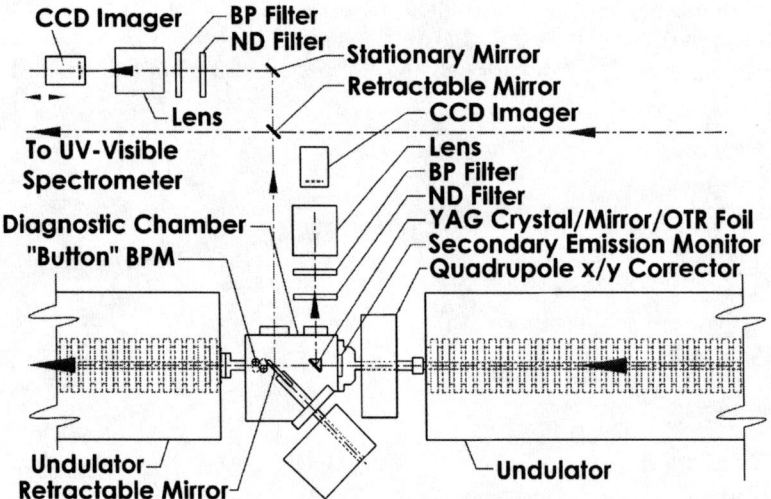

FIGURE 2. A detailed schematic of the intraundulator diagnostic stations showing the YAG actuator/cameras and the downstream 45° pick-off mirror with the visible light detector (VLD) cameras.

EXPERIMENTAL RESULTS AND DISCUSSION

We now present examples of results from the near-field imaging, far-field imaging, and the imaging spectrometer.

Near-Field Imaging

The near-field focus position of the lens provides beam size and profile information. In our geometry we can completely eliminate the competition from the dominant SASE light by using the blocking foil at the upstream position from the 45° pick-off mirrors at each station. Examples of the z-dependent beam sizes are shown in Figure 3. In this case the FEL fundamental was at 530 nm. We note that the needed ND value changed from 0 at VLD1 to 4.5 at VLD5, 6. A significant variation in the observed beam sizes is observed, in contrast to the expected constant beam sizes at the sampling points for a well-matched beam. Possible contributions to this effect are beam match into the undulators, beam image size reductions due to COTR, a transverse dependence of the bunching fraction coupled with the COTR effect, chromatic effects on lens focus for broadband OTR versus narrow-band COTR, and a camera focus error in the early stations. It is noted that we believe the major contributions are the first three. Since we see small beam structures in the VLD0 and VLD1 cameras on some shots out of the 100 images, we believe focus effects are minimal.

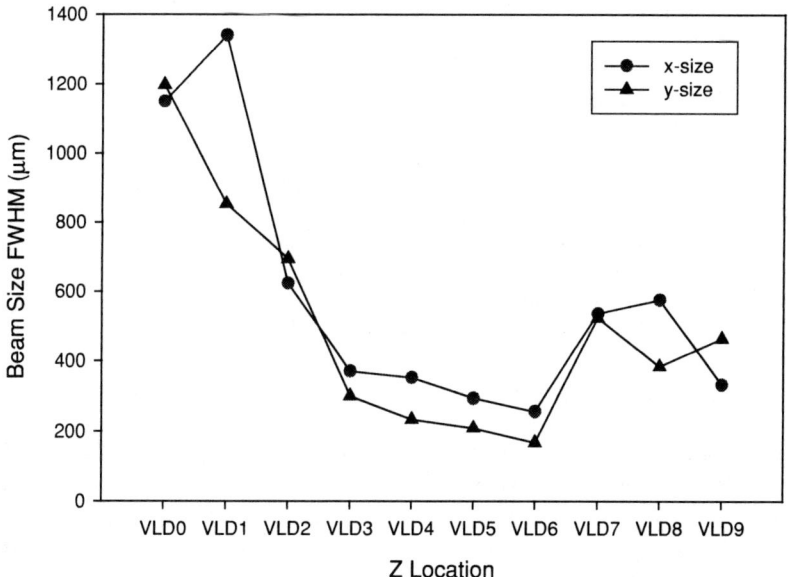

FIGURE 3. An example of z-dependent, x– and y–beam sizes observed in the undulator areas. These data are from March 30, 2001.

We have performed a calculation of the beam spot narrowing that occurs when COTR is used as the conversion mechanism. If the bunching fraction is assumed to be uniform across the Gaussian transverse dimension then the $(b_n N)^2$ term for the peak intensity will approximately give a $\sqrt{2}$ narrowing of the "observed" peak. We have experimentally tested this simple model by taking data without (COTR) and with (OTR) a 500-nm short pass filter, which would attenuate by 100 all wavelengths greater than about 520 nm. An example of this effect is shown in Figure 4. The horizontal beam size observed when significant gain has occurred starting at VLD5 is narrower when the bandpass filter is not used (COTR). The incoherent OTR source provides the actual e-beam size. We note that at VLD8 the single filter is insufficient to block the entire strong microbunching signal, so clean OTR imaging is not achieved for this point.

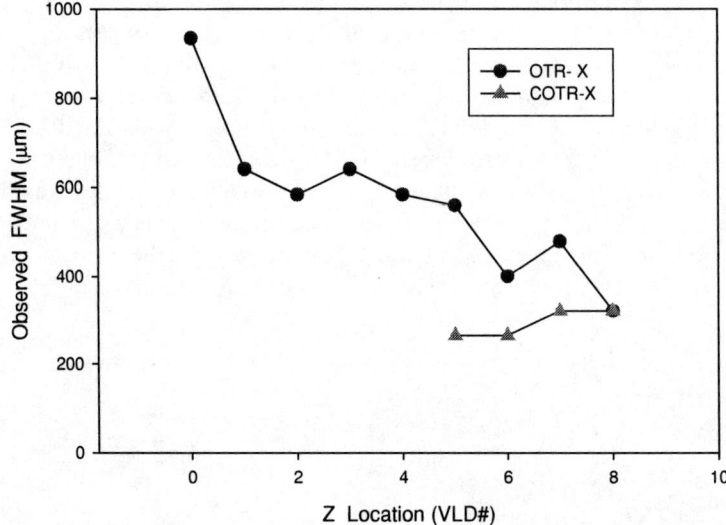

FIGURE 4. An example of z-dependent x–beam sizes observed with (OTR) and without (COTR) the 500-nm short pass filter at VLD5 – VLD8. Data are from December 20, 2001.

Far-Field Imaging

The camera-to-lens z separation can be remotely controlled by the stepper motor. For far-field imaging angular distribution patterns are obtained. Based on the analytical model described in Ref. 13, the single-electron OTR interference (OTRI) pattern for two sources 63 mm apart is first calculated. This includes effects due to beam divergence. For example, we estimate the beam scattering from the first foil as about 0.1 to 0.2 mrad at 220 MeV. The calculated OTRI pattern exhibits fringes at ± 1.9, ± 4.6, ± 6.3, and ± 7.5 mrad. The bunch form factor is multiplied times this pattern. Large beam (> 100 μm) sizes have very narrow form factor functions in θ space. Figure 5 shows an example of the COTR fringe visibility variation with beam size. The COTR second fringe peaks are visible for beam sizes less than 50 μm. The observed first fringe peak position varies rapidly for $\sigma > 100$ μm. An example of an image is given in Figure 6. In this case the initial beam size asymmetry in σ_x and σ_y results in the single peaks in θ_x and multiple peaks in θ_y. By looking at the fringe peak relative intensities for σ_y = 15, 20, and 30 µm, the θ_y pattern was found to be consistent with a beam size of 30 µm and total divergences of about 0.2 mrad. The σ_y = 30 µm result is smaller than the limiting resolution in the optical system of about one pixel at 80 µm/pixel for near-field focus. Further studies are needed in this area to develop accuracy on the beam size. In addition, we have reported elsewhere the sensitivity of the symmetry of the ± θ_y peaks to e-beam steering with the correctors [18].

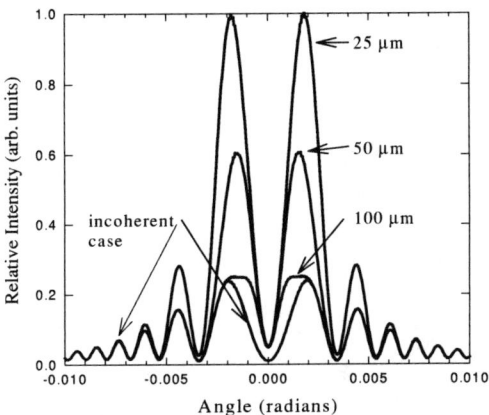

FIGURE 5. A comparison of COTR interference fringe visibility for different beam sizes, σ_y = 25, 50, and 100 µm.

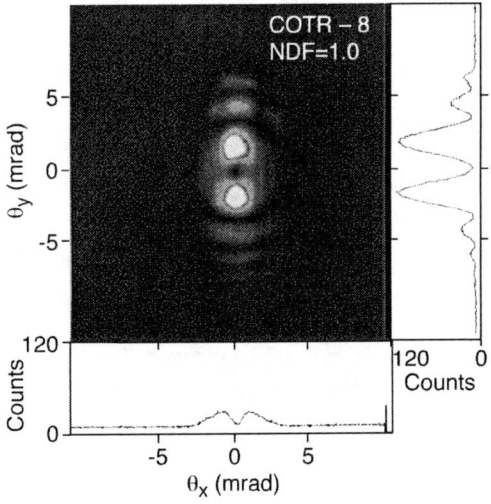

FIGURE 6. An example of a COTR interference image taken after undulator 8 in a case when this was post-FEL saturation. The clear θ_x-θ_y asymmetry and the fringe visibility support corresponding beam sizes of $\sigma_x \approx$ 100 µm, and $\sigma_y \approx$ 30 µm, respectively.

Imaging Spectrometer

The expected microbunching at the fundamental wavelength is clearly illustrated in a spectrum from one of our early studies [19]. As shown in Figure 7, the SASE FEL peak and the COTR peak overlap in wavelength at 530 nm. The COTR is a little broader in this example taken after undulator 5. The electron beam energy jitter can be observed directly in the wavelength jitter of the images. As a reference the GreNe calibration line is shown at 543.5 nm, and its width indicates operational resolution of about 0.8 nm (σ) for these conditions.

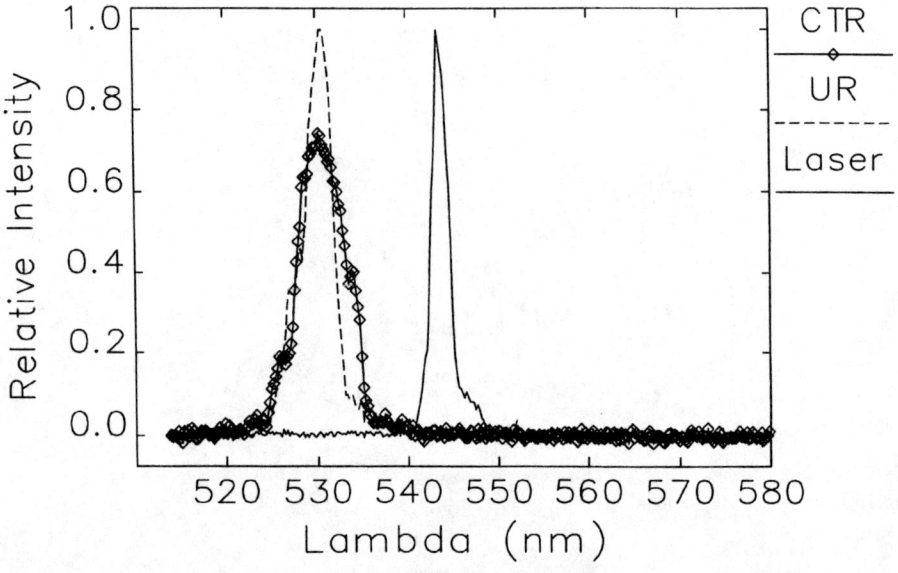

FIGURE 7. A comparison of the SASE FEL spectrum, the COTR spectrum, and the calibration laser signal obtained after undulator 5. The discrete COTR line is quite different from the broadband OTR spectrum.

SUMMARY

In summary, the COTR imaging techniques provide all the advantages of OTR imaging with the additional features of signal enhancements up to 10^5 and monochromatic spectral output. These features provide a near-ideal, intraundulator electron-beam characterization capability. We hope to extend the techniques to the VUV in the coming year as the APS FEL project pushes to operate at 130 nm.

ACKNOWLEDGEMENTS

The authors acknowledge the support of Om Singh, Antanas Rauchas, Efim Gluskin, and Rodney Gerig of the APS.

REFERENCES

1. Wartski, L. et. al., J. Appl. Phys. **46**, 3644 (1975).
2. Rule, D. W., Nucl. Instrum. Methods in Phys. Res. **B24/25** 901 (1987).
3. Lumpkin, A. H. et al., NIM **A296**, 150 (1990), D. W. Rule et al., Idem, p. 739 (1990).
4. Lumpkin, A. H. et al., Nucl. Instrum. Methods in Phys. Res. **A429**, 336 (1999).
5. Derbenev, Y. S., Kondratenko, A. M., and Saldin, E. L., Nucl. Instrum. Methods **A193**, 452 (1982).
6. Orzechowski, T. J. et al., Nucl. Instrum. Methods **A250** 144 (1986).
7. Hogan, M. et al., Phys. Rev. Lett. **81**, 4897 (1998).
8. Milton, S. V. et al., Phys. Rev. Lett., **85** 988 (2000).
9. Tremaine, A. et al., Phys. Rev. Lett. **81**, 5816 (1998).
10. Lumpkin, A. H. et al., Phys. Rev. Lett. **86**, 79 (2001).
11. Lumpkin, A. H. et al., "Evidence for Microbunching "Sidebands" in a Saturated FEL Using Coherent Optical Transition Radiation," accepted in Phys. Rev. Lett. March 2002.
12. Lumpkin, A. H. et al., "Comprehensive z-Dependent Measurements of Electron-beam Microbunching using COTR in a Saturated SASE FEL," Nucl. Instrum. Methods in Phys. Res. (in press).
13. Rule, D. W., and Lumpkin A. H., Proc. of the IEEE 2001 Particle Accelerator Conference, Chicago, IL, pp. 1288-1290 (2001).
14. Lewellen, J. W. et al., Proc. of the 1998 Linac Conference, Chicago, ANL-98/28, Vol. 2, pp. 863-865 (1999).
15. Biedron, S. et al., Proc. of the IEEE 1999 Particle Accelerator Conference, New York, NY, pp. 2024-2026 (1999).
16. Milton, S. V. et al., Nucl. Instrum. Methods in Phys. Res., **A407**, 210 (1998).
17. Gluskin, E. et al., Nucl. Instrum. Methods in Phys. Res., **A429**, 358 (1999).
18. Lumpkin, A. H. et al., Proc. of the 2001 Particle Accelerator Conference, Chicago, IL, pp. 550-552 (2001).
19. Lumpkin, A. H. et al., "Utilization of CTR to Measure the Evolution of Electron-Beam Microbunching in a SASE FEL," Nucl. Instrum. Methods in Phys. Res. (in press).

The LHC 450 GeV to 7 TeV Synchrotron Radiation Profile Monitor using a Superconducting Undulator

R. Jung, P. Komorowski, L. Ponce, D. Tommasini

CERN, CH1211 Geneva 23, Switzerland

Abstract. In LHC it will be important to measure with precision and in a non-destructive way the proton beam profiles from 450 GeV to 7 TeV. The chosen monitor will make use of a 5 T superconducting Undulator with two periods coupled to the D3 bending magnet built by BNL. From the various variants studied, this combination is the only one which could cover the whole LHC energy range. By locating both magnets in the same cryostat, it will be possible to minimise the light source length for best precision. The choice of the undulator parameters and its basic design will be described. The evolution of the synchrotron radiation patterns along the energy ramp will be given, as well as the performance with respect to sensitivity, depth of field and diffraction, with a description of the simulation codes used.

INTRODUCTION

There is a strong need in LHC to measure the beam profiles all along a run. The tight emittance budget asks to measure the emittance at beam injection at 450 GeV to check that the limit of 5% blow-up between the circular machines is respected. A turn-by-turn measurement during the first tens of turns will check that the matching between the accelerators is done properly (1). A relative accuracy of the order of a few percent is requested for the measurement of the turn by turn profile oscillations that are the signature of a mismatch. The beam size evolution has then to be followed through the acceleration cycle from 450 GeV to 7 TeV where the beam size shrinks substantially but for which a normalised emittance blow-up of less than 7% is requested. Finally, the beam profile has to be measured with a relative accuracy better than a few percent to adjust the aperture controlling collimators. During all these phases, there is also a demand to measure individual bunches out of the 2808 circulating bunches, at various locations in a 72 bunches batch in order to identify beam dynamics problems.

An ideal monitor for these tasks is a non-intercepting monitor. One monitor of this kind is a Synchrotron Radiation (SR) monitor. The main candidate was a monitor close to a physics Interaction Region (IR), IR1 or IR5, using the light generated in one of the dogleg bending magnets, D2, bringing the beams back to the nominal LHC separation after the increased separation in the IR. This monitor was in a favourable location where the beam size increases at top energy when the beams are brought into collision optics. Unfortunately, the light production within the spectral range of available detectors, was only sufficient above 2 TeV. From injection energy to 2 TeV, another solution had to be found. Various solutions were looked at in the RF region IR4, using room temperature or superconducting undulators to generate enough light in the neighbourhood of the visible spectrum. These solutions could cover the 450 GeV to 2 TeV region, but were useless above, generating the additional problem of changing monitors during the delicate process of the energy ramp.

An acceptable solution became possible when the IR4 layout was changed for economical reasons and a long dogleg was introduced to go from a separation of 420mm in the IR, dictated by the RF cavities, towards the 194mm in the standard LHC arc dipoles. With this layout, a superconducting Undulator could be introduced in front of the D3 separating magnet, which deflects the circulating beam from the SR generated in the Undulator. A mirror can be introduced to collect and deflect out of the vacuum chamber the SR after a drift of some 10m after D3. Above 2 TeV, the Undulator radiates again mostly outside the detector range, but this time the edge radiation of the D3 magnet will be used as SR source. Finally at top energy, the whole of D3 radiates enough SR, which has this time to be limited to a region close to the entrance edge for limiting the longitudinal acceptance of the imaging optics.

SR CHARACTERISTICS OF THE UNDULATOR

An Undulator is a periodic magnetic structure that concentrates the SR through interference in a cone in the forward direction along the beam path (2). It is characterised by a factor K, with K<1 for an Undulator:

$$K = \lambda_u B_0 \frac{e}{2\pi m_p c} \quad (1)$$

where λ_u is the Undulator period and B_0 the peak magnetic field on the beam axis.

The coherence condition relates the emitted SR wavelength λ_c to a given direction θ with respect to the beam axis and the Undulator characteristics by:

$$\lambda_c = \frac{\lambda_u}{2\gamma^2}\left(1 + \frac{K^2}{2} + \gamma^2\theta^2\right) \approx \frac{\lambda_u}{2\gamma^2}\left(1 + \gamma^2\theta^2\right) \quad (2)$$

The angular spectral energy density in the deflection plane of the Undulator is given in equation (3), with the usual notations, k being a constant and N_u the number of Undulator periods. From this equation, it is clear that the light production will increase as B_0^2. It can also be seen that the light production decreases when going away from the beam axis and that the light spectrum narrows around λ_c when the number of Undulator periods N_u increases:

$$\frac{d^2W}{d\Omega d\lambda} \approx kB_0^2 \left[\frac{\gamma}{1+\gamma^2\theta^2}\right]^6 \left[1-\gamma^2\theta^2\right]^2 \left[N_u \frac{\lambda_c}{\lambda}\right]^2 \left[\frac{\sin\left(\frac{\lambda_c}{\lambda}-1\right)\pi N_u}{\left(\frac{\lambda_c}{\lambda}-1\right)\pi N_u}\right]^2 \quad (3)$$

Based on these considerations, a two period superconducting Undulator, of 28cm period, and with a peak field of 5T was chosen. The relevant parameters of this

Undulator are: K= 0.07 and λ_{co}=608nm at 450 GeV and already down to λ_{co}=55nm at 1.5 TeV on the beam axis. It is only because of the spectral width due to the small number of magnetic periods and to the angular acceptance of –0.5/+1.5mrad, that there will be a reasonable amount of energy available in the spectral range of the detectors.

The evolution of λ_c as a function of beam energy and observation angle is given in figure 1. This situation is acceptable at high energy because the edge of the D3 magnet starts to produce enough SR from 1 TeV onwards.

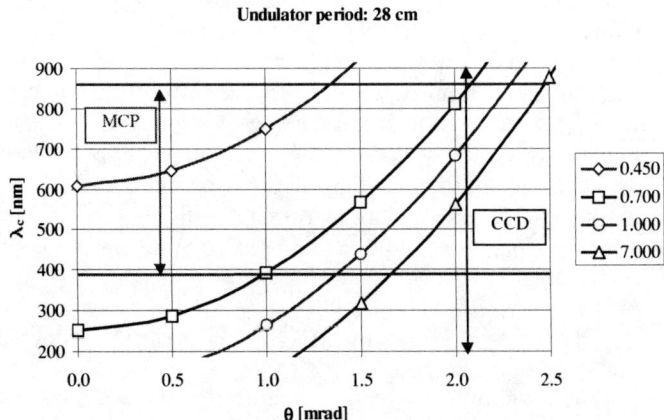

FIGURE 1. Coherence wavelength versus angle to the beam direction as a function of beam energy, with the spectral sensitivity bands of a back-illuminated CCD and a MCP with a SS25 photocathode.

SR EMITTED BY THE D3 BENDING MAGNET

Starting at 750 GeV, the edge of D3 will emit light in the range of interest.

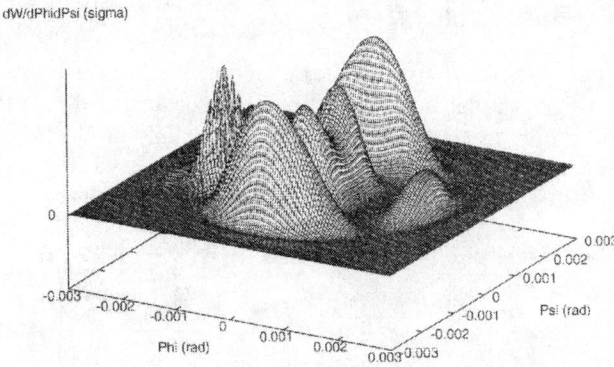

FIGURE 2. Angular light pattern resulting from the combination of the SR from the Undulator (ring pattern with central peak) and of the D3 bending magnet input edge (at the centre of the Undulator pattern) and exit edge (peak at the left) at 1 TeV.

The light is emitted by the entrance edge along the same direction as the Undulator SR, and can hence be extracted under the same conditions. This light will interfere with the light produced by the Undulator. A typical angular light pattern for the intermediate energies is given in figure 2.

Once the energy increases beyond 2 TeV, the whole core of D3 will produce SR.

THE LHC SR PROFILE MONITOR

The principle of implementation of the monitor is given in figure 3. The proton beam leaving Interaction Point 4 at the top right of the figure, enters the Undulator before being deflected by 1.6 mrad by the D3 magnet towards D4.

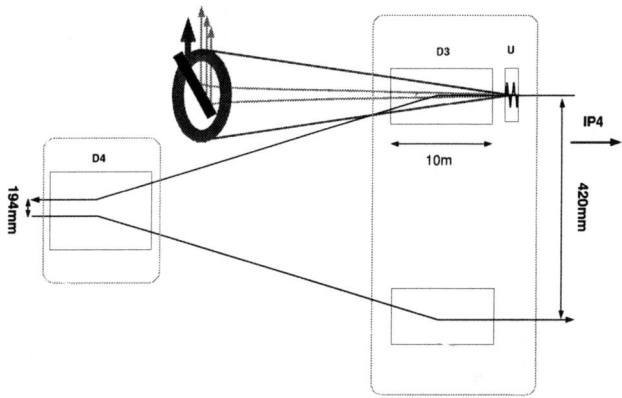

FIGURE 3. Schematic view of the Undulator/D3 SR monitor in IR4 of LHC.

The Undulator and D3 are located in the same cryostat to minimise the distance between them, in order to reduce the extent of the light source. The light generated in the Undulator and at the edge of D3 travels a distance of 23m before an extraction mirror can be inserted at an acceptable distance from the beam, typically 15 σ_H. The beam and the light will travel in an enlarged vacuum chamber with tapered transitions at both ends in order to reduce the perturbation to the beam.

The SR monitor's performance and calibration will be checked at low proton beam intensity with H and V Wire Scanners located at the exit of D3.

Undulator Magnet

The undulator consists of 8 superconducting coils assembled around ferromagnetic iron poles to produce two periods of magnetic field with a sinusoidal shape: figure 4.

To block the conductors during magnet excitation, the coils will be clamped under pre-stress. Vertical clamping will be provided by splitting the magnet into a lower and an upper part, and by closing the structure with spacers between the upper and lower coils outside the beam tube. Horizontal clamping will be provided by retaining blocks fixed by copper/beryllium tie bolts.

The main parameters of the Undulator are listed in Table 1.

TABLE 1. Undulator: main parameters.

Period length	280 mm
Number of periods	2
Iron yoke length	710 mm
Gap	60 mm
Beam tube size	50/53 mm ID/OD
Maximum magnetic field in the gap	5 T
Maximum field error within ± 10 mm from axis	0.25%
Supply current	250 A
Total energy stored at 250 A	150 kJ
Magnet inductance	4.8 H
Coil cross section	36.5 x 42.5 mm^2
Cable size	1.25 x 0.73 mm^2
Overall coil size	140 x 223 x 36.5 mm^3
Operating temperature	4.2 K
Margin to quench on load line	20 %
Main field/peak field ratio	0.83
Hot spot temperature in case of a quench at 5 T	120 K

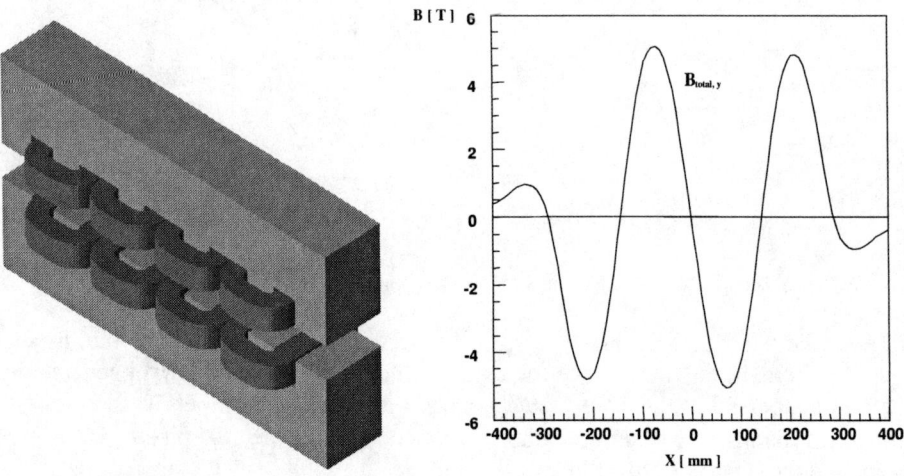

FIGURE 4. Perspective view of the 2 period Undulator with Pole pieces extending beyond the coils (total length 71cm) and Vertical Magnetic Field component along the beam axis.

Telescope

It is intended to re-use the LEP SR telescopes (3) with some modifications. The telescope uses primarily mirrors for folding and focusing. The detectors will be back-illuminated CCDs for highest sensitivity and ordinary CCDs coupled to Multi Channel Plate (MCP) intensifiers for single bunch or single batch, down to turn-to-turn, observations. This telescope has to adapt to changing conditions over a run. At injection energy at 450 GeV, the Undulator is used. The beams are large, σ~1.2mm, and emit little light. At the top energy of 7 TeV, the D3 magnet is used, whilst the Undulator emits in the UV at large angles which can reach the detectors. At that energy the beams are also small, σ~300μm, and D3 emits a large amount of light. For that reason, two detector set-ups are foreseen. As there is enough light available at high energy, a bandpass filter will be used together with a magnifying lens which will image the beam from the first image plane onto the second detector set. This set-up can also take into account the longitudinal separation of the Undulator and the D3 edge, which has to be kept below 80cm. Chromatic, linear density and polarisation filters are installed as well as a slit in the focal plane to restrict the acceptance in D3.

The magnification is determined by the 4m focal length spherical mirror and the 23x23μm² pixel size. It has been set to G=0.2 in order to have 3 pixels per sigma at 7 TeV, which gives then 13 pixels per sigma at injection. One of the limitations of the performance is the distance by which the light extraction mirror has to be retracted from the beam. For the moment, a distance of $15\sigma_H$ has been asked for. It is hoped that with operational experience, this distance can be decreased to come closer to the machine aperture set by the collimators closed to ±7σ. In any case, the extraction mirror will be movable, so that it can follow the $15\sigma_H$ limit to improve the performance at high energy.

Due to the long distance to travel and the small opening of the light cone, proper alignment has to be provided. A set-up using a folding mirror and a laser located close to the SR telescope will be used: see figure 5. A similar set-up has been used in LEP and has proven to be extremely useful. The Undulator itself has to be aligned on the entrance magnetic axis of D3 to a tolerance of the order of ±5mrad.

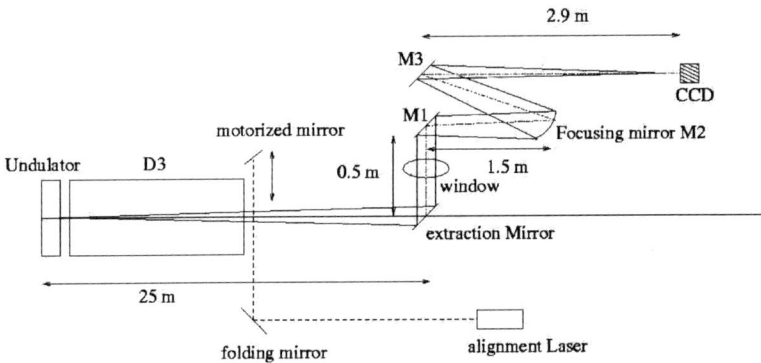

FIGURE 5. Monitor layout with alignment set-up of the optical elements of the SR telescope.

PERFORMANCE ANALYSIS

The photon production has been computed with the ray-tracing code Zgoubi (4). With the optics set-up described there will be a maximum of 200 photons per pixel (px) at injection and $80 \cdot 10^3$ photons/px at top energy for a pilot pulse of $5 \cdot 10^9$ protons in single turn mode. This will be sufficient to observe the beam behaviour in LHC before injecting and accelerating a nominal bunch. It should also be sufficient to check if there are sizeable matching errors. For a nominal bunch of $1.1 \cdot 10^{11}$ protons, there will be $4 \cdot 10^3$ photons/px at injection and up to $2 \cdot 10^6$ photons/px at top energy. This will permit high precision measurements from a statistical point of view. But the imperfections of the LHC monitor are due, to a large amount, to the source length, the interference between the two sources and the diffraction of the SR light cone due to the small opening angle at high energy and the limited acceptance of the extraction mirror. At 450 GeV, the emission pattern is a gaussian like cone with an opening of ~0.8mrad FWHM, within the acceptance of the extraction mirror. At 1 TeV, see figure 2, the light pattern is the superposition of two sources, which will generate a beam broadening through interference and diffraction. Finally, at 7 TeV, where mainly D3 will produce light in the useful spectrum, there is a classical bending magnet SR pattern, with clearly visible edges, cut by the extraction mirror.

The influence of the source characteristics on the performance was evaluated with the program SRW (5). SRW is a numerical code dedicated to the derivation of SR features generated by an arbitrary magnetic field pattern followed by a propagation through an optical chain producing a display of the Point Spread Function (PSF). SRW provides the SR intensity distribution for a "filament" electron beam, i.e. with zero emittance. The electric field in the frequency domain is derived from the Fourier Transform of the retarded potentials, allowing to perform the computing in the far field, as well as in the near field SR approximations. The SR propagation from a transverse plane to another one is implemented using the Fourier optics approach, assuming small angles and large distances compared to the wavelengths. The electric field in a transverse plane after an optical element is derived by applying an operator describing the optical element. The program parameters have been modified to take into account protons and the results have been cross-checked with Zgoubi.

The beam image at the detector will be the convolution of the density distribution of the beam and of the PSF of the optical system.

The images of the filament beam in the detector plane, together with a cut through the horizontal beam axis are given in Fig 6 to 8 for 450 GeV, 1 TeV and 7 TeV. The results for the horizontal polarisation component of the SR are summarised in Table 2.

TABLE 2. Undulator SR profile monitor performance.

Energy / Sizes [μm]	Beam		PSF		Beam Image			
	σ_H	σ_V	σ_H	σ_V	σ_H	σ_V	$\delta\sigma_H/\sigma_H$	$\delta\sigma_V/\sigma_V$
450 GeV	960	1323	159	141	973	1330	1.1%	0.6%
1 TeV	644	888	198	120	674	896	4.6%	0.9%
7 TeV	244	335	156	194	290	387	18%	15%

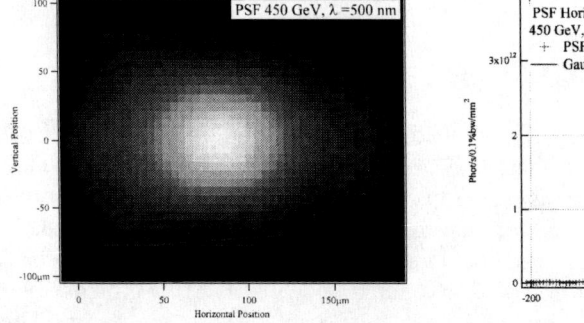
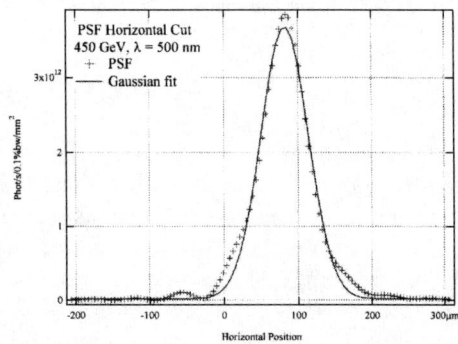

FIGURE 6. 2 D and Horizontal cut of the Point Spread Function of the SR monitor at 450 GeV.

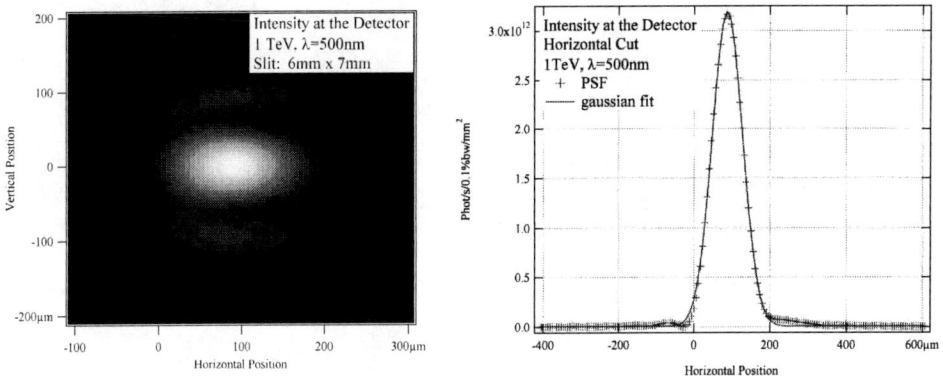

FIGURE 7. 2 D and Horizontal cut of the Point Spread Function of the SR monitor at 1 TeV.

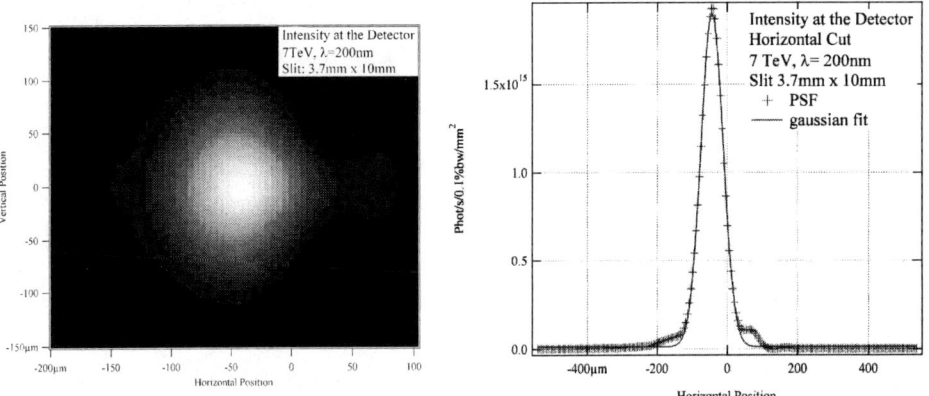

FIGURE 8. 2 D and Horizontal cut of the Point Spread Function of the SR monitor at 7 TeV.

The image broadening introduced by the PSF is small enough to extract the real beam size to the expected accuracy by a simple quadratic subtraction. The broadening will be independent of the beam intensity and will be stable for a given light pattern, i.e. beam energy. The corrections can hence be calibrated with the Wire Scanners.

It would nevertheless be advantageous for the precision of the measurement that the machine optics provides higher βs.

ACKNOWLEDGMENTS

It is a pleasure to acknowledge the help and fruitful discussions with J. Bosser, O. Chubar (ESRF), P. Elleaume (ESRF), A. Hofmann, F. Méot (CEA/Saclay), S. Russenschuck and M. Sassowsky.

REFERENCES

1. C.Bovet, R. Jung, EPAC 1996, Sitgès, June 1996, pp. 1597-1599
2. A. Hofmann, CAS, Grenoble, April 1996, CERN 98-04, August 1998, pp. 1-44
3. G. Burtin et al, CERN SL-99-049 BI, August 1999
4. F. Méot, S. Valero, CEA-Saclay, DSM/DAPNIA/SEA-97-13, 1997
5. O. Chubar, P. Elleaume, EPAC 1998, Stockholm, June 1998, pp. 1177-1179

LHC Beam Loss Monitor System Design

B. Dehning*, G. Ferioli*, W. Friesenbichler*, E. Gschwendtner* and
J. Koopman*

*CERN, CH1211 Geneva 23, Switzerland

Abstract. At the LHC a beam loss system will be installed for continuous surveillance of particle losses. The system is designed to prevent hardware destructions, to avoid magnet coil quenches and to provide quantitative loss values. Over 3000 ionization chambers will be used to initiate the beam abort if the loss rates exceed the quench levels. The time and beam energy dependent quench levels require the acquisition of chamber currents in the range from 50 pA to 0.5 mA and an update of the values every 89 μs. The acquisition and control electronics will consist of a front end electronics near (< 400 m) to the ionization chambers and a threshold controller in the surface buildings. The front end will include a charge balance converter, a counter and multiplexer part. The charge balance converter is most suitable to cover the large dynamic range. The introduced error is smaller than few % in the required dynamic range. Six channels will be transmitted over one cable of up to 3 km length. The threshold controller will issue warnings and dump signals depending on the beam energy and the loss durations.

INTRODUCTION

At the nominal energy of 7 TeV each beam in the LHC stores energy of up to 0.35 GJ. The loss of only a fraction (10^{-8}) of the beam can have a severe impact on the smooth machine operation. Therefore the beam loss detection system must fulfil several requirements, first: Protection: The magnets and other equipments must be protected from damage due to beam losses. Second: Prevention: The super conducting coils of the magnets could be come normal conducting by heat depostion from lost particles. In both cases beam dumps are initiated before a dammage or quench will occur. Thired: Serve as a beam diagnostic tool.

For protection and prevention the beam loss monitors trigger the beam dump via the beam interlock system, whenever they detect beam losses above a certain limit. The quench levels of the super-conducting magnets define this limit.

Detection of shower particles outside the cryostat or near to the collimators will be used to determine the coil temperature increase due to particle losses. The relation between loss rate and temperature increase (quench levels) is based on shower simulation and heat transfer considerations [1]. The expected particle flux outside the cryostat is as well based on shower simulation.

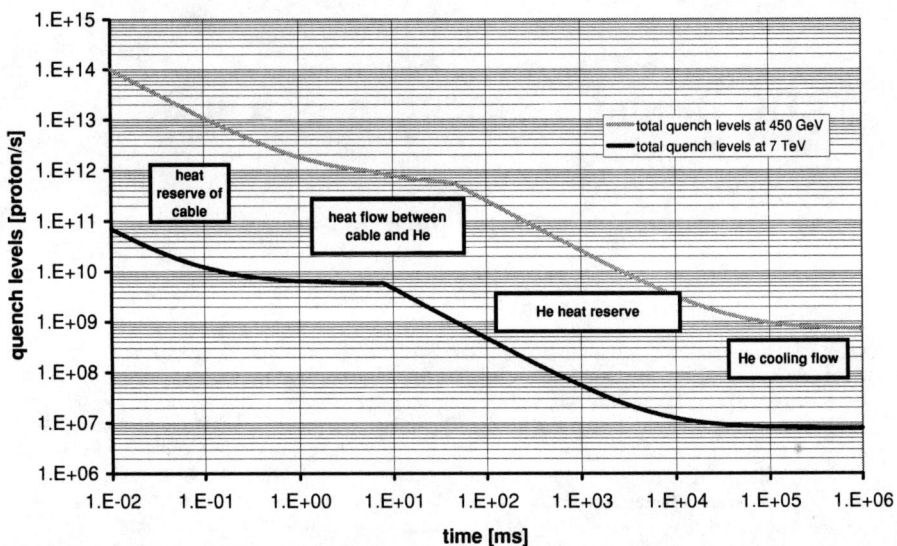

FIGURE 1. The bending magnet quench levels as function of the loss duration for two different energies.

QUENCH LEVELS

The coils of the magnets can quench if a local energy deposition due to beam particle losses increases the coil temperature to a value were the conductor changes from the super-conducting to normal conducting state [1, 2]. In addition to the energy dependence of the quench levels (see Fig. 1), a strong dependence on the duration of the loss is observed. At 450 GeV the quench limit rate is $1\ 10^{13}$ protons/s corresponding $0.9\ 10^9$ protons for a loss duration of 1 turn (89 μs) and the state state rate (t > 100 s) is $7\ 10^8$ protons/s. At 7 TeV the magnet quench rate is $1\ 10^{10}$ protons/s corresponding to $0.9\ 10^5$ protons per turn and the state state rate (t > 50 s) is $8\ 10^6$ protons/s.

The thermal conductivity of the surrounding helium flow determines the maximum loss rate at long time intervals. At short time intervals, the heat reserve of the cables, as well as the heat flow between the superconducting cables and the helium, tolerates much higher loss rates. The heat reserve of the helium determines the quench levels at intermediate time scales.

BEAM LOSS PARTICLE DETECTION

At the positions, where most of the beam losses are expected, simulations of the particle fluences outside the cryostat induced by lost protons at the aperture have been performed with the Monte Carlo Code Geant 3.21. The geometry used in this simulation corresponds to the dispersion suppressor. Calculations for the arc have already been presented in [2]. The simulated shower particles produced by lost protons are counted in two detectors placed all along the cryostat on both sides. Also the energy deposition in these elements is calculated. Figure 2 gives a typical example of detector signals caused

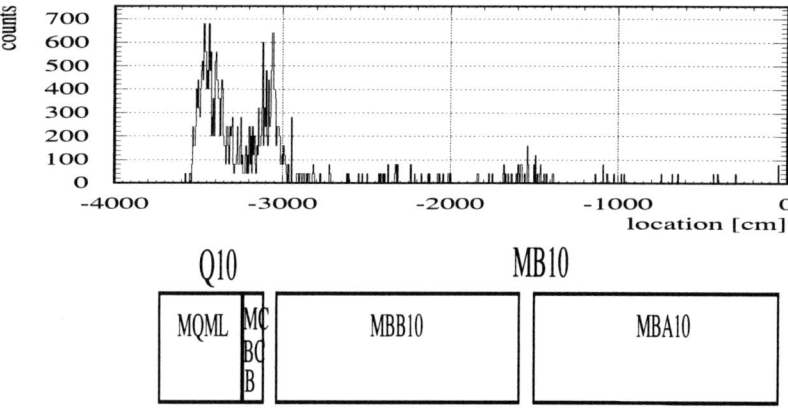

FIGURE 2. Simulated detector signals of shower particles, caused by a point like loss in the middle of the quadrupole magnet MQML.

by the shower particles, which are induced by point like losses of the beam in the middle of the quadrupole MQML (β-function maximum). The shower maximum is about 1 m after the beam loss location. The shower width is 0.7 m. For one lost proton of 7 TeV $1 10^{-2}$ charged particles/proton/cm^2 are observed. The second shower maximum is due to the gap between the quadrupole (MQML) and the dipole (MBB). The signals in the detectors on the opposite side of the cryostat are reduced in amplitude and the beam loss maxima from beam 1 and beam 2 are separated between 1 and 2 m [3]. Positioning beam loss monitors at the shower maxima locations fulfils the requirements for the distinction between the two beams and for localising the beam losses. It is foreseen to place three detectors on either side of the cryostat 3 at the optimal position for the observation of beam losses located in the middle of the quadrupole magnet and located at the transitions between the quadrupole magnet and the bending magnets.

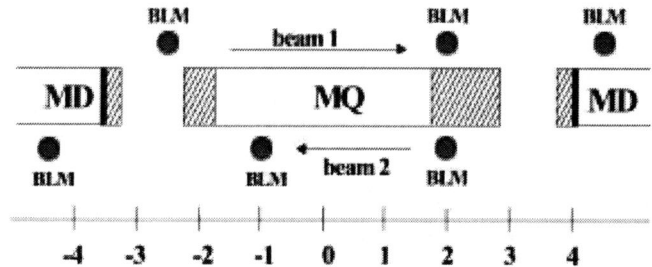

FIGURE 3. Location of the beam loss detectors near to the quadrupole magnets MQ.

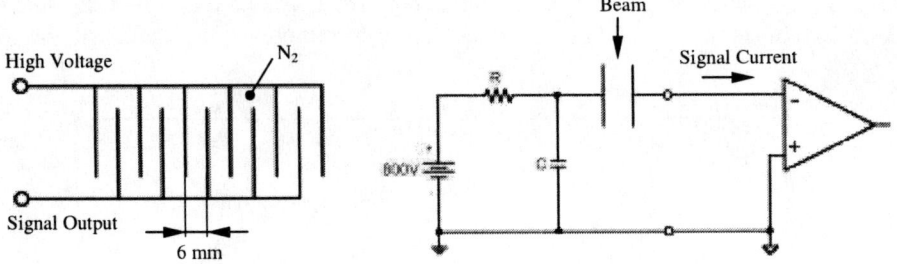

FIGURE 4. Left: Schematic structure of the baseline ionisation chamber. Right: Principle circuit diagram for the frontend electronic circuit.

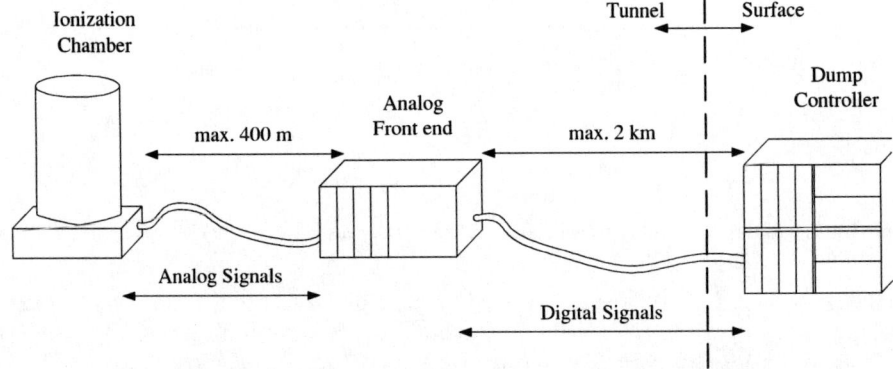

FIGURE 5. Schematic of the beam loss monitor readout chain.

BEAM LOSS DETECTORS

Ionisation chambers will be used as beam loss monitors. The baseline layout is a N_2 filled cylinder with a surface of 80 cm^2, a length of 19 cm and operated with a bias voltage of V = 800 - 1800 V (see fig. 4). The chamber electrodes consist of 31 parallel plates. All the odd and even plates are connected together to increase the electrical field volume, which has a strenght of up to 3 kV/cm. To stabilize the high voltage an low pass filter (R = 1 MΩ and C = 0.5μF) is mounted on the chamber feed throughs.

STRUCTURE OF THE BEAM LOSS MONITOR READOUT CHAIN

The first evaluation of this signal is done by the analog front end. Due to radiation exposure in the accelerator tunnel, the cable length between the chamber and the front end can vary between several meters in the arcs of the tunnel, and up to 400 m in the straight sections (see fig. 5). The increased radiation load in the tunnel (up to 10 Gray

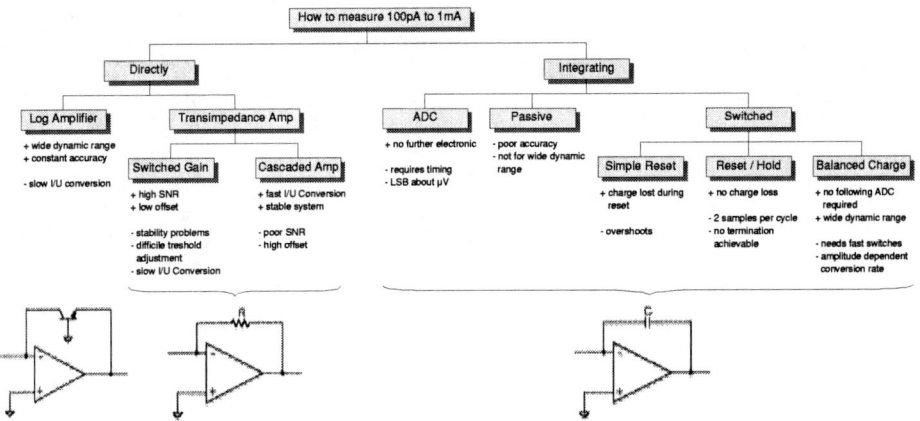

FIGURE 6. Overview of the applicable principles to measure electrical currents.

per year are expected) necessitates a simple and robust design of the front end circuitry. The final evaluation and the beam dump decision is made by the threshold controller, which is located in a surface building. The threshold controllers are located at eight locations where they are connection to the beam abort system. The cable lenght between the frontend electronic and the threshold controllers varries between 1.8 and 2 km The threshold controller itself calculates the beam losses for different loss durations and beam energies and compares them with the appropriate treshold values. A warning or beam dump signal is generated if the tresholds are exceeded.

FRONT END ELECTRONIC DESIGN

The quench levels are varying between 600 pA and 500 μA. To ensure high sensitivity at low losses, the dynamic range must be extended at least one order of magnitude from 600 pA down to 60 pA. Thus the lower limit was set to 50 pA, which results in a dynamic range of 140 dB.

The integrated loss rate has to be acquired in an interval between 89 μs and 100. The minimum time limit is given by the fastest beam dump possibility after one revolution (89 μs) and the longest interval is set by the temperature measurement system of the He of the magnets to 100 s. Simulation studies revealed [1] that the quench level prediction is accurate to ± 50 %. Consequently, the maximum allowable error of the readout electronics was set to ± 10 %.

The usage of the ionization chamber leads to the measurement of a particle loss rate equivalent electric current. Several design considerations to acquire this current have been made [4]. Figure 6 shows an overview of the different methods that have been taken into account.

Direct monitoring techniques provide an output signal that is proportional to the input current. A transimpedance amplifier, e.g., transforms the input current into a proportional voltage. The output voltage can be sampled with an ADC at the required

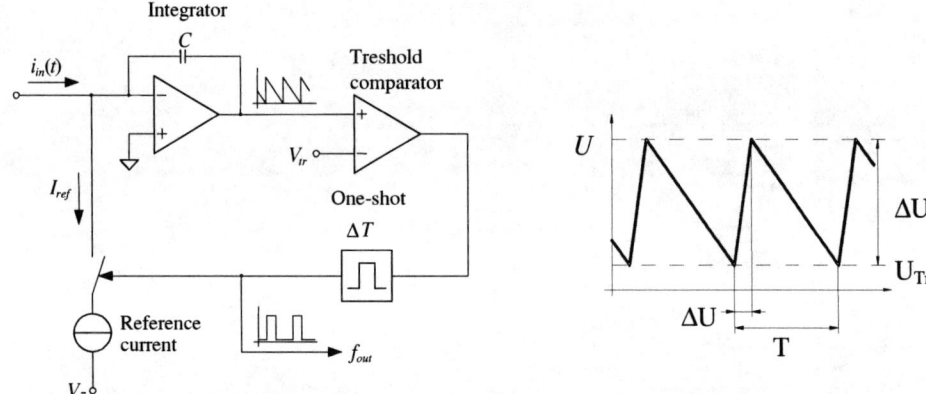

FIGURE 7. Left: Circuit diagram of the current to frequency converter. Right: Output voltage of the operational amplifier versus time applying a constant input current.

frequency, which is determined by the evaluation time of the particle loss rate. It is not possible to achieve with this method a dynamic range of more than approximately 60 dB without introducing large errors. Thus several gain stages have to be incorporated, which have some other features (see pros and cons indicated at the figure: 6).

Integrating techniques compared to direct monitoring provide only average loss rates, but offers unique properties in terms of dynamic range and the signal to noise ratio. The most simple way to realize such a circuit would be an ADC that offers a high dynamic range. Several components, especially for audio applications, have been found but none offered a sufficient low input current. The switched integrator is a widely used circuit to measure small currents. An electric current charges the feedback capacitor of an operational amplifier and the output voltage is equal to the integral of the current over time. However, the dynamic range is only 60 dB. This range can be extended, if a certain treshold voltage is introduced. If the output voltage reaches the treshold within the time interval, the integrator will be reset. At the end of the integration interval, the actual output voltage is sampled as before, but also the number of reset actions within the interval are taken into account. As long as the reset time is much shorter than the integration time, the introduced error can be neglected.

The current to frequency converter functional principle (CFC) is shown in figure 7. It consists of an integrator, whose capacitor is charged by the signal current and discharged by a fixed reference current. The concept of a CFC was successfully improved and used by H. Reeg [5, 6]. The authors investigations are based on the work of E. G. Shapiro [7].

Figure 7, right depicts a typical analog output voltage of the CFC with a constant input current that is almost at the maximum value of the operating range. If the output voltage ramps negative and reaches the treshold voltage U_{Tr}, the one-shot switches the reference current source I_{ref} to the summing node for a fixed time interval ΔT. As the reference current has the opposite polarity and is larger as the signal current, the integrator output voltage is increasing. Later I_{ref} is disconnected, the output ramps again down until it reaches the threshold potential. Assuming an ideal operational amplifier, with no input

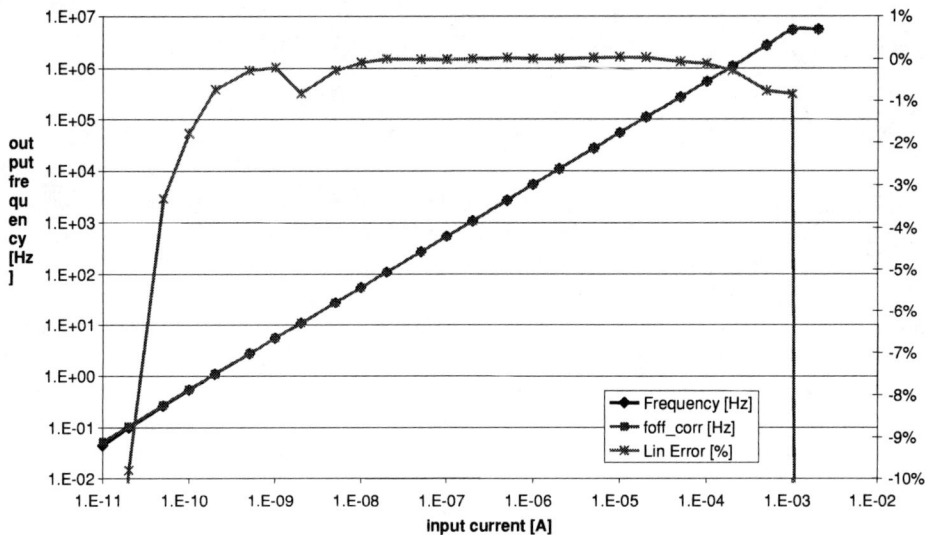

FIGURE 8. CFC output frequency and linearity error versus input current for a maximum current of 2 mA.

current and no offset voltage, a fixed charge $Q_{ref} = I_{ref} \cdot \Delta T$ is extracted from the summing node at the operational amplifier during the reset period ΔT. This charge is equal to the the total signal input charge during the period T. Because the signal charge and the capacitor charge are equal to $Q_{ref} = I_{ref} \cdot \Delta T$ during the time ΔT and since the accumulated charge of the capacitor during the period ΔT and $T - \Delta T$ are equal. By counting the number of reset actions the input current is convertted into an frequency, were one count represents the average integrated charge in the period T. The relation between current and frequency is given by:

$$f = \frac{1}{I_{ref} \Delta T} I_{in} \qquad (1)$$

The output frequency is not depending on the integration capacitor. However, it has to be selected carefully because it determines the amplitude of the integrator output voltage but also the sensitivity to noise and to the charge injection from the current switch. The relative error on the frequency depends only on the square root of the quadratic sum of the relative errors on the reference currents and on the reference time. Since the relative conversion error should be below 0.1 this could be reached without any calibration.

The output frequency of the CFC was measured over the whole dynamic range (see Fig.: 8) [4]. The measured frequency and the linearity error are shown in figure 8. The straight line is set equal to the data at 5 kHz. The CFC shows a deviation from the straight line of less than ± 0.5 % over an in-put range of 500 pA to 100 μA. At low input currents, the error increases mainly because of the total leakage current. Above 100 μA, the one-shot works partly in the instable mode, which leads to a larger error. If the current exceeds 200 μA, the CFC is overloaded and becomes non-linear. The CFC is well within the specified error margin of ±10 %. The absolute error depends on the actual value of I_{ref} and T but is constant over the whole dynamic range.

CONCLUSION

Longitudinal beam loss distribution studies show that losses concentrate on locations with high β-functions (centre of quadrupole magnets) or where mechanical limitations of the aperture can be assumed. From shower simulations at the different loss locations we see that a set of six detectors around the quadrupoles is sufficient for localizing the beam losses and to distinct between the two beams. Ionization chambers will be used as shower particle detectors, which convert the particle loss rate into an electric current. The expected ionization chamber current, equivalent to the quench levels, varies between 500 pA and 500 μA depending on the different loss distributions and detector positions. Introducing a ten times higher sensitivity a dynamic range of 140 dB is needed. Several possibilities to measure the current of the chamber are discussed. The current-to-frequency converter is the most appropriate solution that is able to cover the wide dynamic range with low effort. It produces an output frequency proportional to the input current and simplifies the data transmission from the tunnel to the surface, where the beam loss signal is evaluated. Tests revealed an excellent linearity of the current to frequency converter. The frequency will be counted in intervals of 40 μs and a serial word will be transmitted together with 5 other channel over the cable to the location of the threshold controller.

REFERENCES

1. J.B. Jeanneret et al., Quench levels and transient beam losses in LHC magnets, LHC Project Report 44, CERN, (1996).
2. A. Arauzo-Garcia et al., LHC Beam Loss Monitors, CERN-SL-2001-027-BI, CERN, (2001).
3. E. Gschwendtner et al., The Beam Loss Detection System of the LHC Ring, 8th European Particle Accelerator Conference, (2002).
4. W. Friesenbichler, Development of the Readout Electronics for the Beam Loss Monitors of the LHC, CERN-THESIS-2002-028 (2002).
5. H. Reeg and O. Keller, Linearer Strom-Frequenz-Konverter, Patent No. DE 195 20 315, Deutsches Patentamt, (1996).
6. H. Reeg, A Current Digitizer for Ionization Chambers, SEMS with high resolution and fast response, Proceedings of the 4 th Workshop on Diagnostics and Instrumentation in Particle Accelerators DIPAC, Daresbury, UK, pp. 140-142, (1999).
7. E. G. Shapiro, Linear Seven Decade Current-to-Frequency Converter, IEEE Trans. Nuclear Science, Vol. 17, pp. 335-344, (1970).

Very High Resolution Optical Transition Radiation Beam Profile Monitor*

Marc Ross, Scott Anderson, Josef Frisch, Keith Jobe, Douglas McCormick, Bobby McKee, Janice Nelson, Tonee Smith

Stanford Linear Accelerator Center, Menlo Park, CA, 94025, USA

Hitoshi Hayano, Takashi Naito, Nobuhiro Terunuma

High Energy Accelerator Research Organization, Tsukuba, Ibaraki 305-080,1 Japan

Abstract. We have constructed and tested a 2 um resolution beam profile monitor based on optical transition radiation (OTR). Theoretical studies of OTR [1] show that extremely high resolution, of the order of the wavelength of the light detected, is possible. Such high-resolution single pulse profile monitors will be very useful for future free electron laser and linear collider projects. Using the very low emittance 1.3 GeV electron beam at the KEK Accelerator Test Facility (ATF) [2] (1.4nm εx x 15pm εy), we have imaged transition radiation from 5 micron σ beam spots. Our test device consisted of a finely polished target, a thin fused silica window, a 35 mm working distance microscope objective (5x and 10x) and a triggered CCD camera. A wire scanner located near the target is used to verify the profile monitor performance. In this paper we report results of beam tests.

INTRODUCTION

Measurement of the low emittance beams produced by a linear collider damping ring require a device that can image beam spots as small as 5 microns. Small beam sizes are often measured using wire scanners [3], requiring many machine pulses, often with an over-estimation of the beam size due to beam position and intensity jitter [4]. In addition, wire scanner measurements can take up to half a minute to complete. An optical beam spot size monitor based on OTR can be used to record images from single pulses and, in principle, have a resolution of about one micron.

Optical beam size monitors are typically based on fluorescent screens, OTR or synchrotron radiation. Fluorescent screens are practical when σ > 30 μm [5]. Beam spots with σ < 30 μm are difficult to measure due to effects from phosphor grain size and phosphor transparency, which gives rise to depth of field problems. Synchrotron radiation from a bend magnet is confined to a cone of opening angle $1/\gamma$ (< 0.5 mrad at

---

* Work supported by Department of Energy, Contract DE-AC03-76SF00515

the ATF where $E_b \sim 1.3$ GeV). The diffraction-limited resolution due to this opening angle is typically much larger than the beam size for typical linear collider damping ring parameters.

Transition radiation is produced when relativistic charged particles transit through the surface of a conductor. The backward directed radiation is emitted from the surface at an angle equal to the angle made by the incoming beam. The radiation is emitted primarily in a cone with an opening angle of about 1 mrad for 1.3 GeV. With this opening angle, the expected resolution limit due to diffraction might be expected to be similar to that for synchrotron radiation. However, the transition radiation distribution has large tails that emerge at greater angles, the key feature that makes a very high-resolution OTR monitor possible. Because the radiation is primarily forward-directed, the large angle radiation must be efficiently collected, requiring an optical system that has a larger numerical aperture (by approximately a factor of 2) than would be needed for a diffuse light source.

OTR beam profile monitors will be used in the linear collider or in an FEL light source to check the emittance and beam optics match. Linear colliders are designed with large aspect ratio ($\sigma x/\sigma y$), 'flat' beams. The monitor will be used as part of the skew correction process in which $x - y$ coupling is corrected by a sequence of skew quad magnets. Simulations show that even with ideal skew correction systems, the presence of small, nominal wire scanner errors can make tuning algorithms unstable [6]. OTR has the promise of reducing the error on the beam tilt measurement.

Some controversy exists over the ultimate resolution of an OTR profile monitor[1]. Tests reported here include a search for the minimum observable beam spot. A 5.5 um high, ~200 um wide beam stripe was generated and clearly imaged.

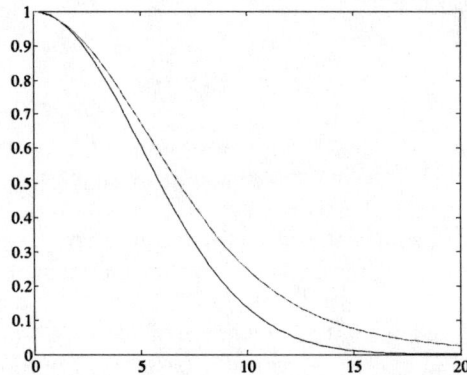

FIGURE 1. A calculated 5-micron spot (lower curve) and its predicted image (top curve).

The OTR can be susceptible to beam related damage when high charge density beams impinge on its polished surface. Since the wide image mentioned above was quite faint, we made a much smaller beam spot and subsequently, over the course of a few minutes operation at 0.75 Hz, noticed damage on the surface of the mirror. The damage quickly ruined the mirror to such an extent that further studies of resolution were not possible.

The design goal of the OTR monitor in the ATF extraction line is intended to measure a (σx, σy) ~ (50, 5) micron beam with better than 10% resolution. The monitor uses a mechanically polished mirror target approximately 500 um thick with an entire field of view of 360 x 250 um. In this paper we discuss the design, installation, and beam testing of an OTR beam size monitor in the ATF extraction line. We also report tests of several different target materials: Cu, Be, glassy Carbon, Ti and Si.

OPTICS

The resolution required to measure the minimum expected beam size in the ATF with a 10% accuracy is about 2 um. Unfortunately, imaging with an ideal lens through a thick vacuum window introduces spherical and chromatic aberrations. Commercial microscope objectives are not compatible with high vacuum systems, and design and construction of custom vacuum compatible optics is time consuming and expensive.

While it is possible to design a lens to compensate for the aberration caused by a window we decided to use a commercial microscope objective to avoid the cost of a custom design. ZEMAX [7] calculations indicated that using a thin (<1 mm) window would reduce the effect of the aberrations to within our design tolerances. Mechanical calculations show that the distortion due to vacuum pressure is acceptable (<$\lambda/4$) for a fused silica window with a diameter of <7 mm.

To meet the high numerical aperture requirements, a long working distance microscope objective manufactured by Mitutoyo was chosen. Their 5X objective has a numerical aperture of 0.14, with a focal length of 40 mm and working distance of 34 mm. The lens is designed for use at infinite conjugate ratio (i.e. focuses a point to a parallel beam), and gives a depth of focus of approximately ± 7 um. The lens has a specified resolving power of 2um, corresponding to diffraction-limited performance. This is roughly equivalent to a resolution of 1 um sigma.

Figure 2 shows a schematic of the OTR monitor. The lens is mounted to a 200 mm tube lens adapter and a 200mm tube that mounts directly to a C-mount camera. The objective and adapter provide a magnification of 5X onto the camera, which is reasonable for a camera pixel size of ~10 μm. For some tests a 10X lens with a numerical aperture of 0.28, but otherwise similar design, was used.

Transition radiation from a target that is not normal to the electron beam is emitted in a direction corresponding to a geometric reflection of the incident beam angle. The microscope axis must therefore be aligned at an angle to the target, resulting in a change in focal depth with position on the target. Tilting the image plane of the camera can in principal compensate this tilted object plane, however for our parameters this was impractical. Instead we oriented the target so that the small vertical spot dimension was aligned with the tilt angle, and accepted a very narrow field of view. The target angle relative to the beam was chosen to be 20 degrees, the smallest allowed by mechanical constraints.

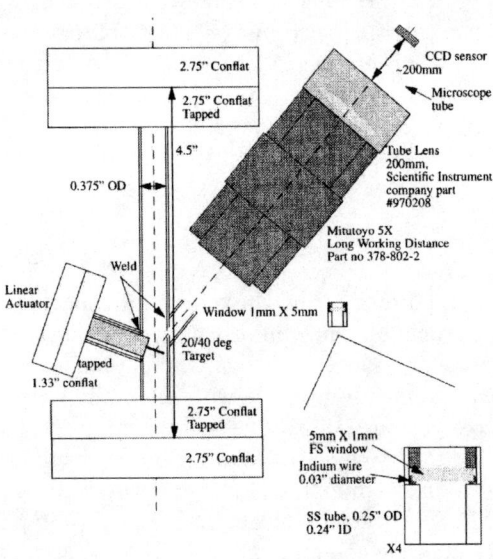

FIGURE 2. Schematic drawing of the OTR monitor with the beam entering from the top of the view. This shows the lens setup of the camera telescope

The optical system should image a point source to a spot with a sigma of ~1 micron; however transition radiation is forward directed, resulting in poorer resolution than for a uniform source. A numerical calculation using the predicted OTR distribution shows that a 5 um sigma spot will be broadened by <10% by the convolution of the transition radiation and diffraction (see figure 1).

MECHANICAL DESIGN

There are two critical mechanical design issues: (1) placing the microscope objective within its working distance, and (2) sealing the thin fused silica window in the body of the device without distorting it.

In order to have the objective within its working distance, the beam pipe would have to be about 1 cm in diameter. Even with the small beams in the ATF extraction line, 1 cm was determined to be too small and a pipe mover was developed, located in the background of figure 3.

The window port was machined to accept an indium seal, and its bore was threaded to accept a screw plug for compressing the window between an O-ring and the indium seal. This seal proved to be problematic and required careful installation to make it vacuum tight.

The target is inserted and removed by a pneumatic actuator that consists of a thin rod to which the target is attached with screws. To make the target insertion reproducible, a stainless steel ball welded on the end of the actuator is designed to seat into a titanium conical receptacle in the wall of the device.

The microscope objective is mounted to an optical translation stage that is remotely controllable using a micro-stepper motor. The stage is used to control the camera focus.

BEAM STUDIES

The purpose of the beam studies was to test the OTR monitor resolution and to begin comparing beam emittance estimates made with the monitor and the nearby wire scanners. First, operational parameters such as focus, depth of field and calibration were tested and a 'minimum' beam size measurement was done. Following that, several emittance measurements were made using the nominal ATF beam.

An alignment test stand was prepared to position the apparatus such that an laser could be shown down the beam pipe onto the target. The target was rotated by loosening the vacuum flange of the actuator and rotating it until the laser light projected out of the center of the window. The target was scribed with four lines using a razor blade to provide position information. Using the laser light, the camera was adjusted until two of the vertical lines were on the edges of the field of view. The apparatus was then installed in the ATF extraction line near one of the wire scanners.

FIGURE 3. The OTR monitor installed in the ATF extraction line.

The extraction line was setup with normal beam optics, and the Cu target was inserted. The light spot was quickly found, steered to the center of the screen, and digitized to obtain FWHM size in digitizer channels.

The camera focus was checked by using a stepper motor to move the camera stage. The depth of focus scan clearly indicated that the beam had to be kept within a ± 25 micron vertical window in order to avoid affecting the beam size measurement due to out of focus effects.

The calibration of digitizer channels to microns in x was calculated by positioning the beam over each of two vertical scribe lines that are 0.5 mm apart. To calculate the calibration for the vertical channels, a reference BPM orbit was saved and both nearby wire scanner y wires were scanned and their centroids noted. Then a corrector was moved by a known amount, another BPM orbit was saved, and the wires scanned again. By plotting the wire scanner centroid and BPM position changes against the predicted changes based on the optics model, the change in the position at the OTR was calculated. This distance was divided by the change between the centroids of the digitized spots (in channels) to obtain the y calibration in microns/channel. The vertical calibration was calculated to be 1.06 µm/channel and the horizontal to be 1.12 microns/channel, accurate to 2%.

For the resolution test, the optics in the extraction line were adjusted to produce a calculated beam size of 1 micron at the OTR. The quadrupole values were implemented in stages to prevent the beam moving out of the camera field of view. At each stage, the beam image was digitized and the y spot size was recorded. Over the course of the scan, an increased tilt was observed in the beam spot; several attempts were made to correct this with skew quads, with only partial success. As the y spot size was reduced (and the x spot size greatly increased), the available light intensity became so small that optimization tuning was difficult as the spot was too dim to see.

At the smallest vertical beam size, we were unable to find the maximum resolution of the device due to decreased light and poor signal to noise ratio. The smallest y spot size measurement was approximately 5.5 microns. Figure 4 below shows OTR images as captured by the digitizer software with projections in x and y.

Next, the beryllium target was installed in the OTR and the nominal optics was implemented in the extraction line. The measurements from the quadrupole scans in the normal optics compare well to five-wire emittance measurements taken after the quadrupole scans were completed. The results are displayed in the Table 1. The slight discrepancies in the results are possibly due to the fact that the five-wire and quadrupole scan measurements were not taken with precisely the same optics setup: while both setups started with the nominal extraction line optics, the extraction line dispersion was measured and corrected in each case to produce two possibly slightly different optics starting points. The quadrupole emittance measurements take into account the dispersion at each monitor, the wire sizes of MW3X and MW4X and the estimated resolution error of the OTR. The OTR is estimated to have a resolution of 2-3 µm and a 2 µm addition in quadrature was used for the quad emittance calculation.

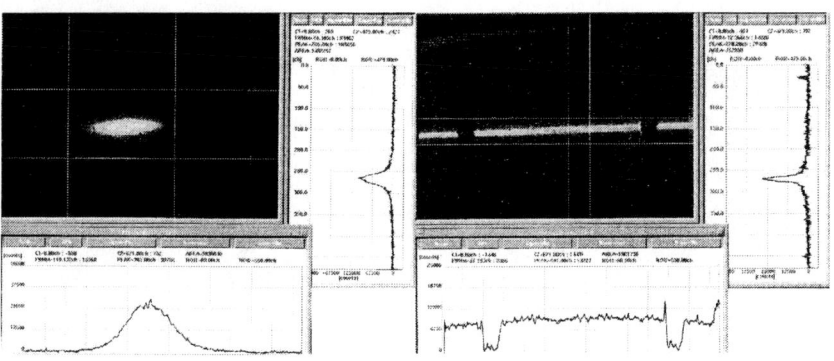

FIGURE 4. Beam spot as seen by Be OTR with normal extraction line optics (left image). The y FWHM is 24 channels, which corresponds to a y sigma of about 10 μm. Beam spot at OTR with "2 μm" optics loaded (right image). Here the y FWHM is 12.9 channels or about 5.8 μm. The gaps in the image are caused by the scribe marks used for calibration.

The wire and uncorrected OTR emittance measurements don't take into account the tilt of the beam, which varied between two and ten degrees for the OTR during the quad scan, and by some unknown amount for the wire scanners for both the quad scan and the five-wire measurements. If the tilt is taken into account for the OTR measurements, the emittance measured by the OTR is reduced by 40%.

TABLE 1. Vertical emittance measurements from OTR, MW3X and MW4X quad scans and five-wire emittance measurement.

Device	ε_y
OTR	2.741×10^{-11} +/- 8.867×10^{-13} m
OTR, tilt corrected	1.779×10^{-11} +- 1.435×10^{-12} m
Wire Scanner 3 (upstream)	2.222×10^{-11} +/- 1.165×10^{-12} m
Wire Scanner 4 (downstream)	2.789×10^{-11} +/- 9.029×10^{-13} m
Five-wire	2.55×10^{-11} +/- 1.2×10^{-12} m

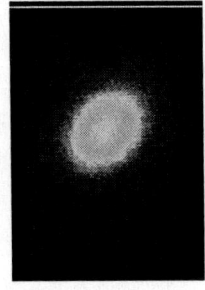 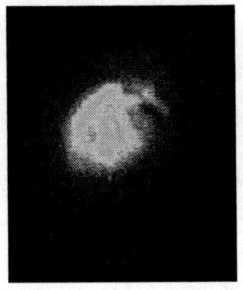

FIGURE 5. Beam on *Be* target before (left) and after (right) five minutes at 5.5 x10^{10} e⁻/bunch train.

TARGET MATERIAL TESTS

The 'energy to break' is defined in equation 1 as the energy at which the material strength is exceeded due to thermal induced mechanical stress:

$$ETB = 2*(1-\mu)*\frac{UTS*C}{\alpha*E} \quad (J/g) \quad (1)$$

where μ is Poisson's Ratio, UTS is the Ultimate Tensile Strength, C is the heat capacity, α is the coefficient of thermal expansion and E is Young's modulus. For a gaussian distribution of N particles with rms σ_r, the normalized probability distribution is p(r):

$$p(r) = \frac{N}{2\pi\sigma_r^2} e^{-\frac{r^2}{2\sigma_r^2}}$$

For energy deposition, $\frac{dE}{dx}$, in units of $\frac{MeV-cm^2}{g}$, q = the charge per electron, σ_r in meters and the peak energy deposition is ΔE_{peak} (in units of J/g):

$$\Delta E_{peak} = 100q\frac{dE}{dx}\frac{N}{2\pi\sigma_r^2} \quad (J/g)$$

The temperature rise is $\Delta T(r)$:

$$\Delta T(r) = \frac{23.9}{C}q\frac{dE}{dx}\frac{N}{2\pi\sigma_r^2}e^{-\frac{r^2}{2\sigma_r^2}} \quad (°C) \text{ and } \Delta T_{peak} = \frac{23.9}{C}q\frac{dE}{dx}\frac{N}{2\pi\sigma_r^2} \quad (°C)$$

where C = the heat capacity in units of cal/(g-°C)

Table 2 summarizes the material damage tests. To test the damage threshold of the beryllium target, the extraction line optics were set up to deliver a 13 by 10 µm of 8 x

10^9 electrons per beam pulse at the OTR. The beam was left on the target over the course of thirty minutes with no damage seen using both the normal x5 lens and a x10 lens. This result is in direct contrast to the damage seen on the copper targets at the same intensities and small spot sizes. Since no damage occurred at the lower current in single bunch mode, the beryllium target was then subjected to 5.5 x10^{10} electrons in 20 bunches per pulse, with the same small spot optics. At this higher intensity, damage occurred after five minutes. Figure 5 shows the OTR beam image at the beginning of this high current destruction test (left image) and the OTR beam image after five minutes (right image). The beryllium target shows noticeable damage that alters the image and would compromise a beam size measurement. The glassy carbon II target was not subjected to a high current destruction test due to time limitations. The sequence of figures, (6a-c), show the progression of damage to the polished Cu surface. At first, a smooth image is seen, shortly after steering the finely focused spot onto a fresh surface. The features above the spot in Figure 6a are from earlier tests. Then, after 3-4 minutes, (Figure 6b), a wrinkled looking image begins to emerge followed by, (Figure 6c), substantially wrinkling. At that point, no further obvious signs of damage appeared after about ½ hour of operation.

TABLE 2: Target material damage test results. The table shows what was observed (D: Damaged, ND: No Damage, I: Inconclusive) and the estimated energy deposition (J/gm). All tests were done with a 1.5 Hz repetition rate.

Target Material	Energy to Break (J/gm)	Single bunch 7.5e9 20x12um	Single bunch 8.9e9 13x10um	Multibunch 5.5e10 20x15um	Multibunch 2.9e10 43x34um	Multibunch 2.8e10 16x10um
Cu	98	D; 118 J/gm				
Be	484	ND; 150	ND; 315	D; 845		
C (II)	580-735	ND; 140				
Si	(?)	ND; 130			ND; 91	
Ti	710	ND; 119				I; 670

At the final stage of damage, the target must have some vertical features or relief, because it became quite difficult to determine when the image was in focus. Different features came into focus at different lens stage positions. Using these expressions a peak deposited energy of 67 J/g and a temperature rise of 174 deg C is expected in the copper for a round spot, beam size σ_{xy} = 20 microns and N = 0.75e10.

In the experiment described here, the beam size σ_{xy} ~ 20 x 12 microns raises the energy deposition by 67% to 112 J/g and the temperature to 290 deg C. While this is well below the melting point of copper (1083 deg C), it is above the temperature where local pressure in the bulk material causes plastic deformation, estimated to be about 180 deg C. The thermal diffusion constant alpha (alpha^2 = 1.16 cm^2/s) is high enough such that at 1 Hz we don't expect any noticeable temperature build up.

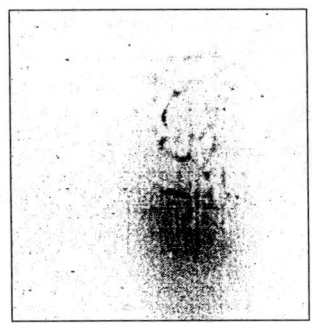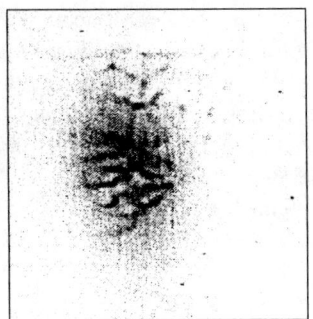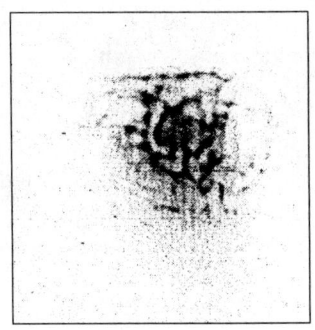

FIGURE 6(a): Negative image of a 20 x 12 um (σ_x, σ_y) beam spot before damage. Some damage is visible above the beam spot from earlier tests. The figure shows the entire field of view (360 x 250 microns). Figure 6(b): Taken a few minutes after the onset of damage. Figure 6(c): Showing damage at a more advanced stage.

The damage to beryllium at 5.5 x10^{10} electrons is not surprising. The energy deposited was roughly 850 J/g, which is well above the energy to break of 484 J/g for beryllium [8]. Glassy carbon type II's energy to break is in the range of 580-735J/g (assuming a 0.29-0.1 µ) and it showed no damage at the single bunch currents up to 0.8 x10^{10} in the small spot optics. The results of the tests are summarized in table 2.

CONCLUSIONS

The OTR was successful in measuring the size and emittance of the beam in the ATF extraction line using two different target materials, beryllium and glassy carbon II. The measurements of beam size versus the predicted size from the five-wire emittance calculation agreed well in the vertical plane and were within 15% in the horizontal plane. The quadrupole scan emittance measurement at the OTR agrees well with the simultaneous scanner downstream wire scanner measurement and with a subsequent five-wire scan emittance measurement. The OTR and the nearby upstream scanner measurements differ by 25%, far outside the error. The tilt of the beam at each of the measuring devices was not corrected online in any of the measurements in this experiment. When the skew was corrected offline for the OTR, the resulting emittance was reduced by 40%, well below the uncorrected wire scanner results. It is presumed that correcting for the beam tilt at the wires would similarly reduce the wire scanner emittance results.

The measurements completed in these experiments suggest that the OTR could be a valuable tool for measuring the beam size and emittance parameters from the ATF damping ring. Given its apparent resolution and its ability to take horizontal and vertical beam size measurements in one beam pulse and to take many measurements quickly, the OTR should be able to measure the beam emittance with high statistics, giving a low error and a good understanding of emittance jitter. Multiple OTR measuring devices located near the wire scanners in the ATF extraction line would be a definitive test of the OTR as a beam emittance diagnostic device.

REFERENCES

1 M. Castellano, V.A. Verzilov, Phys. Rev. ST Accel. Beams 1, 062801 (1998)
2 T.Okugi et al., Phys. Rev. ST Accel. Beams 2, 022801 (1999)
3 H. Hayano, 'Wire scanners for small emittance beam measurement in ATF', Linac 2000.
4 M.C. Ross, 'Wire scanner systems for beam size and emittance measurements at SLC', Second Accelerator Instrumentation Workshop (1990).
5 F. Decker, 'Beam Size Measurement at High Radiation Levels', PAC 1991 (SLAC-PUB-5481).
6 M. Woodley, KEK Accelerator Test Facility Internal Note 99-07, http://atfweb.kek.jp/atf/Reports/ATF-99-07.pdf.
7 http://www.optima-research.com/Software/Optical/Zemax/
8 Private Communication. Material damage calculation from John Sheppard, which is used in explaining damage to e+ target materials.

ELETTRA Photon Source Beam Stabilization [1]

Steve N. Thanos and Rodger H. Hosking

Pentek, Inc.
Upper Saddle River, New Jersey, USA

Abstract. This paper outlines the technical issues involved in configuring an electron beam transverse multi-bunch feedback system for the Elettra Synchrotron in Trieste, Italy. By measuring the electron beam with fast A/D converters and then analyzing the error in beam position using multiple digital signal processors, a corrective kick is applied to the beam to damp its instabilities. Commercial data acquisition and DSP products were successfully deployed to achieve real time operation.

INTRODUCTION

ELETTRA is a third-generation synchrotron radiation facility in Trieste, Italy. The facility provides the scientific community with photons in the range of a few electron volts (eV) to several tens of thousands (keV), the latter corresponding to the spectral domain of soft X-rays. The main characteristics of the synchrotron radiation are very high brightness, polarization and wide range of tunability in wavelength.

Figure 1 is an aerial photograph of the facility. The light source is composed of three parts: a linear accelerator (commonly known as a linac), a transfer line and the storage ring.

Figure 1. The ELLETRA facility in Trieste, Italy (Courtesy of ELETTRA)

The storage ring is filled once a day by the linac with 0.9 GeV electrons whose energy is eventually increased up to 2 – 2.4 GeV. Synchrotron radiation is produced when electrons traveling at relativistic speeds are deflected in magnetic fields. The brightness of the photon beam is derived from the small transverse size and divergence of the electron beam, which is called emittance.

Low emittance, stability, reproducibility and long lifetime of the stored beam are the main requirements for the light source. These properties are a function of the beam environment, e.g. the magnetic field lattice, the accelerating electric fields and the quality of vacuum in the vacuum chamber.

Beam instabilities have different time scales ranging from milliseconds to months and different techniques are used for their suppression and control. Although much is done to passively control instabilities, active feedback systems working at different frequencies and with different bandwidths are needed. Given the small size of the beam (typically tens of microns), stabilizing it requires exceptional performance in position measurement and correction.

SYSTEM OVERVIEW

The electron beam stored in a synchrotron light source is not a continuous beam. It has a "bunched" structure produced by the effect of one or more RF cavities installed in the ring and used to replenish the energy lost by the electrons as electromagnetic radiation on each turn. The reason for this bunch structure is that only electrons arriving at the right time will be accelerated, while the rest are lost.

At ELETTRA, the RF frequency of the four cavities is 500 MHz, so the electron bunches are spaced 2 nsec apart. As currents in the ring are increased, the very high electromagnetic field associated with each bunch interacts with the surrounding vacuum chamber and can start resonating, in one or more storage ring positions, until the next bunch arrives.

As a result, the bunches are no longer independent, but they become coupled to each other by the action of their respective electromagnetic fields. The result is that bunches start oscillating at a characteristic frequency, the betatron frequency (in the order of a few hundreds of kHz) but with different phase relationships among them resulting in different "modes".

There are as many normal modes as there are bunches. In the case of ELETTRA, there are 432. Such coupled-bunch instabilities can be cured by the use of active feedback systems. The approach used at ELETTRA involves a digital feedback system that samples the bunch positions in the storage ring and applies corrections computed by digital signal processors.

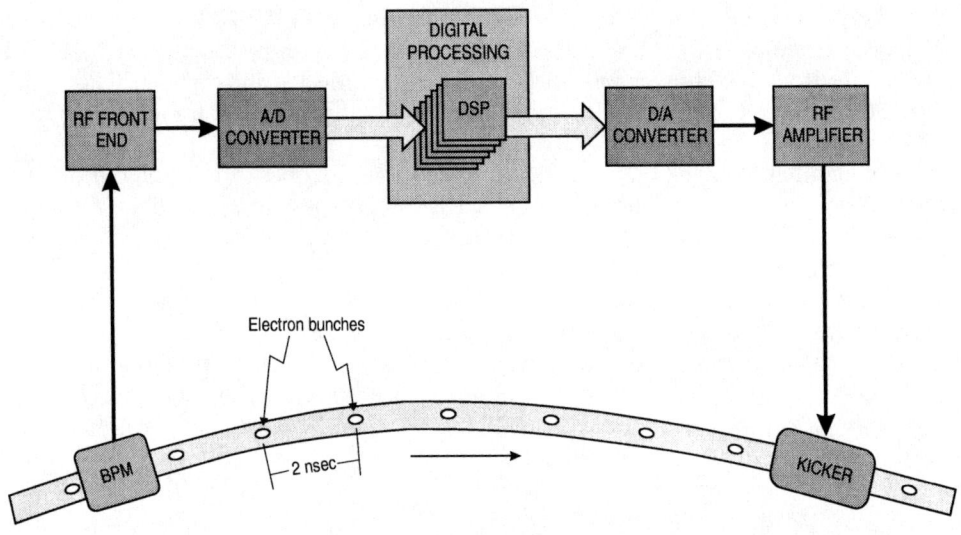

FIGURE 2. Feedback System Block Diagram

As shown in Figure 2, the system includes a 2-axis "Beam Position Monitor" (BPM) and a position corrector, appropriately called a "Kicker". One BPM is used to produce the vertical and horizontal position error signals, while two stripline kickers are used for correcting in the two planes. The wideband signals are demodulated by the RF front end to produce a 0–250 MHz baseband signal that represents the position error of the bunches as they pass through the BPM at a frequency of 500 MHz.

The output analog signal is sampled by a fast 500 MHz A/D converter and passed onto a bank of DSPs that calculate the required correction. The correction is then converted into analog form by a 500 MHz D/A converter. The analog signal represents the correction that has to be applied to the bunches as they pass through the kicker. A RF power amplifier supplies the necessary power to drive the kicker.

SIGNAL PROCESSING

The ELETTRA Transverse Multibunch Feedback consists of a wideband bunch-by-bunch system where the position errors of the 432 bunches, separated by 2 nsec of beam travel time, are individually corrected. After demodulating the wideband signal from the BPM, the baseband x and y signals are sampled by 8-bit 500 MHz A/D converters. One converter is used for the vertical signal and another one for the horizontal.

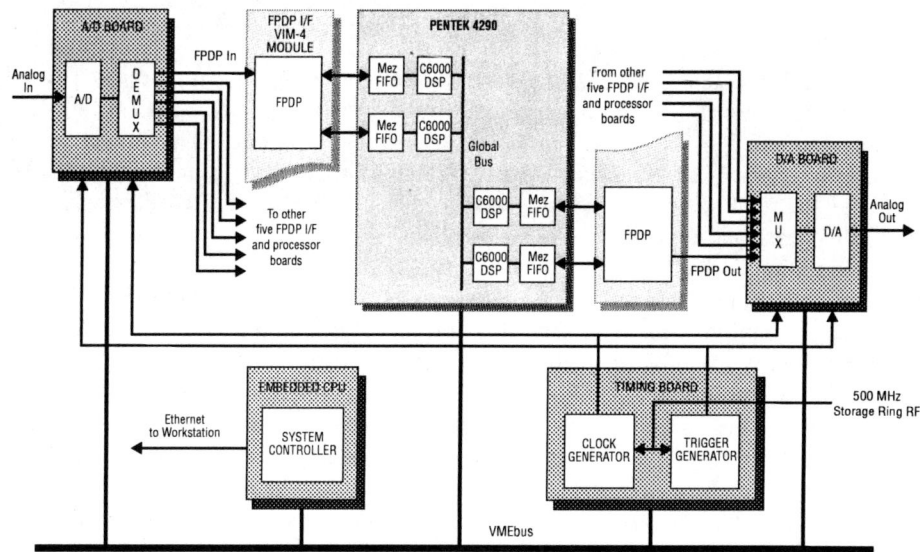

FIGURE 3. Signal Processing Block Diagram

We will address the vertical position signal processing first. As shown in Figure 3, the 500 Mbyte/ sec data flux is first demultiplexed into six 32-bit FPDP (Front Panel Data Port) channels. The data from each channel is then sent through a custom VIM-4 mezzanine module designed by ELETTRA to meet the needs of this project. It provides a FPDP interface with the demultiplexer to distribute the A/D data to the four DSPs of the Pentek Quad 'C6201 board.

The data coming from one FPDP channel, which corresponds to 72 bunches, is passed to one DSP for online diagnostics and signal analysis. Concurrently, the data is split evenly over the remaining three DSPs, each of which executes the feedback algorithm on its respective 24 bunches. Thus, each DSP is responsible for processing the samples of a given group of bunches all the time. To process all 432 bunches, six Model 4290's containing 24 'C6201's are used.

The FPDP input interface receives 32-bit words and writes them to the BI-FIFO of the DSP to which they are specifically assigned for processing. At the same time, the VIM-4 mezzanine reads the BI-FIFO which contains the calculated data words and sends them to the FPDP output interface. The FPDP interfaces act as bidirectional programmable switches for each incoming and outgoing word and the switching rules are downloaded in a table at the beginning of system initialization.

The output data from the six custom FPDP mezzanines are multiplexed by the D/A board and converted to analog form by the 8-bit 500 MHz D/A converter. The entire process is synchronized by the timing electronics. All the electronics required to stabilize the beam in the vertical direction are housed in one VME cage in the service area. One additional VME cage with identical electronics is used for the horizontal direction.

CRITICAL ISSUES

The 'C6201 is a fixed-point DSP clocked at 200 MHz (5 nsec instruction cycle). Its VLIW architecture allows it to execute up to 8 instructions per clock cycle. The requirement here is to execute all the necessary operations in one beam revolution, or 864 nsec (2 nsec x 432 bunches). With a highly optimized code written in assembly language, the time needed to execute the required code of a 5-tap FIR filter, which is used as the feedback element on the 432 bunches, is 600 nsec which is shorter than the beam revolution time.

Another critical issue is the time required for data transfers between the DSP and its BI-FIFO. To minimize the time data stays on the board, thorough use of the Pentek BI-FIFOs, interrupts and DMAs allows very efficient data transfer without interfering with the algorithm execution. The feedback system relies on very strict timing. The A/D converter must sample the analog signal synchronously to the bunch crossing at the BPM and the D/A converter must generate the analog correction signal in phase with the same bunch passing through the kicker.

In addition, the start A/D and D/A triggers must start the conversions in a deterministic and repeatable way with respect to the bunch structure in order to let every DSP work with a known group of bunches.

RESULTS

Figure 4 consists of two synchrotron beam profile images: the one on the left is with the feedback loop open, while the one on the right was taken with only the vertical feedback loop closed. The photographs show how effective the control is at improving beam stability.

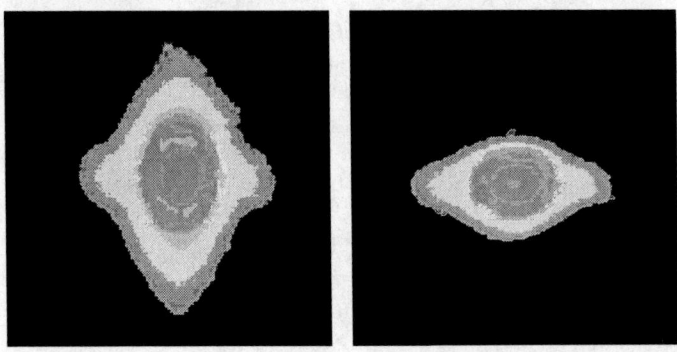

FIGURE 4. Synchotron beam profiles: left, without feedback; right, with only vertical axis feedback stabilization. (Courtesy of ELETTRA)

ACKNOWLEDGEMENTS

The authors wish to express kindest acknowledgements to Dr. Daniele Bulfone and Dr. Marco Lonza of ELETTRA who are responsible for conceptualizing the system. Their dedicated engineering team overcame many system integration challenges which ultimately brought this remarkable system into reality. We are extremely grateful to them for their time in sharing system details and in reviewing and editing the technical content of this paper.

REFERENCES

(1) This article originally appeared in Pentek Pipeline Newsletter, Volume 10, No. 3, Fall 2001 and is reprinted with permission of Pentek, Inc. An on-line version of this article entitled "Pentek DSPs Keep the ELETTRA Photon Beam Stable and Bright", is available from Pentek's website at:
http://www.pentek.com/pildocs/6982/pipelines/PIPE103.PDF

(2) More information about the ELETTRA facility in Trieste, Italy, is available from the facility website at: www.elettra.trieste.it

Posters

10th Beam Instrumentation Workshop
May 6-9, 2002

Progress of Turn-By-Turn System for HLS *

J.H.Wang, J.H.Liu, B.G.Sun, W.M.Li, Z.P.Liu, Z.F.Zhang, P.Lu

NSRL, University of Science and Technology of China, Hefei, Anhui 230029
P.R.China

Abstract. During the PhaseII project of NSRL, a turn by turn system is proposed for storage ring diagnostics which engages log-ratio electronics circuit to measure machine properties of the HLS storage ring. The log-ratio processor works at 408MHz which is 2* RF of HLS. A injection kicker and the stripline resonant exciting methods are used to excite beam for nonlinear beam dynamics studies and phase space of stored beam. Up to 2 seconds data acquisition is ensured. In this paper we present the performance of each components and preliminary test results of the turn-by-turn BPM system.

INTRODUCTION

A 200MeV injected beam and operation at 800MeV and 200-300 mA for HLS storage ring. The multicycle multiturn injection system is used for current accumulation.

In order to monitor the injecting efficiency, damping rate and β oscillation after update of injection and RF system, a turn-by-turn system of HLS has been started[1] and just tested now. The turn-by-turn system based on the Log-ratio technique, which is a fairly new idea for BPM and tends to become mature gradually. Compared with the familiar Δ/Σ and AM/PM method, highlighted features of the log-ratio technique: low noise, high bandwidth and wide dynamic range, as well as responsible linearity [2]. Electronics working at bunch pass frequency can reduce the complexity of deal with signal chain and make it easy to implement.

Because the injection rate of HLS is 0.5Hz, the data acquisition module is designed to capture data and write into disk simultaneously so the max sampling-time is up to 2 seconds. That is one time records 9 million turns. which is flexible for data processing.

Both using stripline exciting controled with a gate in timing system and a injection kicker for exciting and perturbing storaged beam are tested.

---

* Supported by National Important Project on Science- Phase II of National Synchrotron Radiation Laboratory

THE TURN BY TURN SYSTEM ARCHITECTURE AND FUNCTION

HLS turn by turn system consists of front end pick-up electrodes, Log-ratio electronics, timing system, and data acquisition system which will be placed in IPC. We adopt the Log-ratio electronics from Bergoz company. The timing system home made put in NIM Crate. The principle architrave is as Fig 1 shows.

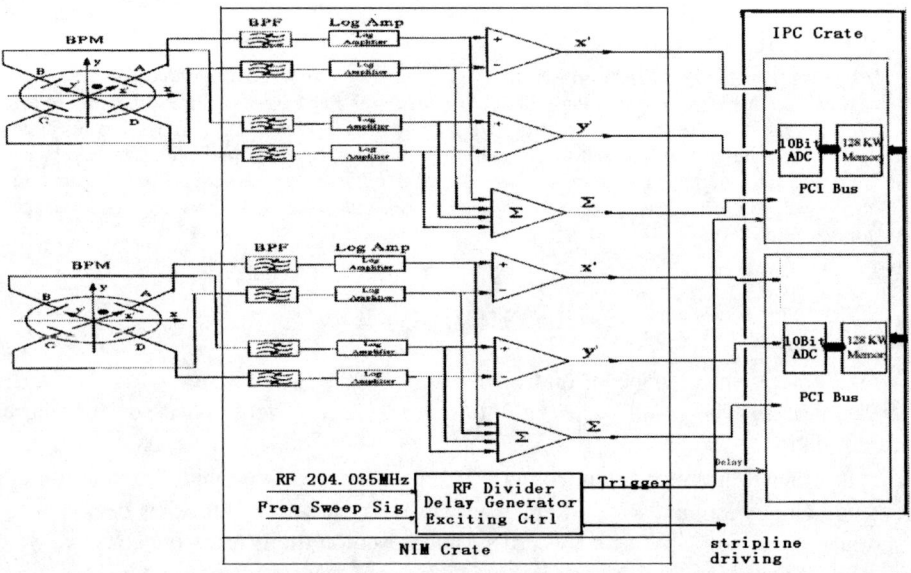

FIGURE 1. Block diagram of the turn-by-turn BPM system

The turn by turn system has two functions. One BPM for turn by turn measurement and two BPM are selected for phase space measurement. HLS ring can be operated in single bunch and multi-bunch mode, with 50mA and 350mA as the achievable maximum storage current respectively now. RF is 204.035MHz, and harmonics is 45. So request the sample rate of timing system is 4.533MHz and the dynamic range no less then 60db. The parameter is as Tab.1 shows.

TAB.1 The parameter of HLS turn-by-turn system

Dynamic range	−50dbm−10dbm
Output S/N	40db
Linearity	<1%
Trigger	
Repeat period	220ns;
Adjust precision	0.1ns
Jitter in 2s	0.2ns

LOG-RATIO PROCESSOR

We utilized a Log-Ratio Module(LR-BPM) from Bergoz company. The LR-BPM processor works at 408MHz and oprates at three modes: Sample & Hold and Track & Hold and Track-Continuous. Ensured dynamic range of LR-BPM is 65dbm.

The beam position calculated with the log-ratio technique is formulated by the following expression[3]:

$$X = \frac{20}{SG_{STM}} \times \log\left(\frac{A}{C}\right) = \frac{1}{SG_{STM}} \times V_{out} \qquad (1)$$

where A and C are the electrode potentials; S[dB/mm] is the detector sensitivity, for circular aperture BPM $S = \frac{80}{\ln(10)} \cdot \frac{1}{a}$ (a-radius); V_{out}[V] is the output from the differential amplifier; G_{STM} (mV/dB) is system gain which includes both log-ratio and differential amplifier gain.

TIMING AND CONTROL SYSTEM

The timing and control system is illustrated in Fig.2. The circuits module includes pulse-shaping, frequency dividing, timing delay and exciting signal switching gate etc. It has two inputs and three outputs. One of the inputs is synchronization RF, another is frequency sweeping signal for exciting the stripline kicker.

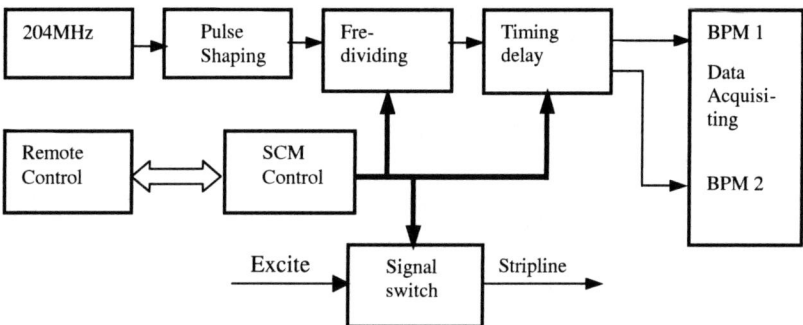

FIGURE 2. Timing and control system of the turn by turn

The rise time of trigger of the timing system and jitter of trigger with RF signal during 2 seconds are shown in Fig.3 . From Fig.3, we can see that the rise time (down curve) is about 2ns and its jitter is better then 200ps during 2 second.

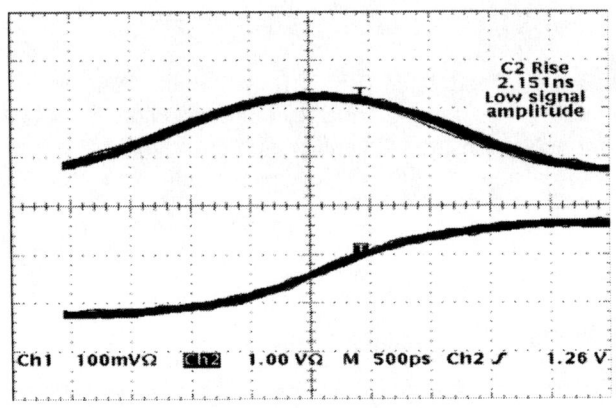

FIGURE 3. The jitter with RF synchronous signal during 2s

DATA ACQUISITION

Revolution frequency of HLS storage ring is 4.533MHz. We choose two 20SPS 10bit ADC for data acquisition of two BPMs. Since the 200MeV injected beam, coherent oscillation damping time is about 1.38s. While for the 800MeV stored beam, it is 22ms. In order to be competent for these two modes, the data acquisition module is designed to capture data and write into disk simultaneously to provide enough details. PCI bus MASTER DMA makes fast data acquisition possible. The max sampling-time is up to 2 seconds.

For different operation mode of ring(GPLS and HBLS), different BPMs are chosen. All modules of the timing system can be controlled via com-ports of IPC.

The acquisited data is transmitted to control room via local area Ethernet network. And another favorable character of this system is that it is designed to be operated under single bunch and multi-bunch modes.

EXCITATION SIGNAL

In order to nonlinear beam dynamics studies and phase space measurment of stored beam, the max sampling length of this system is 2 seconds. First we use a injection kicker, which's repeating rate is 0.5Hz with pulse width 3.5 μ s, perturb storage beam.

We also make experiments to load exciting signal to horizontal or vertical stripline, controled with a gate module packaged in the timing system, can aim for exciting beam. The excitation power and gate width been maked certain according to test and calculate to provide enough for beam motion. After the excitation ends, a delayed trigger is sent to start the turn by turn detect. In our case the delay time(gate width) is about 10ms and need power about 30dBm for excitation beam in operating at 200MeV. Adopted a gate pulse excitation sketch diagram is shown in Fig.4

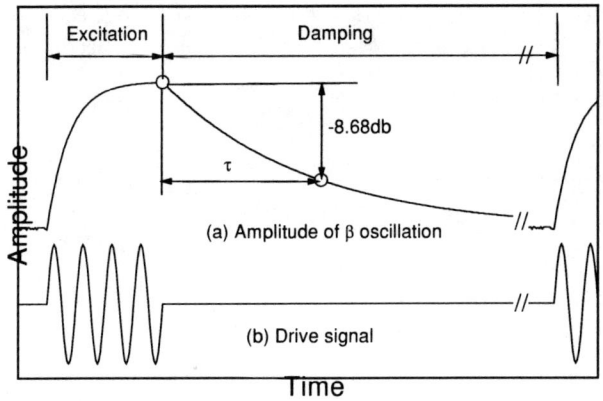

FIGURE 4. Sketch diagram of the excitation system

TURN BY TURN MEASUREMENT

Measasurement result of Turn by Turn

With the LR-BPM module we have made measurements of turn by turn first when there is no any excitation and analyzed beam spectrum from the data when ring is operating at 800MeV.

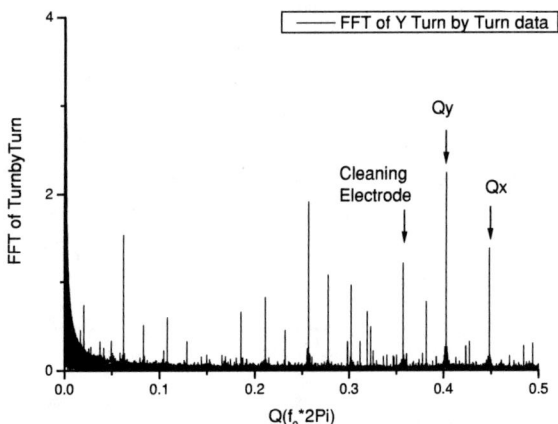

FIGURE 5. The beam spectrum from the turn by turn data

In the beam spectrum include the Qy, Qx and the power spectrum of clearing electrode. Some spectrum are still being investigated. When we using a injection kicker perturb storage beam at 200MeV, the LR-BPM output as below Fig.6. It also clearly shows that the distorting orbit and there's coupling between x and y plane.

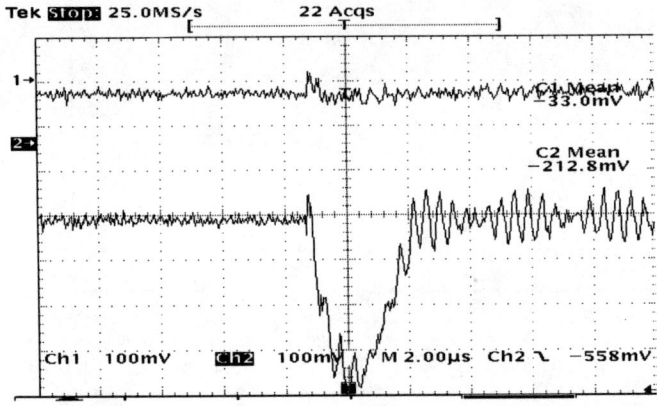

FIGURE 6 Beam position caused by injection kicker

In most cases the beam position variation is transformed to freq field for better understanding. The max available frequency is half the revolution freq. So fraction Q such as 0.54 is leaked to lower frequency 0.46. If we think it as one value close to 0.5, we can also divide it into two columns, so the spectrum line we are interest is the one responding to the fraction part less than 0.1, for instance, 0.51 gives 0.02 in 2 column FFT. Time domain plot of each column is sine like. In a similar way, we can divide data into 5 columns to study the tune close to 0.6, for which the fraction part less than 0.1 is the interested one. One example is the tune y, FFT spectrum of turn by turn and freq spectrograph results are **2.598/3.552** seperately.

Phase Space Measurement

Two BPMs are chosen for phase space tracking and analysis. Different BPM pairs are chosen for different operation mode of ring. They are chosen according as such principles as: large β_x and β_y, phase advance between the BPM pair be close to $\pi/2$ or $3\pi/2$.

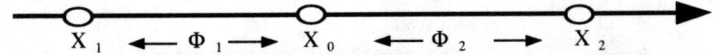

FIGURE 7 Phase space sketch

For Fig. 7, from the transport matrix we can calculate phase coordinates of any point between X_1 and X_2(including these two points), $\left(X_0, \beta_0 X_0' + \alpha_0 X_0 \right)$

$$X_0 = \frac{\sqrt{\beta_1}\sin\Phi_1 X_2 + \sqrt{\beta_2}\sin\Phi_2 X_1}{\sqrt{\frac{\beta_1\beta_2}{\beta_0}}[\sin\Phi_1(\cos\Phi_2 + \alpha_0\sin\Phi_2) + \sin\Phi_2(\cos\Phi_1 - \alpha_0\sin\Phi_1)]}$$

$$= \frac{\sqrt{\frac{\beta_0}{\beta_2}}\sin\Phi_1 X_2 + \sqrt{\frac{\beta_0}{\beta_1}}\sin\Phi_2 X_1}{\sin(\Phi_1 + \Phi_2)} \quad (2)$$

$$\beta_0 X_0' + \alpha_0 X_0 = \sqrt{\frac{\beta_0}{\beta_2}}\frac{\cos\Phi_1}{\sin(\Phi_1+\Phi_2)}X_2 - \sqrt{\frac{\beta_0}{\beta_1}}\frac{\cos\Phi_2}{\sin(\Phi_1+\Phi_2)}X_1 \quad (3)$$

From the phase space ellipse equation:

$$\frac{1}{\beta}[X^2 + (\beta X' + \alpha X)^2] = W \quad (4)$$

To minimum the error of W due to position measurement error:

$$\Delta W = 2X_0 \Delta X_0 + 2(\beta_0 X_0' + \alpha_0 X_0)\Delta(\beta_0 X_0' + \alpha_0 X_0)$$
$$= \frac{2\beta_0}{\beta_2 \sin^2(\Phi_1+\Phi_2)} X_2 \Delta X_2 + \frac{2\beta_0}{\beta_1 \sin(\Phi_1+\Phi_2)} X_1 \Delta X_1 \quad (5)$$
$$- \frac{\beta_0 \cos(\Phi_1+\Phi_2)}{\sqrt{\beta_1\beta_2}\sin(\Phi_1+\Phi_2)}\Delta(X_1 X_2),$$

Then it requires that $|\sin(\Phi_1+\Phi_2)|=1$ and β to be as large as possible. Besides this, to assure the resolution and precision of system, all BPMs should be in sections with no nonlinear devices.

SUMMARY

The newly developed turn by turn system which is based on log-ratio processor working at 408MHz has been implemented for HLS machine study. With the application of high speed data transmission and acquisition technics, the system can accomplish turn by turn measurement of sampling-time up to 2 seconds. And to make use of a injection kicker and horizontal or vertical stripline for excitating beam. Phase space measurement be maked and operation serving the study of parameters of storage ring soon.

ACKNOWLEDGEMENT

Authors would like to thank Dr. K.T. Hsu for the beneficial discussion and communication. Authors thanks Dr. H.L.Xu and L.Wang etc colleague friendly cooperate in machine operating.

REFERENCES

[1] J.H.Wang, W.M.Li etc, *Turn-By-Turn system design of HLS*, APAC01, Sept. 2001, Beijing.
[2] R.E.Shafer, *Log-Ratio Signal-Processing Technique for Beam Position Monitors*, AIP conf. Proc.(1993), pp120-128
[3] G. Roberto Aiello, *Log-ratio technique for beam position monitor systems*, AIP conf. Proc.(1993), pp.301-310

Upgrades to PEP-II Tune Measurements

Alan S. Fisher, Mark Petree, Uli Wienands, Stephanie Allison,
Michael Laznovsky, Michael Seeman,[*] and Jolene Robin[†]

Stanford Linear Accelerator Center
2575 Sand Hill Road, Menlo Park, CA 94025, U.S.A.

Abstract. The tune monitors for the two-ring PEP-II collider convert signals from one set of four BPM-type pickup buttons per ring into horizontal and vertical differences, which are then downconverted from 952 MHz (twice the RF) to baseband. Two-channel 10-MHz FFT spectrum analyzers show spectra in X-window displays in the Control Room, to assist PEP operators. When operating with the original system near the beam-beam limit, collisions broadened and flattened the tune peaks, often bringing them near the noise floor. We recently installed new downconverters that increase the signal-to-noise ratio by about 5 dB. In addition, we went from one to two sets of pickups per ring, near focusing and defocusing quadrupoles, so that signals for both planes originate at locations with large amplitudes. We also have just installed a tune tracker, based on a digital lock-in amplifier (one per tune plane) that is controlled by an EPICS software feedback loop. The tracker monitors the phase of the beam's response to a sinusoidal excitation, and adjusts the drive frequency to track the middle of the 180-degree phase transition across the tune resonance. We plan next to test an outer loop controlling the tune quadrupoles based on this tune measurement.

INTRODUCTION

The PEP-II *B* Factory [1], a 2.2-km asymmetric collider at the Stanford Linear Accelerator Center [2], was built in collaboration with the Lawrence Berkeley [3] and Lawrence Livermore [4] National Laboratories to study *CP* violation by tracking decays of *B* mesons moving in the lab frame. At a single interaction point, 9-GeV electrons in the high-energy ring (HER) collide at zero crossing angle with 3.1-GeV positrons in the low-energy ring (LER). Collisions were first observed in July 1998, during commissioning without the *BABAR* detector, which was installed in May 1999. By the time of this Workshop, peak luminosity has reached 4.6×10^{33} cm^{-2}·s^{-1}, exceeding the design goal of 3.0×10^{33}, and we collide up to 1750 mA of positrons on 1050 mA of electrons.

These high luminosities require high beam-beam tune shifts. In operating the rings, and especially in filling them while colliding, careful attention must be paid to controlling the betatron tunes. However, these conditions also make tune peaks hard to observe. Out of collision, the spectrum analyzer shows tall, narrow peaks at the betatron tunes, but in collision the beam-beam tune spread lowers and broadens the peaks, leaving them at times barely above the noise floor.

---

[*] Present address: Massachusetts Institute of Technology, Cambridge, Massachusetts.
[†] Present address: University of New Orleans, New Orleans, Louisiana.

Two complementary improvements have been pursued. New downconverters for the tune signals have been built on the same principle as the original ones, but with changes to improve the signal-to-noise ratio. In addition, a tune tracker has been built using a digital lock-in amplifier for each of the four tune planes (HER and LER, x and y). The tracker measures the beam's response, via the same downconverter, to sinusoidal excitation at the center of the tune resonance, and follows changes in that frequency by monitoring the phase, rather than the amplitude, of the response.

IMPROVED DOWNCONVERTER

The original tune monitors (one per ring) were described at the 1998 Beam Instrumentation Workshop [5]. They provided strong signals during single-ring commissioning and through the first years of collisions, even with bunch-by-bunch transverse feedback damping betatron motion. However, with increasing luminosity it has become more difficult to see the betatron tunes while in collision, because the beam-beam tune shifts broaden and flatten the peaks, at times bringing them down to the noise floor. An additional concern is that tight budgets during construction left us without a complete spare chassis for quick substitution when occasionally an amplifier, mixer, or power supply fails. To address both of these issues, we recently made a new chassis (Fig. 1) for each ring, with several modifications, rather than making a simple duplicate as a spare. The goal was a 5-dB improvement in the signal-to-noise ratio. The old units have now become the spares.

One might first ask why a downconverter is needed. Our digital spectrum analyzers compute the spectrum (after the downconverter) from a fast Fourier transform (FFT) over a 10-MHz bandwidth, and normally zoom in on a 30-kHz range within one ring revolution frequency (136 kHz) of 952 MHz, twice the ring's RF frequency. If instead we used a broadband spectrum analyzer, it would include downconversion and could

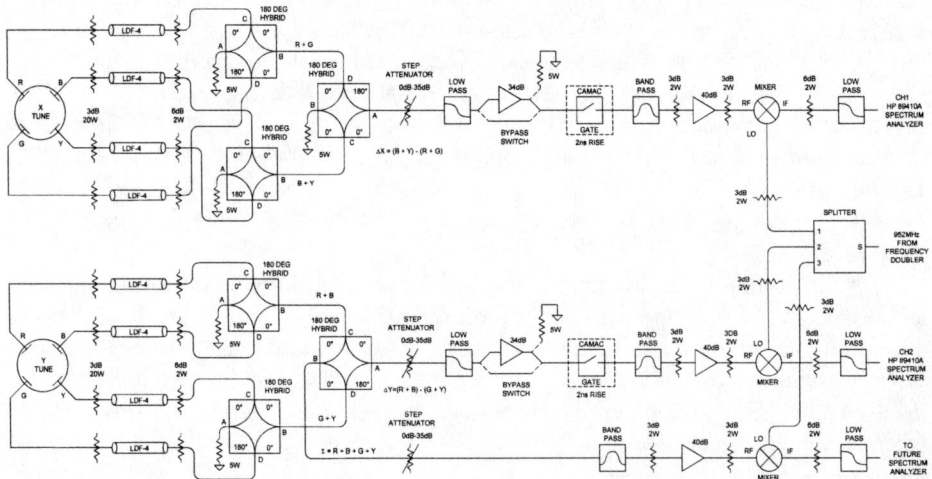

FIGURE 1. The new PEP-II tune monitor, showing the pick-up buttons, the downconverter, and the signals going to the spectrum analyzer.

TABLE 1. Beta Functions at Pickups.

Plane	Quadrupole	β(m)
Original Layout		
LER x	QDP-3014	3.1
LER y	QDP-3014	25.4
HER x	QFP-8062	24.5
HER y	QFP-8062	9.9
New Layout		
LER x	QFP3-3042	23.1
LER y	QDP-3014	25.4
HER x	QFP-8042	37.3
HER y	QDP-8012	36.9

directly display the tune monitor's frequency range. However, the sensitivity is not as good. We have examined the signal directly from one of our 15-mm-diameter pick-up buttons on two broadband spectrum analyzers (a Rohde and Schwarz FSP series and an Agilent PSA series). Both lose the betatron peaks in the noise floor, even when they are plainly visible with our downconverter.

For greater simplicity and fewer long Heliax cables, the old downconverters formed the Δx and Δy signals from one dedicated set per ring of four pick-up buttons (identical to those used for beam position monitoring). Mechanical stability requirements placed all buttons at beampipe supports adjacent to quadrupoles, giving a choice of either a large β_x and small β_y near a QF, or the opposite at a QD. For the new downconverters, we added a second set of cables to get the highest available betas for both planes, and so more signal. Table 1 shows the improvement in beta functions.

The Heliax cables from the tunnel lead to 180° hybrids that form the Δx and Δy signals. Next an amplifier provides gain for small signals obtained when the charge per bunch is small, a condition typical of commissioning or machine development but not of normal operating currents, which can saturate the amplifier. In the old system, we could avoid saturation by first attenuating the signal with a step attenuator at the amplifier input. In terms of gain, this is an adequate solution, but at high luminosity we needed to avoid the additional noise introduced the attenuator as well as the amplifier. Instead, in the new design an RF switch bypasses the amplifier. In addition, the new amplifier offers a lower noise figure and better power handling, as does the second amplifier before the mixer. The step attenuator moves in eight 5-dB rather than 10-dB steps, since we'd prefer smaller steps to the larger range. We also selected a mixer with higher conversion efficiency. Table 2 shows calculated improvements in signal-to-noise ratio for two cases.

Other changes were made for modularity and ease of repair. The main power supply was moved out of the chassis to a separate rack-mounted unit. It provides the input for individual regulators next to each amplifier and frequency doubler (providing the 952-MHz reference). We replaced the single doubler of the old system with one per ring, plus an option to share one if the other fails. The output splits three ways, to provide a local oscillator for mixers for the x, y, and 4-button-sum channels.

There is a phase shifter on the 476-MHz input to each doubler, because the beam's frequency components symmetrically above and below 952 MHz combine with the

TABLE 2. Design Calculations for Small and Large Input Signals.

System	Attenuator (dB)	Amplifier Bypassed?	Gain (dB)	Noise Figure (dB)	Signal to Noise (dB)
Old	0	No	6.7	26.2	37.4
New	0	No	19.1	24.5	40.7
Old	30	No	-22.8	55.7	27.9
New	0	Yes	-8.9	30.9	45.9

reference in the mixer in a phase-sensitive manner, with the output going from zero to a maximum as the 952 phase shifts by 90º. Because the x channel has a different path from the buttons to the mixer, compared to the y and sum channels, the cable length for the reference to the x mixer is adjusted to peak at the same phase as y.

The signal to noise for the four tunes shows improvements of the anticipated magnitude (Fig. 2), although it is puzzling that the biggest improvements were not in the planes with the biggest increases in beta. The figure also shows that the improvement for LER is larger than for HER, probably because LER typically collides with 1.5 to 1.7 times more charge per bunch, and so used a larger step attenuator setting than HER (30 rather than 20 dB) with the old system; it thus benefited more from bypassing the amplifier. Some improvement also came from having the two independent phase shifters, making it easier to peak up all four signals. However, these comparisons were not systematic, because the change-over was made during colliding-beam operation, moving cables during brief intervals without beam. The traces thus show similar, but not identical, machine conditions.

TUNE TRACKER

PEP operators carefully control the betatron tunes to maintaining high luminosity, using suitable linear combinations of the tune-adjustment quadrupoles (the "tune multiknobs") to line up the x or y spectra against reference markers. A feedforward loop [6] also adjusts these multiknobs to compensate for tune variation with current as the rings are filled and the currents decay. The operators frequently consult a history

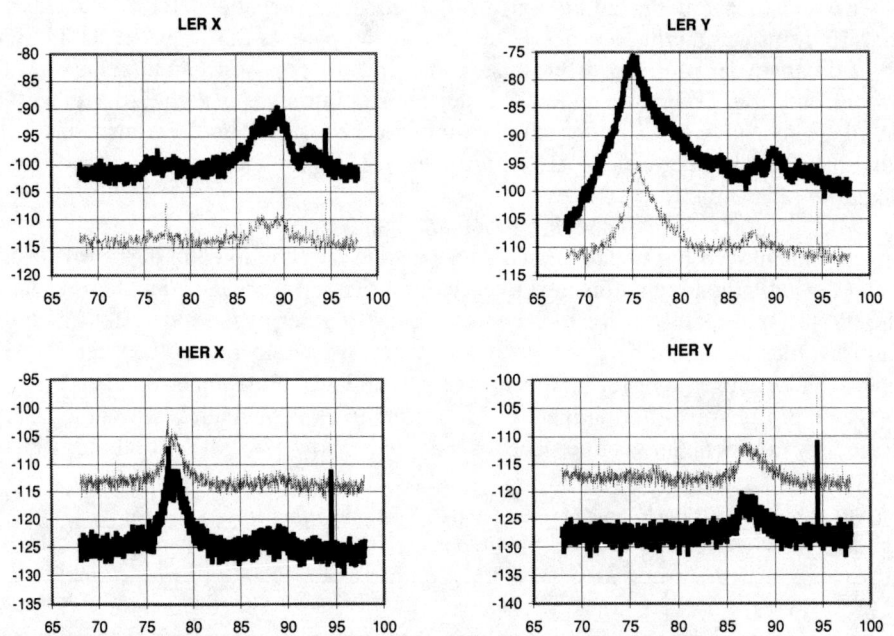

FIGURE 2. Spectra showing typical improvements when the old downconverters (light gray trace) was replaced by the new (heavy black). Amplitude in dBm vs. frequency in kHz.

plot of the knob moves to restore a previously successful configuration.

More complete information would be helpful. These spectra are complex and gradually change shape as the currents decay and the beam-beam forces change. The quadrupoles, the currents, and the beam-beam forces all affect the tunes, but the multiknob history includes only the quadrupole changes made by the operators and by the feedforward, mostly as compensation for the evolution of the beam-beam forces. It would be desirable to plot the history of a single tune value characterizing all the changes affecting the beam, but it is hard to pick out a single number from the amplitude of a broad spectrum. We have tried an algorithm that searches every few minutes for a designated spectral peak, recording its frequency and height, but it generally loses track as the peaks evolve.

The phase shift across the tune resonance provides an interesting alternative. The two channels of the spectrum analyzer for each ring are normally set to display only the magnitude of the Δx and Δy power spectra, as we excite the beam with broadband random noise. To see the phase, we put the excitation into channel 1 and either Δx or Δy into channel 2. The instrument can then plot the magnitude and phase of the transfer function, the ratio of channels 2 to 1. The magnitude is just the familiar power

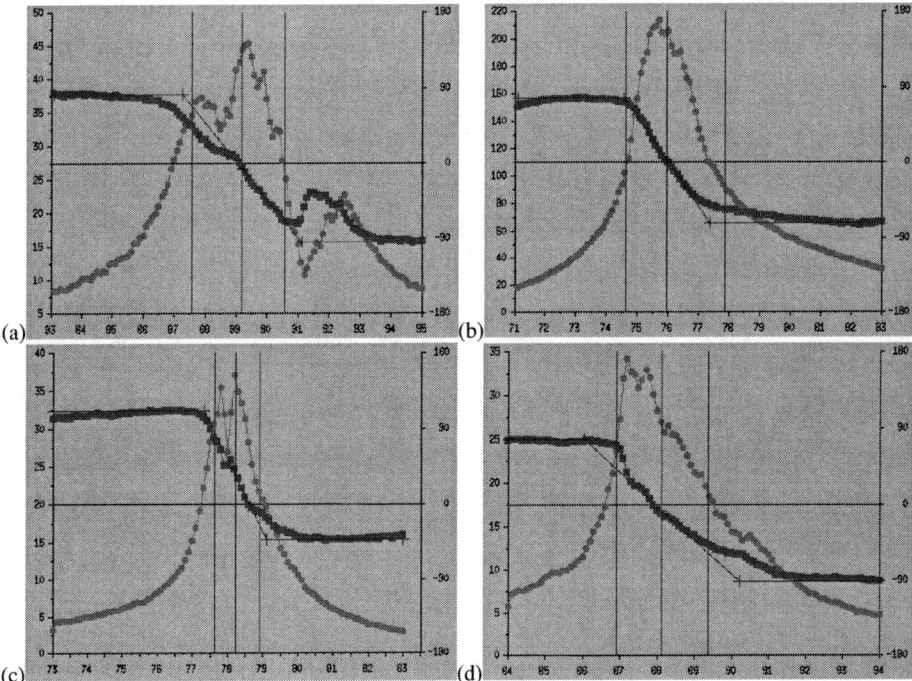

FIGURE 3. Frequency scans of the tune trackers for (a) LER x, (b) LER y, (c) HER x, and (d) HER y, The horizontal scale is in kHz. The vertical scale on the left is the peak's magnitude in µV, while the phase curve uses the scale on the right in degrees. A cross marks the target phase, at the middle of the 180° transition. A slope is computed from a linear fit of points near the target phase, including as many points as possible within the user's cursors without exceeding a chi-squared criterion. Because the beams are colliding, the tunes can have multiple peaks and a non-monotonic phase, as in LER x. Nevertheless, the target phase appears to provide a robust marker to characterize the tune.

spectrum, since the excitation is uniform. The phase drops by 180° (Fig 3) as the frequency crosses the tune resonance. Near the center, the phase can be modeled as a smooth monotonic ramp, making it perfect for a feedback loop tracking the frequency corresponding to this central phase. Also, unlike a peak height, there is little change in this curve with current.

Using the spectrum analyzer, we would have to examine the full trace to extract this frequency. With our two-channel instrument, we would have to forgo separating the beam motion into x and y traces, instead merging both into a single spectrum by using only two diagonally opposite buttons rather than all four.

However, for this purpose we do not need the full spectrum, but only the phase of the response to one frequency at a time. In particular, a digital lock-in amplifier is really a single-frequency equivalent of our spectrum analyzer. Both digitize the incoming signal and translate the desired central frequency down to baseband with the digital equivalent of a quadrature mixer—multiplication by a cosine and sine. The spectrum analyzer then uses a low-pass filter and an FFT to obtain the full spectrum. The lock-in uses a narrow low-pass filter to extract only the DC component, corresponding to the in-phase and quadrature power at the central frequency. Not surprisingly, the resolution is similar, but the lock-in is much less expensive. Also, much less drive power goes into the beam, since we excite only one frequency with a continuous sine, rather than spreading the drive over a band.

Fig. 4 shows the main tune-measuring loop, plus secondary loops to adjust the drive power and keep the instrument in a suitable sensitivity range. The internal sine-wave source of the lock-in amplifier (a Stanford Research Systems SR830) excites the beam by adding its signal to the error signal for transverse feedback (and to the random

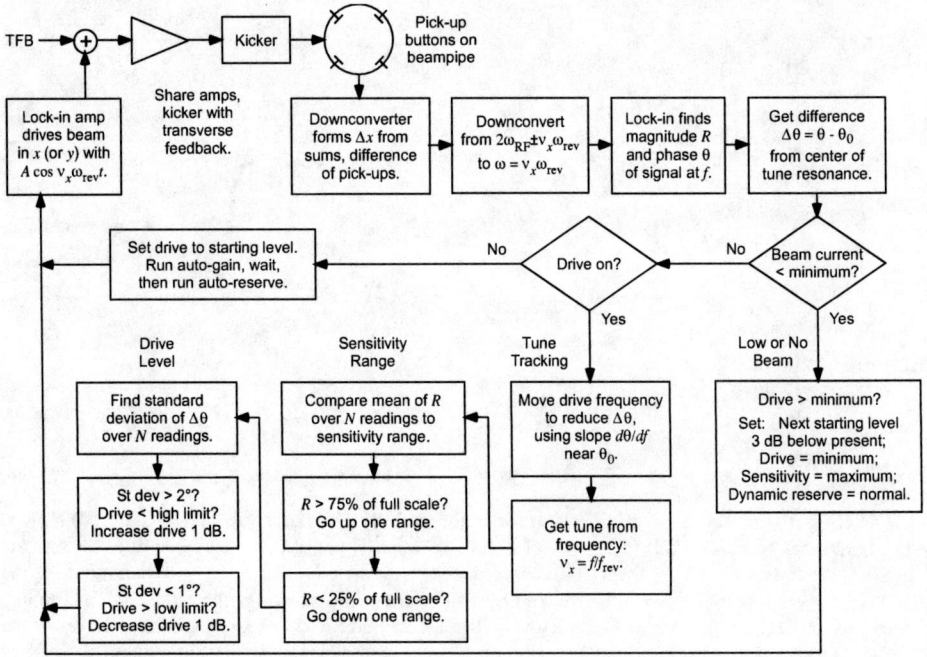

FIGURE 4. Flow chart for the tune tracker.

noise from the source in the spectrum analyzer). The sum goes to the feedback amplifiers and kicker. The input to the lock-in is the same downconverted beam response given to the spectrum analyzer.

A computer running the EPICS control system communicates with each lock-in through GPIB. To set up the tune tracker, the computer first scans the frequency f across the peak and measures the amplitude and phase of the response (Fig. 3). The phase θ_0 at the middle of the transition becomes the loop's target phase. The slope $d\theta/df$ around this phase is also noted. Then the frequency is set to the middle of the transition to start the loop. Every second the computer reads the phase θ and changes the frequency to correct the phase error:

$$\Delta f = g \frac{\theta - \theta_0}{df/d\theta} \qquad (1)$$

where the gain g reduces the step size to avoid overshoot. The frequency gives us the tune. A spike at this frequency from the beam excitation appears on the spectrum, as shown in Fig. 5. The stripchart of Fig. 6 shows the tunes over a 6-hour period, taken from the first 24 hours of tracking in all four planes.

We implemented the tune loop shortly before this Workshop and are now writing code for the drive and sensitivity loops. At present the computer is remote and connects via an Ethernet-to-GPIB interface (Agilent E2050A). It reads all four tunes at 1 Hz. For higher speed or if network traffic eventually becomes a limitation, a local computer with a GPIB card may be used.

We had expected that the effect of the adjustments in the tune multiknobs, by the operators and by the feedforward loop, was to hold the actual tunes constant, with all the knobbing serving to offset tune changes due to currents and beam-beam forces. The stripchart shows that the tunes are not flat between fills, despite these adjustments. With the idea that we might optimize the luminosity by holding the tunes constant, we have begun preparing a true tune feedback that will use the outputs of the tune trackers to slowly adjust the tune multiknobs. To quickly explore this idea in a simpler way, we tried adjusting the tune-versus-current coefficients in the feedforward to flatten the tracker traces. Some flattening may be seen in the last two fills of the stripchart, as yet with little effect on luminosity.

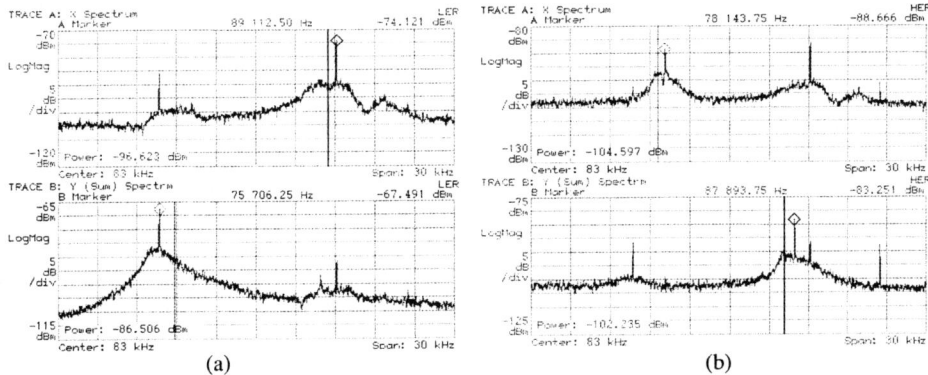

FIGURE 5. Spectrum analyzer traces for (a) LER and (b) HER with narrow frequency spikes from the tune tracker visible on all four peaks. Due to coupling and collisions, a spike can appear on both planes and both rings. The horizontal range is 68 to 98 kHz; the vertical scale is 5 dB/div.

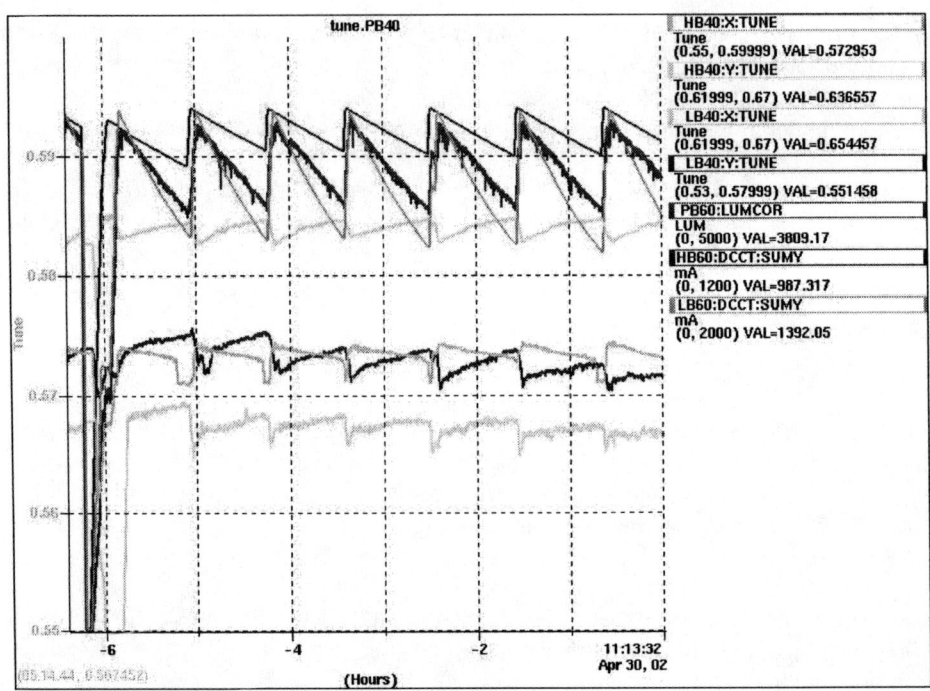

FIGURE 6. EPICS strip chart showing over six hours of tune tracking, through several collide-and-fill cycles and one beam abort. The traces are, from the top: HER current, on a scale of 0–1200 mA; LER current, 0–2000 mA; luminosity (overlapping the LER current and noisier), $0-5\times10^{33}$ cm$^{-2}\cdot$s^{-1}; LER x tune; LER y tune (sloping upward); HER x tune (downward); and HER y tune. The tunes all have a full-scale range of 0.05, but with different midpoints.

ACKNOWLEDGMENTS

We thank Peter Cameron of Brookhaven National Laboratory for suggesting the use of the lock-in amplifier in a conversation last year with Uli Wienands.

REFERENCES

1. *PEP-II: An Asymmetric* B *Factory*, Conceptual Design Report, LBL-PUB-5379, SLAC-418, CALT-68-1869, UCRL-IC-114055, UC-IIRPA-93-01, June 1993.
2. U.S. Dept. of Energy contract DE-AC03-76SF00515.
3. U.S. Dept. of Energy contract DE-AC03-76SF00098.
4. U.S. Dept. of Energy contract W-7405-Eng-48.
5. Fisher, Alan S., "Instrumentation and Diagnostics for PEP-II," Proc. 8th Beam Instrum. Workshop, Stanford, CA, May 4–7, 1998, edited by R. O. Hettel et al., AIP Conference Proceedings 451, Woodbury, New York: American Institute of Physics, 1998, pp. 95–109.
6. Decker, F.-J., Fisher, A. S., Hendrickson, L., Krauter, K. E., Murphy, B., Weathersby, S. and Wienands, U., "Tune Feed-Forward for PEP-II" Proc. Particle Accelerator Conf., Chicago, IL, June 18–22, 2001, edited by P. Lucas and S. Webber, IEEE, Piscataway, NJ, 2001, pp. 3558–3560.

Proposed Profile Monitor Designs for the Advanced Hydrodynamic Facility (AHF)[1]

William. C. Sellyey, James F. O'Hara

Los Alamos National Laboratory, MS H817, Los Alamos, NM

Abstract. The AHF consists of a LINAC, a booster ring, a 50 GeV Synchrotron and an elaborate set of beam-lines for simultaneously delivering a string of 24 proton bunches from 12 different directions to a firing site (fs) chamber in which an explosion is in progress. This paper will discuss profile instrumentation being considered for the fs beam-lines and the Synchrotron. In the beam-lines most profiling devices will probably be harps but some fluorescent screen/camera systems may be used. For the Synchrotron RGIPM's may be used for observing individual bunches. The MCP usually placed in the vacuum for such devices might be replaced by a scintillator viewed by a lens plus multianode PMT. In order to observe individual bunches, it may be necessary to increase the local vacuum pressure using a gas jet, molecular beam or a gas puff. Another option may be to view gas fluorescence with the same optical arrangement as used for the RGIPM. A carbon wire moving with velocities of 1 to 5 m/s is being considered as an intercepting device to observe stored beams consisting of one or two bunches. A quadrupole moment measuring system for determining transverse emittance is being investigated.

Harps in the firing site beamlines

Harps can be used anywhere in the beam transport to obtain projections of beam profiles in the direction perpendicular to the secondary emission (SE) wires. In this section a harp based profile measurement for observing profile projections of beam pulses separated by 200 ns will be described.

All beam bunches in all (six) dimensions will be assumed Gaussian. One way to process the signal from a wire is to send it through a Gaussian filter with rms time width σ_F=50 ns. This is a factor of 5 times longer than the bunch length in the beam lines and thus σ_F dominates. It can be shown that the fractional error in the rms spatial beam width is

$$\frac{\Delta s}{s} = 12.36 \frac{s^{1/2}}{m^{1/2}} \frac{\sigma_F \Delta V}{\varepsilon Q_o \Delta x R} \tag{1}$$

ΔV is the rms amplifier noise, m is the number of wires per mm, ε is the secondary emission coefficient, Q_o is the charge in one bunch, Δx is the width of one wire and R is the signal cable impedance. To get a specific error estimate let ε=0.02, m=3/mm, Δx=0.1 mm, R=50 Ω, Q_o=10^{11} particles = 1.6*10^{-8} C. Take s=1 mm, a typical beam size in much of the transport. ΔV is obtained from a stretcher amplifier designed and constructed for amplifying stripline signals [1]. This amplifier has a gain of 31.5 and a

---

[1] This work was sponsored by the US Dept. of Energy under contract W7405-ENG-36.

36 MHz rms width. The filter considered here has a 20 MHz rms width, and thus will pass less noise than the 30 MHz amplifier. The noise referred to the input of the 36 MHz unit was 25 µV and this is used here as an overestimate of the noise. Inserting these numbers into the above formula gives Δs/s=.006 and this is clearly an adequate result. Temperature estimates for a 100 µm carbon wire indicates that it will never go above 465 °C anywhere in the beam lines.

Viewscreens in the firing site beam lines

Many types of imaging systems have been designed for the beam transport lines from the Synchrotron to the firing sights. The screen materials considered include Cromox-6 ($Al_2O_3(Cr)$), LSO and shiny materials for OTR screens. Cameras investigated include the non-rad hard CCD cameras like the Cohu 2622 and rad-hard cameras manufactured by CIDTEC. Lenses can be off the shelf compound devices, large diameter compound achromats or multiple fused silica. Only one system will be described here.

This system would be used in circumstances where the beam intensity is high at $3*10^{13}$ particles. It will use an OTR viewscreen that may be made out of Molybdenum coated with a shiny material like Ag or Al. If the surface is smooth, the peak backward OTR intensity will be generated in a cone with an opening angle of 0.02 radians. This is an inconvenient situation because it can lead to position dependent imaging on the camera faceplate. However it has been shown that if the surface is rough, the OTR light does not go into a cone [2]. It is assumed here that it can be emitted into 2π (this needs to be demonstrated). The light intensity can be calculated [3]. The result is that there are about 2 photons per 100 nm emitted at the OTR surface for each 1000 protons passing through the surface. A 150 mm diameter achromatic doublet with a 1 m focal length could be used as the imaging lens. The lens would be placed 11 m from the beamline, and the camera would be 12.1 m from the beam line. This results in a magnification of 0.1. The camera is a Cohu 2622 CCD camera. If 100 nm of light is imaged, than $3*10^{13}$ protons will produce $N_{ph}=6*10^{10}$. The number of photons reaching the camera is $N=1.39*10^6$. It can be shown that the fractional error in the rms width of a Gaussian shaped beam is

$$\frac{\Delta\Sigma}{\Sigma} = \frac{\pi^{1/2}\Sigma}{(\Delta x \Delta y)^{1/2}} \left(\frac{\delta N}{N}\right) \quad (2)$$

Δx and Δy are the camera pixel sizes δN is the photon number equivalent of the camera (read) noise. δN =330 at 700 nm, Δx=8.4 µm, Δy=9.8 µm and in the beamlines typically rms beam sizes are 1 mm. The error calculation refers to the camera faceplate and thus Σ=0.1 mm. This results in ΔΣ/Σ=0.05. By using a larger light bandwidth, losses in the lens system can be compensated.

The temperature rise in almost any material used for an OTR screen may be significant but not serious because conduction carries the heat away from the beam impact area during the 25 seconds between beam pulses.

Replacing the CCD camera by a gated, intensified camera will enable the measurement of individual transverse bunch shapes for normal intensity and low intensity bunches.

Gas Fluorescence in the Synchrotron

In a typical gas fluorescence imaging system, protons (or other charged particles) impart energy to a residual gas molecule and some of the molecules will emit this energy as photons. Lenses are used to produce an image of the gas fluorescence. If the gas fluorescence has a sufficiently short decay constant, the fluorescent image is expected to be an accurate representation of the transverse beam profile. Gas fluorescence of the 391.4 nm N_2^+ line [4] has been used extensively to observe transverse beam profiles [5] at low beam energies. Recently it has been shown that this can also be done at 10's of GeV [6] and therefore it will be considered here. The approximate cross section at 50 GeV for producing a 391.4 nm photon is found to be $\sigma_\lambda = 3.3*10^{-20}$ cm^2.

Two types of systems will be describe here, both of which will be designed to operate in a high radiation environment, but one will depend on shielding to protect conventional cameras and optics. The other will be made of radiation hard components. This second system will consist of a fused silica view port to let out the beam produced fluorescence light, a radiation hard lens system and a radiation hard, position sensitive, high gain photon detecting device. One possibility for the photon detector is the Hamamatsu R5900U-07-L16 with a fused silica window. It has 16 strip photocathodes and corresponding anodes. Each strip is 0.8 mm by 16 mm with 0.2 mm space between strips. Typical electron gain is 10^6 for each channel. There is at least one manufacturer who makes a microchannel plate based position sensitive photon detector (Photec Inc.) that might be usable here. They are also capable of manufacturing devices to our specifications. In the shielded system, a gated intensified camera with a lens will be placed in a shielded environment some distance from the beam.

A formula for calculating the statistical error of a measured rms width of a Gaussian charge distribution was derived. It assumes that there is no background signal or other noise sources other than what results from counting statistics and thus gives the lowest error that can be achieved. The fractional rms error is given by

$$\frac{\Delta\Sigma}{\Sigma} = \frac{.866}{N_b^{1/2}} \left\{ 1 + \frac{1}{g} \left[1 + \left(\frac{\Delta G}{G}\right)^2 \right] \right\}^{1/2} \qquad (3)$$

Here N_b is the number of photons produced my the beam in the length of the beam that is imaged, g is the fraction of the photons collected by the lens system and converted to photoelectrons and G is the electron multiplication factor of the photon detector. The rms fractional fluctuation $\Delta G/G$ is typically of the order of 0.5 and thus the coefficient of 1/g is taken as 1.25. A "low" aberration fused silica lens system will be described below. To calculate signal levels, we only need to know that the aperture of the system is 45 mm and that the lens focal length is 500 mm. An inverting, unity magnification system is used. To be specific, it is assumed that an R5900U-07-L16 is used. The fraction of light from the beam entering the aperture is $1.3*10^{-4}$, the product of quantum efficiency (0.2) and fractional strip width (0.8) is 0.16 which results in $g = 2.03*10^{-5}$. Equation 3 can be solved for N_b and if a 5 % measurement of the rms width is to be made, $N_b = 1.8*10^7$ photons will need to be generated in 1.6 cm length of the beam. The number of photons produced in L=1.6 cm of beam per turn is $N_{b0}=\sigma_\lambda\rho_N N_S L=5.61*10^3$ where $\rho_N=3.54*10^9/cm^3$ is the density of gas molecules at 10^{-7} Torr. The number of turns for a 5 % measurements will be $N_b/N_{b0}=3.2*10^3$ and this will take 16.6 ms. Shorter measurement times could be obtained if the local gas pressure can be increased. To observe individual bunches, the local gas pressure would need to be increased to 10^{-2} Torr and it is unlikely that this can be done without adversely affecting the circulating beam.

The measurement error in the rms beam width is also affected by the 1 mm width of the detector photocathodes as well as optical aberrations. The rms width of a 1 mm strip is $W_{rms}=282$ µm. It is straightforward to show that for a circular lens of radius r the rms width of the image of a line source because of chromatic aberration is given by

$$\alpha_{rms} = \frac{1}{2}\frac{I}{f}r\frac{\Delta n_{rms}}{(n-1)} \qquad (4)$$

I is the image distance, f is the focal length, n is the average index of refraction and Δn_{rms} is the rms deviation of the index of refraction. The FWHM line width is estimated as less than 7 nm [4]. Assuming a Gaussian distribution and using a table of refractive index for fused silica [7] $\Delta n_{rms}=3.57*10^{-4}$ mm, n=1.471. Using I/f=2, and r=22.5 mm gives $\alpha_{rms}=17$ µm.

To minimize aberrations, two planoconvex lenses will be use with the convex faces nearly touching. Ray trace equations were derived for this lense arrangement and used to write ray tracing software. The rms image widths of a line source were calculated for various transverse and longitudinal object positions for a 50 mm lens aperture and a focal length of 1 m for each lens. The image position was kept at 1 m. The result is that a 50 mm wide detector would have a 50 mm field of view and a 52 mm depth of field while keeping the line image rms spread below H=0.33 mm. This depth and width of field of view should be adequate for imaging both the horizontal and vertical beams. If both horizontal and vertical profiles are imaged simultaneously and at the same place, the position information from one can be used to correct for spot smearing in the other.

The total rms blurring caused by the three effects discussed above is $[W_{rms}^2+\alpha_{rms}^2+H^2]^{1/2}=434$ µm. The rms width of a typical beam will be observed as $[434^2+1900^2]=1949$ µm and this 49 µm overestimate is less than the 5 % statistics error (1900*.05=95 µm). The most important contribution to the broadening comes from the depth of field and strip width, and both of these can be corrected for.

The second type of system depends on shielding to protect a lens and camera from radiation. To be specific, a Xibion ISG-750 intensified camera could be used as the imaging device, combined with a single 150 mm diameter achromatic doublet with a 1 m focal length (Melles Griot 01 LAO 367). The lens would be placed 11 m from the beamline, and the camera would be 12.1 m from the beam line. This results in a magnification of 0.1. The 12.1 X 9.2 mm^2 intensifier active area will translate into a 121 X 92 mm^2 field of view at the beamline. Equation 3 can again be used to calculate how long it will take to record enough data to make a 5 % measurement in the rms width of a beam. With a 19 % quantum efficiency and an aperture of 135 mm, $g=1.79*10^{-6}$. This gives $N_b=2.1*10^8$ and with L=12.1 cm, $N_{b0}=\sigma_\lambda\rho_N N_S L=4.2*10^4$ at

10^{-7} Torr. Thus it will take 5000 turns or 26 ms to make a 5 % measurement of the rms width. The ISG-750 output signal is 30 frames per second interline transfer. Thus it is limited to making at most 30 measurements per second. It is a gateable camera, so the signal from a single bunch could be integrated over many turns, especially if the local gas pressure is increased.

The rms resolution of the ISG-250 is about 16 microns. Aberration and field of view rms image smearing is less than 1 μm over an 80 mm field of view and an 80 mm depth of field. When added in quadrature to a 1.9 mm beam rms width they have a less than 1 micron effect on the width. The lens focal length could be reduced by a factor of three (say by adding lenses after the primary lens) and this would decrease the observation time to get a 5 % rms by about a factor of 9 if a sufficiently large viewport is available.

In the above discussion, optical losses have been ignored although they may be as large as 50 %. Thus, measurement times may be underestimated by a factor of two.

Residual Gas Ionization profile monitor (RGIPM) in the Synchrotron

For an RGIPM system approximately uniform and parallel electric and magnetic fields will need to be established in the region of the measurement. Many authors have described this type of system and the details of producing the fields will not be discussed here. The electron detection system will be designed so that a 5% measurement can be made in the rms width of a single bunch for each bunch on every turn in the Synchrotron. Additionally, the only detector component in the vacuum will be an LSO scintillator and this will be made as thin as possible to reduce unwanted background signals. The beam transverse distribution will be reproduced on the surface of the LSO by scintillation cased by 5 keV electrons accelerated and guided to the LSO by the E and B fields. This distribution will be viewed by an imaging system outside the vacuum.

Most RGIPM systems presently in operation use microchannel plates (MCP) in the vacuum chamber as a signal amplifier. A problem with this is that the MCP's are limited in total charge they can produce per unit area. In time this leads to uneven gain over the MCP surface because, on average, transverse beam charge distributions are not uniform. The LSO scintillator will be used here in an attempt to circumvent this problem.

The light attenuation of LSO caused by Co^{60} γ's has been measured [8] and is quoted as 7% /(cm*10^8Rad) at the LSO emission wavelength. Assuming an average beam size of 1.9 mm rms, and that 7000 electrons are produced per cm which are then accelerated through a 5 kV potential (see below) it will take about ten days of continuous operation for 8 hr/day to reduce the LSO light transmission by 14 %. It has also been established that heating the LSO to 300 °C for about one day will restore it to essentially its original light transmission characteristic [8]. Thus by providing a means of heating the LSO it may be possible to use the same scintillator indefinitely. If permanent darkening is expected to develop after say 100 heating cycles it would probably not be difficult to design the RGIPM such that the LSO could be easily replaced perhaps once every one or two years. It has been demonstrated that fused silica radiation damage can largely be reversed by heat [9]. Thus the quartz viewport and lenses of this system might also be kept free of radiation darkening by making provisions for heating.

Each 5 keV electron hitting the LSO will produce M=150 photons [11]. These will be observed by the same type of dual silica lens arrangement as discussed earlier with a lens focal length of 150 mm, diameter of 25 mm and the magnification of –1. It can be shown that the fractional error in the rms width of a Gaussian beam is given by

$$\frac{\Delta \Sigma}{\Sigma} = .866 \left\{ \frac{1}{N_{be}} \left\{ 1 + \frac{1}{g} \left[1 + \left(\frac{\Delta G}{G} \right)^2 \right] \right\} + \frac{.0705}{M} \frac{\Delta x}{\Sigma} \right\}^{1/2} \quad (5)$$

Δx is the width of the beam being sampled, N_{be} is the number of electrons produced by the beam and the rest of the symbols have been defined previously. It will again be assumed that the photo detector is an R5900U-07-L16 and that Δx is 1 mm. The value of g is $2.25*10^{-4}$ and is again determined by the aperture and properties of the tube. Assume Σ is 1.9 mm and solve for N_{be} with a 5 % fractional error in Σ gives $N_{be}=1.235*10^4$. Assume that Xenon at a partial pressure of $2*10^{-7}$ Torr is introduced into the beam tube at the measurement location. For Xenon a proton produces 44 (primary) electrons per cm at one atmosphere pressure [10]. If each bunch contains $1.25*10^{12}$ protons, 14000 electrons per centimeter will be produced by each bunch. The length of beam that is imaged is 1.6 cm and thus the total number of electrons is 22400. This is enough to result in a 5 % measurement and also account for a 45 % loss in photons through the optical system.

Using the spectrum of LSO [11] the rms chromatic aberration for this lens system is α_{rms}=63 µm. Restricting the field of view to 40 mm keeps the rms width of a line image below 330 µm. Combining these in quadrature with the rms width of a detector strip results in 439 µm. Finally combining this with the 1900 µm nominal beam size results in 1950 µm.

In order to cover a 40 mm field of view, one can use three lenses each with an R5900U-07-L16 tube for imaging. One would be in the center with its axis perpendicular to the LSO sheet. The other two would have to be at an angle and for these the resolution analysis will be more complicated. Alternatively, it may be possible to have a photon counting MCP device made with 40, 1 mm detection strips.

Quadrupole moments in the Synchrotron

Some very innovative work has been done at CERN for measuring quadrupole moments of beam bunches in a Synchrotron [12] and the CERN design may be adapted for use in the AHF system. However it is worth investigating the possibility of using a conventional stripline because in AHF the quadrupole signal will be relatively large.

Consider a conventional stripline position monitor in a circular beam tube with signal outputs T(top), B(bottom), L(left) and R(right). If a Gaussian shaped beam is centered and the strips are narrow, the quadrupole signal from such a device will be

$$\frac{(T+B)-(L+R)}{T+B+L+R} = \frac{2(\sigma_x^2 - \sigma_y^2)}{b^2}. \quad (6)$$

Here the σ's are the rms beam widths and b is the beam tube radius. Present models of the beam indicates that the largest σ^2 differences will be $(2.68 \text{ mm})^2-(1.12 \text{ mm})^2$ in nondispersive regions at 50 GeV. The vertical beam tube diameter is limited to about 46 mm by the bending magnets. In dispersion free regions, it may be possible to use a circular cross section beam tube with a radius of 23 mm. With these numbers, the quadrupole signal will be 0.026. Figure 1 shows the signal one would get from one bunch at 50 GeV from an electrode of a 0.5 ns long stripline after it is passed through a 10 MHz rms low pass Gaussian filter. If the strip is assumed to intercept 10 % of the wall current the quadrupole voltage signal will be of the order of 2.8*0.4 V*0.026 =29 mV. It is easy to build amplifiers with 10 MHz bandwith that have an input noise of 25 µV and thus signal to noise will be about 800. At injection S/N will be 320, thus amplifier noise will not be a problem.

The proposed processing electronics is shown in figure 1. The most troublesome part of the signal is the monopole part. It is (largely) eliminated by taking a difference between T and R signals and B and L signals with 180° hybrids (MA-COM model HH-108-SMA). The output from the hybrids would go through Gaussian filters, amplifiers and than the signals would be digitized. Current, position and quadrupole information would all be computed digitally.

The striplines would be terminated externally at both ends to minimize reflections and so calibration signals can be injected to measure how well the difference is being taken by the hybrids. It may be useful to add adjustable attenuators and phase shifters before the hybrids so corrections can be made to assure precise cancellation of the monopole signals.

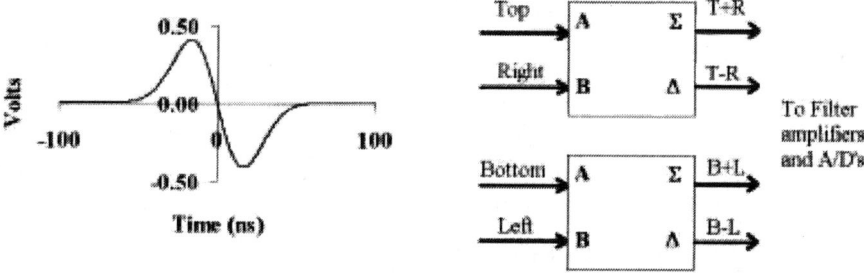

Figure 1. At left, calculated signal from one strip after passing through a Gaussian filter. Right, processing electronics.

Flying wire in the Synchrotron

In one design that has been investigated, a rotating wheel of 30 cm diameter would cause a C wire on a fork to reciprocate through the beam. An 8 µm wire would have a maximum acceleration of a_m=0.38 g (g=9.8 m/s^2), would pass through the beam f=1.6/s and reach a maximum temperature of 1750 °K if the wheels angular velocity was ω=5 rad/s. The effect on the beam with the 8 µm wire will be about the same as if Nitrogen were present in the beam tube at $4*10^{-7}$ Torr. Using a formula for emittance growth [13], the 8 µm wire would cause about a 10 % emittance growth in the 20 s of an acceleration cycle. Thus it would be useful for calibrating the nonintercepting devices under conditions almost identical to an unperturbed beam. With $\omega=15$ rad/s a higher sampling rate of 4.8/s could be achieved with no increase in emittance growth. The peak temperature of the C wire will be 1100 K and $a_m=3.44$ g.

There are two ways the relative charge density being intercepted by the wire can be measured. One is to measure secondary emission and the same electronics may be used as described for the harps. Equation 1 can be used to estimate the measurement error due to amplifier noise. $m=183$/mm, the number of samples per mm, is calculated using the wire velocity of 1.05 m/s and the revolution time of $T=5.2$ μs. Using $s=1.5$ mm, $x=35$ μm, $Q_o=7*10^{-7}$ C and the other quantities the same as for the harps one gets that the S/N=5000.

The second method is to detect radiation by attaching a scintillator like LSO to a photomultiplier. It has been calculated that, on average, each proton from a circulating bunch will deposit 10 μeV of energy per cc of LSO when a 35 μm C wire is in the center of the beam, and the detector is 100 m downstream of the wire and 20 cm from the (straight) beamline center [14]. For a beam bunch of $1.25*10^{12}$ protons, this becomes 125 MeV/cm^3. For an 8 μm wire this results in 6.5 MeV/cm^3. One photon is produced per 340 eV of deposited energy [11], and thus $1.9*10^4$/cm^3 photons are produced. Using the sampling rate $m=183$/mm, one gets that a total of $1.24*10^8$ photons/cc are produced in one sweep through the beam. Equating this to N_b in equation 3 and using a detection efficiency of $g=.05$ results in S/N=835 if the LSO volume is 1 cm^3. Energy deposition results for lower beam energies are not yet available, but it is likely that several cm^3 of LSO will be needed. In the actual system the noise will probably be dominated by beam loss.

References

1) W. C. Sellyey and R. W. Kruse, "Operational Amplifier Based Stretcher for Stripline Beam Position Monitors", 1991 IEEE Particle PAC, San Francisco, California, pp. 1145-1147, May 1991
2) S. Reiche, J. B. Rosenweig, "Transition Radiation for Uneven, Limited Surfaces", Proceedings of the 2001 PAC, Chicago, P-1282
3) D. W. Rule, R. B. Fiorito, "The Use Of Transition Radiation As A Diagnostic For Intense Beams", NSWC 84-134, Naval Surface Weapons Center, Dahlgren, Virginia 22448, Silver Springs, Maryland 20910
4) R. H. Hughes, J. L. Philpot and C. Y. Fan, "Spectra Induced by 200-kev Proton Impact on Nitrogen", Phys. Rev. V 123, Number 6, Sep. 15, 1961 P-2084.
5) J. H. Kampershor, et. el., "Initial Operation of the LEDA Beam-Induced Fluorescence Diagnostic" BIW 2000, AIP Conference Proceedings 546, P-456
6) Burton G., et. el., "THE LUMINESCENCE PROFILE MONITOR OF THE CERN SPS", CERN-SL DIVISION, CERN-SL-031 BI, presented at EPAC 2000-Vienna-Austria, 26-30 June 2000
7) J. A. Melles, editor, Melles Griot Inc., "Optics Guide 5", 1990
8) Masaaki Kobayashi, Mitsuru Ishii, Charles L. Melcher, "Radiation damage of cerium-doped oxyorthosilicate single crystal", NIM in Phys. Res. A 335(1993) 509-512
9) W. Tighe et. el., "Proposed experiment to investigate use of heated optical fibers for tokamak diagnostics during D-T discharges", RSI 66(1), Jan 1995, P-907
10) F. Sauli, "PRINCIPLES OF OPERATION OF MULTIWIRE PROPORTIONAL AND DRIFT CHAMBERS", CERN 77-09, 3 May 1977
11) V. V. Avdeichikof et. el., "Light output and energy resolution of CsI, YAG, GSO, BGO and LSO scintillators for light ions", Nuclear Instruments and Methods in Physics Research A 349 (1994) 216-224
12) A. Jansson, D.J. Williams, "A new optimized quadrupole pick-up design using magnetic coupling", NIM A 479 (2002) 233-242
13) D. A. Edwards, M. j. Syphers, "An Introduction to the Physics of High Energy Accelerators", John Wiley & Sons, Inc.
14) W.P. Lysenko, LANL, private communication

Design and Experiment of 200MeV Energy Spectrum Analysis System of Linac in NSRL

Ping Lu, Yuan Ji Pei, Baogen Sun, Weimin Li, Hongliang Xu, Zuping Liu

Jun Hong, Junhua Wang, Duohui He

NSRL, University of Science and Technology of China, P.O.Box 6022, Hefei, 230029, China

Abstract. Energy Spectrum Analytic System of 200MeV Linear Accelerator in NSRL is used to measure the beam energy and energy spread. This paper shows the status of the system, and analyses a few of its shortcomings. In the NSRL phase-II Project, the system is required to improve. The results of simulation calculation and some preliminary experiment are described.

INTRODUCTION

National Synchrotron Radiation Laboratory (NSRL) uses a 200MeV linear accelerator as the injector of Hefei Light Source (HLS). The LINAC construction began in 1984, and commissioning was successful in 1987[1]. The beam energy spectrum analysis system[2-5] was installed in a beam switch yard area at the end of the LINAC as is shown in Fig.1. There are three ways for an electron beam to go by means of the switch magnet, one way is to an electron storage ring of 800MeV, one to the beam dump and nuclear physics experimental hall, and another to the energy analysis system. The energy analysis system is composed of an analysis magnet of 60°, a fluorescent ceramic plate to image the beam spot, a reflection mirror, and a CCD. The system as mentioned above has run well since 1988 and provided important information about the energy spectrum of the electron beam from the 200 Mev LINAC . But the system is not perfect, because the tunnel width where the analysis magnet was located, is not enough so that the beam target is not located at the focus point.

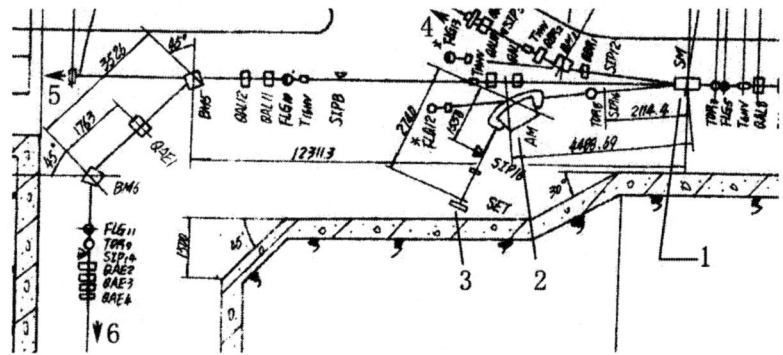

Fig.1. general layout of the transport line after switch magnet. 1. switch magnet, 2. analysis magnet, 3.target, 4.to ring, 5.to beam dump, 6.to nuclear physics experiment hall.

With the NSRL Phase II Project, our machine's performance will be upgraded, this demands that we know the beam energy spectrum accurately, so we plan to redesign the linac's spectrum analysis system.

Spectrum Analysis Principle

The beam motion in the energy analysis system will be described by Beam Transformation Matrix[6]. When we have the initial conditions of the beam and the transfer matrix between the start point and the target, we can get the terminal information. For example, the horizontal motion is as the following[7]:

$$\begin{pmatrix} x_f \\ x'_f \\ \frac{\Delta p_f}{p} \end{pmatrix} = \begin{pmatrix} m_{x11} & m_{x12} & m_{x13} \\ m_{x21} & m_{x22} & m_{x23} \\ m_{x31} & m_{x32} & m_{x33} \end{pmatrix} \begin{pmatrix} x_i \\ x'_i \\ \frac{\Delta p_i}{p} \end{pmatrix} \qquad (1)$$

Then we get:

$$x_f = m_{x11} x_i + m_{x12} x'_i + m_{x13} \frac{\Delta p_i}{p} \qquad (2)$$

We usually use a CCD to get the fluorescent image on the target. The intensity distribution indicates the particle distribution. If we make m_{x11} or m_{x12} equal to 0 and x_i or x'_i can be limited to a certain range, then we can get the energy spectrum by means of the intensity distribution of fluorescent image on the target. According to the principle, we designed the energy analysis system in 1984. The sketch map of the beam line arrangement and its parameters are shown in Fig.2.

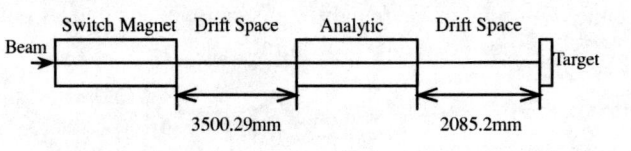
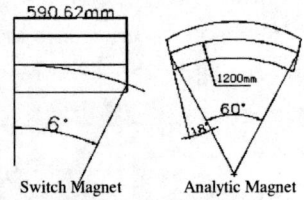

Fig. 2 sketch map of the beam line arrangement.

Fig. 3 sketch map of the switch magnet and the analytic magnet.

The transfer matrix of the horizontal motion from the switch magnet entrance to the target is the following:

$$\begin{pmatrix} 0 & 2.683 & 0.9403 \\ -0.3727 & -0.7541 & 0.1506 \\ 0 & 0 & 1 \end{pmatrix} \qquad (3)$$

In the computation, we use meter, radian as x and x''s units respectively. The half width of the fluorescent target is 5cm=0.05m, the centre energy of HLS Linac electron beam is 200MeV, the half energy offset ΔE displayed on the target corresponds to $\frac{\Delta E}{E} \times m_{x13} = 0.05$ m, thus we get $\Delta E = \frac{0.05 \times 200}{0.9403} = 10.635$ MeV's. In order to make the beam dispersion function less than 1mm on the target, the aperture of the scraper should limit the dispersion angle to $\pm \frac{0.5 \times 10^{-3}}{2.683}$ rad $= \pm 0.186$ mrad.

From the above analysis, we have the following conclusions:
1. The resolution of this system is very low. The fluorescent material granule and the CCD have a definite resolution, the detailed spectrum will be blurred.
2. The demand on the aperture of scraper is high; it will greatly influence the machine operation and storage injection.

As the building's construction didn't match the design well, as mentioned above, the vacuum chamber had to be shortened so that the real length of the drift space downstream of the analytic magnet was 1.422m(see Fig.1). Using this length, the horizontal transfer matrix will be

$$\begin{pmatrix} 0.2469 & 3.183 & 0.8405 \\ -0.3727 & -0.7541 & 0.1506 \\ 0 & 0 & 1 \end{pmatrix} \quad (4)$$

As the matrix shows, $m_{11} \neq 0$, so both x_i and x'_i will influence the target image.

NEW ENERGY ANALYSIS SYSTEM

According to the design requirements, we have tried and computed many schemes and determined a layout for the new energy analysis system as shown in Fig.4. Whereas the useable space in the existing linac tunnel is so limited, a rational idea is

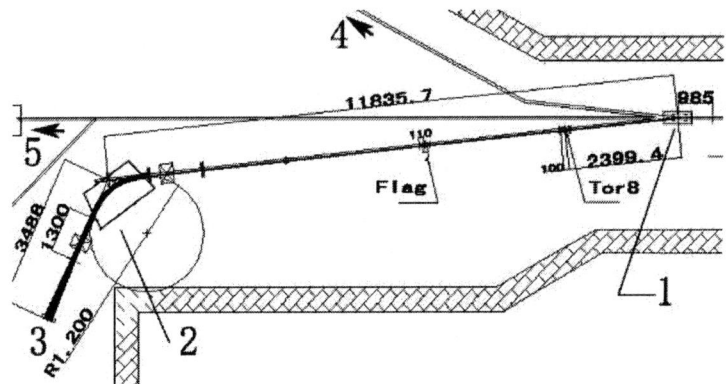

Fig.4 Layout of the new energy analytic system. 1.switch magnet, 2. analysis magnet, 3. target, 4. to ring, 5. to beam dump.

to move the analytic magnet and expand the transfer line to the spacious corner near the nuclear physics experimental hall (see Fig.4). There is a quadrupole magnet both downstream and upstream of the analytic magnet respectively.

Obviously, we can select an appropriate length so that the image of the object can be displayed on the target (make the transfer matrix's element m_{x12} equal zero from the object to the target).

We can adjust the resolution of the spectrum analysis by setting the corresponding exciting current of the quadrupoles. After a simulation calculation, we found a problem. The β_y function is so big that it is difficult to transfer the beam to the target. To solve the possible problem, we add a quadrupole magnet in the middle of the

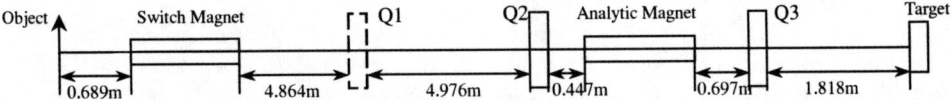

FIG.5. sketch of new energy analytic system. Q1, Q2, Q3 are the quadrupoles magnets, the dashed Q1 is supplemented, the effective lengths of Q1, Q2, Q3 are all 0.28m.

transfer line between the switch magnet and the analytic magnet, then set its K<0, to decrease the β_y, so it is limited to an acceptable range. The dimension parameters of the new energy analytic system are shown in Fig.5.

Calculated Result of Twiss Parameter

The initial Twiss Parameter at the entrance of the switch magnet is: $\beta_x = 48.32974$ m, $\beta_y = 1.35255$ m, $\alpha_x = -2.56926$, $\alpha_y = -0.09582$, $\eta = 0$, $\eta' = 0$, the emittance is 0.5mm.mrad(the horizontal and the vertical are the same) [8]. We calculated the β function along the transport line of energy spectrum analysis system with and without Q1. Fig.6 shows the β function when Q1 is not added, the different curves present different resolutions on the target. Fig.7 shows the curves where Q1 is added ($K_{Q1}=-0.65m^{-2}$). It is obvious that the β_y is depressed, but the β_x is increased.

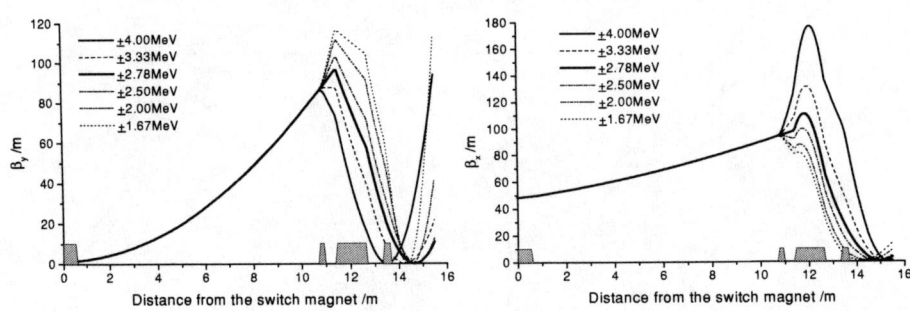

Fig.6. the β function curve when the Q1 is not added

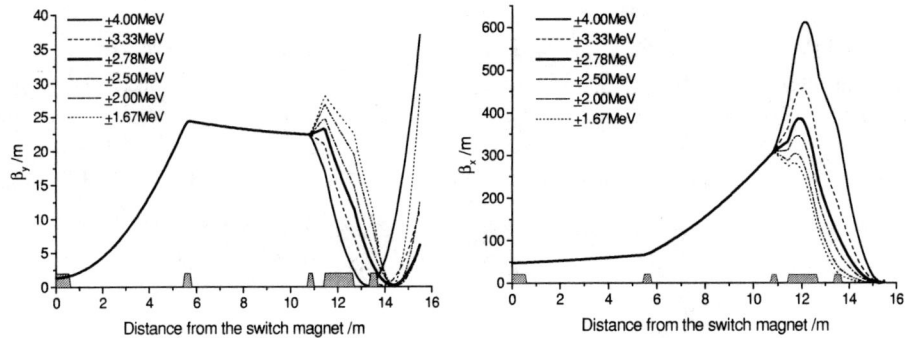

Fig.7. the β function curve when the Q1 is added

Table 1 is shows some parameters of the system according to the simulation calculation. In the table, K values of Q1, Q2, Q3 are for a different energy range on the target, the matrix element m_{x11} is of the horizontal motion, and the element m_{y11}, m_{y12} of the vertical motion from the object to the target.

From the table we can find a stirring advantage. The horizontal matrix element m_{x11} is very small, in the range about 0.1~0.36. We have also observed the beam spot at the flag detector. The diameter is about 4mm, so the blurring effect by the beam size on the target is aproximately 1mm. In the future, the system will usually run at ±2.78 MeV resolution(no current supplied to Q3), so the beam size influence is small and we need not add a gap in the transfer line upstream of the switch magnet. The nearest flag detector can be used for measuring the approximate beam spot size, so we can use it as the object.

TABLE 1. Numerical Values.

ΔE (±MeV)	4.00	3.33	2.86	2.50	2.00	1.67	4.00	3.33	2.78	2.50	2.00	1.67
$K_{Q1}(m^{-2})$	0	0	0	0	0	0	-0.65	-0.65	**-0.65**	-0.65	-0.65	-0.65
$K_{Q2}(m^{-2})$	-0.884	-0.323	0	0.2073	0.4615	0.6109	-0.751	-0.19	**0.1275**	0.3395	0.5939	0.7428
$K_{Q3}(m^{-2})$	1.258	0.6255	0	-0.611	-1.813	-2.981	1.2512	0.6163	**0**	-0.625	-1.831	-3.004
m_{x11}	-0.15	-0.207	**-0.265**	-0.322	-0.437	-0.552	-0.098	-0.136	**-0.174**	-0.212	-0.288	-0.364
m_{y11}	-1.216	-0.716	**-0.534**	-0.499	-0.634	-0.887	-0.688	-0.589	**-0.476**	-0.354	-0.111	0.1266
m_{y12}	-11.029	-5.266	**-3.684**	-4.041	-7.383	-12.22	-7.410	-4.178	**-3.089**	-2.979	-4.137	-6.084

EXPERIMENT RESULT

The new system was installed in August of 2001, some preliminary experiments were performed to test the analysis system in October and November last year. Fig.8

shows some photos. The upper picture was taken by setting the parameters at ±2.86MeV and no exciting current in Q1 and the lower picture was taken by setting the parameters at ±4.0MeV with no exciting current in Q1. We found the system worked well, and all the quadrupoles did not need to be energized during normal machine operation. Now the linac and storage ring have a long shutdown for reconstruction, and further experiments will be conducted in the near future.

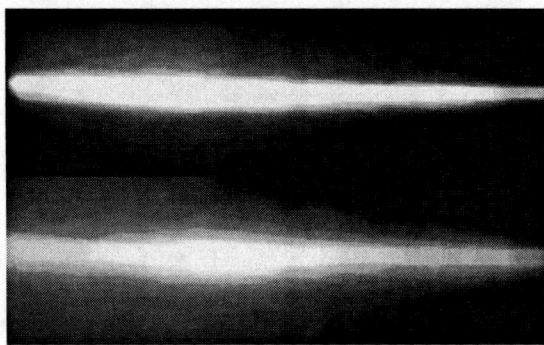

FIG.8. Photograph on the target. Upper picture was taken by setting the parameter as ±2.86MeV and no exciting current in Q1; the lower was taken by setting the parameter as ±4.0MeV when no exciting current in Q1.

REFERENCES

1. Y.J.Pei "200 MeV LINAC – injector for storage ring of HESYRL" Rev. of Scie. & Inst. Vol.60, No.7 P.1701 (1989)
2. Y.J.Pei, M.Bai, G.R.Huang, "Energy stability of 200 MeV electro LINAC" Proc. of the 1994 International LINAC Conference Vol.1, p.196 Tsukuba(1994)
3. SUN Bao-gen, FANG Zhi-gao, et al. Beam Energy Spectrum Monitor for Hefei 200MeV Linac. Nuclear Techniques, Vol.21, No.1, 1998.1, 48-50
4. SUN Bao-gen, LU Ping, et al. Beam Measurement System in NSRL. Journal of Systems Engineering and Electronics, Vol.11, No.3, 2000.9, 9-13
5. Huang Gui-rong, PEI Yuan-ji, et al. NSRL 200MeV Linac Beam Energy Stability System. Nuclear Techniques, Vol.24, No.3, 2001.3, 233-236
6. Zimmerman F. Measurement and Correction of Accelerator Optics. Proceedings of the Joint US-CERN-Japan-Russia School on Particle Accelerators, Montrux, 1998, 21-107
7. LIU Zu-ping. Beam Optics. University of Science and Technology of China
8. ZHAO Ai-hua, LIU Zu-ping. Physical Desigh Calculations for an Electron Beam Transport Line to the Nuclear Physics Experiment Hall at NSRL. Journal of China University of Science and Technology, Vol.28, No.4, 1998.8, 461-465

Enhancements To The Digital Transverse Dampers At The Brookhaven AGS[*]

M. Wilinski, A. Drees, R. Michnoff, T. Roser, G.A. Smith

Brookhaven National Laboratory
Upton, NY 11973

Abstract. Since 1993, a digital transverse damper system has been used at the Brookhaven Alternating Gradient Synchrotron (AGS). The dampers are used to damp coherent oscillations and injection errors in both planes for protons and all species of Heavy Ions. Over nine years, several AGS improvements, the addition of the Relativistic Heavy Ion Collider (RHIC) operations, and our experience, created a critical need to improve the original system. Several enhancements have been made to the digital electronics including compatibility with harmonic numbers up to 24, an increase in the system resolution from eight to ten bits, and the conversion of the system interface to VME. The analog electronics were also modified to appropriately interface with the new digital electronics, as well as to provide an overall functional improvement. The pick-up electrode (PUE) preamplifiers were redesigned to decrease the radiation susceptibility of the electronics. The concepts of the AGS Damper system can be utilized in developing a solution for the damping requirements in RHIC.

INTRODUCTION

The AGS Transverse Damper system was commissioned in 1993 to damp coherent oscillations and injection errors. It consists of beam position pickups, preamplifiers for the horizontal and vertical planes, signal processing electronics, and stripline kickers[1,2]. A simplified block diagram of one plane is shown in Figure 1.

The capacitive pickup electrodes (PUE) located at F20 in the AGS Ring are used to measure the beam position. A four-channel preamplifier conditions the signals before they are sent to the processing electronics located outside the AGS Ring in the F18 House. Hybrid transformers generate horizontal difference, vertical difference, and sum from the four PUE signals. The Buffer Gain Module provides a continuously adjustable gain of 0 to 40dB for each signal independently. Each signal is baseline restored and integrated over one clock cycle. The integrated signals are then digitized by ADCs. Each beam bunch centroid is calculated in a digital processor. Using an established algorithm,[3,4] a correction kick is determined and clocked into DACs. With the use of 500W Kalmus power amplifiers, the correction signal is sent to 50Ω stripline kickers in the AGS Ring to damp beam oscillations and instabilities.

The clock signals for the integrators, ADCs, and DACs are generated from RF sine and cosine signals. They are phased-locked to the frequency sweep of the low level RF used in the AGS. The RF signals are input into two Phase Shifter modules that

---

[*] Work performed under auspices of the U.S. Department of Energy

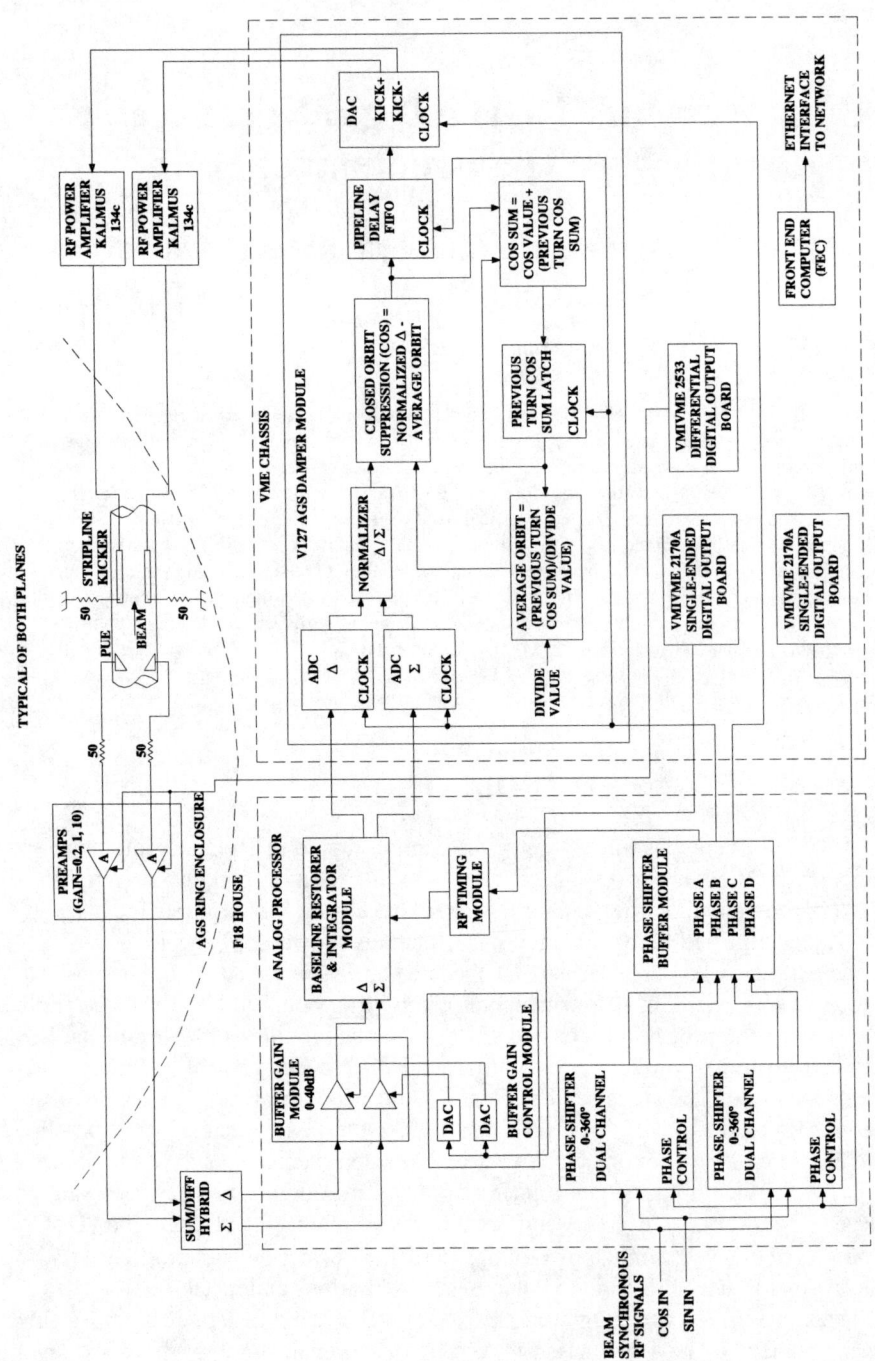

Figure 1. Block Diagram of the Digital Transverse Damper System.

produce a total of four independent phase-shifted TTL outputs. The signals are sent to a Phase Shifter Buffer Module that has two functions: (1) it operates as a buffer and, (2) it introduces a time delay on three of the four channels. The delay compensates for the amount of time it takes for the PUE signals to arrive at the F18 House, as it is necessary to keep the clock signals and beam signals in phase relative to each other over the entire RF frequency sweep. The buffered signals are then distributed to the appropriate device.

Changes and improvements made within the Brookhaven accelerator complex imposed a need for an upgrade to the damper system. The maximum harmonic number for the AGS was raised from 12 to 24. Also, the harmonic number selection range was expanded. Previously, only harmonic numbers 8 and 12 could be selected. The upgrade allows harmonic numbers in the range of 1 to 24.

The addition of RHIC operations also served as motivation for an upgrade. An injection damping system is currently planned for RHIC[5] for which the AGS dampers are a strong foundation. As the amount of time available for processing in RHIC is decreased, an effort was made to increase the processing speed of the AGS dampers for use as a prototype.

There were also several approaches on how to improve the system including increasing the flexibility of the programmed algorithms, increasing the system resolution to ten bits, and providing digital control to the Phase Shifters. The increase in system resolution had an immense impact on the digital portion of the electronics, as it required the processing gate array to be completely redesigned. Therefore, an upgrade was done to make the AGS damper system compatible with the new operational requirements.

DIGITAL SYSTEM UPGRADE

Prior to the upgrade, a BNL Instrument Controller controlled the processing electronics. This interface was outdated and contained several limitations, thereby providing the impetus for conversion to a VME based controller. The VME chassis contains a Front End Computer (FEC) with an ethernet interface to the BNL network, allowing the status to be monitored. In the conversion, the diagnostic memory, which had the capacity to store up to 16,000 turns of data, was removed.

A BNL custom designed VME module, the V127 AGS Damper module, was developed to perform the beam damping algorithm and to produce an analog output signal to the power amplifiers. One module is installed for each of the two planes.

Two Edge Technology, Inc. 700140 ten bit, 20 MHz ADCs are used to digitize the integrated sum and difference input signals provided by the analog processing electronics. The mathematical calculations for the algorithm, as shown in Figure 1, are performed in an Altera EPF10K200 gate array using pure combinatorial logic. As soon as the digitized ADC data becomes available, the calculations begin and propagate to the Pipeline Delay FIFO and the Previous Turn Closed Orbit Suppression (COS) Sum latch. An Edge Technology, Inc. 700145 ten bit, 20 MHz DAC is used to generate the plus and minus outputs to the Kalmus power amplifiers. The minus output is always of the same magnitude but opposite polarity of the plus output, thus doubling the kick strength by providing a push-pull effect on the beam.

The damping algorithm calculates a linear output function that is proportional to the difference between the measured single turn beam position and the calculated average orbit. This calculation is performed independently for each bunch, where the harmonic number defines the number of bunches. The Output Function to the power amplifiers is selected to be one of the following types: Linear Function Direct, Linear Function with Programmable Multiplier, Bang-bang, and Bang-bang with Programmable Deadband. The two programmable methods allow the output signal to change with greater flexibility.

The major system programmable parameters are shown in Table 1. Additional parameters, not shown in the table, allow the Output Function to change at a programmable delay time from the beginning of the AGS cycle.

TABLE 1. Programmable System Parameters

Parameter	Description
Harmonic Number	Number of RF clocks per revolution; maximum number of bunches in AGS Ring
COS Divide Value	Used by COS algorithm to control orbit correction response relative to average orbit; number of revolutions used to calculate average orbit
Sum Cutoff	Output is zero if sum signal is below this value; inhibits output if beam signal is not detected
Output Function Select	Select one of four algorithm types
Output Delay	Delays output value by programmable number of RF clocks for proper bunch synchronization
Preamplifier Gain	Select one of three gain settings
Sum Gain, Difference Gain	Select gain amount between 0 and 40dB
RF Clock Phase Shifter	Select a value between 0 and 360 degrees for critical timing control of the clock signals

ANALOG SYSTEM UPGRADE

To appropriately interface with the improved digital electronics, several of the analog electronics modules were modified. The ADCs were moved out of the analog processor crate and incorporated into the V127 module located in the VME chassis. The remaining analog modules were consolidated into one 19-inch, 6U Eurocard crate, requiring some modules to be repackaged from NIM modules. Several existing modules also had their inputs and outputs transferred from the rear panel to the front panel to accommodate the new analog-digital interconnections. Also, the two integrator modules had signal outputs added to view the integrated sum and difference signals without disturbing the signals used for processing.

Concurrent to the interface alterations, modifications were also made to the analog electronics to provide overall functional improvement. Digital phase control for the RF Phase Shifter module was implemented. Four digital phase controls are sent to the phase shifter by a VME based Front End Computer (FEC) with user control through SpreadSheet, a BNL developed, UNIX based software program. A decimal setpoint value between 0 and 255 is entered into SpreadSheet with an eight bit binary word corresponding to the setpoint sent to the appropriate channel. The conversion factor is 1°/least significant bit. If necessary, the module can be returned to analog phase control by changing a jumper internal to the module.

The digital control for the Buffer Gain Control Module was also changed. Previously, the module had one input of eight bits that was multiplexed among the horizontal difference, vertical difference, and sum signal gain controls. The module was redesigned to have three eight-bit inputs with each signal receiving its own gain control. The SpreadSheet program is used to control the gain by entering a setpoint value between 0, corresponding to a gain of 0dB, and 255, corresponding to a gain of 40dB. The gain conversion is linear between 0 and 40dB.

The final module that underwent a major upgrade was the Phase Shifter Buffer module. Formerly, the module had eight phase inputs and outputs. However, since only four outputs are produced by the Phase Shifters, eight channels were superfluous. The module was reconfigured with four single-ended phase inputs (Phase A, B, C, & D) and five differential outputs (Phase A, B, B, C, & C). The fourth input channel, Phase D, does not have a corresponding output on the front panel. The Phase A output is sent to an intermediary timing module and then used to clock the two integrator modules. The two Phase B outputs clock the horizontal and vertical ADCs, and the two Phase C outputs clock the horizontal and vertical DACs. The Phase A, B, and D channels contain an internal time delay of 200ns.

PREAMPLIFIER REDESIGN

The PUE preamplifiers were redesigned to reduce the damaging effects of radiation and to increase the amount of RF shielding. Previously, the preamplifier's power supplies were attached to its enclosure in the AGS Ring. Radiation damage to the power supplies was a major source of failure. However, the preamplifier cannot be moved from its location since the capacitive PUEs need a high impedance input. After consideration, the power supplies were moved outside of the AGS Ring into the F18 House, while keeping the preamplifier in its current location.

New inner and outer preamplifier enclosures with improved EMI shielding were purchased. The inner enclosure, which directly encloses the preamplifier circuit board, has shielding capabilities at or greater than 80dB up to 20GHz. This inner enclosure is placed inside an outer junction box that has shielding greater than 35dB up to 1GHz. To provide better shielding at the signal connection points, metal circular connectors for the gain control signals and power were used instead of plastic connectors. SMA connectors were used at the signal input and output connections instead of BNC connectors.

The circuit boards were also redesigned to take advantage of surface mount technology. The design is based on the preamplifiers that were developed for the AGS Booster[6]. By using surface mount chips, the four preamp channels were placed on one circuit board. The preamp has remotely selectable gains of one-fifth, one, and ten. Gain changing is accomplished through the use of relays by either introducing a voltage divider in a feedback loop for x10 gain or by loading the input signal with capacitance for x0.2 gain.

RESULTS

The upgraded digital and analog electronics have been installed and in use since February 2001 and performing as expected. There have been no major problems with setup or functioning of the upgraded portions. The enhanced computer control made the setup of the system faster and easier.

Some experimental data was taken to demonstrate the functioning of the dampers with the improved electronics. With the dampers off, a kick was delivered to the beam by the AGS Tune Meter Kicker, as shown in Trace B of Figure 2. Trace A on Figure 2 shows one bunch from the vertical-bottom PUE. As can be seen on the lower portion of Trace A, the kick has caused the beam to oscillate. Figure 3 shows the same two signals with the dampers turned on. The oscillations are damped out after approximately 150us. Figure 4 depicts a close-up version of the same two signals, as well as the Stripline Kicker signal on Trace C. The stripline kicker is kicking the beam in the appropriate direction to move it back toward the central orbit.

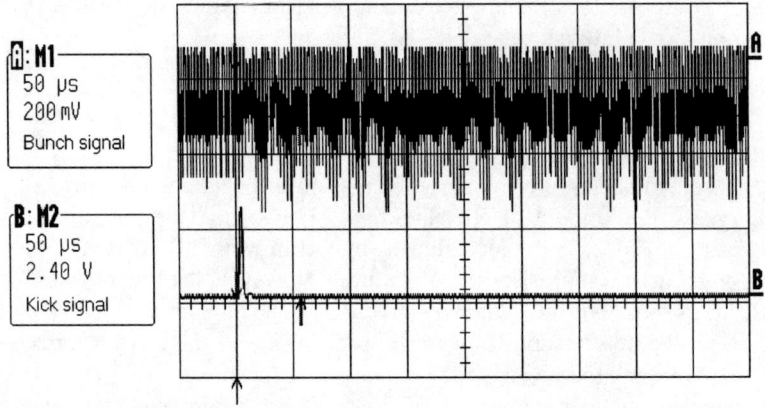

Figure 2. Bunch signal and Tune Meter Kicker signal with no damping.

Originally, the smallest gain of the preamplifier was intended to be one-tenth. During testing, the signal output exhibited large overshoots and undershoots on the rising and falling edges as a result of reduced bandwidth, as shown in Figure 5a. The input signal is on Channel 1 and the output of the preamp at the lowest gain is on Channel 2. Reducing the value of the input signal capacitive load from 2200pF to 680pF resulted in a cleaner signal as shown in Figure 5b. The input signal is on Channel 1 and the output of the preamp at the lowest gain is on Channel 2. The change in capacitor value does not affect the other gain ranges, as the capacitive load is only introduced in the lowest gain range.

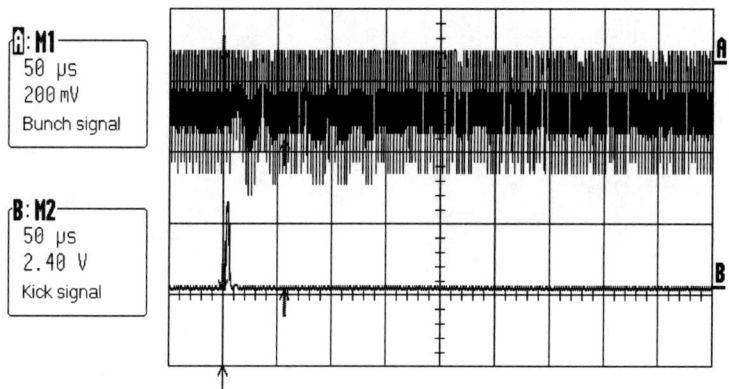

Figure 3. Bunch signal and Tune Meter Kicker signal with damping.

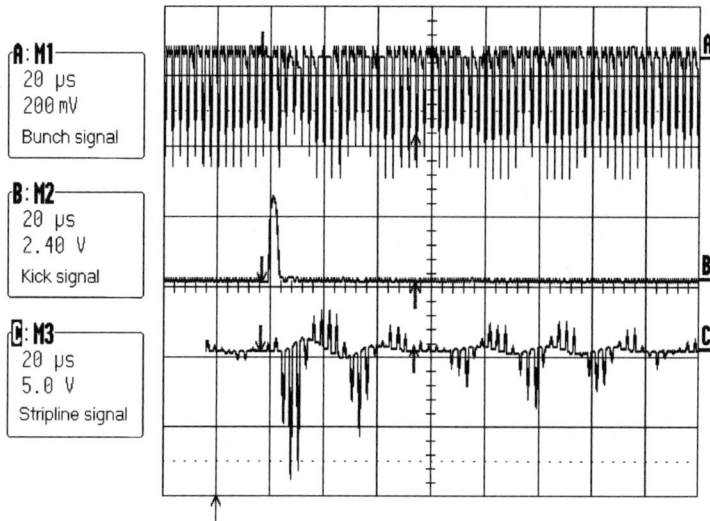

Figure 4. Bunch signal, Tune Meter Kicker signal, and Stripline Kicker signal with damping.

Since preliminary tests of the preamplifier were successful, one was installed at the G7 location in the AGS Ring for tests with beam during the most recent high intensity proton run. The remote power supplies were the only portions of the system not in use. The signals returned from the preamp were examined and were comparable to signals from the previous preamplifier. Therefore, the installation of the system at F20, as well as the implementation of the remote power supplies at G7, will be done over the upcoming shutdown period.

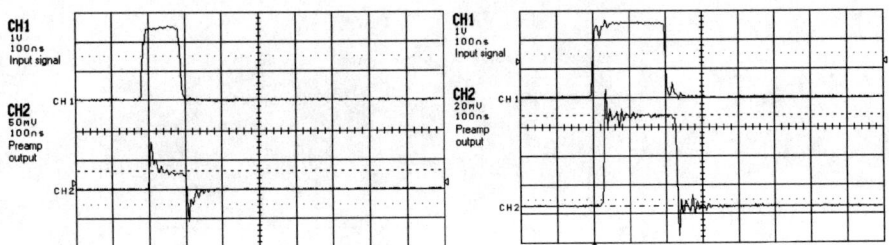

FIGURE 5. (a) Preamplifier output with C=2200pF and x0.1 gain, (b) Preamplifier output with C=680pF and x0.2 gain.

REFERENCES

1. G.A. Smith, T. Roser, R. Witkover, V. Wong, <u>Transverse Beam Dampers for the Brookhaven AGS</u>, AIP Conference Proceedings 319, Beam Instrumentation Workshop, Santa Fe, NM 1993, pp 309-318, BNL-49437.
2. G.A. Smith, V. Castillo, T. Roser, W. Van Asselt, R. Witkover, V. Wong, <u>Digital Transverse Beam Dampers for the Brookhaven AGS</u>, Proceedings of the 1995 Particle Accelerator Conference, Dallas, TX, pp 2678-2680, BNL-61021.
3. T. Roser, <u>Transverse Damping Algorithms</u>, Accelerator Division Technical Note, AGS/AD/Tech. Note No. 377.
4. T. Roser, <u>Recursive Transverse Damping Algorithms</u>, Accelerator Division Technical Note, AGS/AD/Tech. Note No. 398.
5. A. Drees, M. Brennan, P. Cameron, C. Montag, R. Michnoff, G.A. Smith, M. Wilinski, <u>RHIC Transverse Damper</u>, Beam Instrumentation Workshop, Upton, NY 2002.
6. D.J. Ciardullo, G.A. Smith, E.R. Beadle, <u>Design of the AGS Booster Beam Position Monitor Electronics</u>, Proceedings of the 1991 Particle Accelerator Conference, pp1431-1433.

Biasing Wire Scanners and Halo Scrapers for Measuring 6.7-MeV Proton-Beam Halo[+]

J. Douglas Gilpatrick *, Michael Gruchalla †, James Kamperschroer ¶, James O'Hara*

Los Alamos National Laboratory, MS H808, LANL, Los Alamos, NM, 87545
† Honeywell FM&T/NM, PO Box 5250, Albuquerque, NM, 87185
¶ General Atomics, Los Alamos, NM, 87544

Abstract. Wire scanners and halo scrapers (WS/HS) were used to acquire projected beam distributions over a very wide dynamic range in order to determine the extent and study the formation of beam halo at the Low Energy Demonstration Accelerator (LEDA). To detect beam distributions over a large dynamic range, it was necessary to understand the effects of WS/HS biasing for optimizing wire and scraper signal amplitudes. Both wire scanners and halo scrapers were biased with both positive and negative potentials to +/− 200 V. WS/HS depleted-charge data were acquired at these different potentials and the amount of signed charge leaving or accumulating on the wire or scraper was measured. This paper will show these data, and will offer a discussion of an optimal biasing potential for these types of projected beam profile measurement devices.

HALO INSTRUMENTATION

At the LEDA, a 100-mA, 6.7-MeV beam is injected into a 52-quadrupole-magnet lattice (see Fig. 1). Within this 11-m FODO lattice, there are nine wire scanner/halo scraper (WS/HS) stations, five pairs of steering magnets and beam position monitors, five loss monitors, three pulsed-beam current monitors, and two image-current monitors for monitoring beam energy [1]. The WS/HS instrument's purpose is to measure the beam's transverse projected distribution [2]. These measured distributions must have sufficient detail to understand beam halo resulting from upstream lattice mismatches [3,4]. The first WS/HS station, located after the fourth quadrupole magnet, verifies the beam's transverse characteristics after the RFQ exit. A cluster of four WS/HS located after magnets #20, #22, #24, and #26 provides phase space information after the beam has debunched. After magnets #45, #47, #49, and #51 reside the final four WS/HS stations. These four WS/HS acquire projected beam distributions under both matched and mismatched conditions. These conditions are generated by adjusting the first four quadrupole-magnetic fields so that the RFQ output beam is matched or mismatched in a known fashion to the rest of the lattice. Because the halo takes many lattice periods to fully develop, this final cluster of WS/HS are positioned to be most sensitive to halo generation.

[+] Work supported by the US Department of Energy.

FIGURE 1. The 11-m, 52-magnet FODO lattice includes nine WS/HS stations that measure the beam's transverse projected distributions.

As the RFQ output beam is mismatched to the lattice, the WS/HS actually observe a variety of distortions to a properly matched Gaussian-like distribution [3,4]. These distortions appear as distribution tails or backgrounds. It is the size, shape, and extent of these tails that represent specific types of halo. However, not every lattice WS/HS observes the halo generated in phase space because the resultant distribution tails may be hidden from the projection's view. Therefore, multiple WS/HS are used to observe the various distribution tails.

WS/HS DESCRIPTION

Each station consists of a horizontal and vertical actuator assembly (see Fig. 2) that can move a 33-μm-carbon monofilament and two graphite/copper scraper sub-assemblies [5]. The carbon wire and scrapers are connected to the same movable frame. Attached to this movable frame is a linear encoder that provides the wire and scraper edges' relative position to within a typical rms error of 5 μm, and an additional linear potentiometer provides an absolute approximate position for LEDA's run-permit systems. A stepper motor coupled to a ball lead screw is used to drive the moveable frame. A motor brake and microswitches limit the frame's movement.

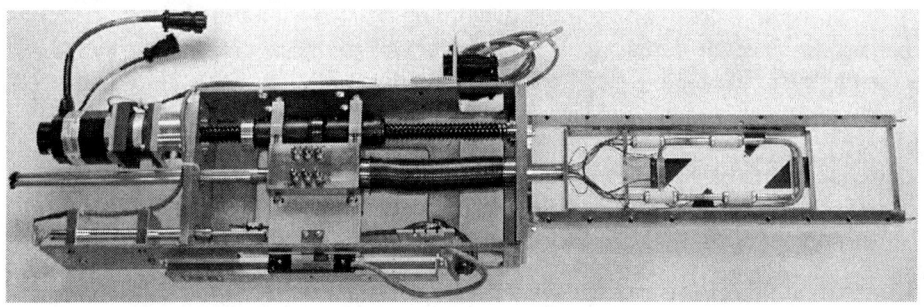

FIGURE 2. The WS/HS assembly contains a movable frame on which a 0.03-mm carbon wire resides between two water-cooled graphite scrapers.

The carbon wire, which senses the beam's core, is cooled by thermal radiation. If the beam macropulse is too long, the wire temperature continues above 1800 K resulting in the onset of thermionic emission [6]. Thermionic emission causes an inaccurate appearance to the distribution by exaggerating the core's current density. To eliminate these effects for the halo experiment, the maximum pulse length and repetition rate is limited to approximately 30 μs and 1 Hz, respectively.

The halo scrapers are composed of a 1.5-mm thick graphite plate brazed to a water-cooled 1.5-mm thick copper plate. Since 6.7-MeV protons average range in carbon is approximate 0.3 mm, the beam is completely stopped within the graphite plate. Cooling via conduction lowers the average temperature of the scraper sub-assembly and allows the scraper to be cooled more rapidly than the wire. The lower average temperature and faster cooling allows the scraper to be driven in as far as 2 rms widths from the beam distribution peak without the peak temperature increasing above 1800 K.

The movement and positioning of each wire and scraper pair is controlled by a motion control system that contains a stepper motor, stepper motor controller, a linear encoder, and an electronic driver amplifier [7]. The controller's digital PID loop controls the speed and accuracy at which the assembly is moved and placed.

The target position, as defined by the WS/HS operator, is relayed from the EPICS control screen via a database process variable to a National Instruments LabVIEW Virtual Instrument (VI). The VI also calibrates the relative position of the linear encoders based on the measured position of the limit switches, and provides some error feedback information [7]. The total error between the target wire position and the actual wire position attained is within a total 4% range of a typical 1-mm rms-width beam.

As the wire is moved through the beam, it senses the projected beam core distribution. A small portion of the beam's energy is imparted to the wire causing secondary electron emission to occur. The secondary electrons leaving the wire are replaced by negative charge flowing from the electronics. This current flow for both axes is connected through a negative bias battery to an electronic lossy integrator circuit and followed by an amplification stage.

The integrator capacitance and amplifier gain are set to allow a very wide range of values of accumulated charge [8]. Data are acquired by digitizing the accumulated charge through the lossy integrator at two different times within the beam pulse. This charge difference, acquired by subtracting the two values of charge, provides a low noise method of relative beam charge acquisition. The wire and scraper accumulated charge signals are digitized using 12- and 14-bit digitizers, respectively. The analog noise floor has been measured to be 0.03 pC, a noise level slightly lower than the scraper digital LSB noise level of 0.15 pC using the highest gain settings within the detection electronics.

The front-end electronic circuitry, mounted on a daughter printed circuit board, is connected to a motherboard that has all of the necessary interface electronics to communicate with EPICS via a controller module within the same electronics crate. A software state machine sequence was written within EPICS to control and operate WS/HS instrumentation [9]. The state machine instructs the VI to move the wire and scraper to a specific location, acquire synchronous distribution data from either the wire or scraper, trigger the IDL routine to normalize the acquired charge with a nearby toroidal current measurement, graph the normalized data, and write the distribution to a file. The sequence also instructs IDL to calculate the first through fourth moments, fit a Gaussian distribution to the wire scanner data, and calculate the point at which the beam distribution disappears into the background noise.

To plot the complete beam distribution for each axis, the wire scanner and two scraper data sets must be joined [10]. To accomplish this joining, several analysis tasks are performed on the wire and scraper data including,

(1) scraper data are spatially differentiated and averaged,

(2) wire and scraper data are acquired with sufficient spatial overlap, and

(3) differentiated scraper data are normalized to the wire beam core data.

The scraper data need only be normalized in the relative charge axis since the distances between each wire and scraper edge are known to within 0.25-mm. In addition, the first four moments and the point at which the beam distribution disappears into the noise are also calculated for the combined distribution data.

WIRE AND SCRAPER PHYSICS

For several different wire scanners and halo scrapers, an emission current curve was generated as a function of device bias potential. This was done to verify the acquisition goals of detecting only secondary electrons for the WS wire and 6.7-MeV protons for the halo scraper. The procedure included acquiring an integrated charge waveform at each bias potential. The emission currents for each point on the following graphs were acquired by fitting a straight line to the integrated charge data and calculated the fitted slope or current. Emitted current data were acquired during the last 10 µs and at least 10 µs after the 30-µs-length beam pulse.

Wire Scanner Bias

Several interesting details show up in these wire scanner bias curves. As the wire bias is positively increased from ~+6 V to > +200 V, the wire secondary electron emission is nearly completely inhibited and the net current flowing on to the wire reduces to very near zero (as shown in both wire scanners #22 and #51 of Fig. 3). At approximately +150 V, the secondary emission is completely inhibited. As the bias potential is reduced from ~-6 V to < -200 V, the net current varies from being relatively stable initially to a slight reduced trend. For the purposes of the halo experiment, the stable or 0-slope area in the –6 V to –12 V region is what was chosen as the optimal operating point and is the region in which the wire emission current does not change with bias. Furthermore, it appears that the wire collects positive ions with < -25 V bias potentials after the beam pulse, as shown by the red crosses in Fig. 3. This ion collection additionally limits the amount of negative bias that is applied to the wire for proper secondary emission operation. As the negatively biased wire's potential is further reduced, the small positive slope of the "during pulse" wire data appears to reinforce the premise that slow ions are being collected.

One rather interesting area of the graph is that of the emitted wire current near the 0 V bias potential. It is peaked at 0 V approximately 15% higher than at either +6 V or –10 V bias potential. One proposed explanation of this 0-bias elevated wire current is the interception of electrons and ions from protons ionizing residual background gas - both of these ionized species creating further secondary emission. As the wire is biased negatively, low energy intercepted electrons are rejected causing a reduction in secondary emission, and as the wire is biased positively, the intercepted ions are rejected causing a reduction in secondary emission. If the intercepted electron-ion pairs are the mechanism for the elevated net current, the wire should be biased to not include this additional current component since this is not directly due to 6.7-MeV beam impingement. Since we did not have clear proof as to what the real cause of the elevated current is, it was decided to bias the wire away from this effect, i.e., a –12 V bias.

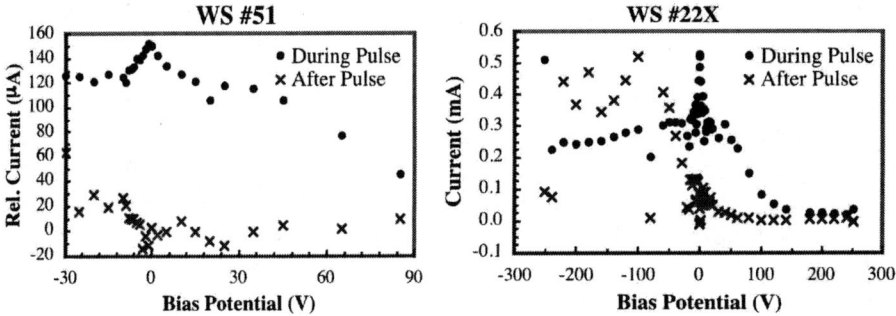

FIGURE 3. The two above graphs of wire scanner #22X and WS #51 show the wire scanner emission current as a function of wire potential. The goal of proper biasing is to optimize the emission current such that the detection current is only due to secondary electron emission. This goal was addressed by biasing the wire with approximately –12 V where the curves slope is approximately zero.

Halo Scraper Bias

The scraper bias data, Fig. 4, show similar results as the wire data. Increasing the scraper bias from ~+ 6 V to > +100 V, reduced the amount of electrons leaving the scraper. At +20 V to +40 V, the total amount of current detected levels out to a minimum and is composed only of deposited protons and is constant with respect to increasing bias. For the purposes of the halo experiment, a scraper bias of +25 V was chosen. Also, note that with this +25 V bias, the data show that no after pulse current flowing. One interpretation of the after pulse current was that it is composed primarily of slow low-energy positive ions. This interpretation seems to hold based both on the temporal waveforms observed during data acquisition and the fact that after pulse current is a positive non-zero value for negative bias potentials.

With 0 V applied to the scraper, the scraper net current is also elevated. As with the wire, it is not clear what is causing this increase in emission current so it was decided to bias the scraper outside this region, i.e. a +25 V bias.

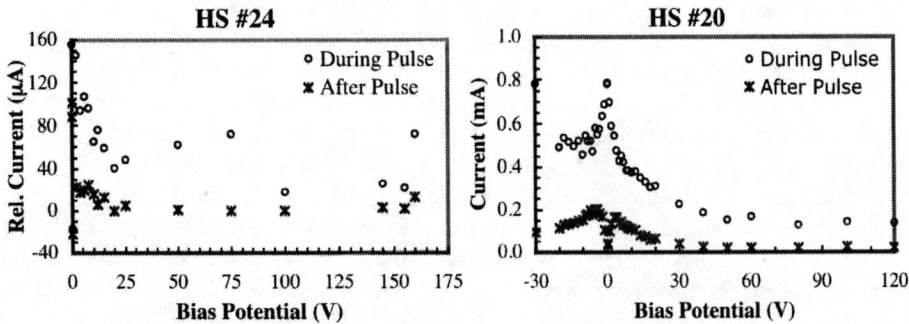

FIGURE 4. The two above graphs of halo scraper #20 and WS #24 show the scraper emission current as a function of scraper potential. The goal of proper biasing is to optimize the emission current such that the detection current is only due to deposited protons. This goal was addressed by biasing the scraper with approximately +25 V where the curve slope is approximately zero.

Secondary Emission Yield

As was shown in the previous section, the WS wire is biased negatively to optimize secondary emission (SE) yield, where these SE yield is defined as the ratio of the emitted secondary electron current and the proton beam current intercepted by the wire. TABLE 1 shows the measured values of a representative sample of the lattice and HEBT WS wires. All of the wires in the halo lattice WS are configured with a 33-μm, carbon monofilament. The HEBT WS is configured with a 100-μm SiC wire. All of the SE currents were acquired with the nominal –12 V bias so that SE was optimized and with the wire placed in the core of the proton beam. In both cases, the 6.7-MeV protons did not stop in the wire but deposited sufficient beam energy into the wire to cause SE.

One particular model for emission of secondary electrons resulting from energetic ions impinging on various materials is described by E. J. Sternglass. In his 1957 paper, Sternglass defined the secondary emission yield, Y, as

$$Y = \frac{Pd}{\varepsilon} \frac{dE}{dx} \qquad (1)$$

where P is the probability that an electron will escape (approximately –0.5), d is the average depth from which electrons escape the material (in this case a small monolayer of ~ 1 nanometer), ε is the average amount of kinetic energy lost by an ion or proton per ionization in material (~ 25 eV), and dE/dx is the stopping power of the proton beam for the wire material (163 MeV/cm for SiC and 162.9 MeV/cm for C) [11, 12]. The resulting calculated yield based on the Sternglass model for both C and SiC is approximately 33%.

However, the above model does not include the geometric factor of a round wire. In this case, a round wire will have areas on both sides of the wire that have a longer distance to deposit the proton beam's energy in the 1-nm annular "electron escape" region of the wire. This "form factor," as Sternglass describes it in his paper, is a value proportional to a $sec(\theta)$ function where θ is the angle between two rays, an incoming proton trajectory ray and a ray between the wire center and the impact point of the proton on the wire perimeter. The "form factor" has been calculated to be ~3.8 and ~4.2, for the 33-μm and 0.1-mm round wires, respectively.

Additionally, Sternglass also mentions a SE yield temperature dependency that can lower the emission efficiency by as much as 50%. If both the temperature dependency effect and the "form factor" are treated as multipliers to the initial theoretical yield, a final yield is calculated to be ~63% and ~70% for the C and SiC wire, respectively. The acquired experimental data, as shown in Table 1, are within a few 10s of percent and approximately agrees with the calculated yields as suggested by Sternglass's model.

TABLE 1. Secondary Emission Yield: SiC and C Wires with 6.7-MeV Proton Impingement

Wire Scanner Number	X/Y Beam Current (mA)	X/Y Rms Width (mm)	X/Y S. E. Current (mA)	X/Y Yields (%)
22	75/76	0.86/0.67	0.63/0.6	55/41
24	76/76	0.78/0.88	0.55/0.61	42/54
47	76/76	0.7/0.75	0.6/0.6	42/47
51	77/77	0.8/0.77	0.65/0.61	51/46
HEBT	75/75	5/8.2	0.30/0.24	51/66

SUMMARY

This paper has described the general operation of the WS/HS combination profile measurement used in the LEDA. In order to operate the WS and HS properly, a series of biasing measurements was performed. The wire scanner and halo scraper overall acted as expected at various biasing potentials and the resulting V-I curves show that the wire and scraper are optimally biased at −12 V and +25 V, respectively. However, an additional effect near 0 V potential of an elevated emission was unexpected and not well understood. A proposed explanation for this elevated emission current was suggested but certainly not proven. The amount of secondary emission yield was measured to be approximately the expected amount based on the Sternglass model of secondary emission of electrons as a energetic ion impinges on a specific material (in this case, 6.7-MeV protons on C or SiC).

REFERENCES

1. Gilpatrick, J. D., et al., "Experience with the Low Energy Demonstration Accelerator (LEDA) Halo Experiment Beam Instrumentation," Proceedings of the *2001 Particle Accelerator Conference*, June 18-22, 2001, pp.2311-2313.
2. Gilpatrick, J. D., et al., " Beam-Profile Instrumentation for Beam-Halo Measurement: Overall Description and Operation," Proceedings of the *2001 Particle Accelerator Conference*, June 18-22, 2001, pp. 525-527.
3. Wangler, T. P., et al., "Experimental Study of Proton-Beam Halo Induced by Beam Mismatch in LEDA," Proceedings of the *2001 Particle Accelerator Conference*, June 18-22, 2001, pp. 2923-2925.
4. Colestock, P. L., et al., "Measurements of Halo Generation for a Proton Beam in a FODO Channel," Proceedings of the *2001 Particle Accelerator Conference*, June 18-22, 2001, pp. 170-172.
5. Valdiviez, R., et al., "The Final Mechanical Design, Fabrication, and Commissioning of a Wire Scanner and Scraper Assembly for Halo-Formation Measurements in a Proton Beam," Proceedings of the *2001 Particle Accelerator Conference*, June 18-22, 2001, pp. 1324-1326.
6. Valdiviez, R., et al., "The High-Heat Flux Testing of an Interceptive Device for an Intense Proton Beam", Proceedings of the *2001 Particle Accelerator Conference*, June 18-22, 2001, pp. 1321-1323.
7. Barr, D., et al., "Design and Experience with the WS/HS Assembly Movement Using LabVIEW VIs, National Instrument Motion Controllers, and Compumotor Electronic Drive Units and Motors," Proceedings of the *2001 Particle Accelerator Conference*, June 18-22, 2001, pp. 794-796.
8. Gruchalla, M., et al., "Beam Profile Wire-Scanner/Halo-Scraper Sensor Analog Interface Electronics," Proceedings of the *2001 Particle Accelerator Conference*, June 18-22, 2001, pp. 2314-2316.
9. Day, L., et al., "Automated Control and Real-Time Data Processing of Wire Scanner/Halo Scraper Measurements," Proceedings of the *2001 Particle Accelerator Conference*, June 18-22, 2001, pp. 1309-1311.
10. Kamperschroer, J., et al., "Analysis of Data from the LEDA Wire Scanner/Halo Scraper," Proceedings of the *2001 Particle Accelerator Conference*, June 18-22, 2001, pp. 1306-1308.
11. Sternglass, E. J., "Theory of Secondary Electron Emission by High-Speed Ions," *The Physical Review*, Second Series, Vol. 108, No. 1, October 1, 1957, pp. 1-12.
12. Borovsky, J. E. and Suszcynsky, D. M., "Experimental Investigation of the z^2 Scaling Law of Fast-Ion-Produced Secondary Electron Emission," *The Physical Review A*, Vol. 43, No. 3, February 1, 1991, pp. 1416-1432.

The Mechanical Design and Preliminary Testing Results of Beam Position Monitors for the LANSCE Isotope Production Facility and Switchyard Kicker Projects

J. F. O'Hara, J. D. Gilpatrick, J. E. Ledford, R. B. Shurter, R. J. Roybal, B. E. Bentley

Los Alamos National Laboratory
Los Alamos, NM 87545

Abstract. The Los Alamos Neutron Science Center (LANSCE-1) Beam Diagnostic Team is providing Beam Position Monitors (BPMs) to the LANSCE Facility for use in two on-going projects: The Isotope Production Facility (IPF) and The Switchyard Kicker Upgrade (SYK). The BPM designs for both projects are very similar. The BPMs are classic, four, micro-stripline units having one end terminated in a 50-ohm load. This paper will discuss the position measurement requirements, mechanical design, fabrication, and alignment issues encountered for both sets of BPMs, as well as report the results obtained from the initial taught wire testing of the IPF BPMs.

INTRODUCTION

The Los Alamos Neutron Science Center (LANSCE) is currently in the process of implementing two different projects to upgrade the 800 MeV accelerator facility's capabilities. The two projects are the construction of the Isotope Production Facility and the Switchyard Kicker Project upgrade. Both of these projects require Beam Position Monitors (BPMs), which will be provided by the LANSCE-1 Beam Diagnostic Instrumentation Team (BDIT). The IPF projects BPMs have been fabricated and tested, four of the eight required units have been installed during the recent LANSCE maintenance outage. The Switchyard Kicker BPMs are in the process of being fabricated. The Switchyard Kicker mechanical design is based on the IPF design with some intended improvements.

Isotope Production Facility

Radioisotopes can be introduced into the body where their absorption by different organs can be detected and used for diagnosis and treatment of diseases [1]. LANSCE has been producing radioisotopes for the nation's health program for over 20 years, and it is essential that Los Alamos National Laboratory along with other isotope production facilities at Brookhaven National Laboratory, TRIUMPF (Vancouver, Canada), Institute of Nuclear Research (Troitsk, Russia), National Accelerator Centre (Faurve, South Africa), and Paul Scherrer Institute (Villigen, Switzerland) continue to deliver a dedicated year-round supply of key medical isotopes. In order to help meet this demand, a new facility is being built at LANSCE that consists of a new beam line leading to a new target irradiation area (located below ground), and a new target equipment handling area (above ground). The new beam line will contain eight BPMs, seven four-inch ID and 1 six-inch ID, whose main purpose is to ensure proper beam/target interaction.

Switchyard Kicker Project

Typical LANSCE operation [2], [3], consists of two simultaneously accelerated beams using alternating cycles of the linac RF. Two 0.75 MeV Cockroft-Walton injectors are used; one supplies protons and the other supplies H^- ions. Beams are accelerated up to 800 MeV. The beam switchyard is used to direct the two beams from the linac to the different experimental areas. One problem with the existing switchyard configuration is that it is not currently possible to deliver simultaneous beam to experimental areas served by lines D and X. It requires a few hours in order to reconfigure beam delivery from one line to the other and during this time delivery of beam to all users is interrupted.

The Switchyard Kicker project involves removing existing unused magnets and replacing them with a kicker system consisting of two C-magnet benders, two pulsed kicker magnets, and assorted beam diagnostics including three, four inch ID, BPMs. The purpose of the BPMs is to monitor and track the kicker system operation by acquiring the H^- beam's position as the beam is switched between lines D and X. The Switchyard Kicker BPMs are placed such that they will only see accelerated H^- particles.

BEAM POSITION MONITORS

Position Measurement Requirements

For both facilities, the overall position measurement requirements are similar with some differences. As Table 1 shows, the beamlines transport H^+ (IPF) and H^- (SYK) beams, their dynamic range, and timing constraints are also somewhat different. The Switchyard Kicker BPM has various chopped beam structures that required the beam position measurement to detect and monitor the chopped beam's position with as few as 10 to 20 beam bunches per chopped beam pulse. Whereas the IPF will be both tuned and operated with no chopped beam pattern. However, due to the tuning and operating procedures both have similar range of detected beam position and dynamic range requirements. Finally, one additional but no less important requirement was added. Both sets of instruments must have an online method of both unambiguously verifying the instrumentation's operational health and maintaining the electronics calibration. For this reason, the calibration and operational verification method, first applied in the LEDA beam position measurements, was adopted. This method allowed the operators to initiate a calibration of the electronics processors from the control room and to verify that the cable and BPM electrode are working in a known and unambiguous manner. This additional cable and electronics verification step is performed by injecting a signal into the downstream termination and acquiring the know attenuation through the electrodes and full cable plant. Further detail of the processing electronics and calibration and verification processes will be discussed in a later paper.

Table 1. Overall position measurement requirements.

Measurement or Beam Parameter	IPF	Switchyard Kicker
Accelerated Beam Species	H^+	H^-
Beam Repetition Rate (Hz)	30 (possibly 1 to 120)	1 to 100
Data Acquisition rate (Hz)	1 to 10	1 to 10
Macropulse length (ms)	0.05 to 1	0.15 to 1.225 ·
Chopped Beam Rate (MHz)	No chopped beam	2.8 and 5.6
Chopped Beam Duty Factor (%)	No chopped beam	22 to 56
Macropulse Beam Current (mA)	16.5 to 0.1	13 to 1
Bunching Frequency (MHz)	201.25	201.25
Position Measurement Range (% pipe radius)	50	50
Position Measurement Dynamic Range (dB)	63	~70
Bandwidth (kHz)	15	~2500
Position Precision (% pipe radius)	0.3	0.25
Beam Pipe Radius (mm)	50.4 and 76.2	50.4

IPF BPM Mechanical Design

The BPMs consist of four micro-stripline units having one end terminated in a 50-ohm load, two LANSCE style modified flanges, and the center-body housing. Figure 1 is a photograph of the fabricated IPF BPMs. Two different size BPMs were built for IPF, a four-inch ID and a six-inch ID. There will be a quantity of 7, four inch ID BPMs and 1, six inch ID BPM used in the IPF line. The six-inch ID BPM is required where the beam is being expanded as it approaches the target.

The BPM is fabricated in two separate sub-assemblies, which are joined at the final assembly step. The first sub-assembly is the housing. This consists of the BPM center-body and flanges, all fabricated from 316-stainless steel. The BPMs are located near quadrupole magnets and therefore are desired to be non-magnetic so as to not interfere with the quadrupoles magnetic fields. 316-stainless steel was chosen because it possess good non-magnetic stability during cold work [4]. The flanges are initially fabricated in a rough machined condition. One flange is slightly different from the other. The BPM has four flats cut into the downstream flange's outer-diameter; each flat also receives a 0.250" hole for an alignment target. The flats and holes are later used in the characterization and alignment of the BPM. Some material is left on the sealing face so that later it can be removed once the flanges have been welded onto the center-body. This final machining step allows the sub-assembly to be brought within dimensional tolerance after any distortion due to the welding operation. The shape in the center-body was cut out using a wire EDM (electrical discharge machining) technique. The flanges themselves are a modified version of the standard LANSCE flange. This style of flange utilizes an aluminum wire sandwiched between mating flanges as the vacuum sealing mechanism.

FIGURE 1. Fabricated 4" ID and 6" ID IPF BPMs.

The center-body housing and stripline electrode geometry were optimized to give a matched 50-ohm impedance [5]. The electrodes are recessed 1 mm from the bore of the flange for two reasons. The first reason is to protect the electrode from stray beam impingement. The second reason is to minimize the signal coupling between adjacent electrodes.

The electrode sub-assembly units consist of a 316 stainless steel electrode 0.030 inches thick connected to two KAMAN [6] UHV microwave feed throughs. The feed throughs are attached to a transition piece that is in turn welded into the BPM center-body. Figure 2 shows the electrode sub-assembly. The process for assembling the electrode is to first weld the transition piece to each of the individual feed throughs. The KAMAN feed throughs have a borosilicate strengthened glass seal that has a temperature limitation of 300 °C. Care must be taken during welding to avoid overheating the feed through, damaging the seal, and causing a vacuum leak. The next step is to attach the feed through center-pins to the electrode. The center-pin material is a molybdenum alloy (TZM) that needs to be secured to the 316-stainless electrode. A torch braze technique has been found to be extremely effective in making this a secure joint. The alloy used is AWS (American Welding Society) Bag-13a alloy with AWS FB3C flux, again care must be taken to avoid overheating the feed through.

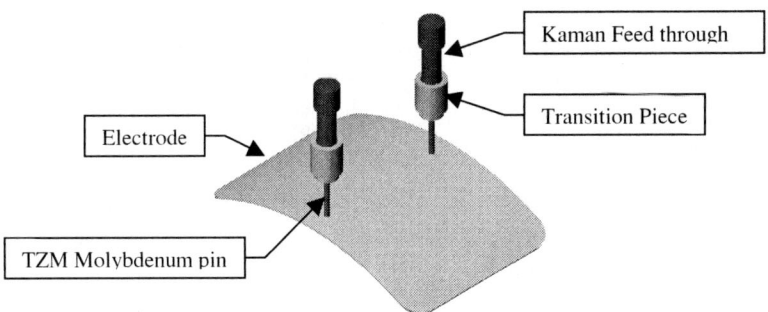

Figure 2. Electrode sub-assembly.

The final step in assembling the BPM is to weld the electrode assembly into the center-body. The electrode assembly is installed from the inside of the center-body. The feed throughs are passed through the holes in the center-body until the tops of the transition pieces are flush with the flats in the center-body. Figure 3 depicts the installation of the sub-assembly into the BPM center-body.

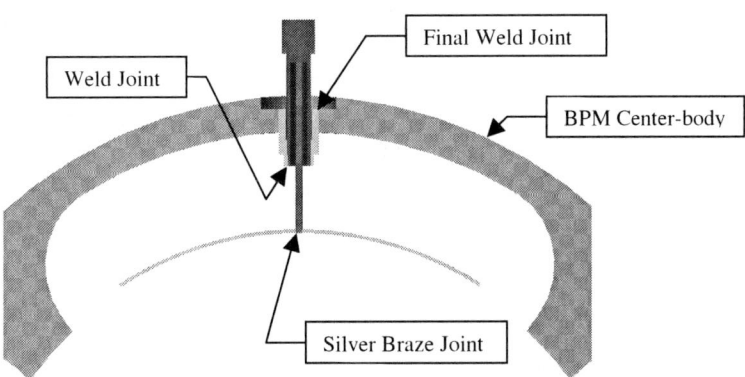

Figure 3. Electrode sub-assembly installation.

IPF BPM Characterization

The fabricated BPMs are then characterized using the BDIT's BPM mapping system [7], [8]. The mapping system consists of a 0.004-inch diameter, stainless steel wire, stretched tightly between two horizontal plates and running through the bore of the BPM. A 201.25 MHz RF signal is injected onto the wire, which induces RF currents on the BPM's electrodes. Figure 4 shows the BDIT mapping system schematically. A stepper motor system is used to move the upper and lower plates, while the BPM remains stationary, such that the wire is positioned at discrete locations throughout the bore of the BPM. The induced RF signal on each electrode is read out and recorded on a Boonton [9], 4300 power meter. The power ratio between opposing electrodes is then calculated for each wire location. This data set of power ratios is fitted to a third order, two dimensional equation using the Levenberg-Marquardt [10] least-squares fit method using IDL programming environment [11].

The reference features (fiducial flats and alignment target holes) on the BPM flange described previously are used to locate the BPM in the mapping system. The offsets determined from the mapping process are based on these reference surfaces. These reference surfaces are also used during BPM installation. The BPMS are installed in the beam line and their locations are measured using a theodolyte system. The theodolyte system uses alignment targets, which are placed in the holes provided in the BPM flange. This way alignment data and mapping data are based on the same reference surfaces. The offsets determined from the mapping and alignment processes can then be incorporated into the measurement software and accounted for during a measurement.

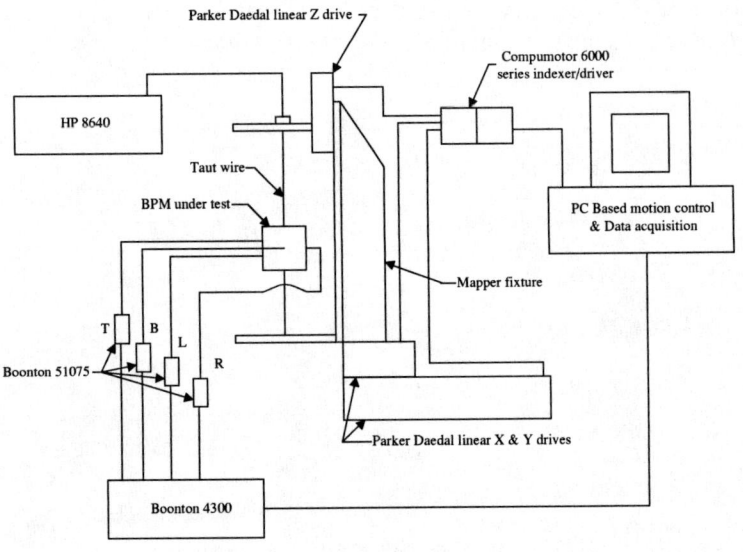

Figure 4. BDIT BPM mapping system schematic.

IPF BPM Mapping Results

Each of the eight (seven required and one spare) four inch ID BPMs has been mapped using the above-mentioned method. Each of the eight BPMs had at least three mapped data sets. Each data set is analyzed and the fit coefficients are generated. Table 1 shows the offsets and sensitivities of each of the eight four inch BPMS. The values listed in table 1 are the mean value of the coefficients based on the number of data sets taken for each BPM. The calculated theoretical sensitivity for these 50-mm radius BPMs was 0.654 dB/mm which is approximately 4% higher than the actual measured sensitivities shown in Table 2.

Table 2. 4" IPF BPM Offset and Sensitivity Data

IPF BPM	X offset (dB)	Y offset (dB)	X Sensitivity (dB/mm)	Y Sensitivity (dB/mm)
BPM #1	-1.9661E-01	-4.9739E-02	6.2674E-01	6.2674E-01
BPM #2	-6.9556E-02	-1.3268E-01	6.2608E-01	6.2595E-01
BPM #3	-1.1288E-01	-2.4562E-01	6.2602E-01	6.2615E-01
BPM #4	2.0089E-02	-1.9652E-01	6.2580E-01	6.2548E-01
BPM #5	-1.2761E-01	-8.1582E-02	6.2588E-01	6.2603E-01
BPM #6	-1.8416E-01	-4.2807E-01	6.2573E-01	6.2587E-01
BPM #7	1.3820E-02	-1.2591E-01	6.2594E-01	6.2591E-01
BPM #8	-1.1351E-01	-1.5327E-01	6.2547E-01	6.2623E-01

The maximum standard deviation of the offsets values for each individual BPM was 0.020 dB (0.032 mm). This value was generated from the numerous data sets related to each individual BPM and gives insight into the mapping system's precision. Other important information can be gained by looking at the standard deviation of the offset values for all of the different BPMs. From this value one can get an indication of the BPM manufacturing tolerances. The standard deviation of the offsets of the entire BPM set was 0.106 mm (0.004 inches).

Switchyard Kicker BPM Mechanical Design

The Switchyard Kicker BPM design is very similar to the IPF BPM. The same modified LANSCE style of flanges, the same recessed electrode design, and the same KAMAN feed throughs are used. There were a few areas where some improvements were attempted. The Switchyard Kicker BPM design calls for the brazing of the flanges to the center-body. This is done to eliminate any potential virtual vacuum leaks caused by trapped volumes due to the outside weld required in the IPF BPM. The manufacturing process for the Switchyard Kicker BPMs has been modified from the IPF plan by including more rough machining steps during the housing sub-assembly fabrication stages. These extra steps allow for the precise location of key features such as the reference flats and alignment target holes to be accomplished after the parts have gone through the braze process, as opposed to the IPF design where potential distortion due to welding could adversely impact the location of these features.

Another improvement is a modification to the electrode sub-assembly. A transition piece was added between the two feed throughs. This added piece keeps the feed throughs supported during the time prior to installation in the center-body. The interface between the transition piece and the center-body becomes the final sealing weld. Plans call for this weld joint to be done with at laser to minimize the amount of heat used to make the joint and also minimize any electrode placement distortion. Figure 5a shows the Switchyard Kicker BPM with the upstream flange removed and figure 5b shows the modified electrode sub-assembly.

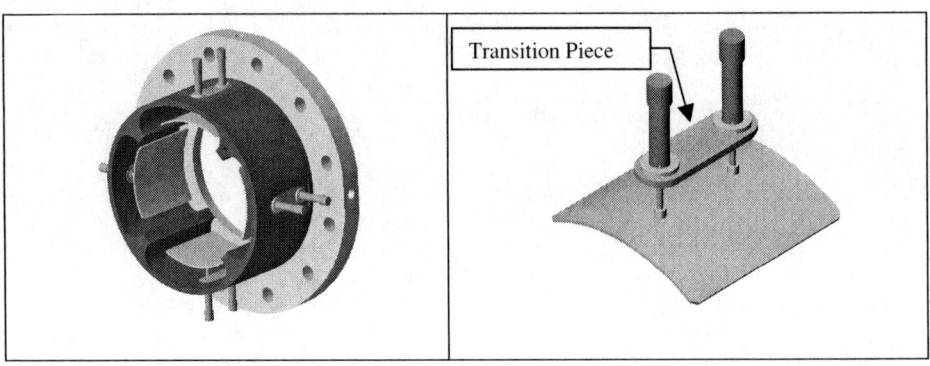

FIGURE 5A & 5B. Switchyard Kicker BPM with upstream flange removed and modified electrode sub-assembly.

Summary

Two similar style BPMs have been designed for two different on-going upgrade projects at LANSCE. The BPMs are classic, four, micro-stripline units having one end terminated in a 50-ohm load. The four-inch ID IPF BPMs have been characterized and the results were presented. The results show that the mapping system has a precision of 0.020 dB (0.032 mm), the BPMs have a sensitivity of 0.626 dB/mm (standard deviation of 3.55e-4 dB/mm), which compared well with the theoretical value of 0.64 dB/mm. Based on the offset data the error associated with the manufacturing process can be described to be within 0.004 inches. Four of the IPF BPMs have been installed in the beam line. The Switchyard Kicker BPM design is similar to the IPF design with some modification that should improve these BPMs overall performance. The Switchyard Kicker Upgrade schedule calls for the BPMs to be delivered and tested by October 1st of 2002.

REFERENCES

1. Heaton, R.C., and Peterson, E.J., "Construction of a New Isotope Production Facility", *LANSCE Division Technology Review*, LALP-01-258.
2. Fitzgerald, D.H., *"LANSCE Switchyard Kicker Proposal Introduction"*, April 2, 2001.
3. Carpenter Steel Company, *"Carpenter Steel Company Data Book"*, 1980, Wyomissing, PA.
4. Garnett R.W., et.al., *"Status of a New Switchyard Design for LANSCE"*, LINAC '98, Chicago, IL, August 23-28, 1998.
5. Kurennoy S.S., Los Alamos National Laboratory, private communication.
6. KAMAN Instrumentation, an operating division of KAMAN Aerospace, Colorado Springs, CO.
7. Gilpatrick, J.D. et al., *"LEDA & APT Beam Position Measurement System: Design and Initial Tests"*, LINAC '98, Chicago, IL, August 23-28, 1998.
8. Shurter R.B., et.al. *"An Automated BPM Characterization System for LEDA"*, BIW '98, Stanford, CA, May 4-7, 1998.
9. Boonton Electronics, Parsipany NJ.
10. Craig B. Markwardt, *"MPFIT2DFUN"*, NASA/GSFC Code 662, Greenbelt, MD 20770.
11. IDL, Interactive Data Language, Research Systems Incorporated a Kodak Company, Boulder, CO.

Electron Beam Diagnostics at the Radiation Source ELBE

P. Evtushenko, U. Lehnert, P. Michel, C. Schneider, R. Schurig,

J. Teichert

Radiation Source ELBE,
Research Center Rossendorf, Postfach 510119, 01314 Dresden, Germany

Abstract. In the research center Rossendorf, the radiation source ELBE, based on a super conducting LINAC, is under construction. In the year 2001 the first accelerating module was commissioned. The electron beam parameters like emittance, bunch length, energy spread were measured. Here we present results of the measurements as well as the methods used to make the measurements. In the ELBE injector, where electron beam energy is 250 keV, the emittance was measured with the aid of a multislit device. Emittance of the accelerated beam was measured by means of quadrupole scan method and is 8 mm×mrad at 77 pC bunch charge. Electron bunch length was measured using the coherent transition radiation technique. At the maximum design bunch charge of 77 pC the RMS bunch length was measured to be 2 ps. A set of online diagnostic systems is also under development. One these include a system of stripline beam position monitors is also described here. A BPM resolution of about 10 µm was achieved using logarithmic amplifier as the core element of the BPM electronics. A system of beam loss monitors based on the RF Heliax cable working as an ionization chamber is intended to be another online diagnostic system.

TRANSVERSALE EMITTANCE MEASUREMENTS

Introduction

The ELBE accelerator is a conventional design with an injector section followed by an accelerator section. The injector itself consists of the thermionic gun and a beam line section wherein two bunchers, focusing and steering elements as well as diagnostics are installed. The bunched beam is injected into the first accelerating module with an energy of 250 keV. After the accelerating section the beam is available with energies up to 20 MeV; in a later stage of improvement up to 40 MeV are planned. For some of the applications at ELBE, the emittance for a specific current or bunch charge is an important input. Two commonly used approaches are applied at ELBE to determine the emittance. The sampling of partitions of the phase space named multislit or pepper pot method. The determination of the beam matrix through the measurement of the beam diameter at different values of a focusing magnetic lens namely a quadrupole or solenoid scan method. In the injector section both methods mulitslit / pepper pot and solenoid scan are used while for the accelerated beam the quadrupole scan method is used to determine the emittance.

The Multislit Method

In the injector section of ELBE different masks can be put into the beam to register slices of the beam on a luminescent screen. In the past we have used slit masks with slit dimensions of 100 µm and slit distances of 1 mm and 3 mm, while now we us pepper pot masks to measure vertical and horizontal emittance in one step. The image on the luminescent screen is recorded by a vidicon camera, which delivers the image to a frame grabber card in a PC. From the projection of the beam through the slits onto the screen the space location and the divergence of the phase space samples can be measured. These parameters are extracted by a multi Gauss fit from a profile slice of the image data. From the fit parameters a RMS emittance is calculated.

Quadrupole Scan Method

A focusing magnetic lens changes the beam diameter at a fixed distance. If the beam diameter on the screen is recorded for different magnetic strengths of the lens the elements of the beam matrix in front of the lens can be extracted. From the matrix elements the transverse emittance can be calculated. In the injector we use solenoids while after the accelerator cavities quadrupoles are used as focusing elements. In the injector the multislit and the solenoid scan method agree to within a 20% percent level, which gives a hint to the precision of the emittance determination.

Results

On Fig. 1 all the relevant emittance measurements are summed as a function of the bunch charge. The squares show older measurements performed with the multislit method in the injector at a micro pulse rate of 13 MHz. They agree well with model calculations of our thermonic gun [1] (solid curve) especially in the range of higher bunch charges. For applications in the radiation physics an attempt was made to reduce the emittance by cutting off the outer part of the beam through apertures, (see the triangles). With a transmission of 10% the emittance can be improved significantly but the micro pulse rate has to be raised to reach the claimed mean current. Repeating the multislit measurements (circles) at the injector in a next stage of extension it was found that the emittance has improved. From an image of the cathode it could be seen that the emission characteristic has changed, now the main part of the electrons come from a smaller region of the cathode surface. These data compared with model calculations performed with a smaller cathode radius (2.5 mm) agree well with this assumption (dashed curve). The measurements after the accelerating section with the quadrupole scan method (stars) show that within the errors of both methods there is no visible increase of the emittance due to the accelerating cavities.

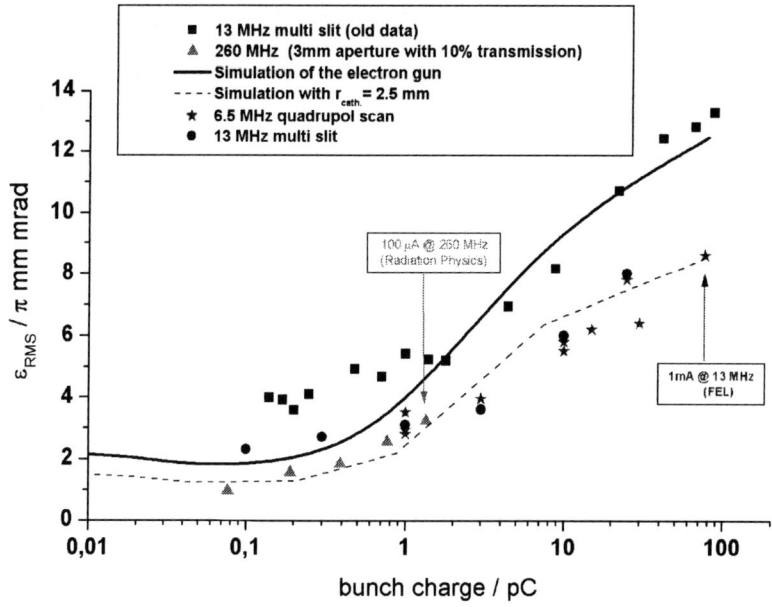

FIGURE 1. Emittance as a function of bunch charge for different settings of ELBE.

BUNCH LENGTH MEASUREMENTS

Introduction

The ELBE electron gun is a thermionic triode, which produces electron pulses with a charge of up to 100 pC and an electron energy of 250 keV. Electron pulse length is about of 450 ps RMS at the gun output. The electron bunch is compressed in the injector down to 10 ps with the help of two buncher cavities. The first sub-harmonic buncher operates at 260 MHz, and the second one works at the fundamental frequency of the LINAC which is 1,3 GHz. The ELBE accelerating module consists of two TESLA type nine-cell cavities. Since the electrons become relativistic in the first cavity, the phase of the RF field there is the key parameter influencing the bunch length at the output of the accelerating module. Bunch length behavior was studied at different bunch charges as a function of RF field phase in the first cavity. The bunch length was measured during commissioning of the first acceleration module using well-known coherent transition radiation (CTR) techniques. The method uses a Martin-Puplett interferometer to measure an autocorrelation function of the CTR pulse. The measurements were done at electron beam energy of about 12 MeV. Here we present results of the measurements, description of the experimental setup, and the data evaluation procedure.

The Coherent Transition Radiation Technique

The coherent transition radiation technique is already well described in a number of publications [2]-[4]. Here we just want to remind one of the key issues of this method. As soon as a charged particle crosses a boundary of two media with different dielectrics or magnetic constants, transition radiation (TR) is produced [5]. If an electron bunch consisting of N electrons crosses such a boundary, each electron of the bunch radiates the TR. For a wavelength shorter than the bunch length, the radiation power is proportional to N, since for every electron there is an electron radiating in opposite phase and the coherent term is equal to zero. Part of this radiation lies in the optical range and is used nowadays very widely for beam profile measurements. For a wavelength longer than the bunch length, all electrons radiate almost in one phase and since the phase difference is constant, the radiation is coherent and therefore the power of the radiation is proportional to N^2. Of course, there is a transition region when the spectral power density goes from N to N^2. Obviously, the position of this transition depends on the bunch length; hence measurements of the transition radiation spectrum can give information about the bunch length. We want to point out here also that for the ELBE bunch charge 77 pC N is about 5.5×10^8, which leads first to a huge difference of N and N^2 and, moreover, almost all power of the transition radiation is in the coherent part. PARMELA simulations of the electron beam transport predict the bunch length to be in the picosecond range.

Experimental Setup

An aluminum foil as thin as 10 μm, stretched to a frame, was used to produce the CTR. The view screen is oriented 45° to the beam direction. Thus the backward CTR part is propagating perpendicular to the electron beam. We use a crystal-quartz window for the output of the CTR from the beam line. A Martin-Puplett interferometer is used to measure the autocorrelation function of the CTR pulses. A parabolic aluminum mirror with focal distance of 200 mm is used to transform the divergent transition radiation into a quasi-plane wave, which then goes to the interferometer. Wire grids are used as a polarizer, analyzer and also as a beam splitter in the interferometer. The grids are made of gold covered tungsten wires, with diameter of 20 μm. The grid period is 100 μm. Another parabolic mirror at the output of the interferometer focuses the radiation on the input windows of the detectors. We have used two Golay cell detectors for the measurements with the interferometer. The theory of the Martin-Puplett interferometer is well developed and shows that an interferogram is the autocorrelation function:

$$V_{detector}(\tau) \propto \int (2f(t)^2 + 2f(t) \cdot f(t-\tau)) \cdot dt \qquad (1)$$

of the incoming into interferometer radiation pulse f(t) [6]. The Wiener-Kintchine theorem proves that the Fourier transform of the autocorrelation function is the power spectrum. Hence we measure the power spectrum of the CTR pulse. The power spectrum defines uniquely the amplitude of the components of the frequency domain representation of the pulse. But information about the relative phases of the different components is lost in the interferometric measurements. This way a direct pulse reconstruction from the power spectrum is not possible.

Bunch Length Estimation From The Interferogram

The following procedure is used to evaluate the measurement data. First of all an assumption about a longitudinal charge distribution is made. Then the power spectrum of the distribution is calculated. A filter function of the interferometer is applied to the power spectrum to get the modified power spectrum. This modified power spectrum is compared with the measured power spectrum, which in turn is the Fourier transform of the interferogram. Finally we have to change parameters of the assumed charge distribution so long, until the calculated power spectrum fits well to the measured one. In practice we assume a Gaussian distribution of the charge, which makes it possible to do the calculations analytically. The filter function we use corresponds to a low frequency cut off which comes from the detector and the wire grid spectral response as well as from the crystal-quartz window transmission. The function was chosen empirically to be:

$$F_{filtrer}(v) = 1 - e^{-\left(\frac{v}{v_0}\right)^4} \qquad (2)$$

here the v_0=0.1 THz. The criterion is that with such a filter function, all experimental data have a good fit. Under these assumptions we have the fitfunction:

$$F(v,Q,\sigma) = \left(1 - e^{-\left(\frac{v}{v_0}\right)^4}\right) \cdot \frac{Q}{\sqrt{2\pi}} \cdot e^{-\left(\frac{v}{\sigma \sqrt{2}}\right)^2} \qquad (3)$$

which was used for the bunch length extraction from the measured power spectrum. Result of a fit is the σ parameter, which is the RMS bunch length. Some examples of the measured data with their fit functions are shown on Fig. 2.

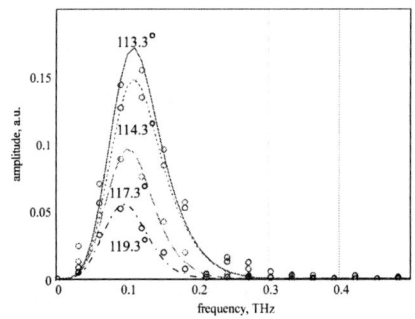
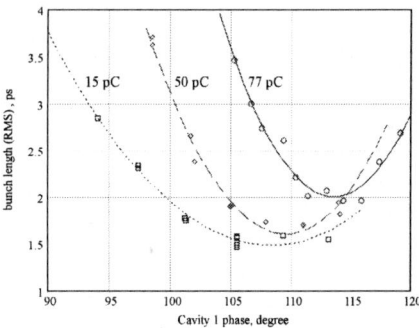

FIGURE 2. Measured power spectrums (points) and their fit functions (curves).

FIGURE 3. Bunch length vs. cavity one phase.

Results Of The Measurements

As was mentioned above, during commissioning of the first accelerating module, it was important to study methodically the work of the accelerator. During this phase we have measured the RMS bunch length as a function of the RF field phase in the first cavity at different bunch charges. The results of the measurements are shown on Fig. 3. The bunch length was measured to be a minimum about 2 ps RMS at 77pC bunch charge. The results are in good agreement with the PARMELA simulations [7].

BEAM POSITION MONITORS

Introduction

ELBE will be used for experiments in radiation physics, nuclear physics and neutron physics. It will also be the driver for the infrared free electron laser (FEL). Obviously, an accelerator needs a system for the beam position measurements. Also the position of the electron beam has to be controlled at the target in any experiment and inside the undulator of the FEL. In the case of the ELBE accelerator, the required resolution of the beam position measurements is about 100 µm. We decided to use stripline BPM, since it is well known that with the BPM one can easily achieve the resolution. The BPM system has to work in all possible modes of the accelerator, which are:

- The FEL mode: repetition rate of 13 MHz, CW or macropulsed beam with macropulse length from 100 µs up to 37 ms. The maximum bunch charge in this mode is 77 pC, which corresponds to an average current of 1 mA.
- The mode for the radiation physics experiments: 260 MHz repetition rate, the bunch charge of about 0.4 pC or average current of 100 µA.
- The diagnostic mode: the repetition rate from 13/128 MHz up to 13 MHz, CW or pulsed beam, maximum bunch charge of 77 pC.

Stripline BPM Design Experience

As reported before [8,9] a system of stripline beam position monitors (BPM) is under construction for the ELBE accelerator. The BPM from the JLab FEL machine was a prototype for the first version of the ELBE BPM. The construction was only changed to be appropriate for the ELBE beam line diameter and for the accelerator fundamental frequency of 1.3 GHz. Two BPMs were manufactured and successfully tested at ELBE. The BPM was used during the injector characterization. A resolution of about 10 µm at 1 mA beam current was achieved which is about ten times better than required.

However, we have faced some problems during the BPMs manufacturing. First of all, the SMA feedthroughs, which have to be electron beam welded to the BPM, are very sensitive to mechanical stress. About half the feedthroughs were broken in the first welding attempt. Another problem is the length of the electrodes. To form the electrodes, a pipe with a diameter exactly as the beam line has to be made with four cuts. Because of mechanical stress in the beam pipe, the electrodes can change their position with respect to the pipe center. This leads to an impedance of the transmission line slightly different from the designed 50 Ω. The electrode displacement also imposes some danger for the feedthrough. Besides all of this we could not put the BPM at some desirable places because of its length and the allowable space. All these reasons required some changes in the BPM design. The length of the BPM electrode in general is chosen such that the BPM has maximum sensitivity to the fundamental frequency of an accelerator. To satisfy this condition the length should be $L=(2n+1)\lambda/4$, where n is integer number and λ is the wavelength of an accelerator fundamental frequency (1.3 GHz in our case). The first ELBE BPM has the electrode length of about $3\lambda/4$. The main idea of the redesign was to make the strip length $\lambda/4$. At first, an electrical model of the $\lambda/4$ BPM was built and tested on a wire test bench. It was demonstrated that the resolution of the model is not worse than the $3\lambda/4$ BPM resolution. After this successful test we have constructed a BPM with the electrode length of about $\lambda/4$. In the construction we used another type of the feedthrough, which is not welded to the BPM but sealed to it with a CF flange. Brazing is used in the construction instead of expensive electron beam welding. The feedthrough can be replaced any time, thus the BPM can be repaired without removing it from the beam line, which can save a lot of time. Total length of the BPM is 85 mm from flange to flange. The $\lambda/4$ BPM is in factor two cheaper than the first version. For beam tests two $\lambda/4$ BPMs were manufactured without any technical and technological problems. The new BPM was tested at ELBE and has demonstrated resolution and dynamic range not worse than the first version of the BPM with an electrode length of $3/4\lambda$. Finally, it was decided to use the $\lambda/4$ BPM at ELBE.

BPM Electronics

Stripline BPMs are similar to directional couplers. Energy from the fields generated by the electron beam is coupled out. The signal power on each of the four ports is proportional to the beam current and the beam position with respect to each "coupler" in a x/y-coordinate system. The stripline BPM electronics has to convert the power signal from the BPM into a voltage, which is then fed into an ADC system. The BPM output power ranges from −80 dBm (lowest detectable beam current with no beam displacement) up to −25 dBm (maximum beam current, 10 mm off axis). The signal flow through the different conversion stages is shown on Fig. 4 (one channel). The 1.3 GHz fundamental signal is selected by an external 5-pol inter-digital filter (ID1300) having a 3 dB bandwidth of 8 MHz. Signal attenuation of the coax cables from the accelerator cave to the diagnostics room plus filter attenuation is about of 5 dB, but exactly measured for each channel and considered in the software for calibration. The first stage inside the electronics housing is an integrated MMIC amplifier with 20 dB gain and a high IP3 level to avoid spurious signal generation. The central part is the AD8313 logarithmic detector made by Analog Devices Inc., converting the RF signal into a DC voltage. The dynamic input range of this detector is −65 dBm to −10 dBm for less than 1 dB conversion error. The two following rail-to-rail operational amplifiers LT1630 from Linear Technology shift and amplify the detector voltage to -

5 V to +5 V which meets the input range of the ADC. Low offset voltage and temperature drift are essential here. In order to minimize crosstalk problems, each PCB-board contains the electronics for two channels housed in milled boxes, ensuring proper shielding against stray signals from the nearby klystrons, which may badly affect the measurement at low input levels (note, that the maximum klystron output is 10 kW, or +60 dBm and the minimum detectable electronic input is –85 dBm, so any leakage path must be prevented). Two such boxes and a power supply make the system for one BPM. Three units have been built and tested during commissioning of the first part of the ELBE accelerator. Sufficient sensitivity along with ruggedness against environmental effects, and long-term stability were proven. Crosstalk between channels was found to be lower than expected. Therefore, the next revision will house the electronics for four channels in one box and 10 such boxes in one 19" rack. It is necessary to include a sample and hold (S&H) stage to make the system work at repetition rates below 13 MHz, which was not intended at the beginning of the development. A version with S&H has been built but not tested yet.

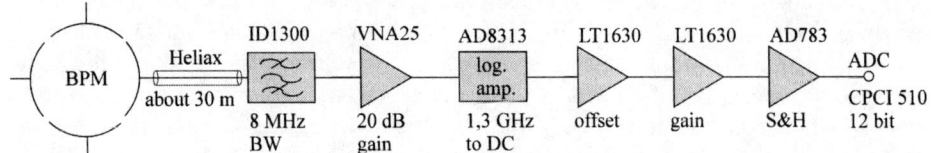

FIGURE 4. The BPM electronics scheme.

ACKNOWLEDGMENTS

We would like to thank Kevin Jordan from JLab for all useful discussions of the electron beam diagnostic at ELBE. Bernt Wustmann did the biggest part of the BPM technical design. We are very grateful to Frank Gabriel for his help with the BPM electronics design. Roland Jainsch and Dieter Proehle helped us a lot with the BPM data acquisition system and the BPM software design.

REFERENCES

1. J. Teichert, et al., FZ Rossendorf, Wiss.-Tech. Ber. 1999-2000, 7-8
2. Happek U., et al., Phys. Rev. Lett. 66, 1967 (1991)
3. Lihn H. C., et al., Phys. Rev. E53, 6413 (1996)
4. Geitz M., et al., Proceedings of the 1999 Particle Accelerator Conference, New York, 2172-2174
5. Jackson J. D., Classical Electrodynamics, Second Edition, New York, 1975
6. Lesurf J., Millimeter-Wave Optic, Devices & Systems, New York, Adam Hilger, Bistrol and New York, 1990
7. A. Buechner, private communication
8. Evtushenko P. et al., FZ Rossendorf, Wiss.-Tech. Ber. FZR-319, 2001, 13.
9. Evtushenko P. et al., Proceedings of the 5-th European Workshop on Diagnostic and Beam Instrumentation, Granoble, 2001.

Cavity BPMs for the NLC*

Ronald Johnson, Zenghai Li, Takashi Naito[†], Jeffrey Rifkin,
Stephen Smith, and Vernon Smith

Stanford Linear Accelerator Center, 2575 Sand Hill Road, Menlo Park, CA 94025, USA

Abstract. The requirements on the Beam-Position Monitor (BPM) system for the proposed Next Linear Collider are very stringent, especially the requirements for position stability. In order to meet these requirements it was decided that cavity BPMs were the best choice. A pair of cavities resonant at 11.424 GHz was designed in a monolithic block. The dipole mode xy-cavity uses a novel coupling scheme that (in principal) has zero coupling to the monopole mode. The other cavity is resonant for the monopole mode and is used to determine the phase. Comprehensive simulations were performed before completion of the mechanical design and production of the first prototype. These results and subsequent tests of the prototype will be presented.

INTRODUCTION

Plans for the Next Linear Collider (NLC)[1,2] that are being developed at the Stanford Linear Accelerator Center and elsewhere will include very stringent requirements on Beam-Position Monitor (BPM) systems. One of the more difficult requirements is that of position stability in the main linacs. This requirement is driven by the necessity to establish and keep precise optics to prevent emmitance growth.

A BPM is placed at each quadrupole along the main linac. There are 1450 of these devices, designated Q-BPMs. They will be rigidly attached to the quadrupole and the whole assembly is mounted on precision movers. Beam based alignment will be used to determine and adjust the centers. But because beam based alignment is an invasive procedure incompatible with colliding for luminosity, the accelerator components must remain stable over a long period of time. The requirements for these Q-BPMs are listed in Table 1.

There are two obvious choices for the Q-BPMs, striplines or cavities. Although striplines have some advantages, they have two major disadvantages. First, the signal of interest is the difference between two large signals from opposing electrodes. Practically, this means that the difference must be obtained with precision analog electronics or with digital electronics with a large number of effective bits. For a cavity BPM there is a null signal when the beam is centered. Second, mechanically the striplines are more complicated than a cavity. The four striplines must be electrically

---

* Work supported by Department of Energy contract DE-AC03-76SF00515
† Permanent address: KEK, High Energy Accelerator Research Organization, 1-1 Oho, Tsukuba, Ibaraki 305-0801, Japan

Table 1. Q-BPM Requirements

Parameter	Value	Conditions
Resolution	300 nm rms	For 10^{10} e⁻ single bunch
Position Stability	1 μm	Over 24 hours
Position Accuracy	200 μm	Wrt the quad magnetic center
Position Dynamic Range	±2 mm	
Charge Dynamic Range	5×10^8 to 1.5×10^{10} e⁻	
Number of Bunches	1 - 190	
Bunch Spacing	1.4 ns	

isolated from each other and ground (at least at one end). For the NLC main linacs the beam tube inside diameter is only 12 mm. Then the signals must be coupled out through vacuum feedthroughs (introducing the possibility of differential expansion). A cavity can be machined out of a single block of metal with the same tolerances as the accelerating structures, 0.5 μm. The mechanical center can be fiducialized to the outside with errors of this order.

Although it was determined that the resolution requirement could be met with a stripline, it was not clear that the mechanical stability of striplines would meet the position stability requirement. The decision was made to begin a research project on a cavity BPM that could meet the NLC requirements.

CAVITY DESIGN

Electrical

The accelerating structures of the NLC main linacs operate at 11.424 GHz. Although other frequencies could be used, the resonant frequency for the cavity BPMs was selected to be the same as the structures for two primary reasons. First, the Q-BPMs are to provide a phase reference signal to the low-level rf control system. Second, this frequency is consistent with a compact design and well established machining techniques.

A simple cylindrical cavity was chosen, but with a novel design for bringing the signals out of the cavity[3,4]. A rectangular waveguide at right angles to the cavity intercepts the cavity only at the corner. The coupling is through the magnetic field and only couples to the TM11 mode. The monopole (TM01) mode does not couple to the waveguide. This is illustrated in Figure 1. Four of these waveguides intercept the cavity symmetrically horizontally and vertically. A 3-D view of the cavity is shown in Figure 2. Also shown in this figure is the signal out of the waveguide for a beam 1.0 mm off center in x. These results are from MAFIA simulations. Note that signals from the vertical waveguides give x-offsets and signals from the horizontal waveguides give y-offsets.

The dimensions of the cavity are 29.426 mm in diameter and 3.0 mm thick. The waveguides are 18.0 mm by 3.0mm by 30.0 mm. The calculated Q of the cavity (with

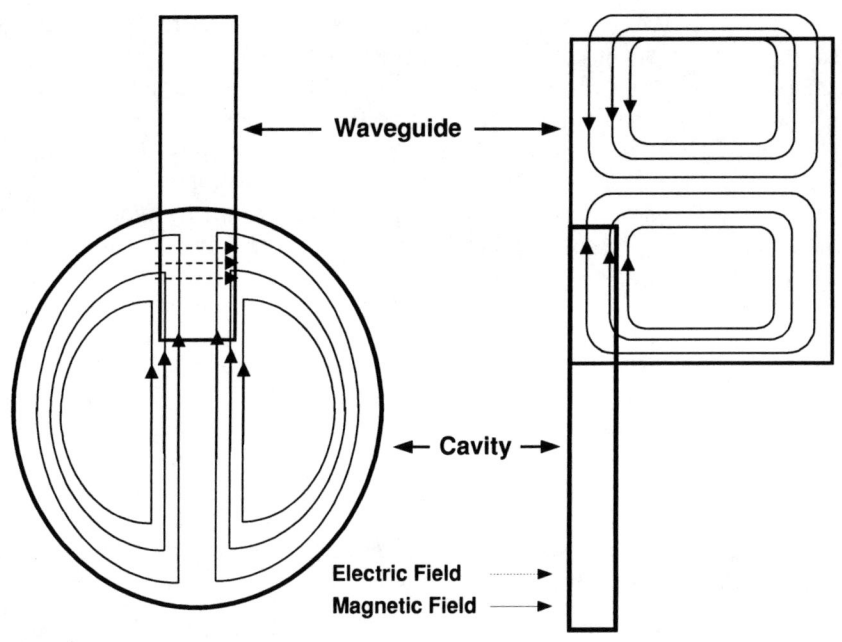

FIGURE 1. Coupling scheme for the cavity BPM. The magnetic field lines illustrate the coupling of the dipole mode to the waveguide.

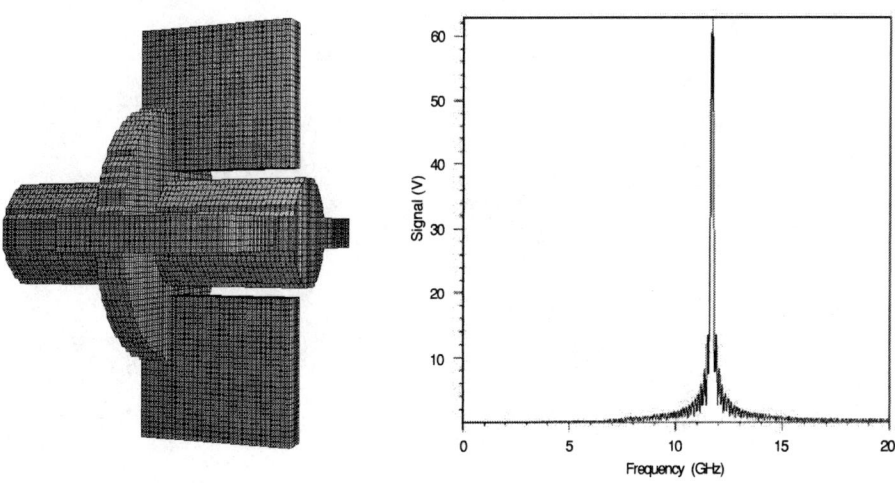

FIGURE 2. 3-D view of the cavity as designed for MAFIA simulations and the signal out of the waveguide coupler. (Only the dipole mode is coupled out.)

the waveguides present) is 1000. Further MAFIA simulations were performed to set tolerances on these dimensions and to determine the sensitivity of the response (especially the suppression of the monopole mode) to errors in machining or construction. For example even a large offset of 0.6 mm (from the ideal radial line) for one of the waveguides introduces a signal from the TM01 mode that is just equal to the TM11 mode. (The beam was offset 1.2 mm for this simulation.)

Also a cavity for phase reference was designed for the monopole mode resonant at 11.424 GHz. The cavity dimensions are 24.711 mm in diameter and 2.0 mm thick.

Mechanical

The body of the cavity BPM was machined out of a single block of copper (OFE Class II) 43.0 mm in length and 34.0 mm in diameter. The xy-cavity and coupling slots were machined at one end and the phase cavity at the other end. Copper end caps 5.0 mm thick were machined and will be brazed on the body to complete the cavities. A 3-D view of the cavity is shown in Figure 3.

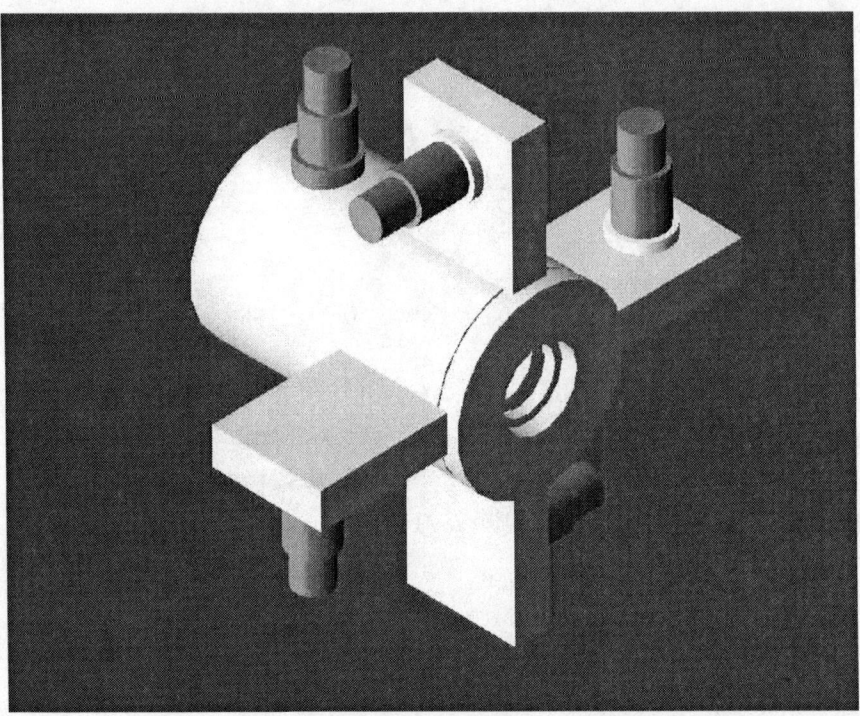

Figure 3. A 3-D view of the cavity BPM as mechanically designed.

The waveguides are formed from two pieces of copper brazed together and then brazed to the cavity body. Although only two waveguides (horizontal and vertical) provide full information for beam position, waveguides at each quadrant were designed to preserve symmetry. This also allows signals from opposite waveguides to be phase shifted by 180° and added to reduce interference from some other modes. In order to couple the signals from the cavity to coaxial cable, vacuum feedthroughs that also serve as antennae are mounted on the face of each waveguide. For the phase cavity a single feedthrough couples to the cavity.

Tolerances for the critical dimensions of the cavity and waveguide were specified to be 1.0 μm. However the actual parts produced for this first attempt did not meet many of these tolerances. For example the diameters of the xy and phase cavities are +8 and -2 μm, respectively, off their specified values. These errors will affect the cavities resonant frequency and other errors may affect coupling. These are achievable tolerances and the actual part was not bad considering that it was a first attempt.

PRELIMINARY TEST RESULTS

Because there was a delay in delivering the vacuum feedtroughs, it was decided to delay brazing of the cavity body and waveguides. Instead a test fixture was constructed in which preliminary tests could be conducted. The test fixture was designed to clamp around the cavity body with slots cut to match the waveguide slots of the cavity. On the outside of the test fixture WR75 waveguide to coaxial cable adaptors were attached. The test fixture also clamped the BPM cavity end plates in place.

This assembly was clamped to an optical bench. An antenna (a quarter wave of the center conductor of a RG141 hardline coaxial cable) was mounted on a xyz stage. The stage was actuated with precision micrometers (sensitivity of 0.07 μm). Measurements were made using a 20 GHz vector network analyzer. The spectra for frequency scans at antenna positions of +1, 0, and –1 mm (in x) are shown in Figure 4.

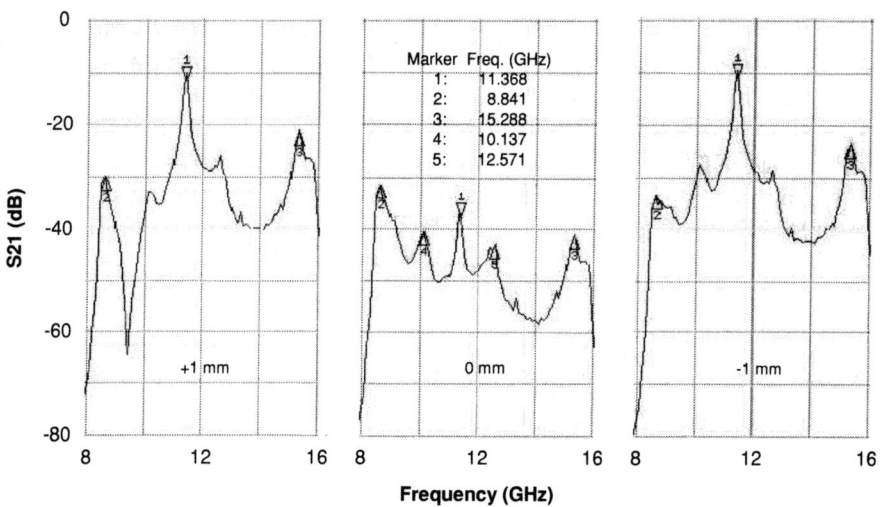

Figure 4. VNA spectra for three x positions of the antenna.

These spectra illustrate many of the features of the cavity BPM. The TM11 mode is at 11.380 GHz, 44 MHz off the design value. However in this test fixture the modes are perturbed differently than in the device as designed. The TM01 mode is at 8.547 GHz and is suppressed by about 20 dB from the TM11 mode for the antenna 1 mm off axis. The peak at 15.388 GHz is due to the TM21 mode. The peaks at 10.137 and 12.571 GHz are predicted by MAFIA to be perturbations due to the antenna. Spectra for coupling from one port to either an adjacent port or the opposite port were also taken. The results are summarized in Table 2.

Table 2. Cavity BPM resonant structure.

Mode	MAFIA	Measured			
	Frequency (GHz)	Frequency (GHz)	Q	Coupling 180° (dB)	Coupling 90° (dB)
TM01	8.724	8.547	~200	-66	-60
TM11	11.43	11.38	~340	-25	-50
TM21	15.92	15.39		-10	-10

Although the Qs are lower than predicted, these are preliminary results from the test fixture where the cavity is simply clamped together. There is good suppression of the TM01 mode, ~30 dB and of xy coupling, ~25 dB. At present no measurements for the phase cavity have been made.

Following the measurements of the mode structure, scans of response as a function of antenna position were taken. In these measurements both the amplitude and phase were obtained so that a full analysis could be performed. The results are shown in Figure 5. This plot is made by fitting the data to a straight line in the three dimensional space of amplitude, phase, and position. This determines the gain and offset (in this case 53.4 µm). Then the data can be replotted as the measured position against antenna position. The deviations about the straight line are the system resolution, i.e., a combination of the antenna positioning resolution and the actual BPM resolution. Since the measured resolution, 230 nm, is near the precision of the micrometer, this number is an upper limit of the BPM resolution. This meets the requirement for Q-BPMs.

An interesting feature of this cavity is that it has response for the TM21 (quadrupole) and TM31 (sextupole) modes. The TM31 mode is not shown on the spectra (Figure 4) but is located at 19.35 GHz. Positions scans for these modes show the appropriate response but were not analyzed in detail.

After the delivery of the vacuum feedthroughs for the waveguides, two were clamped in place and the waveguides were then clamped together. VNA S21 measurements show a loss of only -1.7 dB at 11.424 GHz. The feedtrough for the phase cavity (which is a slightly different design) has just been received.

The next steps in testing this cavity BPM is to complete the assembly and repeat the measurements.

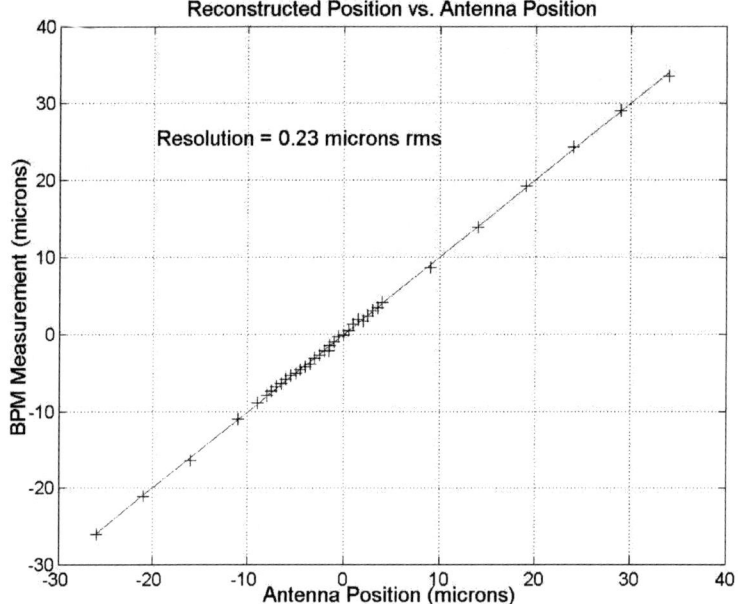

Figure 5. Measured position verses antenna position adjusted for offset and normalized for gain.

CONCLUSIONS

A research program to investigate cavity BPMs for application to the Q-BPMs for the NLC has been started. A prototype cavity BPM was designed and constructed to meet the requirements. The BPM designed for operation at 11.424 GHz has a xy cavity with a coupling scheme that suppresses the monopole mode and a phase cavity. Finished assembly of the device has not been completed, but a test fixture was made so that preliminary measurements could be made.

Results of the preliminary tests show good suppression of the monopole mode, so that it will not interfere with measurements of the dipole signal and little coupling between x and y. The most important conclusion is that an upper limit to the resolution is 230 nm which meets the requirement for the Q-BPMs.

Although the machining of this first cavity did not meet specifications, it is expected that fiducialization to the outside with respect to the cavity center can be done to an accuracy of a few microns. Since the cavity is machined from a single block it should have good mechanical stability.

Assembly of the BPM will be completed and these tests will be repeated. In the future, cavity BPMs of this type will be constructed and tested in an accelerator.

ACKNOWLEDGMENTS

The authors wish to thank Rusty Humphrey, Ray Larsen, and Marc Ross for their encouragement and support of this project.

REFERENCES

1. The NLC Design Group, T. Raubenheimer, ed., "Zeroth-Order Design Report for the Next Linear Collider," LBNL-PUB-5454, SLAC Report 474, UCRL-ID-124161, 1996.
2. The NLC Collaboration, Nan Phinney, ed., "2001 Report on the Next Linear Collider," FERMILAB-Conf-01/075-E, LBNL-PUB-47935, SLAC-R-571, UCRL-ID-144077.
3. T. Slaton, G. Mazaheri, and T. Shintake, "Development of Nanometer Resolution C-Band Radio Frequency Beam Position Monitors in the Final Focus Test Beam," *Proceedings of the XIX International Linac Conference,* Chicago, 1998, pp. 911-913.
4. V. Balakin, A. Bazhan, P. Lunev, N. Solyak, V. Vogel, P. Zogolev, "Experimental Results from a Microwave Cavity Beam Position Monitor," *Proceedings of the 1999 Particle Accelerator Conference,* New York, 1999, pp.461-464.

A Fast VME Data Acquisition System for Spill Analysis and Beam Loss Measurement

T. Hoffmann, D. A. Liakin[*], P. Forck

Gesellschaft für Schwerionenforschung (GSI), Planckstraße 1, D-64291Darmstadt
[*]*ITEP Moscow*

Abstract. Particle counters perform the control of beam loss and slowly extracted currents at the heavy ion synchrotron (SIS) at GSI. For these devices a new data acquisition system has been developed with the main intention to combine the operating purposes beam loss measurement, spill analysis, spill structure measurement and matrix switching functionality in one single assembly. To provide a reasonable digital selection of counters at significant locations a modular VME setup based on the GSI data acquisition software MBS (Multi Branch System) was chosen. An overview of the design regarding the digital electronics and the infrastructure is given. Of main interest in addition to the high performance of the used hardware is the development of a user-friendly software interface for hardware controls, data evaluation and presentation to the operator.

1. INTRODUCTION

For the control of slowly extracted ion currents with energies up to 1 GeV/u from the GSI heavy ion synchrotron into the high energy beam transfer line (HEBT), particle detectors and related technologies are applied. For the low intensities up to 10^6 particles per second (pps) plastic scintillators, for the medium range between 10^4 and 10^9 pps ionization chambers (IC), and for higher ranges secondary electron monitors are used (SEM) [1]. While the scintillators, after discrimination deliver countable logical pulses, the secondary currents from the ICs and the SEMs are converted to logical pulses, close to the detector, by a large dynamic range current-to-frequency converter [2]. For the data acquisition, counting these detector signals, scalers can be used. Scalers are available with high performance at low costs per channel and are installed on densely packed VME-boards leading to a compact and modular setup. By means of voltage-to-frequency converters other parameters, like the stored current inside the synchrotron measured by a dc-transformer, can be taken into account. The main idea of the presented new data acquisition system is digitizing the counts of all detectors, independently from their dedicated application, in a VME system installed in the local-electronics room with a fine time mesh of about 1 ms using only standard hardware components. From the VME processor the reduced data is transferred to a nearby installed Linux PC via a 100Mbit Ethernet connection, where the flexible visualization and data storage is carried out. Using the X capability of LINUX, the information can be exported to any X-terminal located in the main control room. In particular, no additional hardware has to be installed there. This enables a well manageable organization of the required information.

Different types of information can be obtained from the measurement data: For a fast overview the total number of extracted ions per cycle can be displayed. This information may be used as a transmission control from SIS to the experimental area. For the alignment of the accelerator settings, the time evolution during a cycle has to be shown with a time resolution in the order of 1 ms. Comparative studies between different detector types can be used for calibration purposes. A smart selection menu provides a subset of detectors for a particular setting, which then is visualized. More complex analysis algorithms like integration, differentiation or FFT can be applied on demand. An overview drawing is shown in Fig. 1.

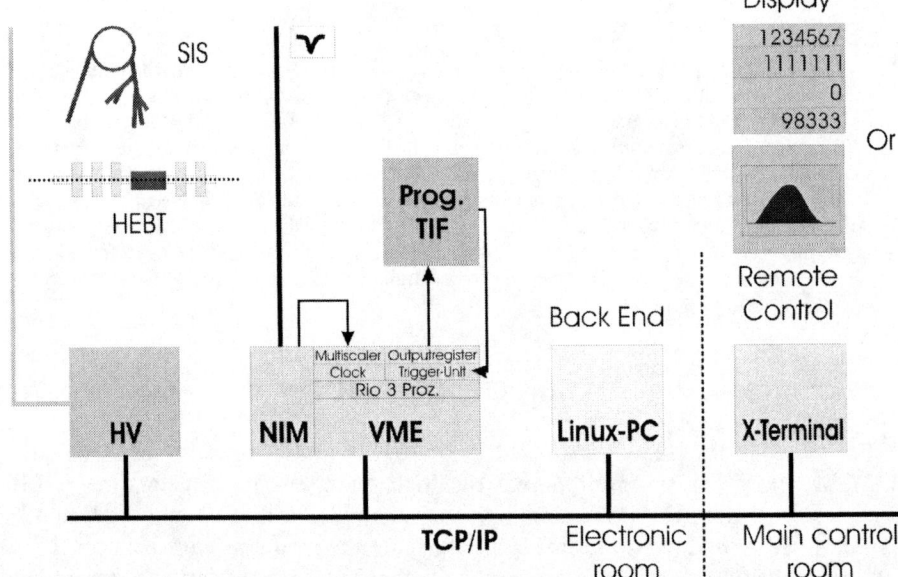

Figure 1. Principle scheme of the new data acquisition system. Analog signals from the synchrotron or the high energy beam transfer line (HEBT) are discriminated and counted in the front-end electronics. Power supply, front-end, back-end PC, and the exported X-display are all connected via 100Mbit Ethernet using the TCP/IP protocol.

2. HARDWARE

On the basis of practical experience a VME system was chosen. In this regime the modular concept and the commercial availability of the components is fulfilled. A CES RIO3 8064 real time processor provided with 128 MB DRAM performs the controlling of the data acquisition. Due to the fact that only count-rate-giving devices shall be observed, four VME SIS3801 32-channels Multiscalers have been installed. These scalers are providing 64 KB FIFO buffers leading to a save and incorrupt data acquisition. The maximum input rate is 200 MHz and the modules provide the possibility for individual channel selection using 32-Bit masking.

To debug and observe bus activities in the VME crate a CES8004AA VME/VSB display unit is used. For beam loss measurement using plastic scintillators the remote controlled VME 16 channel leading edge discriminator CAEN V895 was included into the setup. It provides easy remote threshold modification and adequate countable standard ECL pulses. Different sampling frequencies and precise timing are provided with a programmable VME SIS3807 4-channel pulse generator. To set the start and stop triggers, which means adjusting the acquisition gate depending on the heavy ion synchrotron cycles, a VME SIS3601 32-Bit output register is used. This module controls a programmable GSI timing interface unit, which is interpreting the GSI timing signals. In addition to all these VME devices some standard NIM modules such as Fan In/Fan Out, discriminators and mixed logic units are included. The hardware management and the operation via TCP/IP on the GSI local area network (LAN) is performed with a standard Intel processor based PC using Linux as the operating system.

3. SOFTWARE FRONT-END

The data acquisition software runs on the RIO3 PowerPC-based processor board with booting from a remote LynxOS real-time operating system server. The main idea of the newly developed software is providing a long-term monitoring of various detectors as well as a detailed time or frequency domain analysis of experimental data for short periods of time. It was found that the maximum sampling rate, which may be achieved in the current hardware configuration for continuous mode, is 1Msamples/s for one channel or 7ksamples/s for 128 channels in four modules. These rates are sufficient for the planned applications. Practically the value of 1ksamples/s per channel was selected. The structure of the software running on the processor board is shown in Fig.2.

The MBS (Multi-Branch System) is used as the base software for this application. This software was developed at GSI [3] and is utilized for medium and large sized detector arrays at GSI and a lot of other institutes since several years. It offers a stable and very flexible operation with a relatively small necessity of user programmed software arrangements. As MBS is designed for long period, multi detector data acquisition some modifications have been made to match the requirements concerning online visualization and bi-directional data exchange. Therefore, an additional TCP server thread has been implemented. The MBS controls the GSI VME trigger module, provided for the synchronization between the data acquisition and the accelerator events. The user editable MBS module 'user readout function' (f_user.c) provides all necessary initialization procedures, the basic data acquisition and analysis routines. Measured data from the scalers is summed up in constant time steps and then sent to the network client if it is connected. Meanwhile the original data from the scalers fills a fixed memory segment, organized as a shared memory area to provide access from other processes running on the same processor board. This gives the possibility to connect with other remote clients having access on the same data.

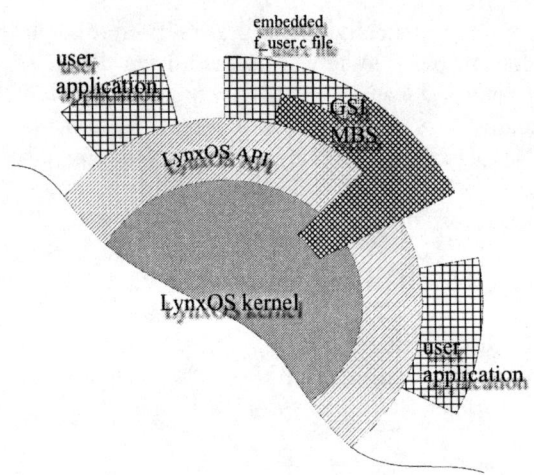

FIGURE 2. The user-readout-function (f_user.c) is part of the MBS (Multi Branch System) having access to the kernel. Being connected to the API, VME-hardware access, network service or inter-process communication is made possible.

To synchronize these processes a special LynxOS system message queue and a semaphore set controlling the shared memory access are initiated. These parallel running processes may be called as separate executables without interrupting the main data acquisition cycle. Assuming no impacts due to network limitations are present, the complete set of raw data may be temporary stored in the local memory for continuative analysis through the PowerPC or any remote Linux computer. While the standard readout process is creating a reduced data stream another user application may use the complete raw data set, which is stored in the shared memory at the same time.

4. SOFTWARE BACK-END

As the client remote computer a standard Intel Pentium 1 GHz CPU was chosen. The operating system is Linux Red Hat 7.2 with an installed KDE 2 window manager. This part of the data acquisition software was programmed in the Borland Kylix development environment. In principle every main function such as beam loss monitoring, matrix switching or spill analysis has to be considered as a user application, which is connecting to the front-end data stream via a TCP/IP client. This client is started inside the main program as a separate thread being responsible for the data transmission between VME crate and the Linux PC, compatible to the TCP/IP server on the front-end. This mechanism is based on a standard Linux socket library and uses network messages with variable data packet sizes.

The evaluator of incoming data at the back-end is implemented as an instance of an Object Pascal class. This class method provides some basic evaluation procedures like calculating basic statistical parameters. The identification data such as physical module, channel number and the device-name etc. is stored in an appropriate data structure. To provide simple manipulation methods like storing data in files, refreshing data presentation, multi-channel data evaluation etc. a special list of objects has been developed.

5. APPLICATIONS

The main advantage of this new data acquisition system is the possibility to include several control operations, such as matrix switching functions and beam loss measurement into one single assembly. The most important functions are presented as follows:

5.1 Matrix Switching

The signals of most of the GSI HEBT (high energy beam transfer) detectors are collected in a designated electronic room. After discrimination the pulses are duplicated and sent to control system scalers and to a matrix switching unit. This crossbar function is used at GSI to display count rates of selected detectors in the main control room. The old system, a standalone 30MHz bandwidth analog solution is still working well. It provides a 64X4 matrix and it is placed in one rack at the SIS control desk. Due to the fact that this widely used device has reached its input channel limit and that more then 4 displayed channels are required a new way had to be found. In addition it is to expensive to update the old analog system to a bandwidth of the required 200 MHz.

The digital data acquisition system, which is discussed here, provides all the functionality of the old system but it is different from its data processing design. With using four 200 MHz SIS3801 multi scalers having 32 channels each, 128 inputs are obtained. Due to the modular design this amount may be increased anytime. The signals of all connected detectors are counted permanently in the VME part using the multi scalers with a sampling rate of 1kHz, but only the data of the selected detectors is used for visual display. The amount of the exported outputs is only limited by the display size. The count rates are displayed on a X-terminal in the main control room. For best view from longer distances within the control room a mode is prepared which displays the count rates in the biggest possible way in digital numbers. For more detailed observation, sitting in front of the terminal, the count rate distribution as a function of the cycle time, the digital value, and the logged end values as a function of cycles are displayed for each selected detector. This is also done in the way that the maximal window size per channel in relation to channel quantity is used.

5.2 Beam Loss Measurement

Having made several experiments [4,5] on beam loss detection to prevent permanent activation in the synchrotron tunnel in exposed areas and to increase transmission to the experimental targets this project had to be moved from the state of improvisation to a regular beam instrumentation system. Blocks of 20x20x75mm³ BC400 standard plastic scintillators coupled to fast multipliers Phillips XP2972 have been mounted at dedicated synchrotron sections such as extraction section and at the electrostatic septum. The amount of nearly 16 units will be installed during this development but further enhancement if necessary was taken into account. All monitors are observed permanently in the time-gate beginning with initialization of the synchrotron and ending with completion of the extraction into the HEBT.

The software provides a drawing of the SIS (Fig. 3) including the installed devices in which the user has to select the desired monitors. In small pictograms the final count-rates after each cycle and a simulated rate meter displaying the variation within the cycle will be shown for each detector.

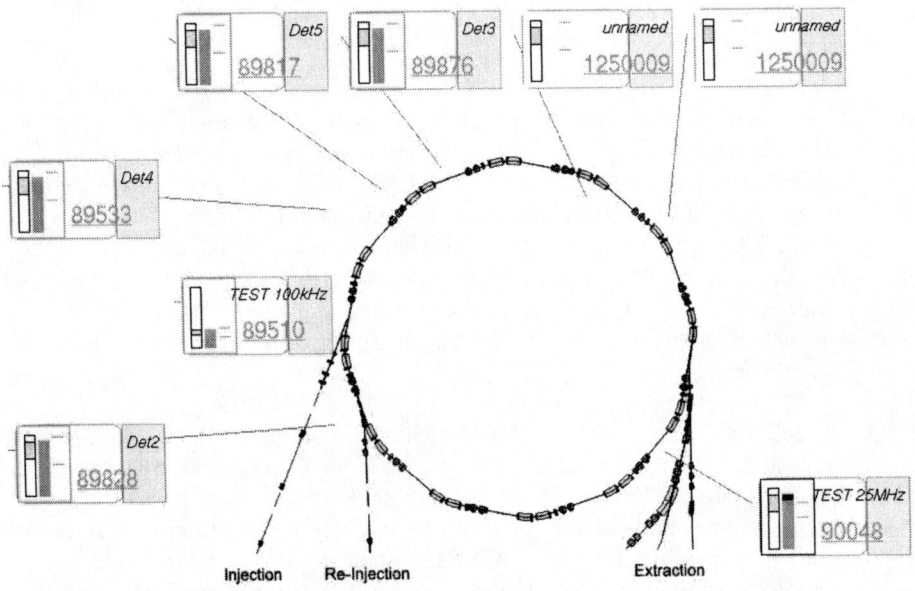

FIGURE 3. Sample drawing of the SIS with included detector icons showing the loss rates.

Due to the fact that these pictograms may be dynamically added, removed or placed on custom position on the program window, the GUI may be easily adapted to the current environment setup. The pictogram set is organized as a list of objects, which properties may be saved and restored by using a separate setup file. These pictograms are based on a common parent class in which main features are defined. The basic class determines the mechanism of synchronization and data receiving from the beam loss data objects. The graphic objects have different shapes to visualize the various detector types. Logging and displaying the end values as a function of cycles while manipulating synchrotron settings provides the possibility to understand and to set the best conditions for transmission or lowest activating rates.

5.3 Spill Analysis and Calibration

For special accelerator experiments a tool was developed to observe the spill as a function of time. The spills of some detectors may be presented as graphs, supplemented with information like FFT, maximal count-rates and statistical parameters. Figure 4 shows in the top curve the currents in the SIS measured with a dc-transformer (a), and then the rate at the experimental target (b) giving information about the transmission, followed by beam loss monitors installed inside the SIS close to the electrostatic septum (c) and at the extraction section (d). During the spill some extraction parameters were not properly set, leading to a decrease in transmission as a function of time.

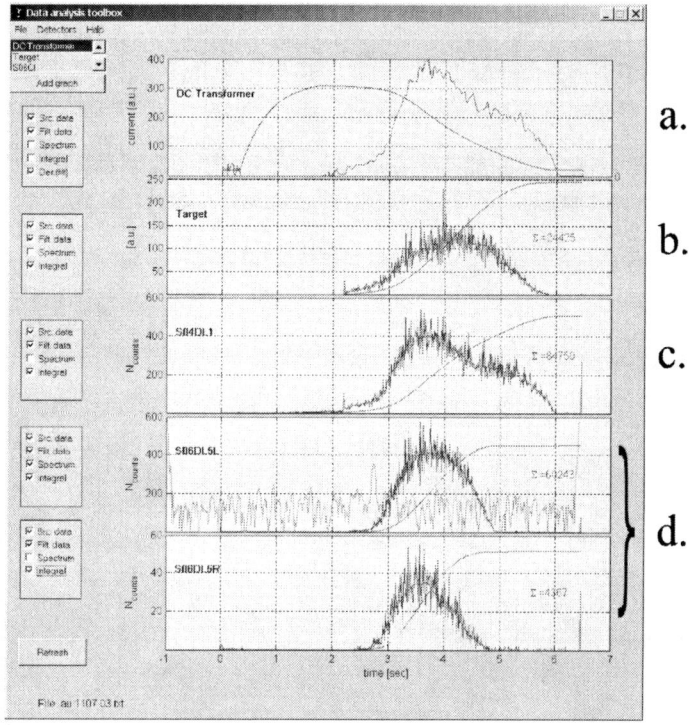

FIGURE 4. The spill analysis function showing selected detectors and mathematical conversions:
a. SIS-DC-transformer, **b.** experimental target behind the HEBT, **c.** scintillator-beam loss monitor at the magnetic septum, **d.** two scintillator-beam loss monitors placed left and right at the extraction tube.

Mathematical routines are used to calibrate the ion chambers with scintillators and secondary electron monitors. In the past this was calculated offline, but having access to every monitor including the SIS beam transformer a correlation between many kinds of detectors is now easily done.

To optimize the extraction parameters this online mode is very helpful. The graphs are getting updated cycle by cycle and are showing the behavior of the extraction while editing the SIS hardware parameters. Comparing the count-rates of the HEBT detectors up to the target with the SIS beam transformer gives the transmission rate, which may then be optimized too.

An interesting use of this function is the observation of the spill time behavior. Due to several effects such as power supply ripples the resonant extraction may be disturbed leading to interrupted extraction behavior. To analyze such effects the sampling rate is increased up to 50 kHz for a maximum amount of 4 detectors. Then a spill structure analysis may be done. This kind of data acquisition was presented in [6]. Introducing the tool into the control system provides the possibility for new data storage and archiving routines.

5.4 High Voltage Power Supply

Since the SIS was commissioned in 1990 the amount of detectors such as scintillators, ion chambers, profile grids etc. has permanently increased. As these devices are all requiring high voltage power supply the maximum load of the existing devices was reached. Looking forward to future plans and particular projects like beam loss measurements a new powerful and expandable power supply unit was purchased. The decision to take the CAEN SY1527 was made due to the modular setup and the remote control function using TCP/IP and OPC server. The software design covers all the required functionalities such as observation and modification of voltage, current, trip-time and ramp-rate. A database for default settings, which can be loaded after shutdown periods, is provided.

6. OUTLOOK

This project has been started on March 2002. The system is yet not ready for implementation into the regular operating, due to debugging, long time tests and further ideas, which have to be added.

It is obvious that other operating features may be implemented into this system as long as they provide countable pulses.

7. REFERENCES

1. P. Forck, T. Hoffmann, A. Peters, Proc. 3rd European Workshop on Beam Diagnostics and Instrumentation for Particle Accelerators, DIPAC, Frascati, p. 165 (1997)
2. H. Reeg, Proc. 4th European Workshop on Beam Diagnostics and Instrumentation for Particle Accelerators, DIPAC, Chester, p. 140 (1999)
3. H.G. Essel and N. Kurz, „The General Purpose Data Acquisition System MBS" IEEE Trans. NS, Vol. **47**, No.2, pp. 337 (April 2000), http://daq.gsi.de
4. P. Forck, T. Hoffmann, Proc. 5th European Workshop on Beam Diagnostics and Instrumentation for Particle Accelerators, DIPAC, Grenoble, p. 129 (2001)
5. E. Berdermann et al., "Diamond Detectors 2001 – Applications for Minimum-Ionizing Particles", GSI Annual Report 2001
6. P. Forck et al., Proc. 7th European Particle Accelerator Conference, EPAC, Vienna, Austria, p. 2237 (2000)

Design of an Improved Ion Chamber for the SNS

R. L. Witkover[†] and D. Gassner
Brookhaven National Laboratory, †TechSource, Inc., Santa Fe, NM

Abstract. Ion chambers are in common use as beam loss monitors at many accelerators. A unit designed and used at FNAL and later at BNL was proposed for the SNS. Concerns about the ion collection times and low collection efficiency at high loss rates led to improvements to this unit and the design of an alternate chamber with better characteristics. Prototypes have been tested with pulsed beams. The design and test results for both detectors will be presented.

BACKGROUND

The ion chambers (IC's) designed for the FNAL Tevatron by Shafer in 1982[1] and built by Troy-Onics[2], have been used at both FNAL and BNL. These are simple in design, consisting of a hollow nickel inner electrode and a nickel foil cylinder outer electrode in a glass enclosure filled with argon. During testing of the chambers for use in RHIC at BNL[3] it was found that detector response varied unacceptably using the preferred bias voltage polarity. Normally, from space charge considerations, electrons are collected on the inner electrode. The IC's were tested with a Cs-137 source which produced 1 Rad/hr at the chamber. The signal current, measured as the bias voltage is varied, is characterized by a rapid increase which quickly rolls over ("knee") into a flat region ("plateau") as the field becomes high enough that all charge produced is collected. Ion chambers operate in this plateau region. At higher voltage, the electric field is sufficient to cause multiplication and the chamber enters the proportional region. Results for the first 40-50 RHIC production IC's varied widely with many having extended "knees" and early onset of the proportional region. Test results obtained from FNAL for the original detectors did not show this behavior. It is possible that the original FNAL detectors, built 15 years earlier, did not exhibit this problem but subsequent vendor technicians may have unknowingly changed details of the construction. However the detectors were not usable as delivered.

At BNL, conditioning using a Tesla coil and "spot-knocking" showed some success but required significant time per unit. It was suggested that the opposite bias polarity would improve uniformity. Tests showed excellent unit-to-unit reproducibility when ions, rather than electrons, were collected on the inner electrode, so it was decided to use the "unconventional" field polarity for RHIC. The major consequences were lower collection efficiency (saturation) for high dose rate losses, and slower signal risetimes, which were not expected to be problems in RHIC. For SNS, however, this drop in efficiency would be significant for the worst-case 1% local loss, and a rise time comparable to the pulse width would be unacceptable.

SIGNAL RISE TIME CONSIDERATIONS

With a 1 msec pulse width, signal rise time is important in the SNS. Electrons are collected in a few microseconds with either bias polarity, but the heavier ions take much longer moving to the inner electrode. At 2000 V the ion collection time would be close to 700 μsec. If the losses were constant the signal would continue to rise over the 1 msec Linac pulse as the slow ions arrived, followed by a tail for another 700 μsec after the pulse. While the electron signal would allow rapid beam abort for large fast losses, the waveform during the pulse would require unfolding.

Electron and ion charges are generated equally, but will produce equal voltages only for a parallel plate geometry. For cylindrical geometry, using the relative capacitance of a line charge to the 2 electrodes, Shafer[4] showed that the fraction of current due to electrons at the inner electrode (anode) is:

$$<F2> = \frac{2\pi \int_b^a F2(R)RdR}{2\pi \int_b^a RdR} = \frac{2x^2 Ln(x) - (x^2 - 1)}{2(x^2 - 1)Ln(x)} \quad \text{where } x = a/b \quad (1)$$

For the FNAL ion chambers, a = 0.75" and b = 0.125" and $<F2> = 0.749$. That is, the fast electron signal will be 3 times the slow ion current with the conventional polarity but only 1/3 for the reverse-bias polarity. This was observed in tests in the RHIC transfer line. Clearly, the FNAL chambers would be unsuitable for SNS unless the conventional bias polarity could be used.

The positive ion transit time is given by:

$$t = \frac{d^2}{\mu_0 V \left(P_0 / P \right)} \quad (2)$$

where:

d = Effective electrode separation [cm] for cylindrical geometry[5]

$$d = \left[(a^2 - b^2) \frac{\ln(a/b)}{2} \right]^{1/2}$$

(3)

μ_0 = Ion mobility at STP [cm²/(V-sec)]
V = Applied Voltage [V]

P_0 = Atmospheric pressure
P = Working pressure

The transit time can be reduced by decreasing the electrode gap but with increased gradient corona may occur. A design was proposed[6] in which the radii were closer together and the chamber lengthened to keep the same volume. Keeping the original outer electrode radius, a = 0.75" but increasing the inner radius to b = 0.5", the ion collection time at 3 kV bias would be reduced from 560 μsec to 72 μsec. The calculated results are shown in Figure 1. "Negative bias" refers to negative voltage on the outer electrode. Clearly the "positive bias" case is not suitable for SNS. The FNAL chamber with negative bias might be acceptable since the electrons contribute 75% of the signal within a few microseconds. The new design provides the best response.

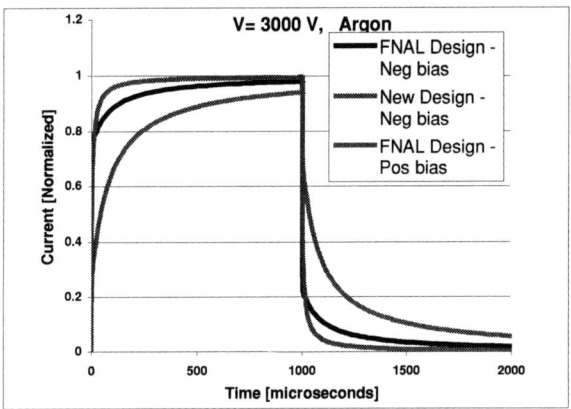

Figure 1. Calculated signal for SNS Linac pulse on original FNAL and new BLM.

COLLECTION EFFICIENCY

While the equations describing the collection efficiency have been known since Thomson described them in 1899, they have never been explicitly solved. Approximate solutions for the simpler parallel plate geometry can be applied to cylindrical if an "equivalent" gap (eq. 3) is used. Boag and Wilson[7] solved the equations assuming the ionization distributions were second and third order polynomials, and that the radiation was constant. Their result, which gives a good fit for efficiency > 0.7 is given by:

$$f = \frac{1}{1 + \frac{\xi^2}{6}} \quad (4)$$

where:

$$\xi^2 = \frac{\alpha}{ek_1 k_2} \frac{P}{P_0} \frac{d^4 q}{V^2} \quad (5)$$

e is the electron charge,
α is the first Townsend recombination coefficient,
k_1 is the electron mobility,
k_2 is the ion mobility,
q is the ionization change density:

$$q = \frac{I_{signal}}{v} \times 3 \times 10^9$$

where I_{signal} is the signal current generated in the ion chamber by the beam loss and v is the active volume of the ion chamber. For the FNAL IC, the sensitivity is 70 nA/rad/sec, and the volume is 110 cm^3 so for the accepted upper limit of 1 % local loss, the dose rate is 9.2 krad/sec during the linac pulse.

Consider a chamber with argon gas:
$\alpha = 1 \times 10^{-6}$ [cm^3/sec]
$k_1 = 1.8$ [cm^2/(V-sec)]
$k_2 = 1.3$ [cm^2/(V-sec)]

$V = 3000$ [Volts] (Negative bias)
$P = 725$ [mTorr]

Then

	FNAL	New Design
Outer Radius [cm]	1.905	1.905
Inner Radius [cm]	.3175	1.27
Active Length [cm]	10	17.4
Volume [cm^3]	110.84	110.21
Effective Gap [cm]	1.991	0.6436
Ionization Density [esu/cm^3-sec]	16850	16850
Ion Transit Time [μsec]	564	72.2
Electron Fraction (neg on outer)	.7495	.5669
Collection Efficiency	.357	.971

While the efficiency calculation for values <0.7 may not be accurate, clearly the new chamber would be superior to the original, both in ion transit time and collection efficiency, but, since the electrode gap is smaller, great care must be taken with the high voltage design. Guard electrodes would be required to decouple the signal electrode from leakage. The design would be more difficult to manufacture than the simple FNAL detector, with higher detector cost. Note that the electron fraction is 0.567, with 43% of the signal coming from the now ions.

Boag[8] also found a solution for the pulsed radiation case when the loss occurs in a time short compared with the ion transit time. For this case and the above parameters the FNAL chamber efficiency is 21% and the new design is 60%. This might apply to the first 5-10 mini-pulses for SNS but the static case solution might better describe the efficiency late in the pulse.

THE NEW ION CHAMBER DESIGN

The design of the new ion chamber is shown in Figure 2. Voltage gradients have been reduced by rounding the electrode ends. Guard electrodes divert leakage from the HV electrode to ground and assure the signal collector is in a uniform electric field region. They also prevent ions at the chamber ends from contributing to the signal resulting in longer transit times. Three prototypes units have been made. The first was limited to 3 kV when filled with argon due to HV feedthrough breakdown, but when filled with nitrogen was able to reach 5 kV. It was then installed in the transport line between BNL Linac and Booster. Units 2 and 3 were built with feedthroughs rated for higher voltage and with improved ceramic design. The radius of curvature of the ends of the guard rings and signal electrodes were also increased. These units have been tested to 4.5 kV with argon. Prototype 2 has been installed in the BNL Booster extraction line for tests with short pulse beams similar to SNS Ring extraction.

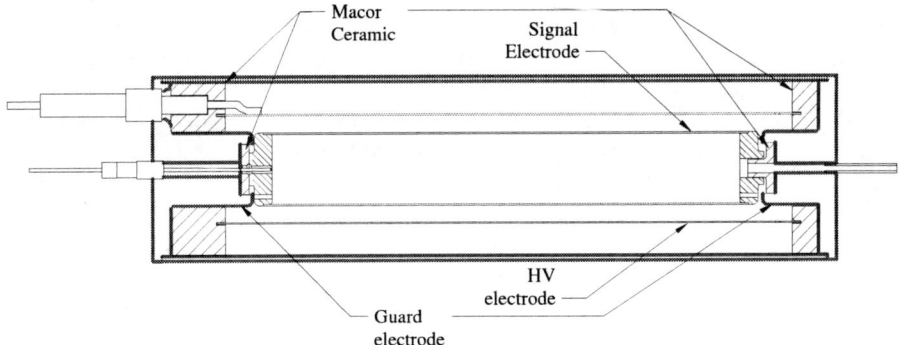

Figure 2. Drawing of new design BLM prototype.

IMPROVEMENTS TO THE FNAL DETECTOR

As an alternative to designing a new chamber, the original was studied to see if modifications could be made to improve its performance. The difficulties with the conventional (negative) bias polarity in the FNAL IC were believed to be due to high field points at the tip of the inner electrode and at the supports for the outer electrode, since these are where breakdown occurs when the voltage is raised. Changes were made to the outer electrode to eliminate crimping at the flared end of the outer electrode at the support pin attachment locations. Figure 3a shows the original outer electrode crimped inward by the support rods. Figure 3b shows a modified chamber with the outer electrode slotted to allow the support to pass the flared end without bending it inward.

Figure 3a. Original FNAL Outer electrode   **Figure 3b**. Slotted outer electrode

So far 5 chambers have been modified in this way. Tests indicate that this has significantly improved the unit to unit response. Figure 4a shows the cesium source tests of 5 of the first RHIC chambers with conventional bias polarity. Figure 4b shows results for 5 units with modified (slotted) outer electrodes. The uniformity is clearly improved with the slotted electrodes.

It was felt that some of the voltage limitation may have been due to sharp edges at the open end of the hollow center electrode. Several chambers were built with the slotted outer and a rounded solid tip inserted in the end of the electrode. Tests of these units did not show a significant improvement from the slotted outer electrode alone. While these modifications seem to have improved the production uniformity, the ion transit time and high dose rate saturation are still a problem for SNS. If the design could be improved to allow higher operating voltage then the ion transit time would decrease and collection efficiency increase. The improvement is somewhat less than linear with voltage, so this might not be very productive.

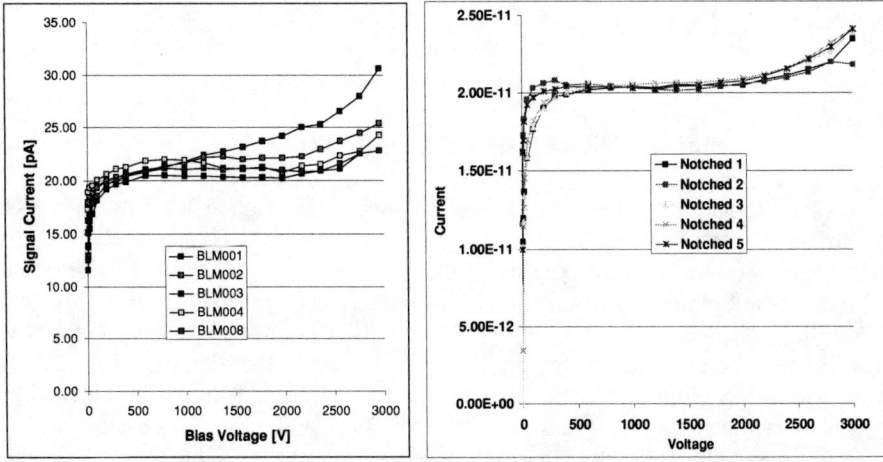

Figure 4a. Original FNAL IC test data **Figure 4b.** Modified (slotted) FNAL data

TESTS WITH BEAM

A non-modified FNAL IC and Prototype 1 were installed together in the BNL Linac-to-Booster (LTB) line, approximated 1 foot from the beam line, where high dose rate losses from the 200 MeV H-minus beam could be observed. Two sets of data runs were made. In the first a high sensitivity multi-channel integrator module of the type used in the source testing of the chambers was used to condition the signals. They were modified to have a 470 Ohm input so rise times of several μsec could be observed on a LeCroy digital scope. The bias voltage was varied to determine the collection efficiency curves for both bias polarities. The beam current waveform was also captured. Figure 5 shows the results of this run. Clearly the positive bias (non-preferred) FNAL response shows low efficiency, especially at lower bias voltage. The data for the negative bias shows large excursions, while summation of the beam current data does not show more than a few percent variation between data points, implying beam motion. The data for Prototype 1 appears fairly flat throughout the entire voltage scan. Calculations indicate that the FNAL chamber should be 77%

efficient while Prototype 1 should be 92% at 3 kV for negative bias at this loss level. Dedicated runs are planned to study the efficiency at various dose rate levels.

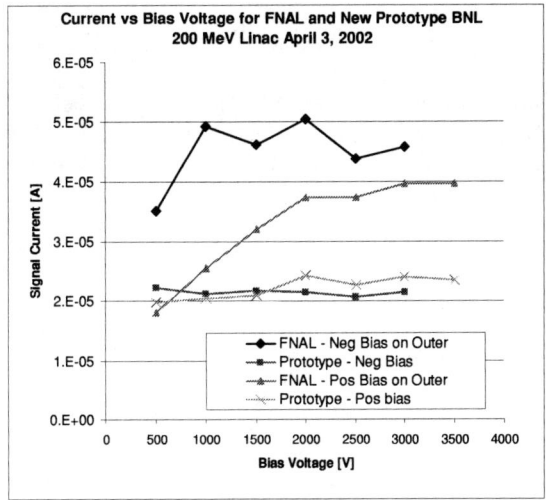

Figure 5. Measured bias curve (collection efficiency) for 200 MeV Linac beam. Aproximately 1000 W/m loss.

A second run was made using the prototype analog front end (AFE) circuitry designed for SNS[9]. This circuit has a current amplifier first stage using a Burr-Brown OPA627BM opamp with outputs from a unity gain (50 kHz BW) buffer, a "leaky integrator" and a filtered (1 kHz) slow amplifier. Data from the fast output and the beam current transformer is shown in Figures 6a and 6b.

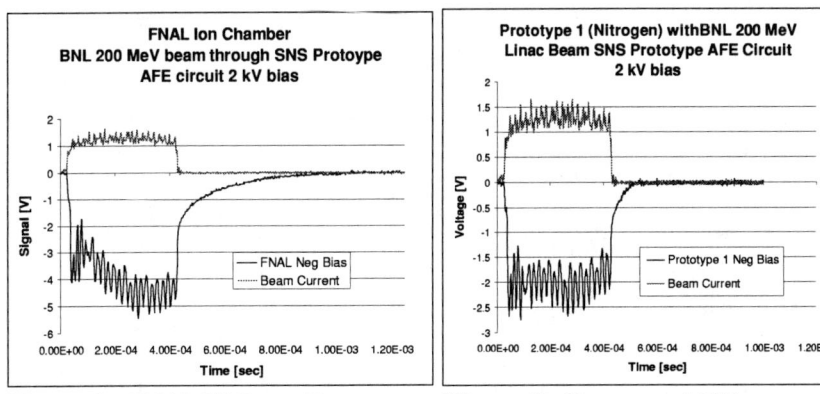

Figure 6a. FNAL IC Beam Test. **Figure 6b.** Prototype 1 IC beam test

The signal from the FNAL chamber (Figure 6a) can be seen to have a very long tail following the pulse, due to the slower ions being collected after the beam. They are also arriving during the pulse, causing the gradual rise rather than directly following the beam current (also see Figure 1). The apparent noise during the pulse is actually modulation of the beam current to tailor the stacking in the Booster. The signal from the Prototype ion chamber has a much faster ion collection time, as expected.

CONCLUSIONS

Improvements have been made to the original FNAL detector design which result in more uniform characteristics of the production units. These changes allow the use of the conventional bias polarity (center electrode is anode) without conditioning of the chambers. However, concern over collection efficiency (saturation) resulting in departure from linearity at high dose rates may preclude its use in SNS. The slow ion transit times are comparable to the SNS pulse width, requiring unfolding to obtain the true time behavior.

The new chamber design appears to yield significantly faster ion transit time and higher collection efficiency at high dose rate, as predicted. However, higher fabrication costs and the uncertainties of bringing a new design to production need to be weighed against the benefits to the SNS application. A firm specification of linearity at high dose rate is needed before the detector choice can be made.

ACKNOWLEDGMENTS

The authors wish to acknowledge the contributions of C. J. Liaw in the mechanical design, Dave Kipp is assembling the prototype chambers and Paul Ziminski in testing the chambers with the cesium source. The prototype circuit used in the beam tests was built and tested by Chaofeng Mi. Anthony Curcio (BNL) was invaluable in the installation and setup of the beam tests.

REFERENCES

[1] R. E. Shafer, et al., The Tevatron Beam Position and Beam Loss Monitoring Systems", Proc. 12th Int'l Conf on High Energy Accel., FNAL (1983) p609
[2] Troy-Ionics, Inc., Kenvil, NJ
[3] Witkover, R. L., Michnoff, R. J. and Geller, J. M., "RHIC Beam Loss Monitor System Initial Operation", Proceedings of the 1999 PAC, NY, 1999, 2247
[4] Shafer, R. E., Private communication
[5] Boag, J. W. in "Radiation Dosimetry, Vol. II", F. Attix, et al, eds.,Academic Press (1966), p22
[6] R. Witkover," Proposal for a Quasi-Parallel Plate, Cylindrical Geometry Ion Chamber for SNS", November 7, 2001, unpublished
[7] Boag, J. W. and Wilson, T., "The saturation curve at high radiation intensity", Brit. J. Phys., 3 222-9, (1952) and Loc. Cit (5), p16
[8] Boag, J. W., "the Dosimetry of Ionizing Radiation", Vol. 2, Ed. By K. R. Kase, B. E. Bjamgard and F. H. Attix, Academic Press, (1987), p 191
[9] Witkover, R. L. and Gassner, D., "Preliminary Design of the SNS Beam Loss Monitoring System", To be published in these proceedings.

Preliminary Design of the Beam Loss Monitoring System for the SNS

R. Witkover*, D. Gassner

*BNL, Upton, NY,*TechSource Inc., Santa Fe, NM*

Abstract. The SNS to be built at Oak Ridge National Laboratory will provide a high average intensity 1 GeV beam to produce spallation neutrons. Loss of even a small percentage of this intense beam would result in high radiation. The Beam Loss Monitor (BLM) system must be sensitive to low level, long term losses, and be capable of measuring infrequent short high losses. This large dynamic range presents special problems for the system design. Ion chambers will be used as the detectors. A detector originally designed for the FNAL Tevatron, was considered but concerns about ion collection times and low collection efficiency at high loss rates favor a new design. The requirements and design concepts of the proposed approach will be presented. Discussion of the design, testing of the ion chambers, and the analog front end electronics will be presented. The overall system design will be described.

BACKGROUND

The Spallation Neutron Source (SNS)[1] is an accelerator based intense neutron source being built at Oak Ridge National Laboratory (ORNL) by a collaboration of 6 national laboratories. The controls and some beam instrumentation were consolidated so that the same detectors and interface hardware would be used throughout the facility to minimize costs and reduce uncertainty due to different instrumentation. BNL has responsibility for the beam loss monitoring system for all of SNS. The SNS H-minus ion source will inject a 60 Hz chopped beam into a 2.5 MeV RFQ. After acceleration to 87 MeV in a Drift Tube Linac (DTL), the beam enters the Cavity Coupled Linac (CCL), leaving at 186 MeV. The Superconducting RF Linac (SRF) will increase the beam energy to 1 GeV. The beam is transported to the Ring via HEBT, converted to protons using a stripping foil, and accumulated in a single 695 nsec bunch over the 1 msec injection pulse. The bunch is transported via the RTBT line to the spallation neutron target. The design peak beam current in the Linac and HEBT will be 38 mA. In the Ring it will increase to more than 40 A (average) at the end of the pulse, for a total of 1.5×10^{14} protons in the single bunch. The BLM system will be designed to accommodate an upgrade intensity of 2×10^{14} protons.

The baseline design will produce 1.4 MW at a 60 Hz rate. This very high average beam power makes it crucial that uncontrolled losses do not produce high activation, preventing hands-on maintenance. This will be done by minimizing losses through

---

Work performed under the auspices of the U. S. Dept. of Energy

careful design and providing beam dumps and collimators where controlled losses can be contained. The BLM system must minimize uncontrolled losses by providing data for tuning the machine and inhibiting the beam when excessive losses occur.

SYSTEM PHILOSOPHY AND DESIGN

The BLM's will be the primary tool for limiting uncontrolled losses by providing sufficient detector coverage in all sections of the SNS. To achieve this the BNL AGS Linac and Ring use extended ion chambers made from hollow argon-filled, large diameter coaxial cables[2]. At ISIS, a machine similar in function to SNS, the same type detectors have been used[3]. A "line" detector has 1/r spatial dependence, but its length often prevents close coupling to the beam line, lowering the response. Since losses predominantly occur at the quad, most of the detector receives a much lower exposure, reducing the benefit of the greater length. The standard connectors for these cables cannot support high bias voltage, limiting the linear range to lower dose rates. For these reasons smaller, sealed glass ion chambers designed for the FNAL Tevatron were used in RHIC[4], but more detectors were required to obtain similar coverage. In the SNS Linac, HEBT, Ring and RTBT, as in RHIC, detectors will be placed at essentially every quadrupole (beta-max) and at other key points. Provision will be made for a number of movable BLMs which can be used for more detailed study.

Ion chambers were chosen as the primary distributed detector over solid state (pin diode) and scintillator-photomultipliers. A commercial pin-diode BLM[5] is available but the 10 MHz maximum output rate would not provide the dynamic range required during a 1 msec beam pulse. Scintillator-photomultipliers do not have the long term stability needed for an accelerator application. However, to be able to see losses within the bunch, scintillator-photomultipliers will be installed at a limited number of points, such as, Ring injection and extraction, and other strategic points. Since only a relative time history is of interest for this application, non-calibrated data is acceptable.

System Requirements

High level losses over several pulses can quench or damage the super-conducting cavities in the SRF Linac. Low level losses can prevent "hands-on" maintenance. To avoid this, losses must be limited to 1 W/m averaged over several seconds duration, roughly equivalent to 10^{-4} of a 2 MW beam lost uniformly around the Ring. Dynamic range must be available to measure such long-term low level losses, yet not saturate under short duration, high level losses. The BLM system will provide hardware and software generated signals to indicate excessive loss and data for tuning the beam to reduce operating losses. These considerations lead to the requirements shown below.

Detailed time history. Data within a macro-pulse will be acquired at 100 kSa/S per channel but will be packed at the front-end computer (IOC) and transmitted at a 6 Hz rate.

Total loss for each BLM per macro-pulse. The data from each BLM will be summed at the IOC and transmitted once per second. This data may be presented as a "waterfall" display[6] which shows the long-term history with detector location as the horizontal axis, time as the vertical axis, and level as color. A simple x-y plot of location versus loss, updated at a few Hz, will be available. The losses for each BLM will be kept in a 1000 point FIFO history at the console level for use in the event of an abort.

Long-term, low-level beam loss alarm and display. The long-term, low-level beam loss (< 1 W/m) will be averaged over 10 seconds in the IOC and compared against a 1 W/m reference. Warnings will be sent as this level is approached and an alarm when it is exceeded. The pre-averaged low-level data will be available as a waterfall display or a strip chart style display.

Fast loss output for the Machine Protect System. Each BLM electronics channel will provide an output signal to the Machine Protect System (MPS) to shut off the beam in the event of a high-level beam loss. This signal will be derived in hardware from the integrated beam loss during the macro-pulse and shall occur within 10 μsec of the loss. Each channel shall have masking capability in the MPS to allow the use of diagnostic hardware which can cause beam loss, such as wire scanners. The mask will also permit studies which might generate a higher than normal local beam loss, and disable malfunctioning BLM channels which would prohibit beam operation. A beam inhibit signal shall also be generated if the integrated loss for any BLM over a 1 second period exceeds a preset limit. This shall be done in software at the IOC and executed through an MPS line in the Utility Module.

Gain setting and readback for each BLM. Each BLM electronics channel will have 3 jumper selectable gain settings to compensate for high radiation locations, such as at dumps, injection, extraction, and the collimators, or lower loss locations such as lower energy regions and well shielded areas. The "Viewing Gain" for the wideband-output can be changed (x1, x10) via the control system without affecting the fast loss trip or the 1 W/m level sensing. Gain setting may be required pulse-to-pulse to allow for different mode cycles, but not during the macro-pulse. Read back of the jumper-selectable gain and Viewing Gain states will be provided.

Bias voltage control. During studies periods, it may be necessary to vary the detector bias voltage. Readback of the voltage and current will be needed. Each rack location will contain at least 2 HV Bias Supplies, powering alternate BLM detectors. The BLM System can be tested by switching the HV bias power supplies off and on again. The resulting transient capacitively couples between the high voltage and signal electrode and through the BLM electronics, checking all components of the system, including the bias and signal cable continuity.

System calibration. Each BLM detector will be calibrated in a laboratory test setup using a static radiation source. Data will be taken throughout the operating voltage range for each detector. Testing will be done at BNL until the production is stable, with subsequent BLMs tested at ORNL. Insitu tests of the system through the installed cabling will be performed using a calibrated 1 G-Ohm resistor and voltage source for use in the calibration database.

Expected Losses

Experience at LAMPF and the PSR, and studies for APT at LANL[7] have determined that losses of 1 W/m will allow hands-on accelerator maintenance (~100 mR/hr at 1 foot), with transverse losses primarily at quadrupoles where the beta-max occurs[8]. The beam-on dose rate can be estimated from the beam-off dose rate by a "rule-of-thumb": multipliers of 500-1000 are typically used. Taking 100 mR/hr at 1 foot as the beam-off activation, the beam-on dose rate may be estimated at 100 R/hr, or 0.46 mR/pulse, which is equivalent to 0.46 R/sec during the 1 msec beam.

The upper end limit has been specified as 1% local loss based on the expected losses in the Ring collimators[9]. The BLM system must generate a beam abort signal for the MPS in less than 10 μsec. This will be done by integrating the beam loss during the macropulse and comparing the result to a programmable reference. Longer term losses at lower level can also do damage to beamline components. The numerically integrated output of each BLM will be compared to settable thresholds in the local processor. Excessive loss will generate an abort signal to the MPS through a VME utility module in the IOC crate. Still lower level losses, corresponding to the 100 mR/hr, 1 W/m loss will be similarly monitored and a software warning provided to the operators.

Detector

SNS will use 255 (Linac 88, HEBT 52, Ring 75, RTBT 40) argon-filled ion chambers as the primary detectors for monitoring beam losses. Argon has the advantage of fast electron transit time compared to slower air filled detectors[10]. The initial choice was an ion chamber designed by Shafer[11] at FNAL in 1982 for the Tevatron, and modified to improve radiation hardness and reduce noise for RHIC at BNL[12], but there was concern about saturation at high dose rate and long ion transit time (~ 700 μsec at 2 kV bias). The ions, constituting 25% of the signal, would appear as a continued rise during the 1 msec pulse followed by a long ion tail after. However, processing could unfold the pulse shape. A new ion chamber, designed to overcome these limitations[13], utilizes a larger inner diameter electrode to significantly decrease the ion transit time and raise the collection efficiency for a 1% local loss.

Figure 1 shows the measured response at the equivalent if a 0.05% local SNS loss in the 200 MeV BNL Linac to Booster beamline, for the FNAL and a new prototype detector with nitrogen fill. The slow rise during the beam pulse, and long tail afterwards due to the ions is clearly seen for the FNAL detector. The "noise" is chopper modulation of the beam.

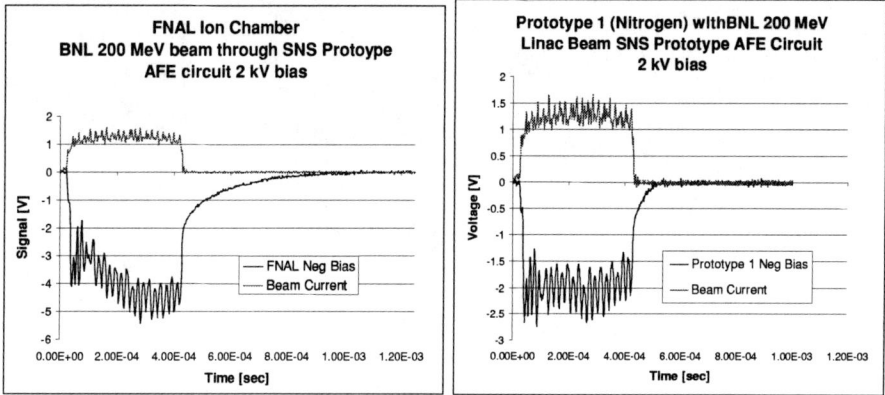

FIGURE 1. Beam Tests of the FNAL and New Prototype Ion Chambers

Installation

The detectors may be mounted directly to the beam pipe, magnet, or moved away from the radiation source to reduce signal output, allow a wider view over a larger area without intervening shielding. A typical SNS installation might raise the BLM 30 cm above the beam line on the upstream side of a quad. BLM's will be placed so that successive units can have overlapping views to allow limited coverage in the event a channel fails. The system will be packaged so that adjacent detectors will not be placed on consecutive electronics channels in the same circuit module, where possible.

Some BLM's may be exposed to X-rays from RF cavities. Measurements at BNL indicate that this may reach 50 R/hr during beam time. A study is needed to determine the type and amount of shielding required that would reduce the X-ray contribution to an acceptable level, and its effect on the beam loss measurement.

Cabling

Low tribo-electric RG-59 type cable such as Belden 9054 or 9224 will be used for the signals. Tribo-electric noise comes from friction between the conductors and insulation due to movement, such as vibration, but a low friction coating between insulator and conductors will reduce this significantly. Two HV cables (RG-59), from separate HV power supplies will connect to alternate BLM's from each rack to provide some coverage in the event of a high voltage short or power supply failure.

ELECTRONICS

The BLM system will be distributed throughout the SNS. The controls electronics will be in VME, with analog signal conditioning housed in a non-VME crate in the same rack. The analog crate will use linear rather than switching power supplies because of the very small signal currents. Figure 2 shows the system block diagram.

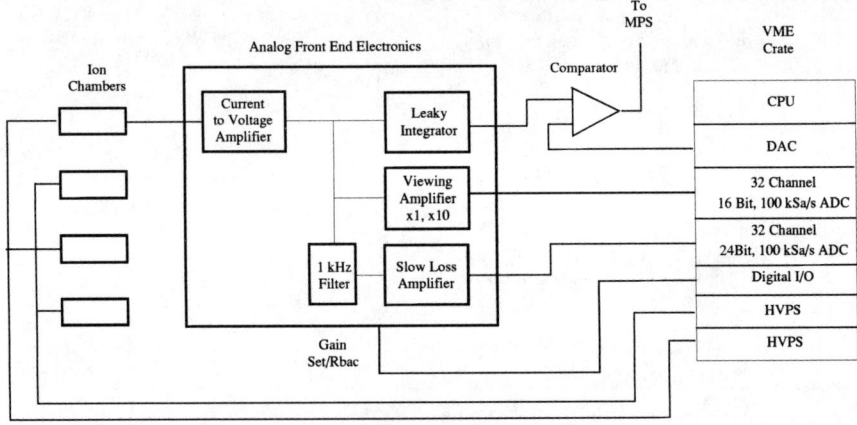

Figure 2. BLM System Block Diagram

System Dynamic Range Requirements

The BLM System must measure beam losses from a 1% local loss down to 1% resolution of a 1 W/m loss, a dynamic range of 126dB, or at least 21 bits. The design exploits the difference in bandwidth between these limits to allow additional resolution at the low end. During the 1 msec beam pulse the 1 W/m dose rate is estimated as 0.46 R/sec at 1 ft. For a typical BLM sensitivity of 70 nA/R/sec, this corresponds to 32.4 nA signal current during the pulse. The upper end loss, allowing for future improvements to 2 MW beam power, corresponds to 20 kW. Since the signal must inhibit the beam within 10 μsec, only the electron signal will be considered, reducing the BLM sensitivity to 35 nA/R/sec. Then $I_{Max} = 16.2 \times 10^{-9} \times 2 \times 10^4 = 0.324$ mA for a loss of 1% uniformly during the 1 msec pulse.

Analog Front End Electronics

Figure 3 shows the AFE electronics schematic. The circuit provides 3 outputs to meet the fast trip, wideband-wide range data, and 1 W/m sensing requirements.

Wideband Signal Circuit

Since the ion chamber is a current source, the input resistor doesn't affect the signal gain but does determine the voltage noise gain and input signal risetime. A 100 m cable and 470 Ω resistor gives about a 5 μsec risetime. A 6.2 kΩ feedback resistor puts the signal mid-range for 5 V ADC for a 1% beam loss. For the 35 kHz SNS BW, the 10 pA at 10 Hz BW noise observed in RHIC would correspond to 3.7 μV. Three jumper selectable gains with readbacks are provided. "Viewing Gain" can be set and read back remotely without affecting the beam interrupt or 1 W/m outputs. The signals of Figure 1 were made using this circuit.

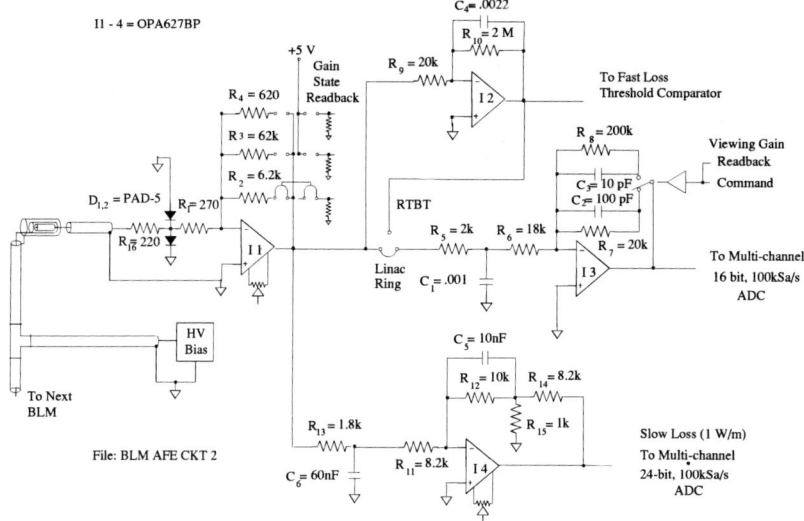

Figure 3. Analog front-end electronics schematic

The 1 W/m loss Circuit

A 1 W/m loss will produce 200 µV out of the input stage, roughly an LSB for a 5V, 16-bit (15-bit plus sign) ADC. Another 7-bits would be needed for 1 % resolution of a 1 W/m loss. Using a 1 kHz low pass filter, the noise would be reduced by the equivalent of almost 3-bits. Having an analog filter precede the output amplifier reduces the peak of short duration losses, allowing the additional gain of 10. A lower cut-off might not allow sufficient time to measure the baseline for offset subtraction. The output goes to a 24 bit, 100 kSa/sec ADC. Tests of one ADC (ICS-110B) indicate it can achieve 18-19 bit resolution. This is not quite enough for the 1% resolution, but sampling over 10 seconds or longer should provide the required sensitivity. Detailed testing of this and other ADCs, including thermal drift of offset and gain, is in process. Using measurements prior to the pulse, offsets of the AFE as well as the ADC, which are likely to be in excess of tens of microvolts, can be measured and subtracted. Dedication of one ADC channel to read a reference voltage may be required to compensate for gain thermal drift.

The Pulse Integrated Dose Circuit

Experience at LANSCE[14] has shown that a beam inhibit signal should be based on integrated dose rather than dose rate. Thus an integrator will be used to provide a signal to a comparator to generate a signal for the MPS when the programmable reference is exceeded (see Figure 2). A gated integrator with a triggered reset would give a precise integral of the loss, but the gating, reset, and charge injection compensation circuitry add considerable complexity to the design. A "leaky integrator", using a large value resistor to bleed the charge, is simple and provides an adequate representation of the pulse dose. Simulations indicate that for an RC time constant of 1/3 the pulse period, the output will reach equilibrium in 3 beam pulses with an error of less than 10% while decaying to under 5% by the next beam pulse. This is a reasonable trade-off to the reduction of complexity and reliability of the triggered integrator.

Since the 695 nsec wide RTBT beam pulse would be too narrow for the 100 kSa/s ADCs to acquire through the wideband output, the signal from this integrator will be jumper selected for input to the Viewing Gain stage.

ACKNOWLEDGMENTS

The authors would like to R. Shafer and M. Plum of LANL for providing valuable insight and discussions regarding the topics in this work. The BNL staff includes Chaofeng Mi who built and tested the AFE circuit. Yongbin Leng evaluated the ADC's, John Smith has been invaluable in supporting the Controls aspect of this system. Tony Curcio, Paul Ziminski and members of the Instrumentation Group provided enthusiastic cooperation in building, testing and installation of the prototype detectors and circuits.

REFERENCES

[1] Alonso, J., "The Spallation Neutron Source Project", Proceedings of the 1999 PAC, NY, 1999, p574. For current status see http://www.sns.gov
[2] Witkover. R. L. ,"Microprocessor Based Beam Loss Monitor System for the AGS", IEEE Trans. Nucl. Sci **26** 3313 (1979)
[3] . M. A. Clarke-Gayther, et al. "Global Beam loss Monitoring Using Long Ionisation Chambers at ISIS", 4th European Particle Accelerator Conference, London, England, 1994, p1634
[4] Witkover, R. L., Michnoff, R. J. and Geller, J. M., "RHIC Beam Loss Monitor System Initial Operation", Proceedings of the 1999 PAC, NY, 1999, 2247
[5] Bergoz Inc, http://www.bergoz.com
[6] Bai, M., et al., "RHIC Beam Loss Monitor System Commissioning in the Year 2000 Run", To be published these proceedings.
[7] Hardekopf, R. A., "Beam Loss and Activation at LANSCE and SNS", LA-UR-99-6825, Los Alamos National Laboratory, Los Alamos, NM. Sept. 1999
[8] Loc. Cit. Reference 7
[9] Y. Y. Lee, BNL, Private Communication, Oct 2001
[10] Plum, M., and Brown, D., "Response of Air-Filled Ion Chambers to High Intensity Radiation Pulses", Proceeding of the 1993 PAC, p2181
[11] . Shafer, R. E., et al., "The Tevatron Beam Position and Beam Loss Monitoring Systems", Proc. The 12th Int'l Conf. High Energy Accel., p609, 1983
[12] Loc. Cit. reference 4.
[13] Witkover, R. L. and Gassner, D. "Design of an Improved Ion Chamber for the SNS", These proceedings
[14] Plum, M., and Shafer, R., Private communications.

Booster Applications Facility Instrumentation*

D. Gassner, S. Bellavia, K. A. Brown, I. H. Chiang, P. Pile, R. Prigl

Brookhaven National Laboratory, Upton, NY 11973, USA

Abstract. A new experimental facility being at built at BNL will take advantage of heavy-ion beams from the AGS Booster for radiation effects studies of importance for the Space Program. A large dynamic range response is necessary to accommodate a wide variety of species (protons to gold) and energies (100 MeV/amu to 1.3 GeV/amu). The instrumentation proposed for extraction control and transport diagnostics will include phosphor screens with video cameras, segmented wire ionization chambers, ion chambers, and scintillators. Design and development of these systems will be presented.

INTRODUCTION

The principle source of ion beams for the US National Aeronautics and Space Administration (NASA) accelerator-based radiobiology program is at the BNL facility[1]. Since 1995, the Alternating Gradient Synchrotron (AGS) has yearly delivered a limited set of ion species and energies for a community of approximately 70 investigators from 15-20 institutions. The focus has been on 600 and 1000 MeV iron, 1000 MeV silicon, and 10,000 MeV gold ions. More than 1000 biological samples were irradiated at the AGS A3 beam line, in addition to physics experiments to establish beam characterization and dosimetry data. Operating time for radiobiology at the AGS is at a premium, as the major running time is dedicated to nuclear physics and high-energy physics research. The NASA program has typically had 1 to 2 running periods of 150 hours duration during a given year. Each period consists of continuous beam operations of 24 hours per day until all the approved experiments for the period are completed.

TABLE 1
Operating Parameters for Slow Extraction Beam for Some Typical Ion Species.

Species	Charge State In Booster	Kinetic Energy Range (GeV/Nucleon)	Estimated Max Intensity (10^9 Ions/Pulse)
H^1	1	0.10-3.07	100
Si^{28}	14	0.09-1.23	4
Fe^{56}	21	0.10-1.10	0.4
Cu^{63}	22	0.10-1.04	1
Au^{197}	32	0.04-0.30	2

---

* Work performed under auspices of the U. S. Department of Energy.

CP648, *Beam Instrumentation Workshop 2002: Tenth Workshop,* edited by G. A. Smith and T. Russo
© 2002 American Institute of Physics 0-7354-0103-9/02/$19.00

The AGS is not designed for providing high quality beams in the energy range of highest interest to the radiobiology research community. The energy range of the AGS Booster, which serves as injector to the AGS, provides a much better match. The Booster Application Facility (BAF), which is scheduled for commissioning in FY2003, will be a dedicated NASA particle beam source that will provide for all ions from protons to gold in an energy range from 40-3000 MeV/nucleon, with beam intensities ranging over 6 orders of magnitude. A pair of octupole magnets in the BAF beam line will provide the flat beam profile required for most irradiations without the need of heavy collimation.

The present NASA planning guidance is for yearly operation of 15 weeks of weekly 5 shift operations. A sample list of available ions with relevant characteristics is provided (Table 1). BAF will operate in slow-extracted[2] mode; the spill can be varied from the Booster uniformly over a 0.5-1 second spill every 3-6 seconds. At the target station the beam size can be varied from 1 cm to 20 cm in diameter for 95% beam intensity and maximum emittance.

BEAM LINE DIAGNOSTICS

A total of 7 instrumentation locations are specified for the 100 meter BAF transport beam line. At these locations a total of 7 phosphor screens (flags), 5 Segmented Wire Ionization Chambers (SWIC's), 5 Ion Chambers (IC's) and 3 Scintillator/PMT's (Scint's) will be installed as shown (Figure 1 and Table 2).

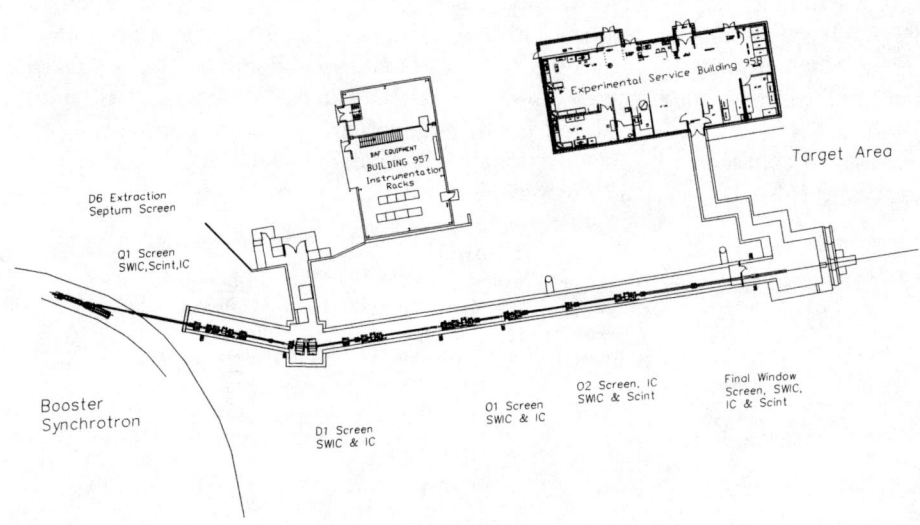

Figure 1. BAF Transport Layout

TABLE 2

BAF Instrumentation Details

Location (just upstream of)	Flag	SWIC Wire Spacing (32/plane)		Beam Size (90% full width)		Ion Cham.	Scint. /PMT	Est. Vacuum (10^n Torr)
		Horiz.	Vert.	Horiz.	Vert.			
1) D6 Septum	Yes	N/A	N/A	25 mm	25 mm	No	No	-11
2) Q1	Yes	6 mm	6 mm	60 mm	120 mm	Yes	Yes	-10
3) D1	Yes	1.5 mm	6 mm	17 mm	75 mm	Yes	No*	-9
4) O1	Yes	6 mm	1.5 mm	90 mm	8 mm	Yes	No*	-9
5) O2	Yes	3 mm	6 mm	20 mm	90 mm	Yes	Yes	-8
Final window				200 mm	160 mm			-8
6) DS window	Yes	6 mm	6 mm	>200 mm	>160 mm	Yes	Yes	Air
7) Target Area	Yes	N/A	N/A	>200 mm	>160 mm	No	No	Air

*Note: All 4 vacuum instrumentation stations are capable of a full compliment of SWIC, IC, and scintillator heads.

VACUUM CONSIDERATIONS

In order to comply with the request from NASA for minimal material in the beam path from the Booster, all the diagnostics stations were designed with plunging capabilities. There is a 6" beam pipe from the Booster to the upstream end of the BAF transport tunnel, then increasing to 8" until the last trim magnet, and a 12" pipe for the remainder of the line to the only transport vacuum window.

An important factor related to this requirement was compatibility with the existing Booster vacuum system. Because it would cause unacceptable beam losses for low momentum heavy ion beams, a vacuum window can not be used to separate the Booster 10^{-11} Torr ultra high vacuum (UHV) system from the BAF beam line vacuum system. A transition vacuum from the Booster ring vacuum to the line vacuum will be provided. Pressures of 10^{-10} Torr and 10^{-9} Torr will be required in the first two vacuum sections of the line respectively. The first section of the line will be bakeable to 150°C. The rest of the line will be a clean all-metal gasket, unbaked vacuum system with ion pumps. Since a robust UHV, bakeable instrumentation assembly had to be designed for the first upstream section, it was decided to use this same design at the 3 other downstream locations. In order to ensure the integrity of the Booster vacuum, the diagnostics chamber design must provide sufficient safety margin. A failure of a vacuum window closer than 45 meters (based on fast valve response time) to the Booster would compromise the Booster vacuum enough to cause a month delay for baking and pumping down.

TRIPLE PURPOSE DIAGNOSTIC STATION

Each stainless steel diagnostic station vacuum enclosure has 8" beam pipe ports and additional ports for vacuum pumps and gauges. The aluminum plunging vessel, which travels about 10" inside the station vacuum enclosure, houses the diagnostics heads, and has 8" diameter aluminum windows machined to a thickness of 10 mils. A mixture of 80% Argon and 20% CO_2 flowing through the plunging vessel at just above 1 atmosphere is used as the counting gas. A 14" O.D. welded bellows is used to plunge the vessel into the beam path as shown in Figure 2. A bolted flange on the bottom of the bellows allows removal and service of the diagnostics heads without disturbing the vacuum system. To incorporate this feature, 2400 lbs of force are needed to retract this large surface area, requiring a reinforced stand and large motor drive. Each vessel has the capability of housing a horizontal and vertical SWIC, dual ion chambers and a scintillator head as shown in Figure 3. The SWIC's will be used to measure transverse beam profiles. For the lower intensity beams, the voltage bias can be increased for operation in the proportional region. The scintillator and ion chamber will measure low and high intensity respectively, as well as time characteristic data.

Figure 2. Plunging instrumentation vacuum assembly and stand. Retracted position.

The SWIC and ion chamber HV bias planes, and ion chamber signal planes are constructed of 1 mil aluminum foil stretched over a G10 frame. The SWIC signal wires, from LUMA Wire, are 0.7mil gold plated tungsten with 3% rhenium. To compensate for the variety (1.5 to 6mm) of SWIC wire spacing, separated HV bias will be used for each plane. The ion chamber electrodes are separated by 6.4mm and an electric field on the order of 1 kV/cm will be applied. The thin scintillator material is mounted downstream of the SWIC and IC. A light guide extends below to the photomultiplier tube and base, which extends below the plunging vessel.

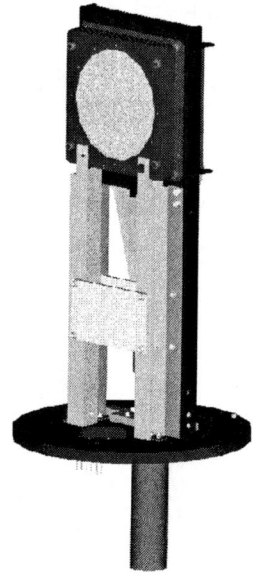 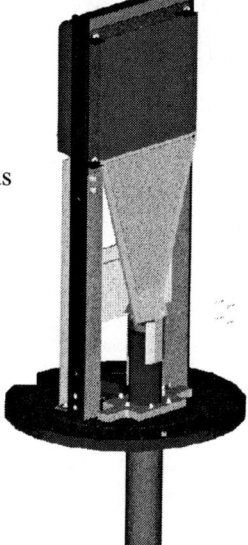

Figure 3. Detail of plunging diagnostics head assembly. Left shows upstream side with SWIC and ion chamber signal and bias planes. Right shows downstream side, with scintillator, light guide and photomultiplier.

ELECTRONICS & CONTROLS

All of the diagnostic electronics (except for the flag upstream of the extraction septum) and controls will be located in the 957 service building. Signals from each of the 32 wires in the horizontal and vertical plane of the SWIC will be processed by a Eurocard chassis using eight, 8 channel Advanced Technology Laboratory model 224900 integrator modules configured with 100pF and 10,100pF capacitors for high and low gain settings. After a programmed time sequence, both sets of 32 channels are scanned and multiplexed into a serial signal path that is digitized by a VME based synchronized 14 bit multiplexed A/D converter designed for RHIC. The profile data is presented graphically by a high level application; data acquisition can be configured to display a mountain range display showing the evolution of the profile over the spill.

Ion chamber electronics will consists of BNL designed current to frequency (I/F) converters that also provide an analog output for monitoring intensity throughout the spill. System calibration is based on calculated pair production in the counting gas, and a precision current source for the electronics. Over all three available gain settings, the module has > 100dB dynamic range. High gain mode calibration is about 7 fC/count. Recycling integrator front-end electronics are under consideration.

Standard Phillips NIM photomultiplier counting electronics that include amplifier, discriminator, and level translator will process the scintillator/PMT signals.

Low numbers of extracted ions are counted directly from the plastic scintillators. When saturation occurs, the ion chambers will be used to measure the higher intensity. The ion chambers and scintillators systems can be cross calibrated when the extracted beam intensity is within the dynamic range of both systems.

A SIS 3808 VME scalar will read in the counts from the I/F converter and scintillator counting electronics; data can be displayed various ways via the high level controls system.

A Bira VME 4500 HV Mini-System will provide high voltage bias for all related systems. This system enables full remote control of all aspects of the bias levels including voltage and current trip thresholds.

Flag illumination lamp controls, electronics gain control, plunging controls, readbacks of limit switches, and system status will be handled by VMIC 1160 and 2170A digital input and output VME boards.

PHOSPHOR SCREENS

To acquire higher resolution transverse beam size at higher intensities, plunging phosphor screens are used which are housed in separate vacuum chambers. The first diagnostic system in the BAF transport is a screen just upstream of the thick ejection septum magnet in the Booster ring. This stepper motor driven 8 position rotating assembly has 1 screen, 3 stripping foils, 3 stripping wires for low intensity experiments, and 1 blank as shown in Figure 4. During Booster slow extraction setup and diagnostics, the Morgan Matroc Chromox 6 aluminum oxide screen will be positioned in the beam path to allow beam position and size measurement of the beam kicked from the upstream thin septum.

Due to high radiation levels produced while running high intensity protons on non-BAF cycles, a rad-hard Dage 70R video camera will be mounted across the aisle in the Booster tunnel. It will gather light from the screen via a quartz vacuum port and mirror mounted above assembly. Between the video camera and lens fixture will be a 6 position rotating neutral density filter assembly that will eliminate saturation problems.

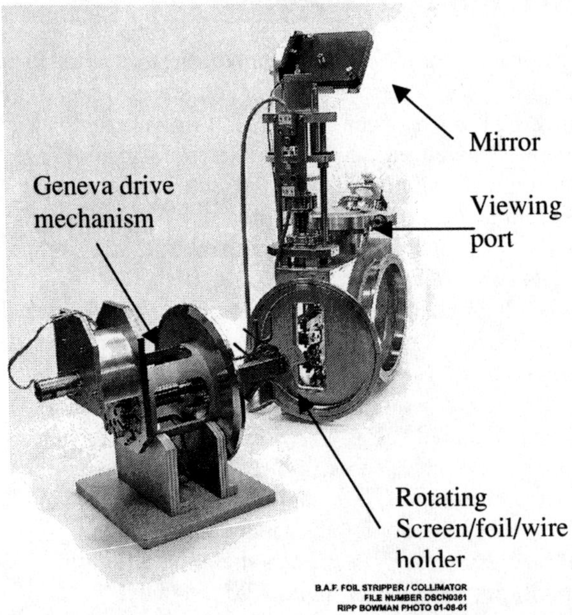

Figure 4. Screen and stripper assembly upstream of extraction septum.

An Imaging Technology Inc.VME based frame grabber will process the analog video signal; a high level application will then display the beam profiles and calculated parameters.

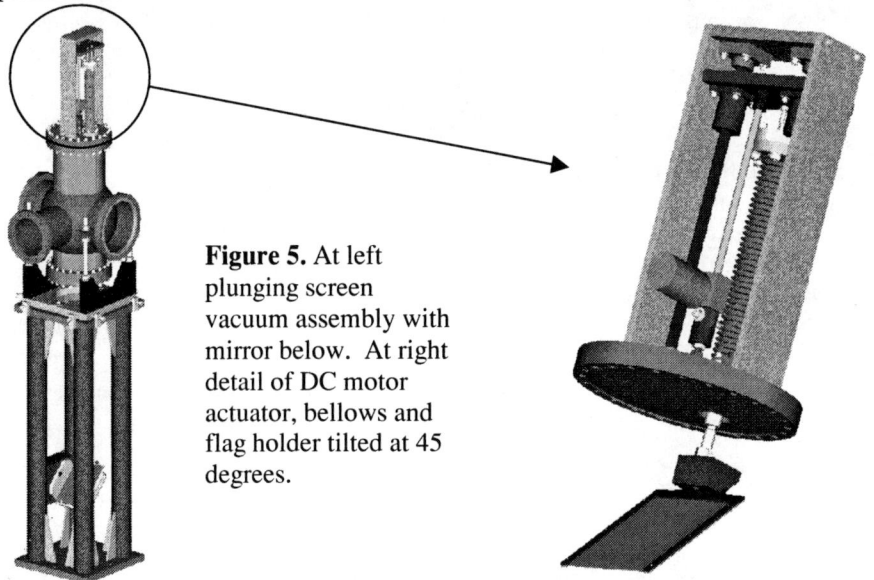

Figure 5. At left plunging screen vacuum assembly with mirror below. At right detail of DC motor actuator, bellows and flag holder tilted at 45 degrees.

There will be 5 plunging screen locations in the 100 meter transfer line, each having a dedicated vacuum chamber assembly which includes a quartz viewing and illumination port, mirror, and a 24VDC motor plunging actuator as shown in Figure 5. The video camera assembly will be mounted on an adjustable stand with drawer slides, and recessed inside the camera cubby as shown in Figure 6. This assembly includes a CCD 1394 Firewire video camera, neutral density filter assembly, lens, and an image intensifier in some locations. The digital video signal from these cameras will be processed in building 957 on a dedicated personal computer running frame grabber analysis software[3].

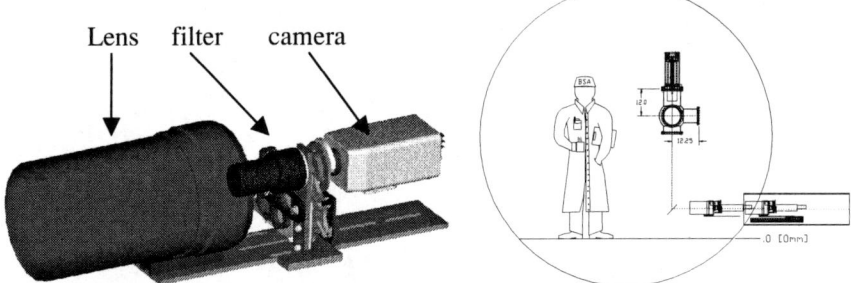

Figure 6. Camera cubby, with CCD camera, intensifier, neutral density filter, and lens on drawer slide. At right, camera assembly orientation to flag vacuum chamber.

At the end of the beam line, after the final vacuum window a full compliment of plunging devices will be installed. Though the size of the SWIC, IC, scintillator and flag will be large to cover the 12" beam pipe aperture, the design is less complex since there are no vacuum issues.

SUMMARY

A total of 20 diagnostics systems will be installed in the BAF transport beam line. All of the system transducers can be retracted during experimental running to minimize the material in the beam path. The mechanical design provides sufficient safety margin to ensure the integrity of the Booster vacuum. The dynamic range of the detectors and electronics is sufficient to provide measurements over most of the planned operating modes. For very low intensity operations, the beam line will initially be set up at higher intensity, and then reduced and the separate experimental target dosimetry system will be used. Commissioning is scheduled during early FY2003.

ACKNOWLEDGMENTS

The authors would like to thank Tony Curcio, Stephen Jao, Peter Oddo, Sal Polizzo, Joe Saetta, Al Weston, and Paul Ziminski for electronics support. We would also like to thank Dave Kipp, Dan Lehn, Al Ravenhall, Craig Rhein, Lou Snydstrup, Don Von Lintig for mechanical support. In addition we are indebted to Larry Hoff, Joe Skelly, and Wes Buxton for handling the Controls aspects of this system. We appreciate the support and advice of Tom Russo.

REFERENCES

1. Lowenstein, D. I., "BNL Accelerator-Based Radiobiology Facilities" First Intl. Workshop on Space Radiation Research and 11the NASA Space Radiation Health Investigators' Meeting, Arona Italy, 5/2000.
2. Brown, K. A., et al., "Resonant Extraction Parameters for the AGS Booster" PAC 2001, Chicago.
3. Brown, K. A., D. Gassner, et al., "IEEE 1394 Camera Imaging System for Brookhaven's Booster Application Facility Beam Diagnostics", EPAC 2002, Paris.

Profile Measurement Of Scanning Proton Beam For LiSoR Using Carbon Fibre Harps

R. Dölling, L. Rezzonico, U. Frei, S. Benz, P.-A. Duperrex, M. Humbel

Paul Scherrer Institut, Villigen, Switzerland

Abstract. Harps using secondary electron emission from 16 carbon monofilament wires have been built to measure the horizontal and vertical beam profiles of an intense 71 MeV proton beam. A very large dynamic range and good time resolution are achieved by a newly developed 16 channel CAMAC read-out module using logarithmic amplifiers. A first test at low beam current is reported.

INTRODUCTION

The LiSoR (Liquid Solid Reaction) experiment, is carried out at PSI within the framework of the MEGAPIE project (MEGAwatt Pilot Experiment), aiming at a liquid metal spallation target based on a lead-bismuth eutectic mixture [1]. PSI's Philips cyclotron is used to irradiate stressed steel specimens in contact with flowing liquid metal with a 71 MeV proton beam. A nearly homogeneous averaged beam current density over a rectangular area is foreseen. This is provided by suitable horizontal and vertical scanning of the beam over the target. A sinusoidal horizontal deflection is combined with a linear vertical deflection. The frequency ratio is 12:1 and the vertical maximum is synchronised to a horizontal zero deflection. Less than 5 % deviation from homogeneity in the 3 x 12 mm^2 central region is expected for Gaussian beam profiles of $\sigma_{hor} = \sigma_{vert} = 0.6$ mm and ±2.75 mm horizontal and ±7 mm vertical amplitudes [2]. The scanning is performed by upstream steering magnets. The power supplies are controlled by a modified PSI CAMAC-ROAD-C-"KOMBI-Controller" [3]. The amplitudes and the horizontal master frequency can be set and interlock signals are generated in case of malfunction, especially if the measured magnet current amplitudes are below a preset safety limit. The working frequency is 15 Hz, as allowed by the power supply response function.

Horizontal and vertical harps are positioned 17 cm in front of the target for the verification of the momentary and time-integrated beam profiles. The harps and the read-out electronics are discussed in the following.

HARP DESIGN

Two harps of 16 wires each are arranged inside a 30 x 30 mm^2 aperture with a separation of 18 mm in the beam direction. The horizontal wire spacing is 1 mm and the vertical spacing is 1.25 mm. A third grid of 13 diagonal wires with 2 mm spacing is mounted in the mid-plane between the two harps. These wires, as well as two 30 x 30 mm^2 electrodes (positioned horizontally 16 mm below the beam axis) in front and behind of the aperture, are biased to +300 V. Thereby, the secondary electrons from the harp wires themselves as well as from the target or from a (15 x 19 mm^2) collimator 21 cm in front of the harps are prevented from reaching the harp wires.

FIGURE 1. Harp assembly. Electrodes for secondary electron suppression not yet mounted.

33 µm diameter carbon monofilament wires (Textron Specialty Materials, Massachussetts) are used. At the working scanning frequency, the wire temperature is estimated to be well below 1000 °C for a 50 µA beam. Hence, no problems with thermionic emission or carbon loss due to its vapour pressure are expected. For a stationary beam, the thermionic emission current would by far exceed the secondary emission current (i. e. the signal) and the wires may be destroyed.

The radiation hardness of the carbon wires is not critical. The same fibre used at the time-structure measurement at the PSI Injector-2 cyclotron [4] has not shown any visible damage after irradiation with 10^{20} protons/mm^2 at 72 MeV.

HARP MANUFACTURE

Due to the small wire spacing a spring tensioning of the wires, as used in [5], seems to be not practical. The elasticity and low thermal expansion of the wires allow a direct mounting as e. g. described in [6, 7].

For reasons of cost and ease of manufacture the carbon wires were glued with conducting epoxy (EPO-TEK H20E) to frames made of 0.5 mm thick ceramic filled printed-circuit board (Rogers RO4350B) with 100 µm copper/gold plating on both sides. Each frame is mounted in a 9 mm thick gold plated aluminium disk (Fig. 2). The stack of four disks (one as a cover) is mounted on a vacuum flange carrying two Caburn 25-pin Sub-D and an MHV feed-through.

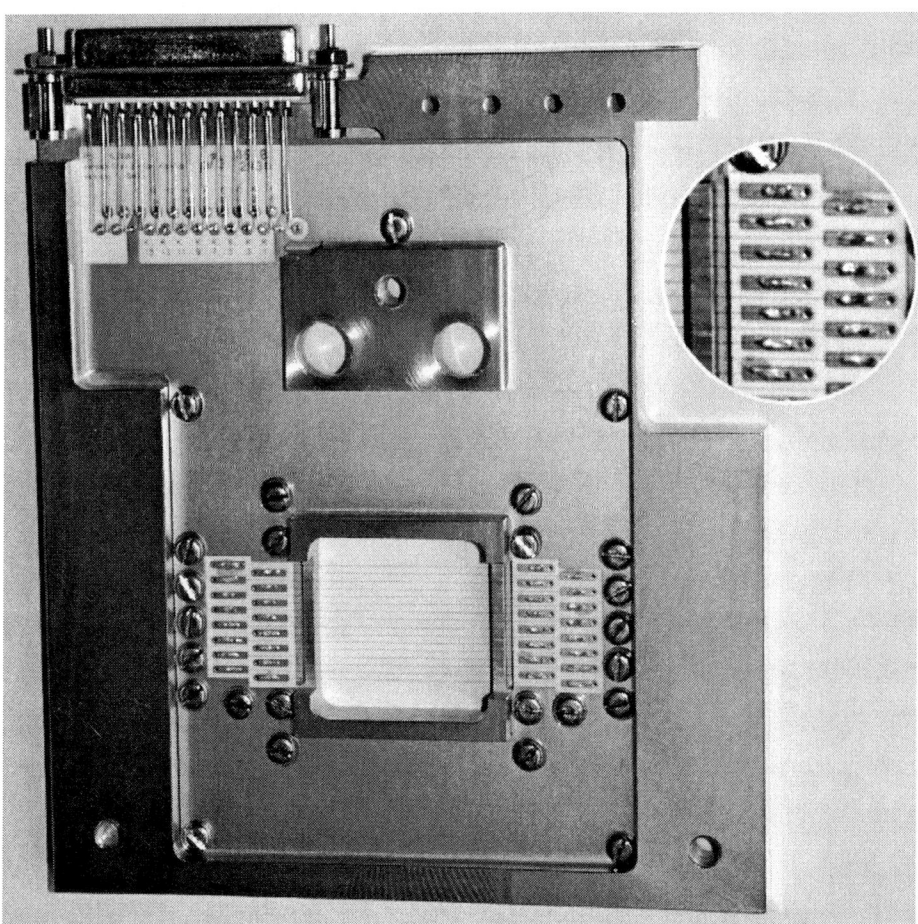

FIGURE 2. Harp for the measurement of the vertical beam profile. The discolouration of the white RO4350B around some of the mounting pads (inset) stems from the solvent of the conducting epoxy and does not deteriorate the isolation.

For the positioning of the wires during assembly, the printed-circuit board is mounted on a support together with two M6 (resp. M8) screws parallel to opposite sides of the aperture. Overlength monofilament with small weights attached with adhesive tape to both ends is then laid over the screws ensuring a defined tension. It is then attached in the threads with cyanacrylat glue. Afterwards, the conducting epoxy is applied with a syringe and cured for one hour at 100 °C and one hour at 160 °C.

This treatment should allow for a permanent operation at 200 °C in vacuum. However, the operation temperature is limited to approximately 170 °C by the Sub-D connectors used. This is well above the expected operating temperature due to heat transfer from the neighbouring lead-bismuth circuit.

READ-OUT ELECTRONICS

The 16 currents of each harp are transferred via 50 m of 10 x 2 twisted-pair cable with double outer shielding to an in house developed CAMAC module (Fig. 3). The 16-channel analogue front-end print was originally developed for the PSI Ultra-Cold-Neutron (UCN) experiment. It uses logarithmic current-to-voltage converters, as our standard LOGCAM and LLCAM modules, but with an extended range. According to first tests the deviation from linearity is within 1 % from 20 pA to 200 µA. The cut-off frequency is approximately 40 Hz at 100 pA, 400 Hz at 1 nA, 4 kHz at 10 nA and 8 kHz above 40 nA with cable and only slightly better without. This measured dependency agrees well with the response to a short current pulse depicted in Fig. 4.

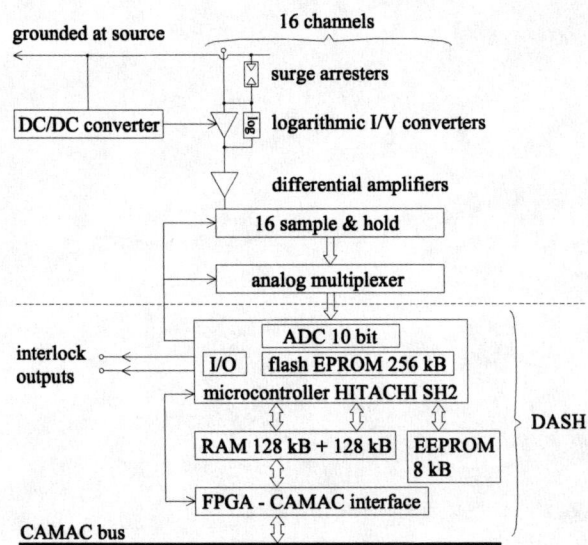

FIGURE 3. Harp read-out electronics. (Unused functionality of DASH not shown.)

Read out and evaluation of the simultaneously sampled currents are performed every millisecond by the DASH back-end electronics. Two modes of operation are implemented: The read out of individual currents (both time averaged and not averaged) via CAMAC and the sampling of up to 4096 profiles every n * 1 ms (with $1 \leq n \leq 65536$) and subsequent read out via CAMAC. Both modes can be used simultaneously.

8 decades of input current range are acquired with the 10-bit ADC of the microcontroller. A single current value is stored in bit position 2 ... 11 of a 2-Byte number according to $nn[\text{digit}] = 4 \cdot ADC_{out}[\text{digit}] = 4 \cdot 128\, \text{digit} \cdot \log(I[\text{pA}]/10\text{pA})$. (The least significant bits 0 and 1 are set to zero. The 2-bit left shift allows the calculation of averaged values with a factor of 4 improved resolution. That way the same inverse transformation for raw data and for averaged data can be applied: $I[pA] = 10\text{pA} \cdot 10^{nn[\text{digit}]/512}$.)

The DASH ("CAMAC <u>D</u>ata <u>A</u>cquisition module with Hitachi <u>SH</u>2 microcontroller") was recently developed as a standardised universal controller, which can support different front ends for beam diagnostic tasks [8]. It includes a programmable interlock and warning logic with watchdog. In the case of the harp front-end, width and position of the beam evaluated from the measured wire currents are monitored as well as the maximum individual wire current. Interlock limits, low pass digital filter and averaging parameters, sampling settings, etc. can be written via CAMAC to an EEPROM (together with the complement as a safety measure). The present module status as well as the status at the last interlock can be read.

FIRST MEASUREMENTS

The presence of the wires can be checked by analysing the response of the wire currents to switching on the secondary electron suppressor voltage (Fig. 4). The signal amplitudes are similar for all 16 intact wires.

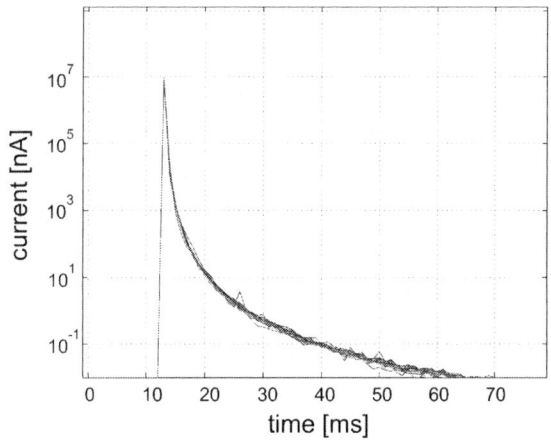

FIGURE 4. Current peak induced to the 16 horizontal wires by switching on the secondary electron suppressor voltage.

Up to now, only a few measurements at low beam intensities (<200 nA) have been performed in order to avoid obstruction of the ongoing installation of LiSoR components by activation. Fig. 5 gives an example of series of horizontal and vertical profile measurements of a scanned beam. (The beam shape was not adjusted to the specifications required for LiSoR in either direction.) Even at the correspondingly low signal levels, the results were very satisfactory. For the given environment, the noise pick-up on the long signal cables was tolerable. Cross-talk was not observed.

Currently the horizontal and vertical series are measured successively. It is foreseen to change the control-system handler to provide nearly simultaneous CAMAC start commands to both CAMAC modules.

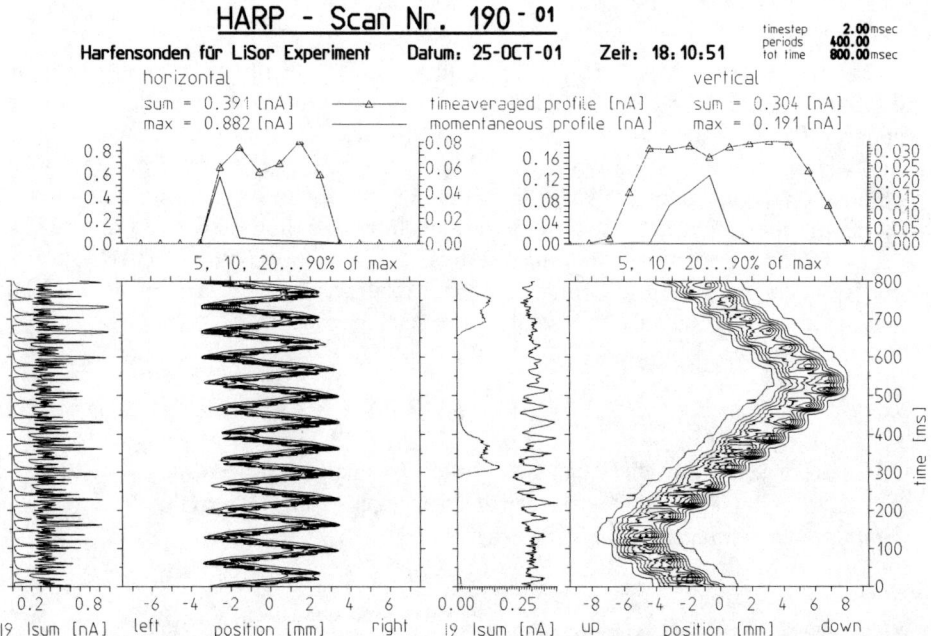

FIGURE 5. 400 horizontal (left column) and 400 vertical (right column) beam profiles. Each series measured (not time-correlated) during 800 ms at a simultaneously horizontally (15 Hz) and vertically (1.25 Hz) scanned beam. In each column: Main graph: time development of profile shown as a contour plot. Top graph: last measured profile (full line) and average of all profiles (broken line with triangles indicating wire positions). Left graph: time development of the sum of the individual currents and of the current on the ninth wire. (Readings $4 \cdot ADC_{out} = 0$ digit which correspond to 10 pA *or less* are depicted as 0 pA.) The temporal fluctuations of the current sums largely stem from the hopping of the beam from one wire to the next. In the present case of a scanned beam, the temporal distribution of the current on a single (here the ninth) wire can give the profile in more detail than the distribution of the momentary individual currents.

ACKNOWLEDGMENTS

We would like to thank D. Anicic for writing the control-system handlers and R. Erne for installing the harps.

REFERENCES

1. Bauer, G. S., Salvatores, M., and Heusener, G., "MEGAPIE, a 1 MW pilot experiment for a liquid metal spallation target", *Journal of Nuclear Materials* **296**, 17-33 (2001).
2. P. Schmelzbach, PSI, internal note, unpublished
3. G. Janser, PSI, internal note, 27.03.2001, unpublished
4. R. Dölling, "Measurement of the time-structure of the 72 MeV proton beam in the PSI injector-2 cyclotron", Proc. of the 5th European Workshop on Beam Diagnostics and Instrumentation (DIPAC 2001), Grenoble, France, 13-15 May 2001, pp. 111-113, http://www.esrf.fr/conferences/DIPAC/DIPAC2001Proceedings.html
5. O'Hara, J. F., et al., "Slow wire scanner beam profile measurement for LEDA", AIP Conference Proceedings 546, New York: American Institute of Physics, 2000, pp. 510-518
6. Mackenzie, G. H., "Beam diagnostic techniques for cyclotrons and beam lines", IEEE **NS-26**, 1979, pp. 2312-2319
7. Fritsche, C. T., Krogh, M. L., Crist, C. E., "High density harp for SSCL linac", Proceedings of the 1993 Particle Accelerator Conference, New York: IEEE, 1993, pp. 2501-2503
8. L. Rezzonico, PSI, "DASH", internal note, 13.11.2000, unpublished

A Digital Signal Receiver VXI Module For BPM And Phase Detection Processing

Brian E. Chase, Keith G. Meisner

Fermi National Accelerator Laboratory, Box 500, Batavia IL 60510

Abstract A VXI module containing eight digital receivers is described for use in the Fermilab Main Injector, Tevatron and Recycler Low Level RF systems. It is used as a phase detector and radial position processor for multi-harmonic RF operation. This module is also slated for use in the Recycler Electron Cooling system as a multiple beam BPM processor. The module and its many operational modes are discussed.

INTRODUCTION

The Fermilab Main Injector RF System supports many RF accelerating frequencies. The 53 MHz RF is currently the main accelerating system, while other harmonics are designed for coalescing multi-bunch protons and antiprotons (pbars) before transfer into the Tevatron for Collider operation. The proposed scheme to decelerate hot Tevatron pbars in the Main Injector for storage in the Fermilab Recycler Ring will require using 2.5 MHz RF in a closed loop feedback mode, with new detectors, preamps, and signal processing for phase and radial position detection at 2.5 MHz. A Digital Signal Receiver (DSR) module has been prototyped that is compatible with the existing LLRF platform and provides the required processing. This module far exceeds the accuracy and dynamic range of the present analog processing equipment. The DSR also fits the dual beam measurement requirements for the BPM system in the Electron Cooling straight section of the Recycler Ring.

DIGITAL SIGNAL RECEIVER

Hardware Description

The DSR VXI module has eight digital radio channels that are controlled and read out by a single Digital Signal Processor (DSP). The channels may be synchronized in pairs, or in groups of pairs to make coherent measurements between multiple inputs. Synchronized DSR channels form the core of any Quadrature Amplitude Modulation (QAM) or vector processing function, while the DSP uses the channel vectors to compute complex functions in real time. This real time data is then presented to off module processes in a variety of high speed data ways.

Each DSR channel receives signal through a daughter card that allows custom analog preprocessing and anti-alias filtering. The daughter card output drives an Analog Devices AD6644 14 bit, 65 MSPS ADC and an AD6620 Digital Receiver Processor. The digital receiver contains a frequency translator and three cascaded filters. It outputs I and Q data that is read out by the ADSP-21062 DSP. The DSP provides additional narrow band filtering of the I/Q data, and calculations of position or phase at data rates of up to 1 MHz. The DSP also programs the AD6620s, drives an eight channel DAC, and provides digital control of the analog processing daughter cards. All ADCs are clocked from a common front panel input. Each AD6620 channel pair is separately programmable to allow processing at different center frequencies, bandwidths, decimation rates and gains.

Noise sources on the DSR module include VXI bus transfers, on-board clocks, DSP internal processes, and data transfers from the AD6620s. This noise must be controlled to take full advantage of the ADC's dynamic range. When all eight channels are synchronized to latch AD6620 output data at the same time, coherent noise from digital power bus currents transfer to the analog power bus and the ADCs. For a data rate of 21.66kHz, this noise signal level measures at about -90 dBc. Careful choices of decimation and clock rates allow for minimum interference to the measurement.

Processor Calculations

Once the I and Q data is acquired by the DSP, it is converted to floating point format to vastly increase the dynamic range of further processing. The DSR module's synchronization between channels allows for measurements of relative phase as well as amplitude, making it effectively a multi-channel Vector Signal Analyzer. The phase measurement is very accurate over a large dynamic range due to the small errors inherent in digital down conversion. The determination of position from split plate detector signals "a" and "b" is solved to first order by the equation (a-b)/(a+b). Maintaining the phase information and solving the equation with the complete vector increases accuracy in the presence of random noise. The phase error between plate signals is measured and zeroed by setting the phase register in the AD6620's NCO. With a zero phase difference between plates some terms may be eliminated from the position calculation as well as removing the computationally intensive square roots.

$$Position = \left| \frac{z_a - z_b}{z_a + z_b} \right| \approx \frac{(I_a - I_b)(I_a + I_b) + (Q_a - Q_b)(Q_a + Q_b)}{(I_a + I_b)^2 + (Q_a + Q_b)^2}$$

The relative phase between two channels is found by dividing the two vectors and taking the arctangent of the resultant vector.[1]

$$\frac{z_a}{z_b} = \frac{(I_a, Q_a)}{(I_b, Q_b)} = \frac{I_b}{I_b^2 + Q_b^2}(I_a, Q_a) + \frac{Q_b}{I_b^2 + Q_b^2}(I_a, -Q_a)$$

$$Phase = \arctan(\frac{z_a}{z_b})$$

Optimizing For Accuracy And Signal To Noise

Any signals other than the signal intended for processing are capable of increasing the variance and or mean of the processed result. These signals may be random noise, out-of-band signals picked up on cables, or distortions of in-band signals caused by non-linearity in the process chain. The sources of random noise are preamplifiers, the quantization noise of the ADC and the digital receiver's filters. This noise is reduced by separately filtering the I and Q outputs to as narrow a bandwidth as practical, and by optimizing the gain at each stage in the signal chain. High over sampling ratios improve SNR by spreading the quantization noise of the ADC over a larger spectrum. This noise may then be largely removed by digital filtering. This "Process Gain" is calculated from the equation:

$$PG = 10 \times \log\left(\frac{Sample\_Rate\_Of\_ADC}{Filter\_Bandwidth}\right)$$

For the case of an output bandwidth of 1 Hz and a sample rate of 65 MHz, the process gain is an impressive 78 dB!

Correlated noise from cable pickup or noise on the DSR module itself presents a more serious problem than random noise. In band correlated signals cause a shift in the mean position output and therefore cannot be filtered out. If these noise sources are identified, then frequency domain filtering can suppress any strong lines that are present. While noise is a major issue for wide band systems, non-linearity and gain stability dominates position accuracy in extremely narrow band systems. Sources of non-linearity are ADC differential non-linearity and round off errors in the AD6620 filters. Adding out-of-band signal or dithering reduces the distortion of the ADC. Operating the AD6620 at the lowest decimation rate possible and setting the filter stage gains as high as possible reduces round off errors. Slight de-tuning of the AD6620 center frequency randomizes the round off errors in the digital receiver's I and Q outputs so that they may be averaged out by the wide dynamic range floating point filters in the DSP.

MAIN INJECTOR LOW LEVEL RF SYSTEM OVERVIEW

The Main Injector Low Level RF (MILLRF) system generates low power signals at several harmonics of the MI beam revolution frequency. Its RF outputs drive High Level systems that develop cavity RF buckets to support RF synchronous transfer, beam energy control, bunch shape manipulations, and preservation of longitudinal emittance. TTL outputs at the revolution frequency trigger local systems and drive the Main Injector Beam Synchronous Clock to support beam transfers between machines, beam diagnostics, and high bandwidth beam control systems.

Beam energy control requires Direct Coupled RF phase and frequency feedback whenever the MI energy (bend buss current) is changing. This feedback is classic RF phase control based on detected beam radial position (RPOS) and the LLRF oscillator to beam Phase Lock Loop (PLL). MILLRF is largely a VXI system that includes seven VXI modules manufactured by Fermilab. A VXI DSR module processes BPM detected signals at 2.5Mhz and 53Mhz, to provide the RPOS signal required for DC beam energy feedback. The DSR also provides 2.5Mhz and 53Mhz I/Q demodulators for the phase detection required by the PLL and frequency feedback loop.

Radial Position (RPOS) Processing in the Main Injector

While beam radial position processing is basically the same as other BPM measurements, there are additional requirements. In the Main Injector, the RPOS system must work well over beam intensities from 1E10 to at least 2E13, from a single bunch to an almost full ring, and with beam bunched at several RF frequencies. [2,3]

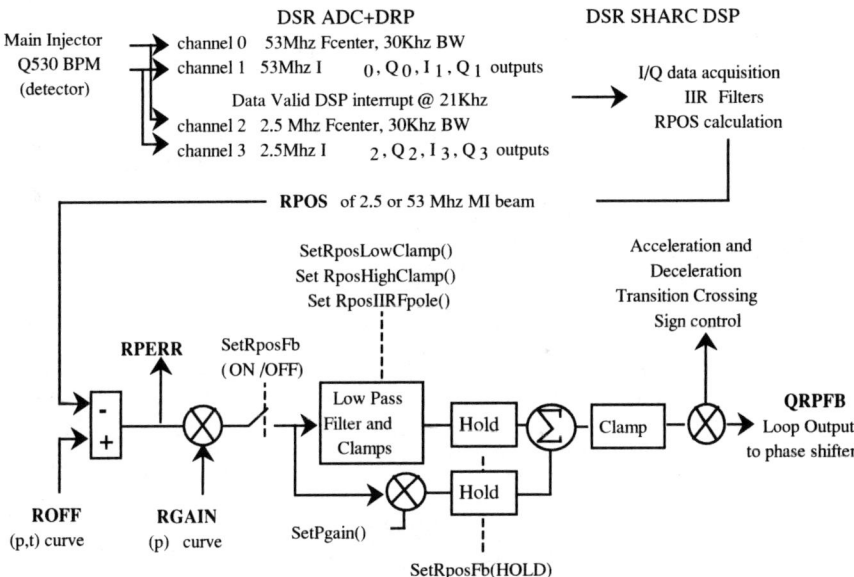

- **NAMES** are DSR DSP sourced data and Acnet Devices updated at 720hz
- FunctionName() describes DSR DSP interface (software) for Rpos feedback processes

Figure 1. MILLRF DSR RPOS Processing Block Diagram

The process must also have short group delay to optimize the radial position feedback loop. For 200 Hz closed loop bandwidth and 30-degree loss in phase margin, the group delay must be less than 400 microseconds. To achieve this delay, the AD6620 RAM Coefficient Filter is programmed with a simulated 2nd order IIR low pass combined with a notch filter centered at the revolution harmonic. This RPOS data is passed to the radial position feedback loop at a 100 kHz rate, as shown in figure 1.

ELECTRON COOLING BPM SYSTEM

The Electron Cooling System for the Recycler Ring will send a low emittance electron beam along the same axis as the 8 GeV Recycler Ring anti-proton beam in a specially designed straight section. The velocities of the beams are closely matched to exchange heat by coulomb scattering and cool the antiproton beam. For cooling to work properly, the relative position of the two beams must be measured and controlled to the 50-micrometer level. Because the detector plates see both beams at

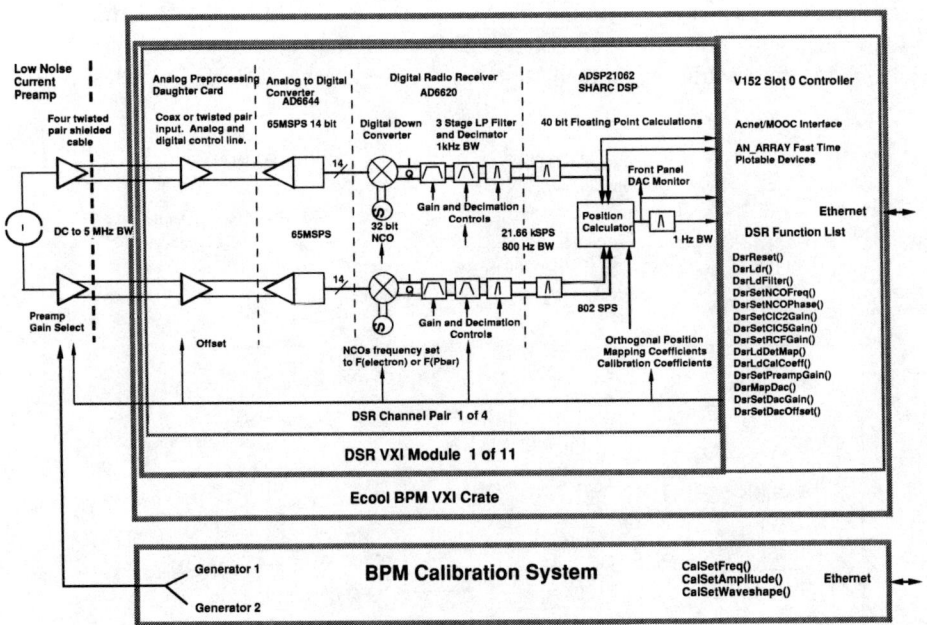

Figure 2.

the same time, their position signals must be resolved in the frequency domain. The anti-proton beam position is processed at the first revolution harmonic signal of 90 kHz and the electron beam is processed at 32 kHz. The 32 kHz signal is produced by a few milliamps modulation on the 1 amp DC beam. By frequency hopping DSR receiver channels between these frequencies, the two beam positions can be measured almost simultaneously. The filter sections of the AD6620s are programmed with an 800 Hz bandwidth (BW) and an output data rate of 21.66 kHz. The DSP filters the I and Q data down to a 100 Hz BW and calculates position and beam intensity at a 800 Hz rate. This data is available to the control system for fast time plots. The position data is further filtered down to a 1 Hz BW and is served to a beam position application program through the ACNET control system.

The AD6620 allows programmable gains in each of the filter stages. The first Combined Comb Integrator filter (CIC2), has a range of 36 dB, the CIC5 filter 120 dB, and the RAM Coefficient Filter (RCF), 42 dB. From the AD6620 data sheet[4,5], the maximum safe gains for our filter design are determined to be:

CIC2 = 6 (-36 dB) for MCIC2 = 15
CIC5 =17 (-102 dB) for MCIC5 = 20
Sout = 7 (-18 dB)
Where MCIC2 and MCIC5 are the stage's decimation rates.

For the best SNR and linearity, the gains in each stage may be increased based on the total signal level in the pass band of that stage. The goal is that each stage works as close to its over load level while maintaining a reasonable safety margin. An AGC is conceived to work in the following manner. Starting at a safe low gain, CIC2 gain is increased while monitoring position. A small calibration signal that is out of band of the CIC5 filter but in band of the CIC2 filter is injected into the preamp. If the gain change or the calibration signal affects position the CIC2 gain is reduced. This procedure is repeated to adjust the CIC5 filter gain. The RCF gain is then set so that the output intensity signal is within –15 to –3 dB of full scale. Gain ranging is of course unnecessary in the DSP filters due to floating point processing. A power sweep of the input power is shown in figure 3. Only two gain changes were made in the AD6620 filter sections during this 100 dB sweep and as a result a systematic error in position is seen around –40 dBm. These errors are expected to be greatly reduced by implementing the AGC algorithm.

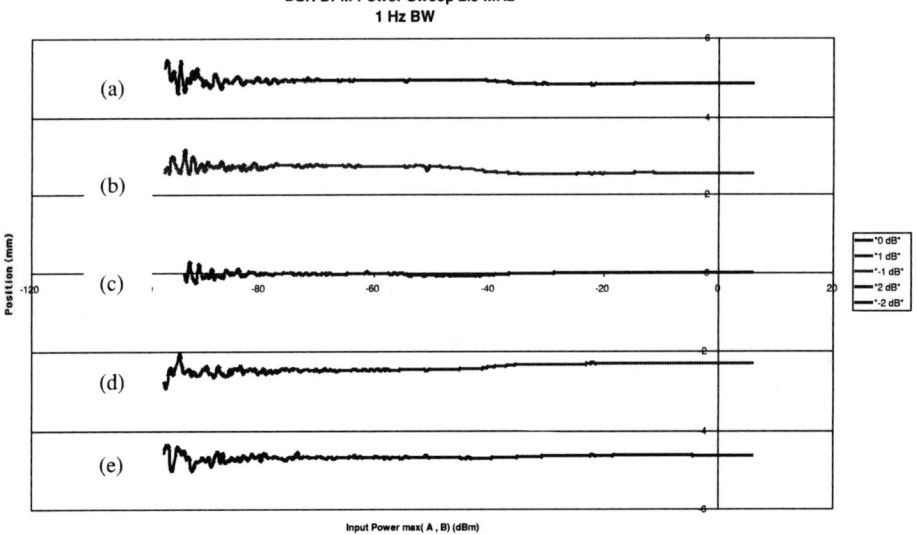

Figure 3. 100 dB power sweep of the DSR Module. The five graphs are produced by changing the power ratio between channels in 1 dB steps: a=2 dB, b=1 dB, c=0 dB, d=-1 dB, e = -2 dB

Calibration And Mapping Of Detector

Each BPM detector in the Electron Cooling section has been carefully mapped over its full range. A 3rd order fit to this data will be applied to the raw data at 800 Hz when the position is calculated. Gain terms determined from calibration are applied here as well. The calibration system will generate arbitrary waveform generator signals that mimic the frequency and wave shape of the beam of interest. This signal is injected through precision resistors to the tunnel preamp. The resistors are of slightly different values to calibrate for a position offset of about 1mm. This offset gives the system the additional information that there are no shorts between channels and the cables are wired correctly. Beam signals may be mimicked as closely as possible in shape and intensity. Calibration may be checked with beam in the machine by moving the calibration signal frequency away from the beam frequency. The calibration system may also be used to generate dither signals to reduce the differential nonlinearity in the ADC.

CONCLUSIONS

Digital Radio technology has advanced to the point where it is a clear winner or at least a strong contender for a wide variety of IF and direct RF processing applications. Using this technology, the DSR module is finding many applications beyond its original designed of radial position and phase processing. The function blocks of programmable analog daughter cards, digital receivers and floating point DSP create a very powerful and flexible processing environment. The bi-directional high-speed digital IO allows the DSR to be used in many real time feedback applications. The ability to program perfectly matched digital filters with over 130 dB dynamic range makes it ideal for processing high precision narrow-band applications such as the Ecool BPM system. At a production cost of less than $2000 per unit, the DSR module is attractive for even high channel count instrumentation projects.

ACKNOWLEDGMENTS

The development of any complex system requires teamwork and the shared experience from an open engineering environment. Special thanks to the LLRF Group members; Barry Barnes, Jerry Cai, Paul Joireman, Dan Klepec and to Duane Voy for continuing improvement of our software development environment.

REFERENCES

[1] Wooton, Beckman, Dolciani, *Modern Trigonometry*, Boston: Houghton Mifflin Company, 1966, pp. 259-280.

[2] Chase, B., Barnes, B., and Meisner, K., "Digital Low Level RF Systems for Fermilab Main Ring and Tevatron" in *Proceedings of the 1997 Particle Accelerator Conference*, edited by M.Comyn et al., Institute of Electrical and Electronic Engineers, Inc, New Jersey: 1998, pp. 2326-2328.

[3] Chase, B.,Mason, A., Meisner, K., "Current DSP Applications in Accelerator Instrumentation and RF" in *Proceedings of the 1997 International Conference on Accelerator and Large Experimental Physics Control Systems,* Editors Jijiu Zhao, Axex Daneels, IHEP Chinese Academy of Sciences, Beijing, China, P230

[4] "65 MSPS Digital Receiver Signal Processor" Analog Devices, Norwood, MA

[5] "Designing Filters with the AD6620" Analog Devices, Greensboro, NC

BPM System for the SNS Ring and Transfer Lines[1]

W. C. Dawson, P. Cameron, P. Cerniglia, J. Cupolo, C. Degen, A. DellaPenna, A. Huhn, M. Kesselman, J. Mead, R. Sikora

Brookhaven National Laboratory, Upton, NY 11973, USA

Abstract. The Spallation Neutron Source Ring accumulates about 1060 pulses of 38mA peak current 1GeV H-minus particles from the Linac thru the HEBT line, then delivers this accumulated beam in a single pulse to the mercury target via the RTBT line. Bunching frequency of beam in the HEBT line is 402.5MHz, and about 1MHz in the Ring and RTBT. Position monitor electrodes in HEBT are of the shorted stripline type, with apertures of 12cm except in the dispersive bend, where the aperture is 21cm. Ring and RTBT electrodes are open striplines, with apertures of 21, 26, 30, and 36cm. All pickups are dual plane. The electronics will be PC-based with the Analog/Digital Front End passing data and receiving control and timing thru a custom PCI interface developed by LANL[1]. LabVIEW will be used to direct the acquisition, process the data, and transfer results via Ethernet to the EPICS control system. To handle the dynamic range required with well over 60dB variation in signal size, the Ring and RTBT electronics will employ a fast gain switching technique that will take advantage of the 300ns tail-to-head gap to provide position measurement during the entire accumulation cycle. Beam-based alignment will be utilized as part of the system calibration.

PERTINENT HEBT-Ring- RTBT REQUIREMENTS

1- Intensity   5E10 – 2E14 protons per pulse
2- Pulse length   .3 – 1000 μSec
3- Accuracy   +/- 1mm
4- Resolution   0.15mm
5- Turn by turn data desired

---

[1] Work performed under the auspices of the U.S. Department of Energy

CHALLENGES

1- Single turn (commissioning) vs. multi-turn operation
2- Dynamic Range
3- Resolution and system band-width selection
4- Cable reflections due to out-of-band mismatch
5- Method of position calculation
6- Linearity over a large range of aperture
7- Beam-based alignment
8- Possible use of the PUE's as electron clearing electrodes
9- Monitoring a tune measurement stimulus signal at ~ 50 MHz

GENERAL DISCUSSION

The beam in the HEBT is primarily composed of short microbunches. These give rise to a large RF component at the 402.5MHz bunching frequency and it's harmonics. The character of the pick-up signals in this region of the SNS is the same as that of the Linac section, and it is intended to use identical electronics to that of the Linac in this region. The Ring accumulates the injected bunches from the HEBT, and after a number of turns the charge diffuses and the RF character is quickly lost. With a millisecond of injection and about a 1 MHz revolution frequency, the charge will grow by a factor of about 1000. Early injected turns will have RF structure, while turns that have been accumulated for many turns will have the character of a 645ns pulse with a repetition rate of 945ns. The pulses with little RF structure will provide a base-band signal, while those that have recently been injected will provide an RF character. Therefore, to obtain position information two types of electronics are required. For early turns something similar to the Linac electronics is necessary, while for the later turns a base-band approach is required. Position monitoring in the SNS Ring has been addressed in two previous publications [2,3].

The RTBT has a beam that displays primarily the character of a single bunch of particles for 645ns. Therefore, this is similar to the Ring base-band signal, and electronics based upon this design would be most appropriate. During commissioning and studies, however, one could expect to see a single mini-pulse of microbunches transported through the entire system. To observe this, electronics similar to that used in the Linac are required.

To measure the tune, the plan is to excite the beam with a stimulus signal in the range of 40 to 60 MHz and observe the response with a resonant pickup. Modification of the input filtering and mixing scheme to accommodate this measurement is being considered.

Electron cloud production is also expected to be a problem in the SNS. The use of the 44 Ring BPM pickup electrodes as clearing electrodes is being considered. This will involve applying a potential difference of 2KV to the electrodes to sweep out electrons that originate at the walls during the later portion of the 645ns bunch. This DC signal will be applied through the signal cables and must be isolated from the input to the electronics (Ring BPMs are of the open stripline design).

A beam based alignment technique will be employed in the Ring to establish the location of the BPM electrical center relative to the adjacent quadrupole magnetic center. A processing scheme employing normalization will eliminate gain from the results, and the calculated position will become dependent upon the basic sensitivity of the BPM.

BACKGROUND

This system is presently in the preliminary design stage. The original thought was to utilize a number of Linac BPM electronic systems [1] to get a general handle on beam position for early turns. For position measurement through the accumulation cycle the base-band concept was pursued and a basic design scheme was developed. This included a switched gain concept that follows that used in the beam current monitor system [4]. The basic concept provides different paths for the signal to follow, each with different gain. The paths are summed in a sum stage. Each path is capable of being switched, in or out, by a logic signal to a switched amplifier (OPA3680). In so doing, the gain can be switched within 100ns, sufficient for switching during the beam "gap" time in the Ring.

To permit a simpler and more general solution to commissioning the Ring and in the interest of project-wide commonality, an approach that utilizes the electronics of the Linac is under consideration, with some modifications to permit it's use with base-band signals. This basic concept uses the RF sensitive electronics for early turns and during the commissioning process, while allowing for switching the mixer off, and providing base-band filtering only, to permit the base-band signal to propagate through the system to the ADC. Processing will continue in the normal manner, and the data will be analyzed for position. The Ring and RTBT do not require the phase information, and the I/Q demodulation present in the Linac electronics need not be used. In addition, it is possible that some of the calibration-switching circuitry may be replaced by local oscillator switches and front end filters to conserve circuit board real estate. There will be some calibration switching necessary to characterize each gain path of each channel.

A digital front end (DFE) board and PCI interface board similar to those used in the linac design will be used to digitize the signals, and allow transfer of information to and from a PC. LabVIEW™ will perform control, signal processing and analysis as well as final communication, through Ethernet, with the control system.

DETAILS

The basic design centers upon the Bergoz® BPM Analog Front End (AFE), designed for the Linac. A simplified block diagram is shown in Figure 1. This diagram shows an example of a 402.5MHz AFE. The pickup signals are passed through a switching matrix that permits calibration information to be placed on the inputs to the electronics or the BPM pickup elements. These signals are band pass filtered to allow only the 402.5MHz information to pass. They are amplified as necessary, and mixed with a local oscillator of 352.5MHz. This yields a difference frequency of 50MHz, which is selected by another band pass filter. The result is digitized by a 40MHz digitizer. This digitizer brings the resulting signal to base-band, and the information of interest is a 5 MHz band after I/Q demodulation of the 10MHz down-converted signal.

The modified design would eliminate some of the calibration switching, since the Ring will be calibrated by beam-based alignment. Some switching will be required to establish filtering and channel gain. The initial band-pass filtering will be switch selected to be either at RF or a 5MHz low pass. For base-band operation the mixer will be biased to avoid mixing. The 50MHz filter already designed for this application can be removed for the base-band application.

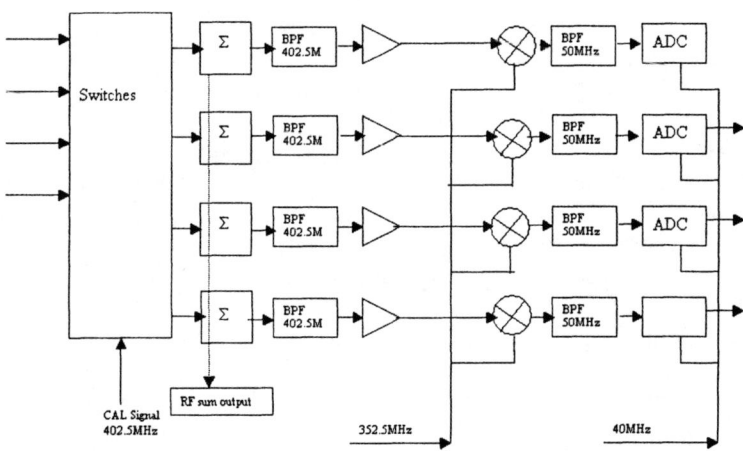

FIGURE 1. Block diagram of basic Linac BPM Electronics

Since I/Q demodulation is not required (not interested in phase in the Ring and RTBT), the available ADC bandwidth is 20MHz when clocked at 40MHz. The required bandwidth is a trade-off between settling time in less than 300ns, noise considerations, signal sensitivity, and signal level at high current. A bandwidth of 5MHz is the preliminary selection to meet these considerations.

In addition to the dual operating modes mentioned above, it would also be desirable to examine the spectral lines of the tune system. This could be accommodated with another modification, now under investigation, that allowed a spectrum of 40MHz to 60MHz to be processed. If this was delivered to the ADC, and sampled at 40MHz, the spectrum would be brought back to base-band and occupy a 20MHz bandwidth. This could be analyzed by an FFT algorithm to obtain the spectrum.

Dynamic Range and Bandwidth

One of the design challenges is to deal with a very large dynamic range in the Ring and RTBT. The signal is expected to change by a factor of about 1000 due to the storage of 1060 turns. In addition the requirement of 0.15mm resolution will necessitate an additional 1000 for a total of 6 orders of magnitude. It would be nice to adjust gain in such a way that we lose no turn information. Input signal voltages could be quite large directly out of the BPM (very wide-band, 20mV to 200V peak). When filtered to 5MHz, signals will range from 1.8mV peak to 10 Volts peak, for 15ma and 100 amp beams with no allowances for terminations and matching circuits. (When filtered to 10MHz signals will range from 3.2mV to 15.3V.) An accuracy of 1mm and a resolution of better than 0.15mm is the present requirement. To achieve this a signal to noise ratio of 48dB is required [5]. This is 3.2nV/rtHz for a 5MHz bandwidth. A 50 Ohm resistor (0.91nV/rtHz) and a typical 2.4nV/rtHz amplifier will yield 2.57nV/rtHz, allowing for little additional resistor noise and signal attenuation due to cabling and "real" filters. A single bunch with RF structure will provide near a 100dB signal to noise ratio. Therefore, for the single turn study case, using the RF detection scheme of the Linac is suggested. A single full current injected turn will yield a signal of about 4.5mV (38mA). This is 8 db more signal than the study current, providing the much needed additional signal to meet the resolution requirement. If the mixer switching speed permits, additional noise margin can be achieved by selecting baseband operation at turns 2 or 3. The system bandwidth clearly affects the noise and in-turn the resolution. The wider bandwidth improves signal almost proportionally, due to the high harmonic content in the signal, while increasing noise by the square root. However, the higher signals need to be dealt with by protecting the amplifiers carefully at the other end of the signal envelope. A compromise to 5MHz was established.

Cable Reflections

A good match at the input is required to assure minimal reflection back to the BPM PUE. In addition to the undesired beam kick which would result, the striplines are unterminated and will reflect signal back yet again. This could interfere with data from the next revolution for cables 100m to 200m long. Three methods have been investigated to match the cable over a wide band; a diplexer, an attenuator, or a combination of both. A simple two-pole diplexer can be constructed using a resistively terminated inductor and capacitor in a low-pass configuration, in parallel with a resistively terminated capacitor and inductor in a high-pass configuration. If the L/C ratio is 2 times the square of the load impedance, such a configuration will have constant input impedance. Difficulties arise using real components, and a better match may be achievable by using a simple resistive attenuator at the expense of signal. If enough signal is available one could use both.

Signal Processing

The standard difference over sum or log ratio method of processing the data will provide information weighted near the beam edges for signals sampled near their peak. To obtain a position more indicative of the average position over the entire mini-pulse a mean square approach is under investigation. With the beam near the center, signals on opposite elements are proportional to the current (I) times (1+ar) for one element, and (1-ar) for the opposite element. The constant a is related to the BPM dimensions, and r is the radial beam position along the axis of the elements. Therefore, the difference over the sum yields a sensitivity of approximately $2ar/2 \cong ar$ for $ar<<1$. Processing the data as mean squared power instead of a peak signal measurement can double the sensitivity of the BPM. That is; $[(1+ar)^2 - (1-ar)^2]/[(1+ar)^2 + (1-ar)^2] \cong 4ar/2 = 2ar$ for $ar<<1$.

To a first order approximation, gain-error affects only a DC error. When employing beam-based alignment the DC term is eliminated, leaving residual second order affect to the linearity with powers of r. In addition, since these signals are averaged over an entire mini-pulse ($\sim 1\mu s$), or a number of mini-pulses, noise term contributions can be reduced. Some of the noise terms are estimates of the noise statistics that do not change with time, and can be characterized for partial reduction, while other noise terms involve the average of the product of noise with signal. This average tends to zero as the averaging time is increased. With increased sensitivity and noise reduction, an improvement in resolution over the standard difference over sum approach is expected. Work continues along these lines to better understand the process.

Linearity

The linearity has been shown to be improved by using log ratio processing. By doing the log ratio analysis of the mean squared power, calculated similarly to that indicated above, the BPM response can be made to be more linear over a larger range of the pipe radius. This calculation need be made only once per position analysis (after calculating a right and left mean square with noise subtraction).

Beam-based Alignment

The location of the HEBT, Ring, and RTBT BPM PUEs relative to the magnetic centers of the adjacent quadrupoles will be determined by beam-based alignment[6,7]. Trim windings with 30A current capacity are available on all quadrupoles, resulting in a 2% modulation of field strength. These winding can be driven with a frequency of about 10Hz. Beam which passes off-center through a modulated quadrupole will experience a transverse kick. This position oscillation can be detected by any BPM of the proper phase relative to the kick. One possibility is to use the acquisition system of the tune meter [8] to detect this motion. If the beam is excited by the ~ 50MHz tune kicker simultaneous with the quadrupole modulation, this modulation will appear as ~10Hz sidebands on the tune signal. This approach benefits both from the sensitivity of the resonant pickups used for the tune system and the signal enhancement of the carrier which results from driving the beam on resonance with the tune system. During SNS commissioning dedicated time will be required to accomplish the initial measurement of the Quadrupole-BPM offset. During normal operations beam-based alignment can also be accomplished continuously and parasitically by taking data over many accumulation cycles if sufficient care is given to maintaining phase coherence between cycles. It is estimated that the precision of the determined offsets should be about 100 microns.

SUMMARY

A description of the SNS Ring BPM electronics, presently in the preliminary design phase, has been presented. The proposed system incorporates a single versatile electronics scheme to accommodate both signals with RF structure and signals with primarily base-band structure. This will permit sensitivity for single turn studies as well as accumulated stored beam operation. A protected switched gain path scheme, similar to the SNS BCM system will provide gain changes to handle the large dynamic range. In addition, the electronics will be made flexible to permit other uses for the BPM signals (tune measurement diagnostic). To make the position calculation more indicative of the average beam position over an entire mini-pulse a mean square approach to data processing is under investigation. The system will be aligned using beam-based alignment. The Ring BPMs are of the open stripline design and may be used as electron clearing electrodes in the future.

ACKNOWLEDGMENTS

The authors would like to acknowledge Tom Shea of ORNL for his assistance with the basic concepts.

REFERENCES

1. J. Power et al, "Beam Position Monitors for the SNS Linac", PAC2001, NY.
2. J. Beebe-Wang et-al, "Simulations of SNS Accumulator Ring Beam Position Monitor Signals", BNL/SNS Tech Note 38, November 1997.
3. T.J. Shea et-al, "Design Study of Position Monitoring in the SNS Ring", BNL SNS Tech Note (unpublished).
4. M. Kesselman et-al, "Spallation Neutron Source Beam Current Monitor Electronics", these proceedings.
5. R. Shafer, "Beam Position Monitoring"; AIP Conference Proceedings 212, p 46 (1989).
6. B. Dehning et-al, "Dynamic Beam-Based Calibration of Beam Position Monitors", CERN SL-98-038 BI.
7. P. Tenenbaum and T. Raubenheimer, "Resolution and systematic limitations in beam-based alignment", Phys Rev Special Topics – Accelerators and Beams, V3, 052801 (2000).
8. P. Cameron et al, "Tune Measurement in the SNS Ring", these proceedings.

The Log-Ratio Beam Position Monitor

A.Kalinin

Bergoz Instrumentation, Espace Allondon Ouest, Saint Genis Pouilly 01630, France

Abstract. Bergoz Instrumentation has designed a new Log-Ratio Beam Position Monitor (LRBPM). It can be used to measure position of a single short bunch, a bunch train, successive and repetitive (circulating) bunches/trains. The monitor has four parallel channels with band pass filters and logarithmic demodulating high dynamic range amplifiers. The amplifiers detect the envelopes of the RF-bursts brought about by the pickup signals in the filters. Log-ratio processing and conversion of the pickup axes to X,Y is done using a broadband analog technique. The monitor can be used either in a continuous mode, or Sample&Hold or Track&Hold modes. In the last two modes, the LRBPM can be triggered by the beam signal itself. The accuracy limits coming from an inherent demodulator noise, logarithmic nonlinearity, speed of response, etc. of the amplifiers (Analog Devices) are discussed. An accessory developed to determine the LRBPM center offset and resolution with the present beam signals is described.

INTRODUCTION

Logarithmic-ratio signal-processing technique [1] is presently in common use for beam position monitoring. It is based on an integrated circuit family of demodulating logarithmic amplifiers, manufactured by Analog Devices, Inc. The family is intended mainly for telecommunication and some characteristics of the amplifiers, sufficient for communication tasks restrict an ultimate accuracy in beam monitoring. This manifests itself mainly as a beam intensity dependence of monitor's readings.

The advantages and disadvantages of the amplifiers available outline the beam diagnostic tasks where the logarithmic-ratio technique is now fruitful. It is mostly beam trajectory measurements in linacs and transfer lines and a similar task of first turn trajectory measurements in circular machines. Besides, coherent betatron and synchrotron oscillations of the circulating beam can be measured and the relevant tasks of betatron dynamics can be solved. Interlock systems for dumping a beam should be mentioned as well. The monitors can perform well with various beam patterns as a single short bunch, a single train of bunches, successive bunches/trains and repetitive bunch/trains in a circular machine. Some of the amplifiers offer a possibility to realise a fast and high dynamic range beam trigger which can be used in a LRBPM to provide it with self-triggering which is especially useful in single pass measurement.

A new Logarithmic-Ratio Beam Position Monitor developed and manufactured [2] at Bergoz Instrumentation and described below, answers to the requirements of the beam diagnostic tasks mentioned above. A number of the LRBPMs have been used on Linacs, Transfer Lines and Storage Rings, mostly for single bunch position measurements and betatron oscillation measurements. The LRBPM is made as an independent card with analog outputs. A beam trigger is built in to sample and hold the X,Y signals and to trigger external ADCs. To eliminate a center offset of the LRBPM, an independent device is developed. For a beam as successive/repetitive bunches/trains, the center offset can be measured using the beam signal, and then subtracted from the readings. A noise resolution at the present beam intensity as a dispersion of the center offset can also be measured.

For a single bunch measurement with resolution $\pm 3 \cdot 10^{-3}$ of pickup's radius, the LRBPM's dynamic range extends 40dB starting from bunch charge 0.1nC.

CONDITIONING OF PICKUP SIGNALS

In a LRBPM, a pickup signal with a short time span is to be converted either to some signal of a band which is within the logarithmic amplifier's base band, or to a RF-signal with an envelope which would have same band. This conversion can be done using correspondingly a low-pass filter or a band-pass one. It is a well-known method.

The band-pass conversion was preferred for the LRBPM to have it applicable to various beam patterns, from a single bunch to a continuous train of bunches. Such a LRBPM can be used, for instance, on circular machines where the beam pattern may be varied just in such limits.

To detect a continuous train, the band-pass filter center frequency is to be chosen equal or multiple of the bunch repetition rate. Reaction of the filter on a single pulse is a RF-burst. Filter's half-band is to be taken within the amplifier's base band.

While design of the LRBPM whose filter center frequency is specified in range (360-500)MHz, different approaches to arrangement of its input network were considered.

The band-pass conversion can be done immediately on the pickup output, using a filter which would include the pickup electrode as one of its elements. In this scheme, a pickup electrode is loaded on some complex impedance.

To provide loading the pickup electrodes on a resistive impedance in the bandwidth of the beam signal, some passive circuit can be used inserted between the electrode and the narrow-band filter which forms the envelope. A circuit as a buffer broad-band filter with bandwidth around 1/10 of the center frequency is developed for the LRBPM. It has 50Ω input impedance, the SWR is less than 1.6 in range 100MHz-3GHz. In addition to purpose to load the pickup electrode, the buffer attenuates high frequency (>1GHz) components more than 40dB. Insertion loss measured on the center frequency is lower than 3dB.

With use of a buffer filter, the basic narrow-band filter can be replaced to amplifier's input. That is convenient from practical point of view. Note that such buffer filters can also be used for increasing dynamic range of narrow-band BPMs used for orbit measurements in circular machines with a high intensity single bunch. Some BPMs may have on the inputs active circuits as switches, amplifiers, etc. Their performance may degrade if a pulse signal with amplitude which may become extremely high with increasing beam intensity is directly applied. The buffer converts the pulse to a RF-burst. Its amplitude is substantially lower that the pulse amplitude. Other properties of the buffer filter may be useful for these BPMs as well.

The same functions of loading the pickup electrode on a broad-band resistive impedance has the device designed for elimination of the center offset, mentioned in Introduction. In the range 30MHz-2.8GHz, the device has input/output impedance 50Ω with SWR less than 1.4 and 1.3 correspondingly. Designed for subnanosecond bunches, it like the buffer filter, converts a pickup pulse to a burst which is now a single sine wave of period 2ns. The central frequency is 500MHz, insertion loss 10dB, bandwidth is (200-800)MHz, attenuation in range (1-3)GHz is more than 15dB.

The LRBPM has the narrow-band filters on the inputs of the logarithmic amplifiers. The buffer filters and the devices are used at user's option.

Designing the input network, it is necessary to take into account signal reflection in the cables. If multiple reflections of the signal occur, a signal on the amplifier's input is a superposition of the bursts caused by the primary signal and by the reflected signals. In two channels, the secondary bursts have different phase relations to the primary bursts, if the cable lengths are not equal and/or the filter center frequencies are different. This manifests itself as a difference of envelope shapes on the channel outputs, which may result in some extra (and uncertain) monitor's center offset. For the case of a continuous train, this effect of reflected signals is characteristic of any BPM if narrow-band conversion has been used. For a single bunch/train, the effect does not manifest itself in a LRBPM if the first reflected signal comes to the input after having the envelope sampled and held. Note that for any length, the reflected signals can cause false triggering in LRBPM's beam trigger.

LOGARITHMIC PROCESSING

For frequency up to 500MHz, two groups of the amplifiers are available: DC-500MHz and (0.1-2.7)GHz. The former has higher accuracies and dynamic ranges. In this group, the only AD8307 [3] has such a converter of the demodulator's signal to the output voltage, that its speed can be increased by use of elements external to the chip. This feature is important practically because first, increase of the base band yields increase of output signal-to-noise ratio, second, requirements to input filter parameters relax.

The possibility to increase the speed was investigated. A RF-burst 500MHz was used as a test signal. Its envelope close to (1-exp)-shape had rise time around 10ns and reached its apex at 15ns. Converter's speed was varied by varying its equivalent load.

The converter response time versus the load was measured for two settings of RF-burst's amplitude: 0dB and (-50)dB. The setting 0dB corresponded to the upper end of amplifier's input dynamic range. It was seen that demodulator's speed and converter's speed change with burst's amplitude in opposite ways. It was found that for some load, when each response time becomes equal to approximately 30ns, the speed changes counterbalance each to other. The response time 30ns can be considered as an optimal one. The amplifier's base band can be estimated, correspondingly, at 5MHz.

Then, an interesting feature of the amplifier was discovered. The logarithmic amplifier output noise consists of a demodulator-converter noise which does not vary with increase of amplifier's input signal (a noise floor) and an input resistor-amplifier noise which comes to the base band and manifests itself increasing linearly with decreasing the input signal. A demodulator-converter noise on the floor can be easily observed, if some DC-bias is applied to amplifier's input, which blocks amplification of the input noise. It was discovered that for settings in range (10-40)ns, the time of reducing of the output noise to the floor value which would correspond to the burst's amplitude, is considerably longer than the response time set. By other words, the noise is still high when the output envelope is reaching its apex.

The signal and noise transients become close each to other, if the RF-burst envelope rise time is twice as much as the response time 30ns. Thus, the filter half-bandwidth is to be taken at least half as much as amplifier's base band width 5MHz.

In the LRBPM, a compromise solution is taken. The response time is set to an optimal one 30ns, but the envelope rise time is less than the optimal one above. It is that time which can be achieved with use of commercial band-pass filters. A filter available (so called a double-tuned helix filter) has the rise time 10ns. A simple way to achieve some stretch of envelope is to cascade two such filters. The rise time comes to 25ns, and a resulting rise time on the amplifier's output is around 40ns.

LOGARITHMIC-RATIO PROCESSING

An analog logarithmic-ratio processing in the LRBPM is done by taking difference of the output signals of the logarithmic amplifiers. Preliminary, the output signals are amplified. The buffer amplifiers are made with built-in gain and zero offset adjustments for equalising the logarithmic slopes and intercept parameters. For obtaining an accurate difference, a fast OA AD8138 is used.

The difference signal is filtered by a LPF to suppress the logarithmic amplifier's output noise beyond a band required for the measurement task. For instance, the bandwidth is to be set to a maximal one 5MHz for a single bunch measurement. For turn-by-turn measurements of a long train, the cut-off frequency can be reduced and be of the order of the revolution frequency.

Signals of the LRBPM are shown in Figure1. The oscillogram was taken with input signals differing each from other by 3dB and falling approximately in the middle of LRBPM's dynamic range.

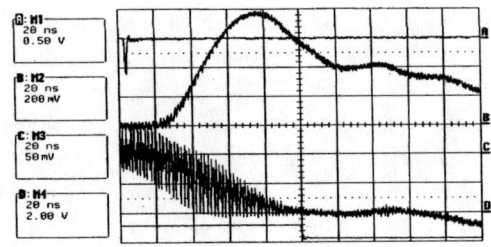

FIGURE 1. Signals of the LRBPM. Trace A shows one of the input signals. It is a 0.5ns-pulse which imitates a single bunch. Trace B shows one of the output signals of two logarithmic amplifiers, and trace C shows the difference signal on the LPF output.

On the picture, one can see that a noise in the difference signal, reducing, reaches some floor value 25ns later then the amplifier's output signal reaches its apex. This moment has happened to be a particular point: in this point, the difference error due to mismatching of transitional characteristics of a pair of logarithmic amplifiers has been minimal as well. It can be clearly seen, when the input signals are equal. Typically just in the vicinity of this point, the transitional zero shift has been passing zero.

Thus, at this moment, the difference signal is to be sampled and then held to be used for further processing. A fast Track&Hold IC AD9101 is used which has Track-to-Hold Settling Time 4ns. In Figure1, trace D shows its CLK signal with a Track-to-Hold transient.

Optionally, the Hold signal which lasts 50ns, can be used for turn-by-turn measurements on small machines with revolution frequency up to 5MHz. For other tasks, is is convenient to have a longer Hold time. This is achieved by use of a Sample&Hold IC AD783 which samples the Hold signal. Its Hold time is set to 100ns for maximal sampling rate 2MHz. For a lower rate, the Hold signal keeps lasting till the next bunch/train/external trigger comes (but not longer than 100ms).

X,Y PROCESSING

By logarithmic-ratio processing, a pair of signals U,V is yielded which represent a position of the beam center of mass with respect to the pickup's own coordinate system. Each axis of this system can be defined (and determined as well) as a population of such positions where the other signal of the pair is equal to zero. The axes intersect in the pickup center. Let the axis of the pair of opposite electrodes A,C/B,D be U-/V-axis correspondingly, and an angle (from U-axis to V-axis) between the tangents in the center be α. Near the center one can use just the U, V-tangents as own pickup axes. Introducing scale coefficients ζ,η as U[mm]=ζU[volt] and V[mm]=ηV[volt], and some output coefficient M[volt/mm], the conversion to X,Y-axes can be written for the LRBPM output signals U,V:

$$X[\text{volt}] = M[\zeta U\cos\theta + \eta V\cos(\alpha+\theta)], \qquad (1)$$
$$Y[\text{volt}] = M[\zeta U\sin\theta + \eta V\sin(\alpha+\theta)].$$

The parameters ζ,η and the angles α and θ can be determined by bench measurements of signals of the electrodes and applying the logarithmic-ratio algorithm. Another way is to determine these parameters from the data obtained on a bench with use of the difference-by-sum algorithm. Assume that the scale coefficients δ[mm] and γ[mm] are available. Using $X=\delta[(A-C)-(B-D)]/(A+B+C+D)$ and $Y=\gamma[(A-C)+(B-D)]/(A+B+C+D)$ which are for the case of a symmetrical about X,Y-axes positions of the electrodes, $2\theta=\pi-\alpha$ ($\theta\neq 0$), and noting that B=D on U-axis and A=C on V-axis, one can obtain the angles as:

$$\alpha = \text{arctg}[2(-\gamma/\delta)/(1-\gamma^2/\delta^2)], \qquad (2.1)$$
$$\theta = \text{arctg}(\gamma/\delta).$$

For a symmetrical pickup, $\zeta=\eta=K$. Again on U-axis, taking A=1+u, C=1−u, u<<1, one can obtain: $X = \delta[1/(1+D/C)]u$, $Y=\gamma[1/(1+D/C)]u$. From other side, U[mm]= Klog(A/C)$\approx$2Ku. Using a segment U, $U^2=X^2+Y^2$, and letting its length tend to zero, one can obtain the scale coefficient K:

$$K = (1/4)\text{sqrt}(\delta^2+\gamma^2). \qquad (2.2)$$

Note, that for some pickup configurations, the logarithmic-ratio algorithm proves to be the least non-linear. [1] With (2.1, 2.2), it can be adopted in BPMs instead of the difference-by-sum algorithm used.

For a symmetrical pickup, (1) yields:

$$X[\text{volt}] = MK\sin(\alpha/2)(U-V), \qquad (3)$$
$$Y[\text{volt}] = MK\cos(\alpha/2)(U+V).$$

Thus, for conversion to X,Y signals, one has to obtain a difference signal $\Delta=U-V$ and a sum signal $\Sigma=U+V$. The coefficients can be realised by attenuation/amplification of the signals. In the LRBPM, for a circular pickup with $\alpha=\pi/2$, $\theta=\pi/2$, the outputs 0.245V are factory-set for displacement by 6dB, i.e. when the ratio of the signals of one pair of opposite electrodes is equal to 2. With such a scaling, if the pickup were linear, the output would reach ±2V for the full radius displacements. This conditional radius is used below as a LRBPM's characteristic measure.

RESOLUTION AND ACCURACY

A residual X,Y zero offset of the LRBPM can be set in the limits $\pm 1 \cdot 10^{-2}$ of pickup's radius (i.e.$\pm$20mV). This can be achieved typically in the range (55-60)dB without selection of pairs of logarithmic amplifiers, by adjustment of the mean slopes and the intercept parameters only using a sine input signal. A zero offset measured with a single pulse, may exceed the limits above. There is no means in the LRBPM to adjust each zero offset in the limits.

The LRBPM's resolution is measured in a full bandwidth 5MHz. The input signal is 0.5ns-pulses with repetition rate around 600Hz. The LRBPM is triggered by the input signal through its built-in trigger. Rms and peak-to-peak values of the Hold signal are measured by a digital oscilloscope for number of the pulses around 150. An oscillogram of the X,Y noise on the floor is given in Figure 2.

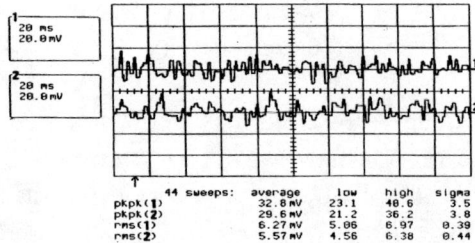

FIGURE 2. An oscillogram of the X,Y noise on the floor.

The rms value is typically less than 7mV, i.e. resolution is $\pm 3 \cdot 10^{-3}$ of pickup's radius. The peak-to-peak value is around 40mV, i.e. the fluctuations may reach $\pm$20mV. That is three times more than the rms value. The floor stretches from the upper end of the LRBPM's dynamic range to approximately (-40)dB, where the rms and peak-to-peak values get increasing due to the logarithmic amplifiers' input noise. Thus, for a single pass measurement, the LRBPM's dynamic range with a fixed resolution is 40dB.

A noise density on the floor is $2.2 \cdot 10^{-6}$ of radius/sqrt(Hz). A rms noise in some abstract bandwidth 100Hz would be $2.2 \cdot 10^{-5}$ of radius. For radius 40mm, the resolution would be 1µm. It is comparable with the resolution of most of the BPMs used for orbit measurements in same bandwidth. This evaluation being loose, however indicates that the inherent demodulator-converter noise has not been extreme and the LRBPM's resolution of single pass measurement can not be much enhanced.

A beam intensity dependence of the LRBPM output signals manifests itself mainly when the beam is displaced. The first effect which causes variation of the output signal is that logarithmic amplifier's response waves in the limits $\pm$3% with a period around 12dB. It is a known effect. [4] One more, a bump effect of beam intensity, manifests itself on the upper end of the dynamic range. A change of the intercept parameter occurs which results in a shift of the logarithmic response for the input signals higher than (−10) dBm. It is a local effect, on the interval 10dB only, but it is quite strong. The bump reaches 25%. In the LRBPM, this local non-linearity is compensated by non-linear dividers located ahead of the logarithmic-ratio processor. Totally, in the range from (-5)dBm to (-55)dBm, the LRBPM's error is around ±3% if the frequency is 500MHz, and comes to ±4% for 360MHz.

For single bunch measurement, a beam intensity dependence manifests itself in other way. No waving effect is seen. In the middle of a 40dB range, a response to beam displacement may be few percent higher or lower than it is for the case of a sine input signal. On the lower end the response gradually lowers by (-15)% in the interval 20dB. The same bump effect is there on the upper end. The gradual effect would be less, if the envelope apex would be sampled. Generally, it is the penalty for the compromise of the envelope rise time against the amplifier's response time, discussed above.

A typical X/Y LRBPM's response to "beam displacement", measured on a bench in a full dynamic range, is shown in Figure 3.

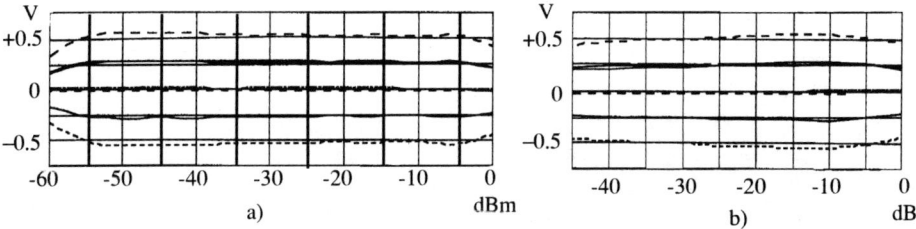

FIGURE 3. The LRBPM's response to beam displacements by 0, ±6.5, ±13dB versus the input signal level. a) The input signal is a 500MHz-sine wave. b) The input signal is a single 0.5ns-pulse.

LRBPM CENTER ACCURACY ENHANCEMENT

To enhance a LRBPM accuracy in the vicinity of the pickup center, especially in single bunch measurement, an optional device can be used mentioned above. The device is based on an idea to measure a BPM center offset caused by its electronics, and its resolution as well, using the same beam signals which are used for the position measurement. The use of real beam signals at any present beam intensity provides a high accuracy of the center offset measurement.

The device intended for subnanosecond bunches, consists of a pair of 50Ω irregular directional couplers which convert the short pulse to a single sine wave, and a switch between their outputs. A pair of opposite electrodes is connected to the inputs. To measure the offset, its output signals, i.e. the BPM input signals, are made equal by the switch short. The position is measured when the switch is open. Then the offset can be subtracted from the BPM readings. To measure both the X,Y offsets, two devices are to be used. To include the BPM cables into the process, the devices should be placed close to the pickup outputs.

When the beam is on the pickup's center, a residual unequalness of the device's output signals is few units by 10^{-3} of pickup's radius. When the beam is displaced, an additional error appears proportional to the displacement. So, for 6dB-displacement, this error reaches $3 \cdot 10^{-3}$ of the radius.

A pin-diode is used as a switch. The switch is controlled by an external signal via the LRBPM and a corresponding pair of its input cables. Note that the device has two optional lossless outputs of the primary electrode signals and can be used as a precise splitter of the signals which have been used in some BPM, to a LRBPM as well.

SUMMARY

Features of a Logarithmic-Ratio Beam Position Monitor developed at Bergoz Instrumentation have been described. Problems and solutions concerning its input network, logarithmic and logarithmic-ratio processing, conversion of the signals to the X,Y-axes are considered. A residual X,Y zero offset, a noise resolution and an error due to beam intensity dependence of the readings are discussed. A beam-based method of enhancement of monitor's center accuracy is proposed.

ACKNOWLEDGMENTS

We would like to thank R.Lubes for his assistance with the design and test measurements, and M.Baffou, P.Kyriacou and A.Murtro for their carefulness in the fabrication. We are grateful to J.A.Hinkson for collaboration and Dr.K.-T.Hsu for the first and supporting beam test.

REFERENCES

1. Shafer, R.E., "Log-Ratio Signal-Processing Technique for Beam Position Monitors" in *Accelerator Instrumentation - 1992*, edited by J.A.Hinkson and G.Stover, AIP Conference Proc. 281, New York, 1993, pp.120-128.
2. "Log-ratio Beam Position Monitor", datasheet and User's manual, Bergoz Instrumentation, http://www.bergoz.com.
3. "Low Cost DC-500MHz, 92dB Logarithmic Amplifier AD8307", datasheet rev.A, Analog Devices, Inc., 1999, http://www.analog.com.
4. Aiello, G.R., Mills, M.R., "Log-Ratio Technique for Beam Position Monitor Systems" in *Accelerator Instrumentation - 1992*, edited by J.A.Hinkson and G.Stover, AIP Conference Proc. 281, New York, 1993, pp.301-310.

Design and Upgrade of a Compact Imaging System for the APS Linac Bunch Compressor

B. Yang, E. Rotela, S. Kim, R. Lill, and S. Sharma

Argonne National Laboratory, 9700 South Cass Avenue, Argonne, IL 60439

Abstract. We present the design, performance, and recent upgrade of a high-resolution, high-charge-sensitivity imaging camera and beam position monitor (BPM) system for the APS linac beam profile measurement. Visible light is generated from the incoming electron beam using standard YAG or optical transition radiation (OTR) converter screens. Two CCD cameras share the light through a beam splitter, each with its own imaging optics. Normally, one camera is configured with high magnification and the other with large field of view. In a different lens configuration, one of the cameras focuses at the far field, allowing the measurement of beam divergence using an OTR screen, while the other camera simultaneous measures the beam size. A four-position actuator was installed recently to provide the option of two screens, a wakefield shield, and an *in situ* calibration target. A compact S-band beam position monitor electrode was designed to mount directly on the flag. The BPM rf circuit was fabricated from a machinable ceramic (MACOR) cylinder substrate, and the copper electrodes were deposited on the substrate. The new design and precision fabrication process make it viable to explore more complex microstrip components printed on the substrate and higher frequency applications. The proximity of the BPM and the camera (< 5 cm) will provide a precise calibration platform to study shot-to-shot jitter, long-term stability of both systems, and the dependence of BPM signal on beam properties (size, charge distribution) due to nonlinearity.

INTRODUCTION

A chicane bunch compressor was designed and implemented at the Advanced Photon Source (APS) in 2000 [1] to increase the peak current of the bunch and improve the performance of the low-energy undulator test line (LEUTL) free-electron laser (FEL). It is expected to operate at ~200 MeV with normalized emittance in the range of 1 π to 4 π mm·mrad. Coherent synchrotron radiation (CSR) effects are expected to be significant at these emittance levels. Their study calls for accurate emittance measurement at the level of several percent or better.

TABLE 1. Electron Beam Parameters at the Three Screen Section

ELECTRON ENERGY (MEV)	200 ($\gamma = 400$)	
Single bunch charge (nC)	0.2 -- 1.0 x	
Normalized emittance (π mm·mrad)	4.0	1.0
Emittance ε (π mm·mrad)	0.010	0.0025
Beta function at beam waist β (m)	1.00	
Rms beam size, $\sqrt{\varepsilon\beta}$ (µm)	100	50
Rms beam size, $\sqrt{\varepsilon/\beta}$ (µrad)	100	50

A compact, modular imaging system [2] was designed and implemented to support studies of the low-emittance beams, with the high resolution, reliability, reproducibility, and accuracy needed. In this paper, we report several implemented and planned upgrades to the system.

A new screen mount was designed and installed. It includes a wakefield shield, a YAG screen mount, an optical transition radiation (OTR) screen mount, and a calibration target. A new set of optics was designed to focus in the far field, or image the angular divergence of the electron beam when the OTR screen is employed.

A compact electronic beam position monitor (BPM) was designed to mount directly on the flag. The BPM design uses a new fabrication process, which is expected to make it more viable to explore complex electrode configurations and for higher frequency applications.

FOUR-POSITION SCREEN MOUNT

Due to high-energy spread used for chirping the beam, the rms beam size in the horizontal direction can reach several millimeters in the midsection of the chicane. This is comparable to the aperture of the vacuum vessel of 25 mm. A smooth transition of the beam pipe is desired to reduce wakefield effects to the electron bunch. Figure 1 shows the design of the wakefield shield. Figure 2 shows an ACAD rendering of the screen mount assembly.

The wakefield shield / transition is mounted at position no. 1 on a four-position vacuum feedthrough, powered by a pair of pneumatic actuators connected back-to-back. The stroke of the short actuator is 38.4 mm and that of the long actuator is 76.4 mm. Four screen positions (38.4 mm spacing) can be reached by extending or contracting these two pneumatic actuators.

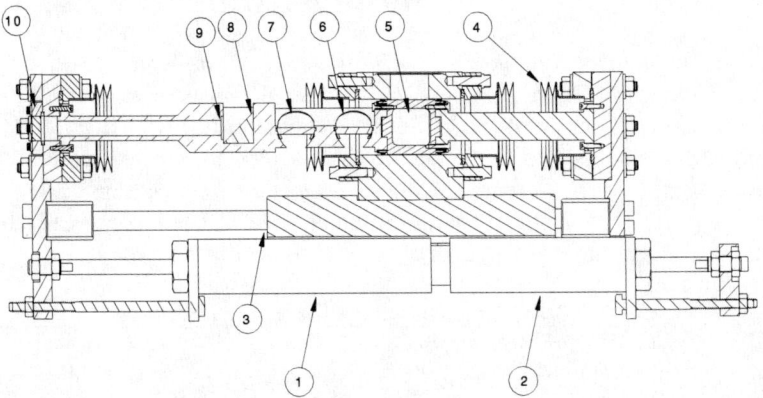

FIGURE 1. Four position screen mount assembly: (1) Long-stroke actuator, (2) short-stroke actuator, (3) mounting block and linear bearing housing, (4) vacuum bellow, (5) wakefield shield / transition, (6) OTR screen mount, (7) YAG screen mount, (8) mirror, (9) calibration target, and (10) viewport for target illumination.

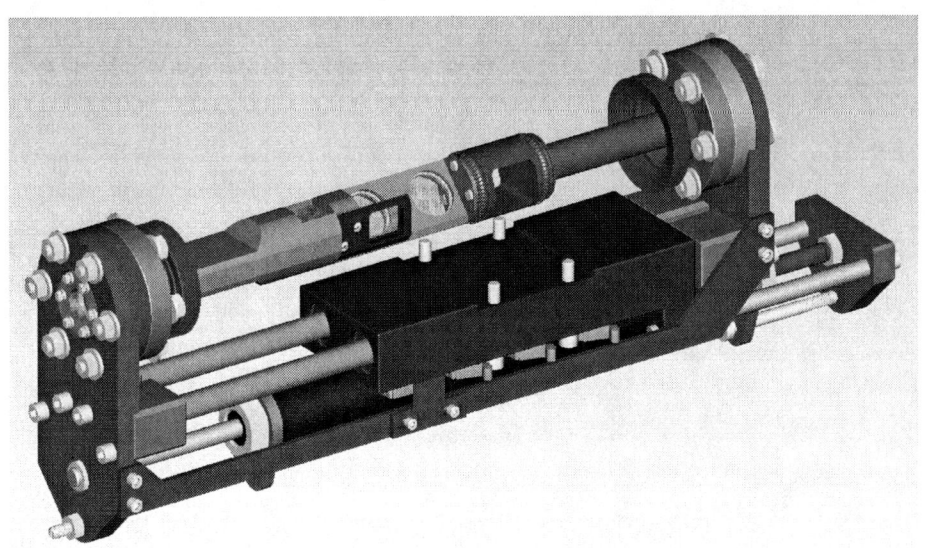

FIGURE 2. ACAD 3-D rendering of the four-position screen mount assembly.

At position no. 2, a 45-degree mirror is mounted as an OTR screen. It is used for high-intensity (charge density) electron beams. At position no. 3, a YAG scintillator screen (20 mm × 15 mm × 0.1 mm) is mounted with a mirror. It is used for low-intensity electron beams.

At position no. 4, an aluminum-coated prism is used as a first surface mirror, in conjunction with a calibration target. The target is designed to have a dark metallic coating with an etched dot matrix at 0.25 mm spacing. The illumination light is fed from outside of the vacuum through a viewport. Images of these bright dots in the dark background are convenient for the video digitizer to pick out and perform profile analysis. The center coordinates of the dots are used to calibrate pixel sizes, while the rms image size of these 10-μm-diameter dots are used to obtain optical resolution of the camera. Since the optical distance from the calibration target is the same as from the YAG screen, the target is also used as a focusing aid.

OPTICS FOR BEAM DIVERGENCE MEASUREMENT

By design, the camera module has two cameras sharing the light with a beam splitter. Normally, each camera has its own imaging optics, one is used for high-resolution imaging but covers only a part of the screen, while another is used for low-resolution imaging and covers the full view of the screens. By reconfiguring the optics, however, one camera could be equipped with optics for the far field, i.e., with object space at infinity.

Measurement Using OTR Angular Distribution

To measure relatively large beam divergence ($\sigma_{x'} \sim 0.1/\gamma$), we can use the features in the angular distribution of the optical transition radiation. A linear polarizer installed in the turn table can be used to select the direction of polarization to be along the x or y-axis. When the x-polarization is selected, the OTR angular distribution is,

$$I_x(\theta_x, \theta_y) = \frac{2A}{\pi} \frac{\theta_x^2}{\left(\gamma^{-2} + \theta_x^2 + \theta_y^2\right)^2}, \qquad (1)$$

where (θ_x, θ_y) is the angle from the specular direction, and A is a constant. For an electron beam with Gaussian divergence distribution, with rms divergence of ($\sigma_{x'}$, $\sigma_{y'}$),

$$g(\theta_x, \theta_y) = \frac{1}{2\pi\sigma_{x'}\sigma_{y'}} e^{-\frac{\theta_x^2}{2\sigma_{x'}^2} - \frac{\theta_y^2}{2\sigma_{y'}^2}}, \qquad (2)$$

the observed spatial distribution is

$$\int_{-\pi}^{\pi}\int_{-\pi}^{\pi} g(\theta_x - \theta_x', \theta_y - \theta_y') I_x(\theta_x', \theta_y') d\theta_x' d\theta_y'. \qquad (3)$$

If we integrate over the y-coordinates to improve statistics, the resultant profile is

$$F_x(\theta_x) = \int_{-\pi}^{\pi}\int_{-\pi}^{\pi}\int_{-\pi}^{\pi} g(\theta_x - \theta_x', \theta_y - \theta_y') I_x(\theta_x', \theta_y') d\theta_x' d\theta_y' d\theta_y. \qquad (4)$$

For $\gamma \gg 1$, we could change the limits of the integration to infinity. Inserting Eqs. (1) - (3) into (4), one obtains the following expression,

$$F_x(p) = \frac{A}{\sqrt{2\pi}\sigma_p} \int_{-\infty}^{\infty} \frac{q^2}{(1+q^2)^{\frac{3}{2}}} e^{-\frac{(q-p)^2}{2\sigma_p^2}} dq. \qquad (5)$$

where

$$p = \gamma\theta_x, \text{ and } \sigma_p = \gamma\sigma_{x'}. \qquad (6)$$

For very small beam divergence, Eq. (5) gives

$$F_x(0) \square A\sigma_p^2, (\sigma_p \ll 1). \qquad (7)$$

The entire profile can be calculated with a given beam divergence $\sigma_{x'}$ (Fig. 3). However, for a quick estimate of the beam divergence, the ratio of the center minimum and the maximum of the angular distribution can be used,

$$\eta \equiv \frac{F_{min}}{F_{max}} = \frac{F(0)}{F_{max}}. \qquad (8)$$

The ratio for Gaussian beams with $\sigma_p \ll 1$ can be easily modeled. The results can be summarized in the following relations,

$$\gamma\sigma_x \equiv \sigma_p \cong \sqrt{0.387\eta + 0.467\eta^2 + 1.135\eta^3}, \quad (\eta \leq 0.7). \qquad (9)$$

This fitted expression has an accuracy of 1% or better for $\sigma_x < 0.8/\gamma$ (Fig. 4).

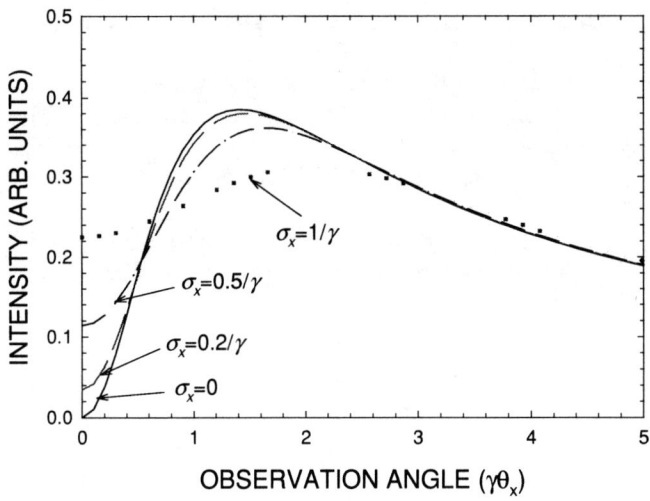

FIGURE 3. Minimum/maximum ratio of the 1-D OTR profile for a Gaussian electron beam.

For our application at the APS bunch compressor, the dynamic range of the system is limited by the video digitizer, which has an 8-bit resolution at zero noise. If we set the noise level at 2 bit, the usable dynamic range is about 6 bit. Best measurements can obtain a ratio of $\eta > 2^{-6}$, hence the lowest measurable beam divergence is given by

$$\sigma_x \geq \frac{\sqrt{2^{-6}}}{\gamma} = \frac{1}{8\gamma}, \tag{10}$$

or 0.3 mrad for $\gamma = 400$. Comparison with Table 1 shows that higher resolution will be needed for the measurements at the APS bunch compressor.

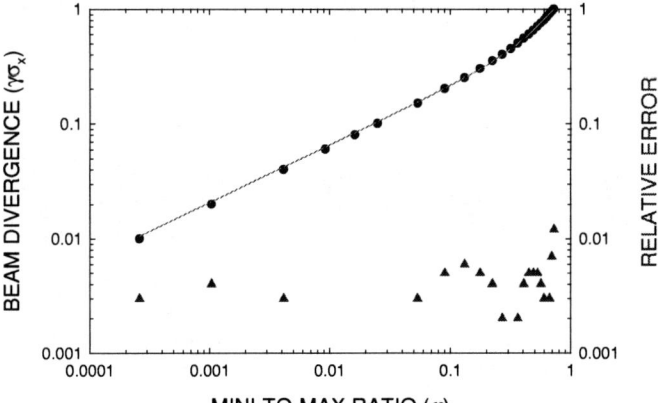

FIGURE 4. The minimum/maximum ratio of the 1-D OTR profile for a Gaussian electron beam, circles: calculated ratio for a given beam divergence, curve: fit by Eq. (9), and triangles: the relative error in beam divergence between the exact calculation and the fit.

Measurement Using OTR Interferometer

To improve the resolution of the beam divergence measurements ($\sigma_{x'} < 0.1/\gamma$), we could use the OTR interferometer [3]. One thin foil is to be inserted at the first station of the three-screen section, and an OTR screen/mirror is inserted at the mid-station, 1.0 m downstream from the foil. The far-field camera located at the second station records the interferogram.

To illustrate the resolution of the interferometer at this geometry, we calculated the OTR intensity distribution for circular electron beam $\sigma_{x'} = \sigma_{y'} = \sigma$. The result is given by a simple integral,

$$K(\theta) = 4A \int_0^\pi \int_0^{2\pi} \frac{q^3}{(1+q^2)^2} \cos^2 \frac{\pi(1+q^2)}{p_\lambda^2} G(\gamma\theta, q) dq, \quad (11)$$

where $\sigma_p = \gamma\sigma$, $p_\lambda = \gamma\sqrt{2\lambda/L}$ and

$$G(p,q) = \frac{1}{\pi\sigma_p^2} \int_0^\pi \exp\left\{-\frac{1}{2\sigma_p^2}(p^2 + q^2 - 2pq\cos\phi)\right\} d\phi. \quad (12)$$

Assuming we have 1 meter foil-mirror spacing, and an observation wavelength of 0.63 μm, we show several sample calculations in Fig. 5, including the beam conditions listed in Table 1. From the simulated interferogram, we conclude that the geometry is suitable for the measurement of beams with divergence less than $0.1/\gamma$.

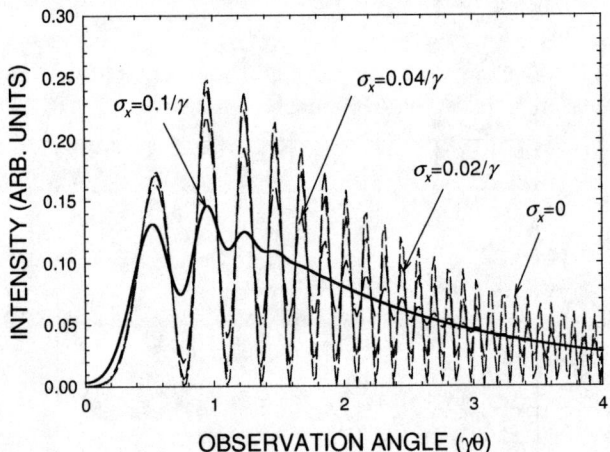

FIGURE 5. The OTR interferometer light intensity as a function of observation angle (in far field). Beam with lower emittance ($\sigma_x=0$) generate intensity oscillations that extends to higher angles, while beam with higher emittance ($\sigma_x=0.1/\gamma$) generates oscillations that damps in a short range.

COMPACT ELECTRONIC BEAM POSITION MONITOR

The shorted $\lambda/4$ wave S-band BPM electrode shown in Fig. 6 was designed as a test vehicle for evaluating the performance of a new compact BPM electrode. The goal was to reduce the size and assembly complexity for the BPM, so a flag and BPM combination can be used in a limited length of beam line. This concept of combining diagnostics also has the advantage of using a cross calibration scheme that can be implemented to verify the performance of each of the instruments. The proximity of the BPM and the camera (< 5 cm) will give us a good opportunity to study, to micron precision, the dependence of BPM signal on beam properties (size, charge distribution) due to nonlinearity. It also provides a precise calibration platform to study shot-to-shot jitter, and long-term stability of both systems.

Several design features can be seen from Fig. 6: (1) The rf microstrip circuit and vacuum feedthrough connectors are replaceable. This is achieved by welding the feedthrough connectors to a standard removable vacuum flange. The center conductors make contact to the stripline via a beryllium copper spring contact. This design simplifies assembly of the BPM and enables the replacement of a connector in the event of damage. (2) The BPM electrodes are fabricated on a machinable (MACOR) ceramic cylinder substrate with a dielectric constant of 4.88. The rf circuit is etched on the copper/ceramic substrate with a process similar to printed circuit-board technology. The ceramic cylinder is captivated in the BPM housing by a threaded transition sleeve. The new design and precision fabrication process make it viable to explore more complex microstrip components printed on the substrate and higher frequency applications. The fabrication costs are expected to be much lower than the 34 mm BPM currently in use in the APS LINAC.

The connectors followed by a bore in the metal housing form 50-ohm transmission lines to each of the four blades. The wall current is intercepted by the 26.24-mm long electrode that forms 50 ohms impedance with the vacuum chamber. The gap that is formed between the vacuum chamber and the stripline is 3.48 mm. The total length of the test BPM stripline assembly is 54.61 mm and the inner electrode radius 35 mm, which is the same as that for the linac vacuum chamber. The downstream end of the transmission line is electrically shorted, which also provides the mechanical strength and rigidity to support the striplines.

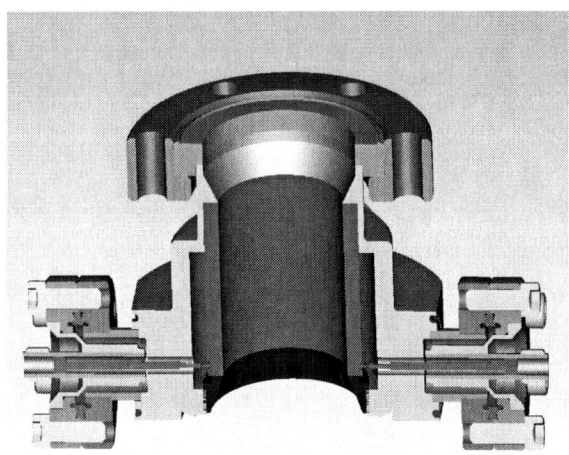

FIGURE 6. Test BPM assembly.

The preliminary electromagnetic modeling indicates that the ceramic actually improve pincushion distortion performance as compared in Fig. 7. The thickness of the ceramic has not been optimized and is presently set to 3.4 mm for mechanical reasons. The overall performance is expected to be similar to the standard S-band [3], which is presently used in all APS linac applications.

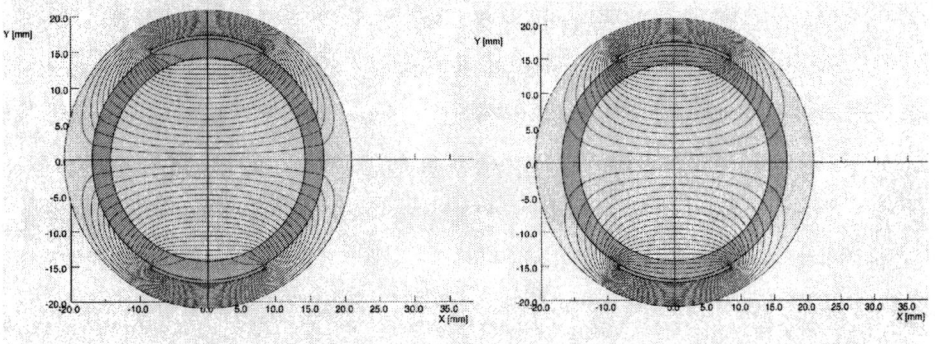

FIGURE 7. Vertical plane equipotential lines with ceramic (left) and without ceramic (right), showing larger linear region in the case of ceramic housing (field lines being parallel and equal spacing).

ACKNOWLEDGMENTS

We wish to thank G. Decker, S. Milton, M. Borland, and O. Singh for helpful discussions, encouragement, and support.

REFERENCES

1. Borland, M et al., A Highly Flexible Bunch Compressor for the APS LEUTL FEL, *Proc. LINAC 2000*.
2. Yang, B. X., et al., Design and Performance of a Compact Imaging System for the APS Linac Bunch Compressor, *Proc. Particle Accelerator Conference 2001*, 2335 (2002).
3. R. B. Fiorito and D. W. Rule, "Optical Transition Radiation Beam Emittance Diagnostics", in AIP Conference Proceedings No. 319, R. Shafer, editor (1994).
4. A. Gorski, R. Lill, "Construction and Measurement Techniques for the APS LEUTL Project RF Beam Position Monitors," Proceedings of the 1999 Particle Accelerator Conference, New York, pp. 1411-1413 (1999).

New Beam Position Monitor System Design for the APS Injector*

R. Lill, O. Singh, N. Arnold

Advanced Photon Source, Argonne National Laboratory
9700 South Cass Avenue, Argonne, Illinois 60439 USA

Abstract. Demands on the APS injector have evolved over the last few years to the point that an upgrade to the existing beam position monitor (BPM) electronics is required. The injector is presently being used as a source for both the low-energy undulator test line (LEUTL) project and the top-up mode of operation. These new requirements and the fact that many new rf receiver components are available at reasonable cost make this upgrade very desirable at this time. The receiver topology selected is a logarithmic processor, which is designed around the Analog Devices AD8313 log amplifier demodulation chip. This receiver will become the universal replacement for all injector applications measuring positions signals from 352 to 2856 MHz with minimum changes in hardware and without the use of a downconverter. The receiver design features integrated front-end gain and built-in self test. The data acquisition being considered at this time is a 100-MHz, 12-bit transient recorder digitizer. The latest experimental and commissioning data and results will be presented.

INTRODUCTION

The Advanced Photon Source injector has taken on a more demanding role over the last few years. The new demands involve operating in top-up mode [1]. In this mode of operation the injector refills the storage ring every two minutes in order to maintain a current of 102 ± 1.0 mA. In order to inject efficiently into the targeted storage ring bucket, the injector beam position becomes critical. The new beam position monitor (BPM) will provide improved dynamic range, resolution, and reduction of other systematic errors to ensure reliable injector operation.

The low-energy undulator test line (LEUTL) project [2] has also created a new set of demands on the injector. The Linac and several transport lines are shared by both the injection process and LEUTL, requiring new and improved diagnostic capability. These new requirements and the fact that many new rf receiver components are available at reasonable cost make this upgrade very desirable at this time.

---

*Work supported by U.S. Department of Energy, Office of Basic Energy Sciences under Contract No. W-31-109-ENG-38.

SYSTEM DESIGN

The system design philosophy was to simplify the topology so the same basic design can be used as the universal replacement for all injector position monitoring applications. The other general design constraint was to make it as reliable and maintainable as possible. The design choice was a log-ratio system with the subtraction function being done in software. The basic expression for the log-ratio beam position [3,4] is:

$$Np = [(\log(A) - \log(B)] = \log(A/B) = \tanh^{-1}\left[\frac{(A-B)}{(A+B)}\right] \approx 2\frac{A-B}{A+B}, \quad (1)$$

where Np is the normalized position and A and B is the induced voltages on the stripline pick-up. This technique adds some flexibility to the system design. The algorithm to calculate the beam position can be changed or adjusted in the future as new techniques develop.

The BPM system was partitioned into subsystems as shown in Figure 1. The stripline detector, bandpass filters, and front-end board make up the major differences between applications. This partitioning allows the use of most of the components for operation at either 352 MHz or 2856 MHz. The stripline sensitivity and system calibration factor can be calculated by the following equations [3,4]:

$$S \approx \frac{80}{\ln(10)} \times \frac{1}{b}, \quad (2)$$

where S = stripline sensitivity (dB/mm) and b = half aperture (mm), and

$$X(mm) \approx \frac{1}{S \times G_{system}} \times V_{out}, \quad (3)$$

where X(mm) = normalized position (mm), G_{system} = system gain (which is set to 36mV/dB), and V_{out} = log amplifier module output (mV).

Table 1 describes the location and number of BPMs that will be upgraded. These applications are transport lines between the linac and the storage ring. This table also describes the type of detector, sensitivity, and system calibration factor.

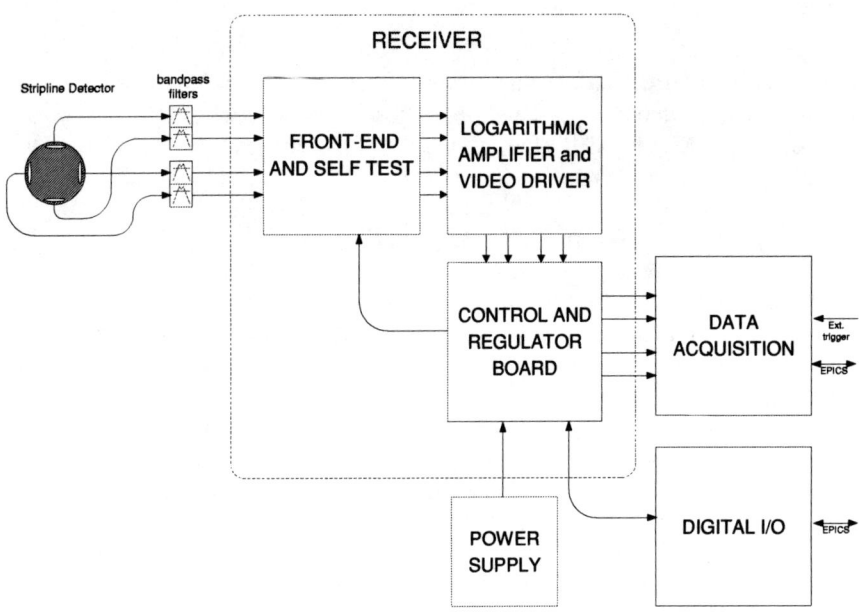

FIGURE 1. System block diagram.

TABLE 1. System Applications.

Location	Number of BPMs	Frequency (MHz)	Half Aperture (mm)	Stripline Sensitivity (dB/mm)	Normalized Position (μm/mV)
Linac	15	2856	17	2.0	$13.6 \times V_{out}$
LEUTL	20	2856	17 and button type	2.0	$13.6 \times V_{out}$
Linac to PAR	4	2856	17	2.0	$13.6 \times V_{out}$
PAR to Booster	5	352	25	1.4	$20.0 \times V_{out}$
Booster to Storage Ring	8	352	25	1.4	$20.0 \times V_{out}$

RECEIVER DESIGN

The receiver is partitioned as shown in Figure 1. There is a front-end board, four log amplifier boards, and a control and regulation board for each receiver. The boards are housed in an EMI-shielded aluminum case. The receivers are installed in a 19-inch-wide, 4-U height card crate where up to eight receivers can be installed.

Front-End and Self Test Board

The rf front-end board provides the gain and self-test capabilities for the system. The four signals from the input bandpass filters are sampled via 15-dB directional couplers, which are printed on the circuit board. This provides the ability to trouble-shoot the system without disconnecting any cables. The directional couplers also serve as feeds for the self-test oscillators. In the self-test mode the coupler is switched from a 50-Ω termination to a voltage-controlled oscillator. The oscillator drives a two-way equal power divider that is also printed on the circuit board and provides an equal input to one plane of the BPM. This same layout is duplicated for each plane. The board employs a selectable gain stage to shift the operating range by 20 dB. Presently we are using a low-noise Stanford Microdevices SGA-3586 cascadeable gain block with a noise figure of 2.5 dB. There are other amplifiers in this series with different gain and noise parameters that could be implemented in the future. This amplification will shift the input operating range while maintaining the same system gain (G_{system}). This feature will be used in some applications and has the effect of extending the dynamic range.

The front-end and self-test rf board was designed using microstrip transmission lines and components laid out on Rogers RO3006 microwave board materials. The impedance was calculated using Eq. (4) and was optimized in the lab [5]. The dielectric constant of the ceramic PTFE composite material is 6.15 and the loss tangent is 0.0025 @ 10 GHz. One of the design goals was keeping the rf board construction process as simple as possible by avoiding bonding substrates. This equates to a two-layer board with a thickness of 0.025 inch to insure the trace width of 0.036 inch for 50-Ω lines. The boards are designed to be drop-in replacements for 352- or 2856-MHz applications.

$$Z \approx \frac{377}{2\pi\left(\frac{\varepsilon_r+1}{2}\right)^{1/2}} \times \left[\ln\left(\frac{8h}{W}\right)+\frac{1}{8}\left(\frac{W}{2h}\right)^2 - \frac{1}{2}\frac{\varepsilon_r-1}{\varepsilon_r+1}\left(\ln\frac{\pi}{2}+\frac{1}{\varepsilon_r}\ln\frac{4}{\pi}\right)\right], \quad (4)$$

where Z=microstrip impedance, ε_r=substrate dielectric constant, h=substrate thickness, and W=microstrip width.

Logarithmic Amplifier and Video Driver Board

The block diagram in Figure 2 shows one of the four logarithmic amplifier video driver boards. The input broadband matching network enables us to use the same hardware for both the 352- and 2856-MHz applications. The AD8313 is one of the main motivations for upgrading the existing 2856-MHz BPMs that require downconverters. The AD8313 uses a cascade of eight amplifier/limiter cells, each having a gain of 8 dB typically and a −3 dB bandwidth of 3.5 GHz, for a total gain of 64 dB.

The ADC video driver is the AD8002 dual current feedback amplifier that provides offset and gain adjustment to the log amplifier. In order to evaluate an off-the-self ADC board, the output was designed to drive a 0-2 volt range with a 1-MΩ load impedance. The board is a two-layer microstrip design with a thickness of 0.025 inch to insure the trace width of 0.036 inch for 50-Ω lines.

The log linearity plot shown in Figure 3 illustrates that each gain range has a dynamic rage of about 60 dB with an overlap of 40 dB. The combination of both ranges enables operation over an 80-dB dynamic range with a maximum of ±0.6 dB errors in log linearity for 352-MHz applications. The S-band application is expected to be similar with some roll-off at the band edges. The overall pulse rise time for 10 to 90% of the input is 45 ns without the lowpass filter implemented for the overall system.

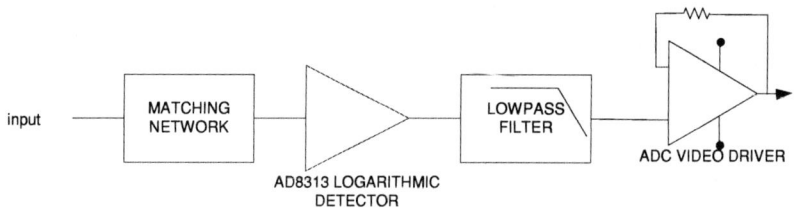

FIGURE 2. One channel log amplifier block diagram.

Control and Regulator Board

The control and regulator board supplies and distributes regulated power to the five boards in the receiver. It also provides the common point for the control interface. The board employs a hybrid coaxial connector that allows a blind mate connection to the main frame.

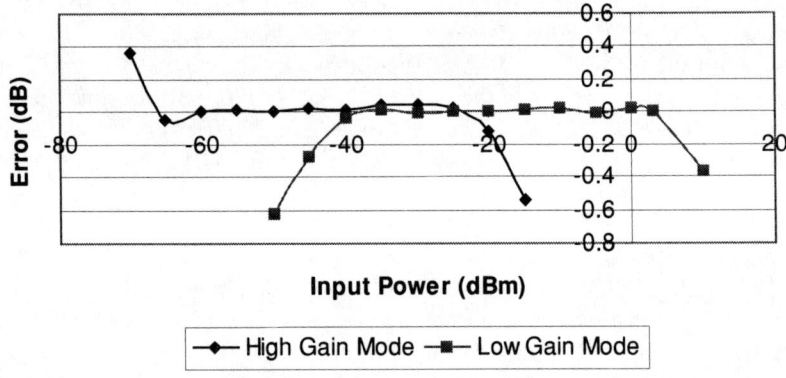

FIGURE 3. 352-MHz receiver logarithmic error.

Striplines and Bandpass Filters

The 2856-MHz BPMs employ shorted S-band λ/4 stripline pickup electrodes, which were recently installed [6], and have a 34-mm-diameter vacuum chamber aperture. The bandpass filters being considered for this application are three-pole λ/4 combline filters. The 3-dB bandwidth is 3 MHz with an in-band insertion loss of 6.5 dB.

The 352-MHz BPM applications also utilize a shorted λ/4 stripline detector with a vacuum chamber aperture size of 50 mm. They differ from the S-band design not only in blade length but also in width (1.4 rads). The system input bandwidth is limited to a 3-dB bandwidth of 3.0 MHz by Surface Acoustic Wave (SAW) bandpass filters. The SAW filters have matching networks on the input and output with an in-band insertion loss of 7.5 dB. The filter sets are matched to < 0.2 dB across the 3-MHz band centered at 350 MHz. The SAW filter is a very economical ($7/filter) solution since it is a standard RFM product.

DATA ACQUISITION

We are presently evaluating several options for data acquisition at this time. One option is a 12-bit, 100-MHz transient recorder. The VME module contains eight channels, each having 256K of memory. Most of the data quoted in this paper was recorded with the 12-bit, 100-MHz digitizer. The other option is a 14-bit, 50-MHz digitizer. The general approach for using the transient digitizer is to reduce the error or standard deviation by $\frac{1}{\sqrt{N}}$, where N is the number of samples collected during the video pulse. In the 352-MHz application we can take as many as 30 samples or 300 ns from the start of the event.

PERFORMANCE

We have built and are currently testing prototypes for a 352-MHz receiver and a 2856-MHz receiver at this time. The plot shown in Figure 4 is the output of the 352-MHz system operating with the 12-bit, 100-MHz digitizer. The input to the BPM system is a simulated beam pulse with the data acquisition set to acquire 30 samples from a single pulse. The top trace is an overlay of all four-blade signals as a function of input power. The second trace shows the position intensity dependence with error bars representing the noise. The third trace shows the peak-to-peak noise of each of the four blades as a function of input power.

The 352-MHz system was also tested in the lab with a 50-mm stripline chamber installed in a wire test set-up. The wire test accuracy was 19 µm rms over a ± 5mm range and 8 µm rms over a ± 1 mm range. The lab testing indicates that the overall electronic system drift stability is <15 µm over a 48-hour period.

The 352-MHz system testing is currently moving from the lab to the accelerator for further evaluation. The single-shot resolution for this system installed in booster-to-storage ring (BTS) transport line is 15 µm over a 30-dB dynamic range. Further testing will continue after the May shut down.

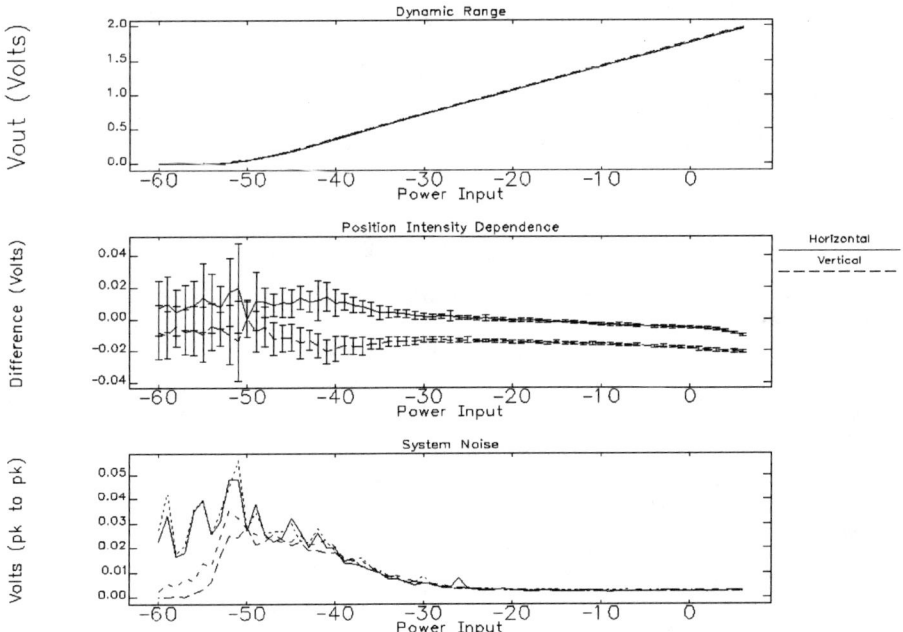

FIGURE 4. 352-MHz performance.

TABLE 2. Design Specifications.

Parameter	PC Gun	RF Gun
Dynamic Range	0.1-2 nC (26 dB)	0.1-10 nC (40 dB)
Single Shot Resolution	15 μm rms	100 μm rms
Drift	15 μm rms	100 μm rms
Accuracy	100 μm ± 5mm	100 μm ± 5mm

DISCUSSION

The BPM system was designed to meet the parameters given in Table 2. The 352-MHz system satisfies the given requirements; however some enhancements warrant investigation. We will explore reducing the single-shot noise by implementing a differentially driven log amplifier output. This will reduce the common mode broadband noise.

The S-band receiver will be more challenging. However, if the signal-to-noise ratio remains relatively constant, we should see about a 32% improvement in the resolution due to the fact that the chamber aperture is smaller.

The plan is to complete testing on the two prototypes by mid-summer, followed by ramping-up production of the first-article units by late summer.

ACKNOWLEDGMENTS

The author would like to acknowledge Glenn Decker for many helpful discussions, Tom Buffington for the mechanical design work, and Robert Keane and Chuck Gold for providing valuable electrical measurement and assembly expertise.

REFERENCES

1. Emery, L., "Recent Operational Data on Continuous Top-Up Operation at the Advanced Photon Source," PAC 2001 Conference Proceedings, Chicago, pp. 2599-2601 (2001).
2. Milton, S., "Measurements of Exponential Gain and Saturation of SASE at the APS LEUTL," PAC 2001 Conference Proceedings, Chicago, pp. 236-240 (2001).
3. Shafer, R., "Beam Position Monitoring," AIP Conference Proceedings No. 212, Upton, NY. pp. 26-58 (1989).
4. Aiello, G. and Mills M., "Log-Ratio Technique For Beam Position Monitor Systems," AIP Conference Proceedings No. 281, Berkeley, CA, pp. 301-310 (1992).
5. Gupta, K. C. et. al., *Microstrip Lines and Slotlines*, Artech House Publishers, Boston, London 1996.
6. Gorski, A. and Lill, R., "Construction and Measurement Techniques for the APS LEUTL Project RF Beam Position Monitors," Proceedings of the 1999 Particle Accelerator Conference, New York, pp. 1411-1413 (1999).

A Transverse Injection Damper at RHIC

A. Drees, M. Brennan, P. Cameron, R. Connolly, R. Michnoff,
C. Montag

Collider-Accelerator Department, Brookhaven National Laboratory, Upton NY 11973

Abstract. During the RHIC Au-Au runs in 2000 and 2001 as well as in the polarized proton run in 2001 no transverse damping system existed. To overcome residual injection jitter and to maintain transverse beam emittances a transverse injection damping system is foreseen for the 2002/2003 run. This report describes the realization of the injection damping system and outlines the future upgrade into a bunch-to-bunch transverse instability damping system, which is expected to be required for increasing bunch intensities.

INTRODUCTION

Since it is planned that the injection damping system evolves into a transverse damper in future years, design and outline considerations will focus on this upgrade as much as possible to avoid duplication of efforts. For a bunch-to-bunch transverse damper a linear amplifier instead of a power supply and fast pulsing switches will be used. The choices of BPMs to be used for a transverse damper are much more constrained. It has to perform under rapid-changing conditions such as the acceleration ramp and with increased bunch intensities. RHIC accelerates from 10 GeV/u to 100 GeV/u for Au and from 23 GeV/u to 250 GeV/u for protons and allows a β-squeeze from $\beta^* = 10$ m at injection down to $\beta^* = 1$m at flattop at certain interaction regions causing significant changes in local phase advance and β-function. Other scenarios as un-squeezing IRs is on the list for future runs as well. Therefore, different power supply and BPMs are required for injection damping and bunch-to-bunch feedback. However, the signal conditioning for the chosen BPM, the trigger and the damper module can be used in either system. The damper module to be used for the injection damper is based on a VME board designed for the AGS damping system with a programmable FPGA (Field Programmable Gate Array) [1]. For the future damper the existing transverse kickers, which are currently used for a tune measurement system [2] and will serve for the injection damper, will be replaced by several dedicated 1m stripline modules in the same area to avoid conflict with the tune measurement system. For the upcoming year, since only used at injection, the kicker hardware including power supply and fast

HV switches will be shared.

I KICKER AND BPM

Each ring has two kicker modules with four 2m-long stainless steel striplines mounted on ceramic stand-offs spaced 1m apart allowing both, horizontal and vertical kicks. The two kickers are connected in series to provide 4m of stripline kickers. Each stripline subtends an angle of $70°$ at an aperture of 7 cm. The assembly is designed to give 50Ω impedance when opposing lines are driven in the difference mode. Figure 1 shows one kicker module in the assembly area. Each of the four

FIGURE 1. *One kicker module on a work bench in the assembly area.*

planes can be powered independently. So far only pulsed power has been used. The kick pulses are generated by fast FET switches [3] producing an approximately 140 ns long pulse. By centering this pulse on the measured bunch single bunch excitation is possible with 60 (RHIC design) and even up to 120 bunches (RHIC upgrade) per ring where the bunch spacing is about 110 ns. Only one out of the 60 or 120 bunches respectively is kicked. All switches for all striplines in both rings are charged by one 5kV/2A power supply. Most BPMs used in RHIC are realized by short circuited transmission lines of 23 cm length, with a design impedance of 50Ω, and an aperture of 7 cm [5]. The selection of BPMs listed in tables 1 to 4 below is based on devices with analog signals available in the 1002 service building, close to the 2 o'clock interaction region (IR2).

II KICKER LOCATION

As can be seen in table 1 and 3 the actual kicker location discriminates the vertical plane by more than a factor of 10 and 20 for the $\beta^* = 3m$ and $\beta^* = 1m$ lattices. Thus, it would be favorable to move the kickers such that during the ramp, when β^* is changed from 10 m up to 1 m, the kicker efficiency would not drop in the vertical plane. To do so, the kickers have to be closer to Q3. However, because of the increasing β-functions, the kicker aperture (7 cm) becomes a concern. Figure 2 shows the vertical and horizontal beam profiles in terms of σ around IR2 for beams at 100 GeV with a normalized emittance of 40 πmm mrad. In order to avoid the kickers being the limiting aperture, they have to be at $\geq 6\,\sigma$, allowing them to move as close as -52 m from the IR. This move by approximately 10 m is actually planned for the next run and would reduce the vertical β-function by 25% at injection while increasing it by a factor of 5 to 10 during the ramp and at storage. The vertical kicker- strength at injection is affected by 10% only and therefore not a problem.

The required kick strength g for a linear amplifying system can be defined as:

$$\frac{2}{tf} = g = \sqrt{\beta_{s0}\beta_s}\,\frac{\Delta x'}{x} \tag{1}$$

with $f = 78$ kHz being the revolution frequency, t being the required damping time and β_{s0}, β_s being the β-functions at the location of the kicker and at the location of the BPM respectively. Equation 1 can be used as a conservative estimate for the required bang-bang kick strength. After a certain number of turns, however, the bang-bang damper would become counter-productive. Therefore the injection damper will be limited to a few hundred turns for damping. For linear damping, a given amplitude should generate a kick $\Delta x'$ of:

$$\Delta x' = \frac{2x}{t\sqrt{\beta_{s0}\beta_s}f} \quad \text{and} \tag{2}$$

$$x = \frac{\Delta x'}{2}tf\sqrt{\beta_{s0}\beta_s}. \tag{3}$$

The achievable kick angle for a single 3 kV pulse is about 11 μrad at injection for Au and p [6]. Using the β-functions at the kicker and the Q3 (Q1) BPMs (see table 1 and 3) and a damping time of 200 turns, i.e. 2.5 msec, results in an approximate orbit amplitude of 32 mm vertically (Q1) and 75 mm horizontally (Q3). The largest observed amplitudes at injection in the straight sections were ≤ 25 mm in both planes. Therefore, the existing kick strength should be sufficient to antagonize injection oscillations in both planes.

III SELECTION OF BPMS

The kick angle after one pulse with 3 kV received by an ion going through the kickers is approximately 10 μrad at injection energy ($\gamma \approx 10$) [6]. The effect of such a kick translates into a beam offset given by:

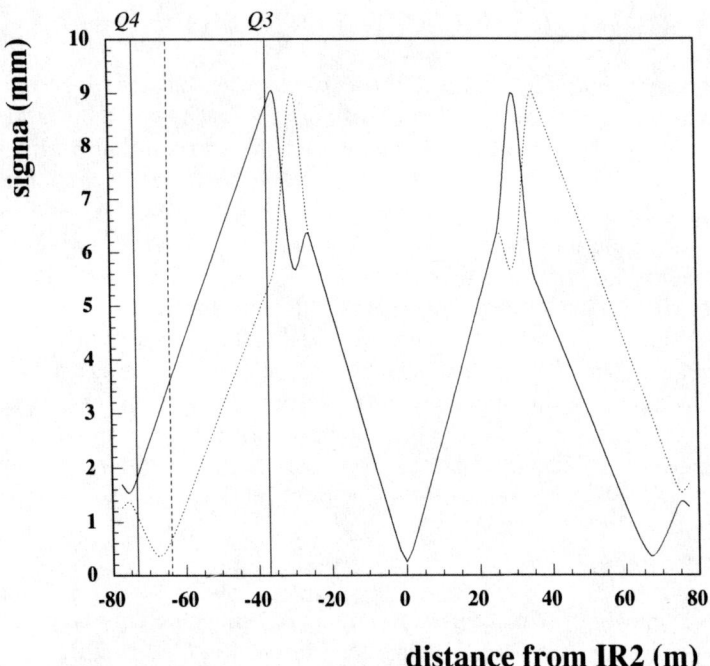

FIGURE 2. Horizontal (solid line) and vertical beam profile (dotted line) in units of sigma around IR2. The dashed line indicates the current kicker location.

$$\delta_{x,y} = \Theta \sqrt{\beta_{x,y}(s_0)\beta_{x,y}(s)} \sin \psi_{x,y}(s), \qquad (4)$$

where $\beta_{x,y}(s_0)$ is the β-function at the location s_0 of the kicker and $\beta_{x,y}(s)$ is the respective horizontal or vertical β-function and ψ the relative phase advance between location s and s_0. Table 1 and 3 summarize the values of the β functions at the location of the kickers and BPMs for the blue and yellow ring respectively. Phase advances, relative to 6 o'clock (IR6) for the two rings are listed in table 2 and 4. The actual lattice strongly favors the horizontal plane when resulting beam displacements are calculated.

The typical phase advance in one turn in RHIC is approximately $0.21/2\pi$ to $0.23/2\pi$ for both planes, corresponding to about $80°$. During the ramp the tune would vary from 0.2 to 0.25 at most. In general, the tunes are fairly close and separated by some 0.01 to 0.02 only. Therefore, after one full turn, a local phase advance of close to $0°$ or $180°$ between BPM and kicker is most suitable for the damper. With this configuration a total shift of close to $90°$ will be kept. The neighboring Q3 BPMs offer $0°$ relative to the new position in the horizontal plane. The Q1 BPMs on the other side of the IR provide an approximate phase advance of close to $180°$ relative to the new kicker location in the vertical planes.

TABLE 1. Approximate β functions at the location of the blue kicker and blue BPMs around IR2 for various lattices. The DX BPM are closest to the IR, indicated by the double line.

device	plane	s(m)	β^*10m		β^*3m		β^*1m	
			β_x (m)	β_y (m)	β_x (m)	β_y (m)	β_x (m)	β_y (m)
kicker	HV	2491.7	34	16	86	6	118	5
Q3 BPM	HV	2519.0	141	48	426	153	633	231
Q1 BPM	HV	2530.9	73	72	211	210	314	314
DX BPM	HV	2547.6	17	17	26	26	37	37
DX BPM	HV	2564.2	17	17	26	26	37	37
Q1 BPM	HV	2580.9	73	73	211	210	314	314
Q3 BPM	HV	2592.8	49	140	154	424	230	634
Q4 BPM	HV	2629.1	36	20	18	32	17	38
Q5 BPM	V	2636.9	8	56	8	59	11	62
Q6 BPM	H	2651.7	42	15	38	18	37	20
Q7 BPM	HV	2668.9	13	49	11	48	10	47
Q8 BPM	HV	2682.0	45	13	46	11	47	10
Q9 BPM	V	2699.4	12	43	12	42	12	43
Q10 BPM	H	2710.8	44	11	44	11	44	11

TABLE 2. Approximate phase advances relative to IR6 at the location of the blue kicker and blue BPMs around IR2 for various lattices. The DX BPM are closest to the IR, indicated by the double line. Values for the new kicker location are in brackets.

device	plane	s(m)	β^*10m		β^*3m		β^*1m	
			$\mu_x/2\pi$	$\mu_y/2\pi$	$\mu_x/2\pi$	$\mu_y/2\pi$	$\mu_x/2\pi$	$\mu_y/2\pi$
kicker	HV	2491.7	18.52 (18.55)	18.97 (19.13)	18.55	19.05	18.53	19.05
Q3 BPM	HV	2519.0	18.58	19.24	18.58	19.25	18.55	19.21
Q1 BPM	HV	2530.9	18.60	19.26	18.58	19.26	18.55	19.21
DX BPM	HV	2547.6	18.68	19.34	18.62	19.29	18.58	19.24
DX BPM	HV	2564.2	18.90	19.56	19.01	19.68	19.00	19.66
Q1 BPM	HV	2580.9	18.98	19.64	19.05	19.72	19.03	19.69
Q3 BPM	HV	2592.8	19.00	19.66	19.05	19.73	19.03	19.69
Q4 BPM	HV	2629.1	19.34	19.78	19.43	19.78	19.43	19.73
Q5 BPM	V	2636.9	19.39	19.83	19.51	19.81	19.51	19.76
Q6 BPM	H	2651.7	19.71	19.90	19.71	19.87	19.67	19.82
Q7 BPM	HV	2668.9	19.82	20.02	19.85	19.98	19.82	19.92
Q8 BPM	HV	2682.0	19.91	20.11	19.95	20.08	19.92	20.03
Q9 BPM	V	2699.4	20.06	20.23	20.09	20.21	20.06	20.17
Q10 BPM	H	2710.8	20.14	20.32	20.18	20.30	20.15	20.26

TABLE 3. Approximate β functions at the location of the yellow kicker and BPMs around IR2 for various lattices. The order is defined by the direction of the beam. The DX BPM are closest to the IR, indicated by the double line.

device	plane	s(m)	β^*10m		β^*3m		β^*1m	
			β_x (m)	β_y (m)	β_x (m)	β_y (m)	β_x (m)	β_y (m)
Q10 BPM	V	2710.8	12	43	12	42	12	42
Q9 BPM	H	2699.4	44	11	44	11	44	11
Q8 BPM	HV	2682.0	13	45	11	46	10	47
Q7 BPM	HV	2668.9	49	13	47	11	47	10
Q6 BPM	V	2651.7	15	42	18	38	20	37
Q5 BPM	H	2636.9	56	8	59	8	62	11
Q4 BPM	HV	2629.1	20	36	32	18	38	17
kicker	HV	2620.1	34	16	85	6	118	5
Q3 BPM	HV	2592.8	141	48	426	153	633	231
Q1 BPM	HV	2580.9	73	72	211	210	314	314
DX BPM	HV	2564.2	17	17	26	26	37	37
DX BPM	HV	2547.6	17	17	26	26	37	37
Q1 BPM	HV	2530.9	73	73	211	210	314	314
Q3 BPM	HV	2519.0	49	140	154	424	230	634

TABLE 4. Approximate phase advances relative to IR6 at the location of the yellow kicker and BPMs around IR2 for various lattices. The order is reverted in s to reflect the direction of the beam. The DX BPM are closest to the IR, indicated by the double line. Values for the new kicker location are in brackets.

device	plane	s(m)	β^*10m		β^*3m		β^*1m	
			$\mu_x/2\pi$	$\mu_y/2\pi$	$\mu_x/2\pi$	$\mu_y/2\pi$	$\mu_x/2\pi$	$\mu_y/2\pi$
Q10 BPM	V	2710.8	8.54	8.37	8.57	8.33	8.60	8.36
Q9 BPM	H	2699.4	8.62	8.49	8.65	8.41	8.68	8.45
Q8 BPM	HV	2682.0	8.74	8.60	8.78	8.56	8.82	8.60
Q7 BPM	HV	2668.9	8.83	8.70	8.88	8.66	8.93	8.70
Q6 BPM	V	2651.7	8.95	8.81	8.99	8.80	9.03	8.85
Q5 BPM	H	2636.9	9.02	9.13	9.05	9.00	9.09	9.01
Q4 BPM	HV	2629.1	9.07	9.18	9.09	9.08	9.12	9.09
kicker	HV	2620.1	9.12 *(9.16)*	9.25 *(9.4)*	9.12	9.26	9.14	9.33
Q3 BPM	HV	2592.8	*9.18*	9.51	9.14	9.46	9.16	9.49
Q1 BPM	HV	2580.9	9.21	9.54	9.15	9.46	9.16	9.49
DX BPM	HV	2564.2	9.29	9.62	9.18	9.50	9.19	9.52
DX BPM	HV	2547.6	9.51	9.84	9.57	9.89	9.61	9.94
Q1 BPM	HV	2530.9	9.59	*9.92*	9.61	9.92	9.64	9.97
Q3 BPM	HV	2519.0	9.61	9.94	9.62	9.93	9.64	9.97

IV TRIGGER AND DATA ACQUISITION

Figure 3 sketches the signal processing and triggering of both, the BPMs and the kickers, for a bang-bang injection damping system. The damper module is based on the existing AGS module and needs adjustments for the RHIC damper in the I/O area.

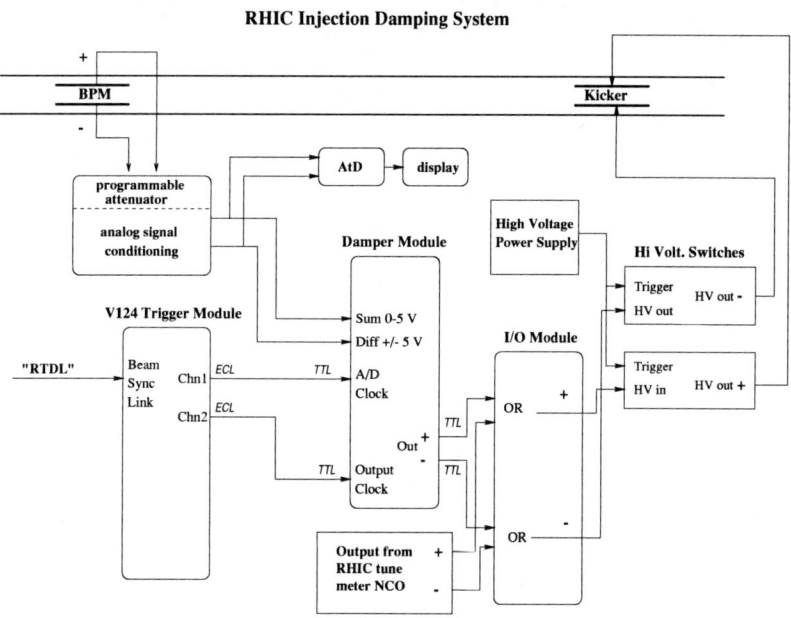

FIGURE 3. *Block diagram for the RHIC injection damper.*

The V124 module [7] receives and decodes the beam synchronous event link [8]. The raw data acquisition from the two BPM planes will be triggered by two channels of the V124 board where a total of 8 channels is available. Each channel for BPM readout and the kicker trigger has the appropriate delay so, on turn-by-turn acquisition, the same bunch will be observed on the BPM and then kicked. Start turn number, total number of turns for acquisition and damping as well as time delays are all parameters which can be remotely set from a console level computer. In general, the V124 allows the system to be triggered by any event broadcasted on the beam synchronous link such as the injection-event, start-acceleration-event or on demand. However, to damp injection oscillations only the injection-event is going to be used.

The raw (bipolar) signal from the BPM will be attenuated by a programmable attenuator of 0-30 db. In the signal conditioning version as used by the ARTUS tune meter two signal processing modules compute the two difference signals and the sum of all stripline signals for each ring. We are currently working on an upgrade of the signal conditioning, which will then be used by both systems, the tune meter and the transverse injection damper. It will also be used for signal conditioning in the future bunch-to-bunch transverse damper.

V CONCLUSION

The existing kicker modules, once moved by approximately 10 m, are suitable to act as a transverse injection damper. A basic design of a signal conditioning exists and is currently in use for the ARTUS tune meter system. However, it is planned to upgrade this version for both, the damper and the tune meter, for future runs starting in FY'03. The actual damping module is based on the existing AGS damper module and needs some modifications. This module, together with the signal conditioning design, will be kept for the future bunch-to-bunch transverse damper. Relative to the new kicker location, the existing Q3 and Q1 BPMs provide a suitable phase advance after one turn of about $90°$ and $270°$ respectively. The high β-function at the BPMs of 140 m and 71 m eases an amplitude measurement with good signal to noise ratio in both planes. For the future feedback system, the injection damping kickers will be replaced by several dedicated 1m stripline modules in the same area to avoid conflict with the tune measurement system.

REFERENCES

1. Michelle Wilinski et al., "Enhancements to the Digital Transverse Dampers at the Brookhaven AGS", these proceedings.
2. A. Drees, R. Michnoff, M. Brennan, J. DeLong, "ARTUS: The Tune Measurement System at RHIC", Proceedings BIW2000, Boston, 2000.
3. Behlke Electronic GmbH, http://www.euretek.com/
4. J. Xu et al., "The Transverse Damper System for RHIC", Proceedings of the Particle Accelerator Conference (PAC) in San Francisco, 1991.
5. P. Cameron et al., "RHIC Beam Position Monitor Assemblies", IEEE Proc., 1995 PAC.
6. P. Cameron, R. Connolly, A. Drees, W. Ryan, H. Schmickler, T. Shea, D. Trbojevic, "ARTUS: A Rhic TUne Measurement System", RHIC/AP/98-125, internal note.
7. H. Hartmann, T. Kerner, "RHIC beam synchronous trigger module", Proc. PAC 1999 (p. 696).
8. T. Kerner, C. R. Conkling Jr., B. Oerter, "V123 Beam Synchronous Encoder Module", Proc. PAC 1999 (p. 699).

Spallation Neutron Source Beam Current Monitor Electronics[1]

M. Kesselman, W. C. Dawson

Brookhaven National Laboratory

Abstract. This paper will discuss the present electronics design for the beam current monitor system to be used throughout the Spallation Neutron Source (SNS) under construction at Oak Ridge National Laboratory. The beam is composed of a micro-pulse structure due to the 402.5MHz RF, and is chopped into mini-pulses of 645ns duration with a 300ns gap, providing a macro-pulse of 1060 mini-pulses repeating at a 60Hz rate. Ring beam current will vary from about 15ma peak during studies, to about 50Amps peak (design to 100 amps). A digital approach to droop compensation has been implemented and initial test results presented.

BACKGROUND

This is the fourth[1,2,3] paper presented on this subject. The earlier papers describe the decision process to attempt to use identical electronics throughout to provide a compatible approach to system diagnostics. A PC based instrument design philosophy was adopted for the diagnostics wherever it would apply, and a compromise in transformer frequency response, droop, and standardization resulted in a decision to use Bergoz® FCT (Fast Current) transformers. Individual areas within the SNS will have current transformers that suit the dimensional requirements, while maintaining electrical performance compatible with required performance and the system electronics. All electronics would be identical and could be placed anywhere in the system. Therefore, the electronics has been designed with flexibility in its configuration.

CHALLENGES

There were a number of major challenges to be addressed:
- Measurement of chopper characteristics vs. Ring turn by turn current.
- Goal of a single design to minimize design cost, and provide interchangeability
- Large dynamic range in Ring and RTBT
- Baseline restoration and droop compensation
- Integration accuracy vs. sampling rate
- Noise, response characteristics, filtering and digitizing aliasing
- Testing
- Calibration

---

[1] Work performed under the auspices of the U.S. Department of Energy

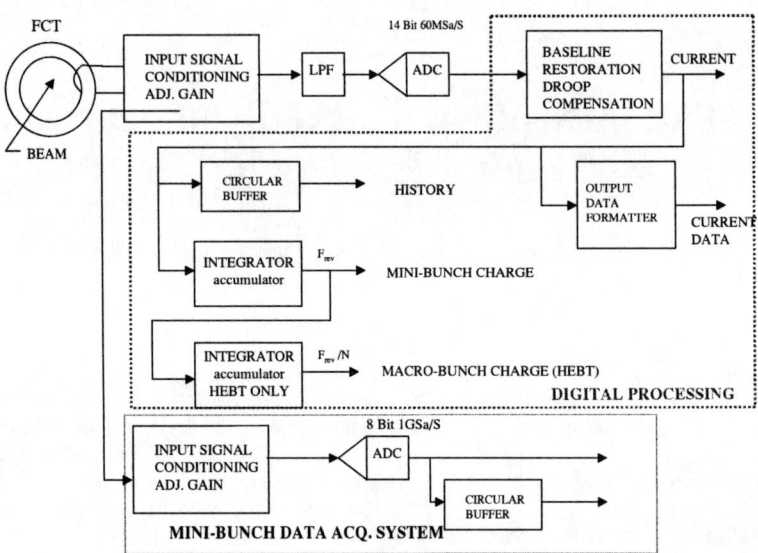

FIGURE 1. Block diagram of BCM electronics. This shows two digitizing systems. The faster digitizer is a separate system used for analysis of chopper characteristics and is intended for short time samples. The slower speed digitizer is intended for general current monitoring and has a reduced bandwidth. Baseline restoration and droop compensation are accomplished digitally in the slower system.

DETAILS

The selection of the FCT transformer permits high speed response, while allowing droop to be compensated digitally. The electronics will provide a broad-band output suitable for observing the chopper characteristics. Provisions for jumper selected configurations for the lower current MEBT, Linac, and HEBT and the higher current Ring and RTBT allow the same electronics to be used throughout the SNS. The large dynamic range in the Ring and RTBT is handled by employing different gain paths that are selected by switchable amplifiers. These amplifiers (OPA680 series) switch in less than 100ns, allowing them to be switched during the "gap" time in the Ring. The large voltages expected (25 to 50 volts) are sufficient to cause sensitive amplifiers to fail. Therefore, the use of protected amplifiers is necessary in the Ring and RTBT. A system block diagram is shown in figure 1. The input signal conditioning circuitry will separate the signal into two paths. The broadband path will be sent to a separate high-speed digitizer capable of analyzing the chopper characteristics. The second path will be channeled to a signal processing system capable of amplifying the signal adequately for digitization by a 14 bit 68MSa/s digitizer. This digitizer is synchronized with the revolution frequency of the Ring ($64 \times F_{rev}$).

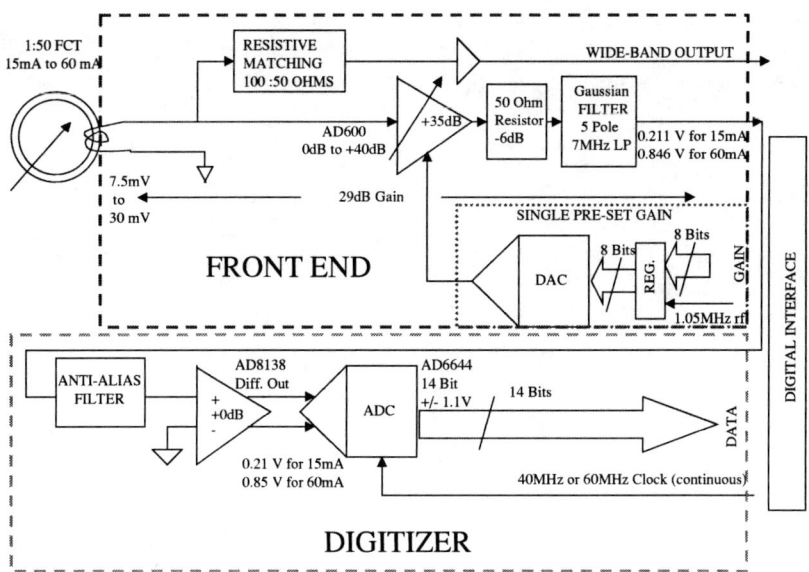

FIGURE 2. Block diagram of BCM electronics used for the MEBT, Linac and HEBT. A 100Mhz bandwidth output is provided for monitoring purposes. Jumper selection configures this processing configuration.

MEBT, Linac, and HEBT Analog System

A high gain configuration is jumper selected to accommodate the lower currents expected in the Front End, MEBT, Linac and HEBT, and is shown in figure 2. A variable gain controlled amplifier (AD600) is used to amplify signals in the 0 to 25mV range (for 0 to 50ma) to a voltage range of 0 to 500mV for the ADC. A factor of two in headroom has been reserved for current peaking. The wideband transformer signal is buffered by an amplifier with bandwidth >100MHz (OPA3680).

The Ring rise time is expected to deteriorate to about 50ns. This requires a 7MHz bandwidth. Increasing the bandwidth increases the noise and the resolution degrades. Therefore, a 5 pole - 7MHz Gaussian filter was chosen to minimize overshoot, and provide significant filtering for aliasing considerations. This filter will provide about 42dB attenuation at 34MHz. Additional attenuation of 34dB by a 5 pole – 17MHz, 0.01dB, Chebyshev filter and two additional amplifier stages limited to 34MHz provides >80dB attenuation at the Nyquist frequency.

Ring and RTBT Analog System

To accommodate the large dynamic range of the Ring and RTBT, the configuration of figure 3 is employed by a jumper selection. This configuration employs protected high gain stages and a number of lower gain paths that are digitally selected to establish a variable gain capability. The paths are switched by fast switching amplifiers. Switching during "gap" time will allow for no loss of turns.

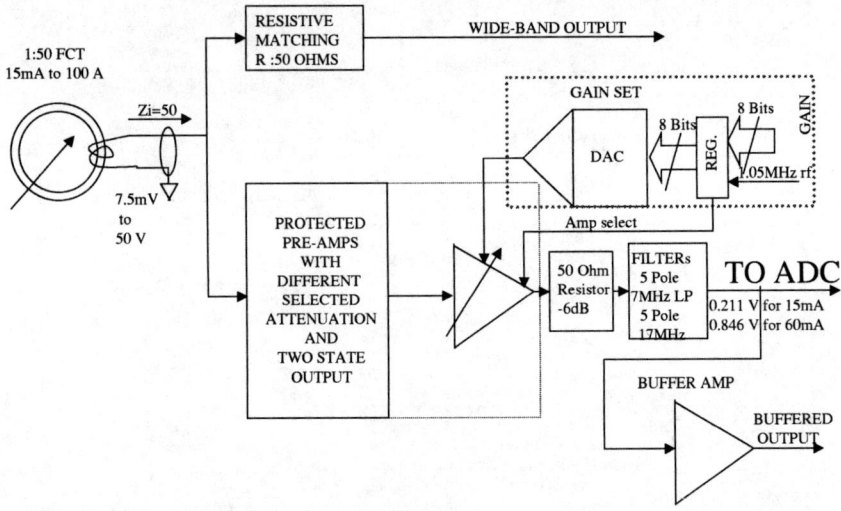

FIGURE 3. This shows the general block diagram for the jumper selected configuration for the Ring and RTBT BCM electronics. A protected amplifier system handles the high voltage levels expected in the Ring and RTBT. Different amplifier gain paths are selected digitally to permit handling the 1000:1 dynamic range.

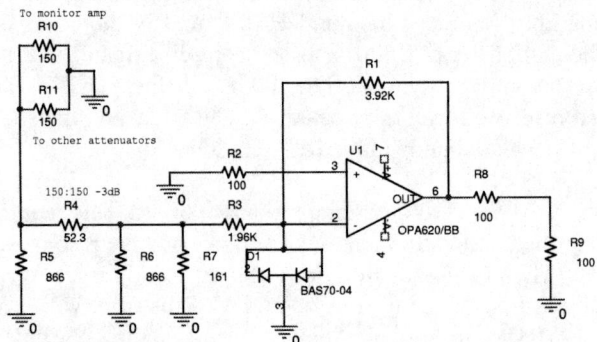

For the attenuator:
$\Delta Z_i / Z_i = 2*(\Delta Z_L / Z_L)/[2*N+(N-1)* (\Delta Z_L / Z_L)]$; N= power ratio=$P_i/P_o$
$Z_L = 161 \| Z_i$; $Z_i = V/I = V/(V-V_d)/1.96K \sim 1.96K*(1+ V_d/V)$
For $V_d/V = 1/50 = 0.02$; $\Delta Z_L = 149.000648 - 148.7788 = .2218$ and $(\Delta Z_L / Z_L) = 0.001491$
For the 3dB pad N=2 and $\Delta Z_i / Z_i = .074\%$

FIGURE 4. A protected amplifier must both protect the amplifier against high input signal voltage as well as assure that input impedance is maintained constant.

Protected Amplifier

Signal levels in the Ring can get as high as 25 Volts for 50 amps. If one considers doubling this for headroom, the signals are clearly too large for the amplifiers to handle. Therefore, the high gain paths must be protected. The amplifier shown in figure 4 provides both protection and assures a near constant input impedance. The input impedance must not vary by more than 0.1% to assure proper accuracy. This is achieved by providing an input resistance of more than 1K Ohm before the protection diodes. In so doing the diode distortion for high amplitude signals is not reflected to the other gain paths, however, noise is increased. One of the 150 Ohm input resistors represents a 150 Ohm attenuator chain. The attenuation is selected such that after the first stage of attenuation the signal is maintained below +/- 2.0 V (max input signal for the AD600).

Adjustable Gain Stage

The gain changing configuration is shown in figure 5. The attenuation provided is also shown. The amplifiers are part of a triple op-amp (OPA3680). This amplifier is a voltage feedback type amplifier with a digital control input permitting it to switch "off" within 100ns. The amplifiers of figure 5 are all of this type. The switched amplifier is configured as a +2 gain amplifier, providing no signal punch through when switched "off". The summer, adds the outputs of each path (all paths "off" except one). The system resolution has been established at 0.5% of full scale, allowing 4 gains. A table of gains and expected resolution is shown in Table 1.

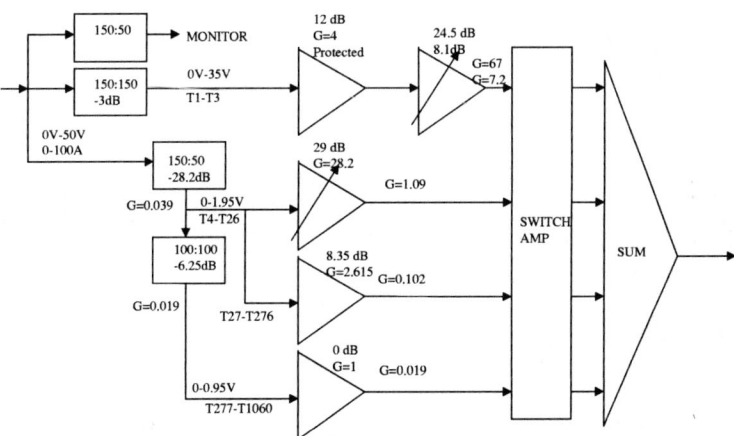

FIGURE 5. Gain Changing configuration

TABLE 1. Gain Switching						
Turn	Beam Current	Est. Input Signal	Gain	Est. Output Noise	Output Voltage	Resolution
1 test	15mA	7.5mV	67	0.65mV	0.5	0.13%
1	38mA	19mV	7.2	0.65mV	0.136	0.47%
4	152mA	76mV	7.2	0.65mV	0.547	0.12%
5	190mA	95mV	1.09	0.485mV	0.103	0.47%
26	988mA	0.494V	1.09	0.485mV	0.538	0.09%
27	1.026A	0.513V	0.102	0.246mV	0.052	0.47%
276	10.488A	5.244V	0.102	0.246mV	0.535	0.046%
277	10.526A	5.263V	0.019	0.246mV	0.1	0.246%
1060	40.28A	20.14V	0.019	0.246mV	0.382	0.064%

Digital control of gain is accomplished via separate DACs to set gain voltage levels on the AD600 variable gain amplifiers, and a gain storage register that establishes which of the gain paths is made active. The gain storage register is updated by a sequence generator on the digital interface section. The gain register information is stored in a FIFO, along with data to tag each data point with a gain path.

ADC Driver Stage

The driver for the ADC has been selected as per the manufacturer's recommendation. An AD8138 wide-band differential ADC drive amplifier is employed to shift the reference to match the reference of the AD6645. The summer feeds the 5 pole Gaussian filter as described earlier. This is buffered by an amplifier and fed to the 17MHz Chebyshev filter.

Digital Processing

The key to the system is the digital processing. This permits us to use transformers that have a very wide bandwidth, and compensate for the droop. The digitized data are transferred to a FIFO for DMA transfer to the PC. A LabVIEW® program processes the data and interfaces the PC with the network.

The DC offset is calculated by averaging points prior to the arrival of the beam. This is subtracted from the data to provide a data set that has been corrected to a zero baseline. The data is then compensated for droop with a digital IIR filter algorithm that cancels the transformer low frequency pole and establishes a new low frequency pole of 1 rad/sec. To cancel the transformer pole, it is necessary to calculate the droop time constant. This is accomplished by providing a calibration pulse and computing the exponential time constant during the transformer recovery time. The sensitivity of resulting compensated droop to errors in this calculation requires a good estimate. An analysis of this indicates a sensitivity of −0.6%/% (error in droop/error in transformer time constant). It is, therefore, necessary to average many data points or calculations.

The sensitivity of the droop to sampling time has a similar sensitivity, +0.6%/% (error in droop/error in sampling time).

The droop compensation formula is:

$$Y(n) = \{1/(2/T + 1/\tau_2)\}\{y(n-1)(2/T - 1/\tau_2) + x(n)(2/T + 1/\tau_1) + x(n-1)(-2/T + 1/\tau_1)\}$$

Where: T is the sampling period, $1/\tau_1$ is the transformer lower cut-off radian frequency, and $1/\tau_2$ is the desired new transformer lower cut-off radian frequency.

The data are integrated to determine total charge and filtered for a comfort display. An analysis of integration errors due to insufficient samples indicates a sampling frequency of more than 25MSa/s is required for a 0.1% error in the integral. This analysis was carried out using both a simple sum and a Simpson's rule algorithm. It is interesting to note that the two methods converge at 64MSa/s, and differ only slightly at lower sampling rates. Therefore, there is little advantage to using the more computer intensive Simpson's rule. An example of a simulated beam processed by this software is shown in figure 6.

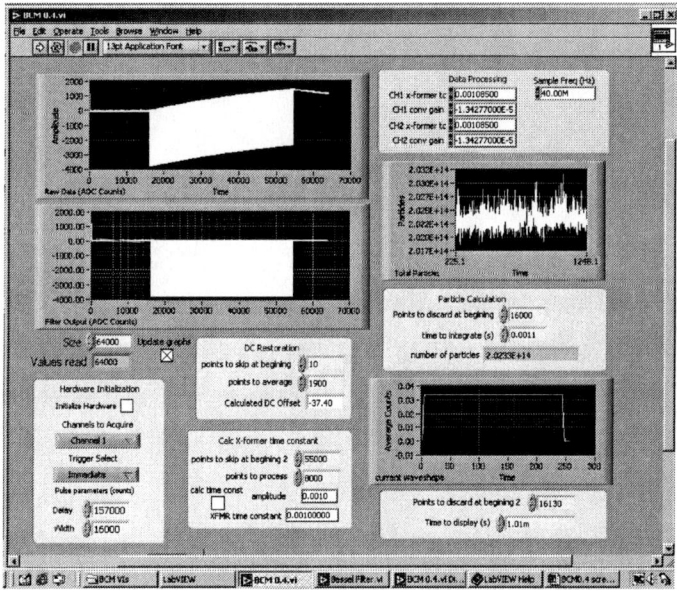

FIGURE 6. Screen dump of a simulated beam processed by the BCM electronics. The upper left graph is raw data showing a simulated 1ms, 645ns pulse with a 945ns period, pulse train. Droop is obvious, and shown compensated in the graph below. The filtered comfort display is shown with 256 points in the lower right, and the integrated charged particle count calculation shown above it.

Testing

A test apparatus was constructed using an 8 inch 50 Ohm coax line. The outer conductor was cut, insulated, and a shroud built to allow the transformer to measure the current in the center conductor. This was found to have a single resonance near 500MHz and provided a good 50 Ohm match for excellent current transient response measurements.

Calibration

An isolated dual, current output, DAC (DAC2902) provides the fundamental source for the calibrator. This is a fast current output device that settles quickly, and can deliver up to 20mA. The calibrator is isolated to avoid ground loops, and is AC terminated to back terminate the calibration winding, while allowing a DC current previously measured by an accurate DMM to flow into the winding. To simulate larger currents for the Ring, an additional current amplifier will be used.

ACKNOWLEDGMENTS

The authors would like to acknowledge Chris Degen for his assistance with the LabVIEW software, and Richard Witkover and Julian Bergoz for their technical assistance during system development.

REFERENCES

1. Kesselman, M. et-al., "SNS Project-Wide Beam Current Monitors", *BIW 2000*, May 8-11, 2000, Cambridge MA.
2. Kesselman, M., et-al. "SNS Project-Wide Beam Current Monitors", *EPAC 2000*, Vienna, Austria, June 26-30, 2000
3. Kesselman, M., "Spallation Neutron Source Beam Current Monitor Electronics", *PAC 2001*, Chicago, Il., June 18-22.

Beam Based Calibration of BPM Position Sensitivity at SPring-8 Storage Ring

S. Sasaki*, K. Soutome* and H. Tanaka*

SPring-8/JASRI, Kouto 1-1-1, Mikazuki, Sayo, Hyogo 679-5198, Japan

Abstract.
A method for beam based calibration of BPM position sensitivity is proposed, and preliminary result at SPring-8 Storage Ring is presented.

Beam based alignments of the BPM offsets were performed in various accelerator facilities. We extended this kind of beam based alignment techniques to calibrate the position sensitivity of the BPM.

When the strength of one quadrupole magnet is changed slightly, a certain amount of extra COD arises, depending how much the offset from the quadrupole center the COD has. The amplitude of COD and the distance from the quadrupole center has a certain functional relation which is defined by the quadrupole magnet field distribution and the storage ring optics parameters. By comparing the COD amplitude and the BPM position reading, we can re-scale the position sensitivity of the BPM.

Recently, a preliminary test of the method was performed. In this paper the obtained result are presented.

INTRODUCTION

The beam based calibration of beam position monitor(BPM) offsets are applied in various accelerator facilities.[1-5] The common method is a minimum search of closed orbit distortion(COD) occurred by a small change of the focusing strength of a quadrupole magnet with scanning the orbit around the quadrupole axis. In performing this kind of measurement, the amount of extra COD occurred by the quadrupole strength change has a relationship to the offset from the quadrupole axis. This relationship can be used for calibrating the position sensitivity of the BPM adjacent to the quadrupole magnet whose strength is changed.

We describe the principle of the method first, the preliminary measurement performed at the SPring-8 storage ring next

PRINCIPLE OF THE METHOD

First, we pose the following conditions for the method to be valid.
1. BPM are linear, which means the measured position (x, y) of a BPM is expressed as

$$x = x_0 + \frac{u}{S_x}, \text{ and, } y = y_0 + \frac{v}{S_y}, \tag{1}$$

where, x_0 and y_0 are offsets of horizontal and vertical directions, S_x and S_y are position sensitivity coefficients, and u and v are defined as

$$u = \frac{(A_1+A_4)-(A_2+A_3)}{A_1+A_2+A_3+A_4}, \quad v = \frac{(A_1+A_2)-(A_4+A_3)}{A_1+A_2+A_3+A_4},$$

or, in another definition, which is used for SPring-8 Storage Ring,

$$u = \frac{1}{2}\left(\frac{A_1-A_2}{A_1+A_2}+\frac{A_4-A_3}{A_4+A_3}\right), \quad v = \frac{1}{2}\left(\frac{A_1-A_4}{A_1+A_4}+\frac{A_2-A_3}{A_2+A_3},\right),$$

where A_1, A_2, A_3, A_4 are the signal amplitude of the electrodes from 1 to 4, shown as figure 1

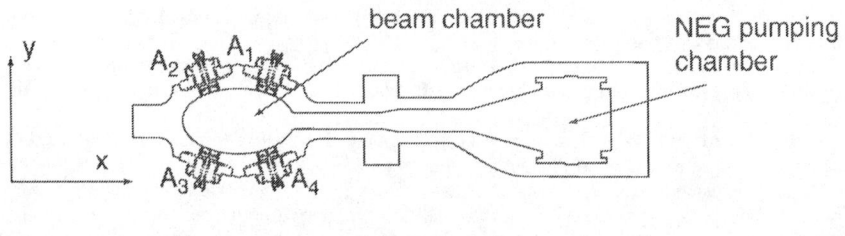

FIGURE 1. Cross section of the BPM part of the vacuum chamber. The coordinates (x, y) and electrode No. and its amplitudes $(A_1, \cdots, A_4)$ are defined as shown in the figure.

2. The optics of the storage ring is linear, which means the COD caused by a single kick at $s = s_0$ is expressed as

$$COD(s) = \frac{\theta(s_0)}{2\sin(\pi v)}\sqrt{\beta(s)}\sqrt{\beta(s_0)}\cos(\pi v - |\mu(s)-\mu(s_0)|), \quad (2)$$

where, $COD(s)$: COD at position s along the storage ring, $\beta(s)$: beta function at s, $\mu(s) = \int_0^s \frac{ds\prime}{\beta(s\prime)}$, v: tune $\left(=\int_0^C \frac{ds\prime}{\beta(s\prime)}\right)$, where C is the circumference of the storage ring. $\theta(s_0)$: kick angle at $s = s_0$.

3. The kick angle generated by changing the quadrupole strength is expressed as

$$\theta(s_0) = \delta K \cdot L \cdot \delta X(s_0), \text{ for } x \text{ direction, and} \quad (3)$$
$$\theta(s_0) = \delta K \cdot L \cdot \delta Y(s_0), \text{ for } y \text{ direction,} \quad (4)$$

where $\theta(s_0)$: kick at $s = s_0$, δK: amount of the change of the quadrupole strength, L: quadrupole length, $\delta X(s_0)$, $\delta Y(s_0)$: offset of the orbit from the quadrupole axis.

If the x_0 and y_0 in the equations (1) coincide with the quadrupole axis, $\delta X(s_0) = u/S_x$ and $\delta Y(x_0) = v/S_y$.

Since we performed measurements for only y direction, because of the analysis convenience, we describe only y-direction case hereafter. The equation (2) becomes

$$COD_y(s) = \frac{\delta K \cdot L}{2\sin(\pi v_y)} \cdot \left(\frac{v}{S_y}\right)\sqrt{\beta_y(s)}\sqrt{\beta_y(s_0)}\cos(\pi v_y - |\mu_y(s)-\mu_y(s_0)|). \quad (5)$$

The equation (5) suggests that the S_y can be calibrated if all the storage ring optics parameters are accurate and the $COD(s)$ are measured without any error. This means that the COD generated by quadrupole strength change is used as a reference for the BPM position reading to be calibrated.

The method is applicable for obtaining the deviation of the coefficients from the common factor to all the BPM. If all the BPM has same amount of factor difference of the position sensitivity coefficients from the true value, the exact value of the coefficients cannot be obtained since the $COD(s)$ themselves are measured with the BPM.

MEASUREMENT

SPring-8 storage ring and its BPM

Before proceeding to the experimental method, the SPring-8 storage ring and its BPM are briefly described. The storage ring is a third generation synchrotron radiation source with the stored electron energy of 8 GeV. It consists of 40-normal and 4-long-straight cells. Normal cell is a double bend achromat type. Each normal cell has 6 BPM, and the long straight cell has 8 BPM.

A BPM consists of 4 button pickups as shown in figure 1. The electrodes are welded directly on a long vacuum chamber with the lengths up to over 5 m. The chamber is a ante-chamber type one, as shown in figure 1. The tuning frequency for detection of electrode signals is same as RF acceleration frequency, which is 508.58 MHz.

Experiment

We adopted the following procedure for the measurement of the beam based calibration of the BPM position sensitivity.

1. Make a bump across one of the quadrupole magnet and its adjacent BPM. This BPM is the target for calibration.
2. Change the quadrupole strength slightly.
3. Measure COD all around the storage ring before and after the change of the quadrupole strength.
4. Repeat the step from 1 to 3 for the predefined set of bump heights.

We adopt the amplitude of the tune component of the COD fourier spectrum for the indicator of the amount of the extra COD caused by the single kick generated with quadrupole strength change.

After transformation from $(y(s_j), s_j)$ to the normalized coordinate ($\frac{y(\phi_{y_j})}{\sqrt{\beta(\phi_{y_j})}}$, ϕ_{y_j}), fourier-cosine and -sine components were calculated, where y_j is the position reading value of j-th BPM. The definition of the cosine($C_y(n)$) and sine($S_y(n)$) components of the n-th fourier harmonics are

$$C_y(n) = \int_0^{2\pi} \frac{COD_y(\phi_y)}{\sqrt{\beta_y(\phi_y)}} \cos(n\phi_y) d\phi_y , \; S_y(n) = \int_0^{2\pi} \frac{COD_y(\phi_y)}{\sqrt{\beta_y(\phi_y)}} \sin(n\phi_y) d\phi_y, \quad (6)$$

where $\phi_y(s) = \frac{2\pi \mu_y(s)}{\nu_y}$, and the amplitude is $A_y(n) = \sqrt{C_y^2 + S_y^2}$.

We approximated the fourier components by the summations of measured values as,

$$C_y(n) \approx \sum_{BPM_j} \frac{1}{2} \left(\frac{y_{j+1}}{\sqrt{\beta_y(\phi_{y_{j+1}})}} \cos(n\phi_{y_{j+1}}) + \frac{y_j}{\sqrt{\beta_y(\phi_{y_j})}} \cos(n\phi_{y_j}) \right) (\phi_{y_{j+1}} - \phi_{y_j}), \quad (7)$$

$$S_y(n) \approx \sum_{BPM_j} \frac{1}{2} \left(\frac{y_{j+1}}{\sqrt{\beta_y(\phi_{y_{j+1}})}} \sin(n\phi_{y_{j+1}}) + \frac{y_j}{\sqrt{\beta_y(\phi_{y_j})}} \sin(n\phi_{y_j}) \right) (\phi_{y_{j+1}} - \phi_{y_j}). \quad (8)$$

The sign to the amplitude was applied as $A_y(n) \leftarrow \text{sgn}(C_y(n)) \cdot A_y(n)$, for the analysis convenience, where $\text{sgn}(x) = \begin{cases} +1 & x \geq 0 \\ -1 & x < 0 \end{cases}$.

Further modification to the $A_y(n)$ was made for normalization to the kick angle as

$$A_y^N(n) = \frac{A_y(n)}{\delta K \cdot L \cdot \sqrt{\beta_y(s_0)}} \propto \frac{v}{S_y}.$$

Taking the n as the integer part of the vertical tune($n = 18$), the position reading values of the target BPM were plotted against the normalized fourier amplitude of the extra COD-difference occurred by change of quadrupole strength for each bump height. The plot is expected to be on a straight line, and the slope has the information about the BPM position sensitivity.

preliminary result

We made measurements on 8 BPM-quadrupole sets for vertical direction. They are listed in table 1.

The quadrupole strength was changed by changing the current applied to the magnet. The change of the current(δI) were the same for all the measured quadrupoles, which was 10 A. The amount of strength change δK was obtained as $\delta K = K \cdot (\delta I / I_0)$, for the magnets operated in the linear range of excitation curve, where K is the nominal value of strength, I_0 is the nominal current applied in usual operations. Corrections were applied for the magnets whose nominal current were non-linear range of the excitation curve. Relative changes of the strength($\delta K/K$) are also listed in table 1.

In figure 2(left), an example of the extra COD caused by 10-A change of the quadrupole current is shown. The calculated value from the designed optics parameter and the unit kick angle at the same quadrupole was plotted on the same graph. The

TABLE 1. Measured set of BPM and quadrupole magnet and relative variation of quadrupole strength

set No.	BPM(serial No. / cell-No.in the cell)	Q magnet(cell-No. in the cell)	$\frac{\delta K}{K}(\%)$
1	23 / 4-5	4-8	2.6
2	87 /15-3	15-4	2.0
3	119 / 20-5	20-8	2.6
4	122 / 21-2	21-3	2.6
5	144 / 24-6	24-10	4.5
6	192 / 32-6	32-10	4.5
7	227 / 38-5	38-8	2.6
8	266 / 45-2	45-3	2.6

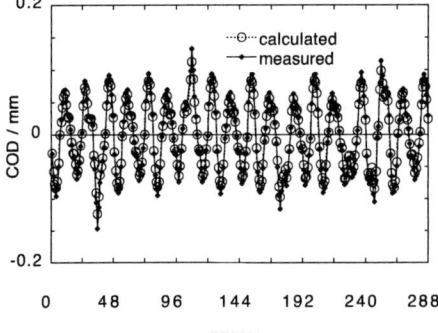 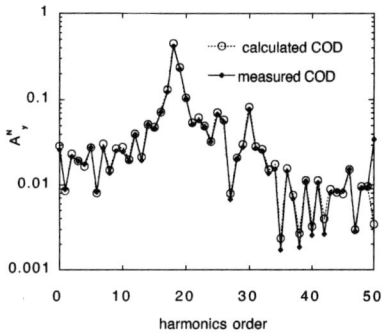

FIGURE 2. Calculated(open circles) and measured(closed diamonds) COD(left) and their fourier spectra(right) for the 7-th set of BPM-quadrupole(serial No. 227 BPM) in Table 1. Calculation was made using the linear part of the designed optics parameters. Fourier components were calculated up to 50-th order harmonics.

agreements of the COD pattern were quite good for all the measured set of quadrupole and BPM. It shows that the nonlinearity of the optics parameter was small enough so that the condition 2 was fulfilled.

An example of fourier spectrum of the extra COD is shown in figure 2(right) with the spectrum of the calculated COD. The spectrum also agrees well. The $n = 18$ component of the fourier spectrum was used for the estimation of the sensitivity coefficient, since the integer part of the vertical tune is 18 for normal operation condition of the SPring-8 storage ring.

The BPM reading values are plotted against the normalized amplitude $A_y^N(18)$ as shown in figure 3. The graph shows a linear dependency, and the data were fitted with a straight line. The slope, which we express as α here, is related to the sensitivity coefficient.

To compare the sensitivity obtained with the designed optics parameters, the COD was calculated in the case of the 1-mm offset of the target quadrupole, and the fourier amplitude of $n = 18$ component was obtained for the calculated COD, which we express as $A^N_{y_{calc}}(18)$ here. Since the calculated data make a line connecting the origin (0,0) and the point $(A^N_{y_{calc}}(18), 1 \text{ mm})$, the slope of this line is $1/A^N_{y_{calc}}$.

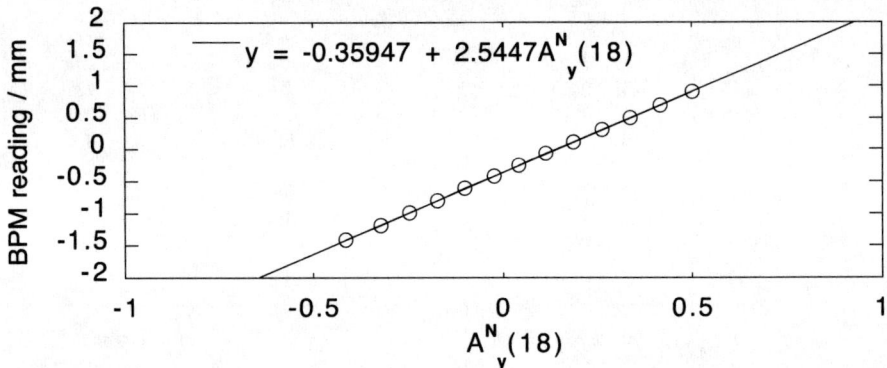

FIGURE 3. BPM reading values of serial No. 227 BPM vs. 18-th order harmonics of normalized fourier amplitude for each bump height (open circles). The line indicates the fitted straight line.

The slope of the line consists of the measured data should be equal to $1/A^N_{y_{calc}}(18)$, if the sensitivity derived from the measured COD is accurate. Here we introduce the ratio of the slope of measured data(α) to the slope of the calculated line as r;

$$r = \frac{|\alpha|}{\left(\dfrac{1}{A^N_{y_{calc}}(18)}\right)} = |\alpha| A^N_{y_{calc}}(18).$$

The obtained values of α, $1/A^N_{y_{calc}}(18)$, and r are summarized in table 2.

TABLE 2. Summary of fitted slope, the inverse of the calculated fourier amplitude, and the ratio of them, for the measured set of BPM and quadrupoles

Set No.	BPM (serial No. / cell-No. in the cell)	$|\alpha|$	$\dfrac{1}{A^N_{y_{calc}}(18)}$	r
1-1	23 / 4-5	2.43	2.31	1.05
1-2 *	23 / 4-5	2.42	2.31	1.05
2	87 / 15-3	2.74	2.32	1.18
3	119 / 20-5	2.28	2.24	1.01
4	122 / 21-2	2.55	2.34	1.09
5	144 / 24-6	2.29	2.20	1.04
6	192 / 32-6	2.31	2.18	1.06
7	227 / 38-5	2.54	2.33	1.09
8	266 / 45-2	2.57	2.36	1.09

* The measurement for BPM 4-5 was made twice with the interval of about 1 month, from which the reproducibility of the measurement is estimated to be 0.4%

DISCUSSION

The reproducibility of the measurement was estimated to be less than 1% from the difference of two measurements on the same BPM (BPM 4-5), as noted in the table 2. The slope fitted with the measured data agree withe calculated values within 10% except one BPM(serial No. 87: 15-cell No.3 BPM). The values of r are different from unity above the reproducibility level. Suppose the accuracy of r is the level of reproducibility, corrections to the sensitivity coefficients(S_y) must be applied, and the serial No. 87 BPM has the correction as large as 18%.

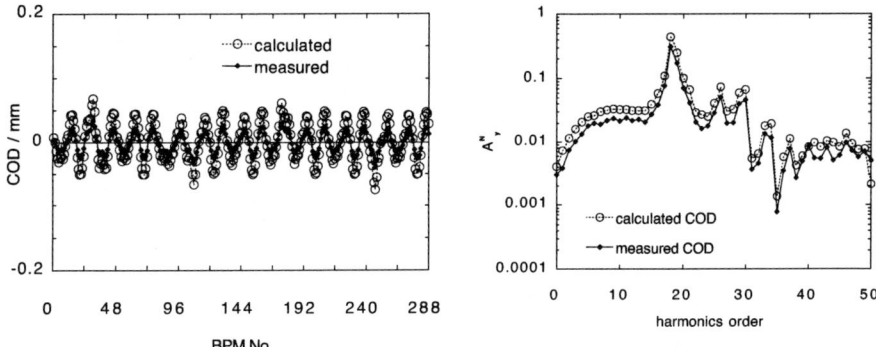

FIGURE 4. Calculated(open circles) and measured(closed diamonds) COD(left) and their fourier spectra(right) for the serial No. 87 BPM. Calculation was made using the linear part of the designed optics parameters. Fourier components were calculated up to 50-th order harmonics.

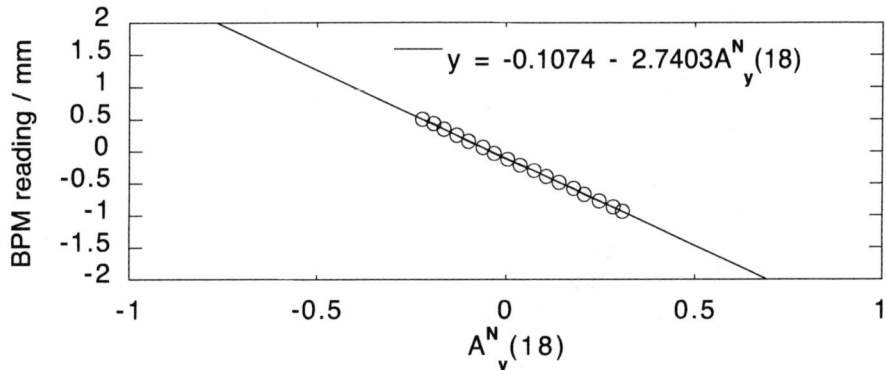

FIGURE 5. BPM reading values of serial No. 87 BPM vs. 18-th order harmonics of normalized fourier amplitude for each bump height (open circles). The line indicates the fitted straight line.

In figure 4, the COD and its spectrum are shown for the serial No. 87 BPM. The COD and the fourier spectrum have the similar patterns of the calculated ones. The plot of BPM reading valued vs. fourier amplitude is shown in figure 5. The data are on a straight line. From these figures, we can judge that the cause of 18% difference from the unity of the r value is not the non-linearity effect.

For the SPring-8 Storage Ring BPM, the sensitivity coefficients were obtained by scanning an RF antenna. The antenna was supported by the single end and attached to an $x-y$ stage for scanning purpose. The antenna consisted of a supporting rod and a semi-rigid coaxial cable with center conductor disclosed at the tip of the cable. The total length of the antenna (rod, cable) from the top end to the supporting end was about 2 m.

The measurement was done for all the 288 BPM of the storage ring before installing the vacuum chambers. The sensitivity values obtained the RF antenna scanning method varied from BPM to BPM by several %. If this variation of the values of S_y came from the measurement accuracy of the RF antenna method, the several % of correction to S_y is plausible. However, the 18-% correction, as in the case of serial No. 87 BPM, exceeded our expectation. Thus, the estimation of the accuracy of this method must be investigated to decide whether the 18-% correction should be applied or not.

SUMMARY

We propose a method for beam based calibration of BPM position sensitivity, and made preliminary measurements for some of the BPM in the SPring-8 storage ring. The reproducibility of the measurement of this method is better than 1%. The level of accuracy of the measurement should be investigated further.

REFERENCES

1. Peter Röjsel, "A beam position measurement system using quadrupole magnets magnetic centra as the position reference", Nuclear Instruments and Methods in Physics Research A343 (1994) 374-382
2. A. Jankowiak, C. Stenger, T. Weis, K. Wille, "The DELTA Beam-Based BPM Calibration System", Proceedings of The 8th Beam Instrumentation Workshop(BIW98), SLAC, 1998
3. B. Dehning, G.-P. Ferri, P. Galbraith, G. Mugnai, M. Placidi, F. Sonnemann, F. Tecker, J. Wenninger, "Dynamic Beam Based Calibration of Beam Position Monitors", Proceedings of the Sixth European Particle Accelerator Conference, Stockholm, 22-26 June 1998, p430-432
4. K. Soutome, H. Tanaka, M. Takao, H. Ohkuma, N. Kumagai, "Estimation and measurement of effective beam position monitor offsets by using a stored beam", Nuclear Instruments and Methods in Physics Research A459 (2001) 66-77
5. Mitsuhiro Masaki and Storage Ring Commissioning Group, "A Method of Beam-based Calibration for Beam Position Monitor", Proceedings of The 11th Symposium on Accelerator Science and Technology, Harima Science Garden City, Hyogo, Japan 1997, pp83-85

Nonintercepting Imaging Diagnostics for the APS Injector during Storage Ring Top-Up Operations[*]

A. H. Lumpkin, W. J. Berg, and B. X. Yang

Advanced Photon Source, Argonne National Laboratory, Argonne, IL 60439

Abstract. The recently implemented top-up operating mode of the Advanced Photon Source (APS) storage ring has motivated an emphasis on nonintercepting imaging diagnostics in the injectors. We present the upgrades to the optical synchrotron radiation (OSR) monitors on the accumulator ring and injector synchrotron as well as the plans for a new OSR monitor on a chicane dipole in the linac and for an optical diffraction radiation (ODR) monitor for the 7-GeV transport line to the storage ring. Two key issues are signal strength for a single macropulse in the chicane and discriminating key transverse information from the visible light ODR, respectively.

INTRODUCTION

The Advanced Photon Source (APS) is a third-generation x-ray synchrotron radiation user facility [1]. We have recently implemented a top-up operating mode that involves single-shot injection into a targeted rf bucket of the fill pattern to maintain the 100-mA stored beam current. When the fill pattern is 23 singlets and a low-emittance (~3.5 nm rad) lattice is used, injection efficiency must be high on each shot (which occurs every two minutes). The sequence involves about 20 seconds of checkout time, which has been found to be impractical for tuning the injector between shots using intercepting beam profiling screens in the linac and transport lines. This scenario resulted in a plan to upgrade the existing optical synchrotron radiation (OSR) monitors on the accumulator ring (AR) and the injector synchrotron (IS) and to evaluate other nonintercepting (NI) diagnostic imaging techniques for the linac and various transport lines between the accelerators. In particular, we are exploring the addition of one OSR station using a dipole source in the chicane bunch compressor located at the 150-MeV point in the linac and an optical diffraction radiation (ODR) monitor on the 7-GeV transport line between the IS and the storage ring (SR). The status and feasibility aspects of these initiatives will be presented.

---

[*] *Work supported by U.S. Department of Energy, Office of Basic Energy Sciences under Contract No. W-31-109-ENG-38.

EXPERIMENTAL BACKGROUND

The APS facility includes the three electron guns (two rf thermionic [2, 3] and one rf photocathode [4]), an S-band linac with acceleration capacity up to 600 MeV, the accumulator ring, the injector synchrotron that ramps the beam energy to 7-GeV, the 1104-m circumference 7-GeV storage ring, and the transport lines between these accelerators [5]. The facility is schematically shown in Fig. 1. There are rf beam position monitors (BPMs) installed in all accelerators and transport lines that provide the beam position noninterceptively. However, aspects of the beam transverse or longitudinal parameters are evaluated by both intercepting screens and nonintercepting imaging techniques during standard operations. In this case, the fill-on-fill interval may be 12 or 24 hours depending on the stored beam lifetime. There are OSR ports on both the AR and IS, and there are both an OSR port and an x-ray synchrotron radiation (XSR) port on the storage ring. The SR OSR is typically used for bunch-length measurements with a streak camera while the XSR is used with a pinhole camera to provide ~ 22 µm (σ) resolution beam size imaging [6].

In the linac, intercepting beam profile screens are used at a number of locations and several have been upgraded to utilize optical transition radiation (OTR) converters or YAG:Ce single crystals instead of the original Chromox scintillation screens [7]. For top-up mode, the single shot of charge is injected every two minutes. The rf beam is only available in the linac, transport lines, AR, and IS for about 20 seconds preceding the single shot. It has been problematic to tune the linac rf phase using intercepting screens to optimize injection into the AR in this time period. It is also useful to operations to be able to observe simultaneously the beam in all three rings of the facility and verify the basic injector functions. In support of these needs the upgrades to the suite of nonintercepting imaging diagnostics were planned as described in the next section.

DIAGNOSTIC UPGRADES

The initial upgrades and proposed upgrades on the injector will now be discussed. Since OSR is a major feature, Table 1 summarizes the expected signal strengths from various points in the machine, and Fig. 1 shows schematically where the sources are. In the case of the single-pass mode of the linac beam, signal strength is one of the challenges for a linac macropulse. In the rings, thousands of turns might be integrated into a 30-ms image.

TABLE 1. Estimates of Total Visible Photons per 1-nC Charge in a Single Pass (λ = 400-700 nm).				
Accelerator	Beam Energy (GeV)	B – Field (T)	Angular Width	Integrated Flux
Linac OTR	0.20	Thin film	2 π solid angle	10.6×10^7
Chicane	0.15	0.6	20 mrad	4.1×10^7
Accumulator Ring	0.375	1.2	10 mrad	3.1×10^7
Injector Synchrotron	7	0.7	8 mrad (16mm @ 2m)	8.4×10^7
Storage Ring	7	0.6	3 mrad (35mm @ 12m)	4.4×10^7

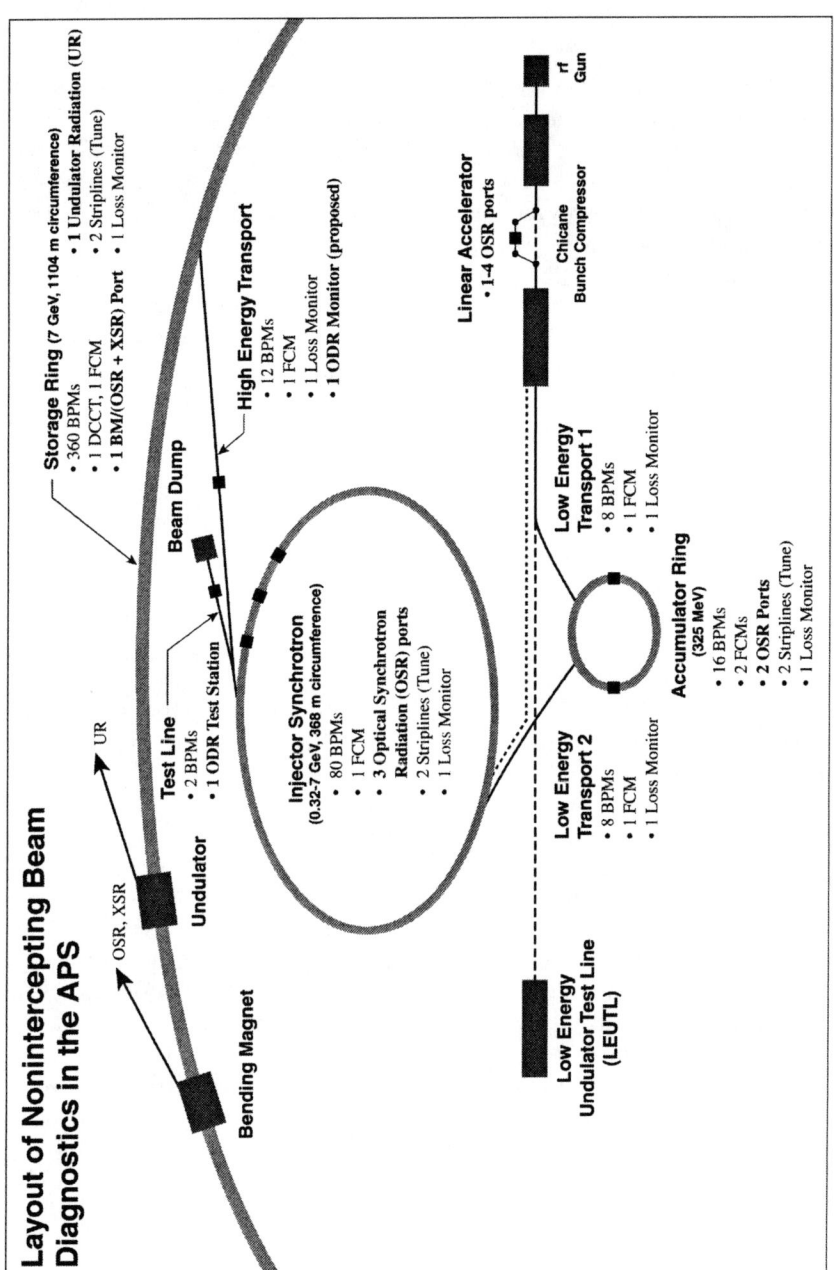

FIGURE. 1. Schematic of the APS injector facility and storage ring with list of the nonintercepting imaging diagnostics.

Linac

At the 150-MeV location, a four-dipole chicane has been installed to provide bunch compression of the rf photocathode (PC) gun beam [8]. We propose to add an OSR monitor after the second dipole, which would be a dispersive point in the lattice. Then transverse (horizontal) beam centroid and size measurements would yield relative information about beam energy and energy spread, respectively. These latter two parameters are directly related to the upstream rf phase and amplitudes for the accelerating structures. Operationally, this could be one of the key tuning aids since we have experienced rf phase shifts sufficient to impact injector efficiency in the past. The projected OSR strength is given in Table 1, and this may require some sensitivity boost with an intensified camera for successful imaging.

Accumulator Ring

Typically, the linac injects a 325-MeV, 8-ns-long macropulse consisting of 25 S-band micropulses. Although the two OSR ports have been functional and were used extensively during the commissioning period several years ago, they have been used in later years for machine studies. The video was selectable in the APS-wide video multiplexor. As a first phase, a dedicated fiber line was added that provides the video at all times to a dedicated quad view with monitor in the main control room (MCR). This will be followed by upgrades of the cameras and optical table components with EPICS control capabilities. The present video clearly shows the top-up sequence of injection of beam, storing of beam, damping of beam size, extraction, and then the absence of beam after the SR injection shot.

Injector Synchrotron

This ring is used to ramp the beam energy from the 325 MeV out of the AR to the full energy of 7 GeV. The ramping of the rf cavity power/phase and the power supplies for the dipoles, quadrupoles, sextuples, and correctors is a delicate balance that occurs over a 200-ms period at a 2-Hz repetition rate. The OSR images of the two-dipole sources reveal the damping of the beam transverse size, a centroid motion, and then the absence of circulating beam after extraction. The beam image centroid at the end of the ramp should be stable for top-up. A dedicated fiber line was also implemented to transport the images to the MCR, and the ramping phenomena are monitored. Ultimately, a dedicated video digitizer with video processing will be added to track the final beam centroid and size.

7-GeV Transport Line

The transport line between the IS and SR at 7 GeV provides another opportunity to track the beam quality. Besides the intercepting screens, we propose complementary information on transverse parameters may be obtained from an ODR monitor [9]. In this case of high gamma, a 1-mm aperture or slit will result in visible diffraction radiation (DR) being emitted. Basically, appreciable DR will be generated as the charged particle beam passes through the aperture when gamma times the reduced wavelength is equal to or larger than the aperture radius (a); i.e., $\gamma \lambda \geq a$. Backward ODR will be emitted around the angle of specular reflection as in OTR. So slits placed at a 45° angle to the beam direction will result in ODR emitted around 90° to the beam direction as illustrated in Fig. 2. Our elliptical beam size should result in a contribution to the asymmetry of the angular distribution pattern. Calculated signal strengths for horizontal and vertical polarization components are about 25% and 80% of OTR for $R = a/\gamma\lambda = 0.5$ but for a lower beam energy example [9].

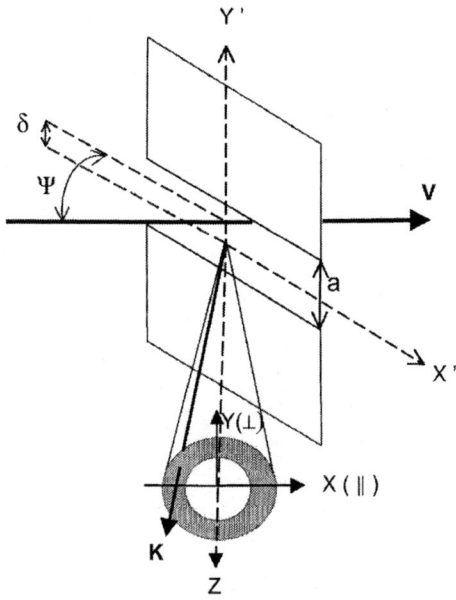

FIGURE 2. A schematic representation of the backward diffraction radiation emitted when a charged particle passes through a slit. The conducting plates are at 45° to the beam direction. (Based on Fig. 1 of Ref. 9).

SUMMARY

In summary, we are in the process of augmenting our NI imaging diagnostics at APS to support the top-up operating mode. In addition, we will be evaluating other locations to employ OSR or ODR imaging in support of interleaving operations where the linac will be used to inject another beamline for free-electron laser (FEL) experiments.

ACKNOWLEDGEMENTS

The authors acknowledge the engineering assistance of Steve Shoaf, Rich Diviero, Richard Koldenhoven, Jim Stevens, and Ned Arnold of the ASD Controls Group and the support of Om Singh, Glenn Decker, and Antanas Rauchas of AOD.

REFERENCES

1. Galayda, J. N., Proc. of the 1995 Particle Accelerator Conference and International Conference on High-Energy Accelerators, Dallas, TX, pp. 4-8 (1996).
2. Borland, M., Proc. of the 1993 Particle Accelerator Conference, New York, pp. 3015-3017 (1995).
3. Lewellen, J. W., et al., Proc. of the 1998 Linac Conference, Chicago, ANL-98/28 Vol. 2, pp. 863-865 (1999).
4. Biedron, S., et al., Proc. of the IEEE 1999 Particle Accelerator Conference, New York, NY, pp. 2024-2026 (1998).
5. Lumpkin, Alex H., Proc. of the Seventh Beam Instrumentation Workshop, Argonne, IL, AIP Vol. 390, pp. 152-172 (1997).
6. Yang, Bingxin, "Optical System Design for High-Energy Particle Beam Diagnostics," these proceedings.
7. Lumpkin, A. H., et al., Nucl. Instrum. Methods in Phys. Res. A429, 336 (1999).
8. Borland, M., Lewellen, J., and Milton, S. V., Proc. of the 2000 Linear Accelerator Conference, SLAC-R-561, pp. 863-865 (2001).
9. Fiorito, R. B., and Rule, D. W., Nucl. Instrum. and Methods in Phys. Res. B173, 67 (2001).

Simple "Package Design" Ion Chamber Monitors for TRIUMF's Proton Beamlines

Daniel Gray and Brian Minato

TRIUMF, 4004 Wesbrook Mall, Vancouver, B.C. Canada, V6T 2A3

Abstract. In the beam line designed to supply 100 µA of 500 MeV protons to the two ISAC production targets at TRIUMF, 13 profile monitor stations were required. The design allows each station to be fitted with either an air driven wire scanner module for high currents or an ionization chamber for low currents. Ring shaped multilayer G10 circuit boards were designed for the latter to enable a simple modular "gas package" that is easily serviced and aligned. These gas packages have only five basic parts, two outer window frames with 0.010 in. thick E-beam welded Al windows, two ring shaped circuit boards with 2 mm wire spacing and edge card connectors (X and Y use the same design of board) and one center frame for mounting to the inserting mechanism and holding a .001 in. Al foil. The circuit boards are critical components due to the necessity to hold vacuum along their edges. Signal traces pass from the inner part of the ring that is gas filled to the outside of the ring that is in vacuum. The windows and center foil frame are at -300 V bias. This gas package design led to a similar design used to upgrade the existing (1970's vintage) proton beamline ion chamber monitors.

PROFILE MONITORS FOR THE ISAC BEAMLINE

2A Beamline Standard Profile Monitors

From the TRIUMF 500 MeV cyclotron, the 2A beamline [1] supplies up to 100 µA of protons to the ISAC targets. Nine profile monitors are required for the main 2A beamline and 2 additional monitors in each leg supplying the 2 target stations.

Two types of profile monitor are used in the 2A beamline. Both use the same drive mechanism, but they can be assembled as either an ion chamber or a wire scanner Fig. 1. The drive mechanism utilizes an air cylinder with a 6 in. stroke, end cushions and air speed controls. Motion is guided by a linear slide fitted with two guide blocks and a 6 in stroke edge welded stainless steel bellows. This monitor drive mechanism is a modification of a prototype designed and tested in 1993 [2].

Monitor boxes are installed at a 45° angle and have been precision manufactured to allow changing of monitor drives without realignment. Hand operated toggle clamps and alignment dowels simplify monitor removal and replacement.

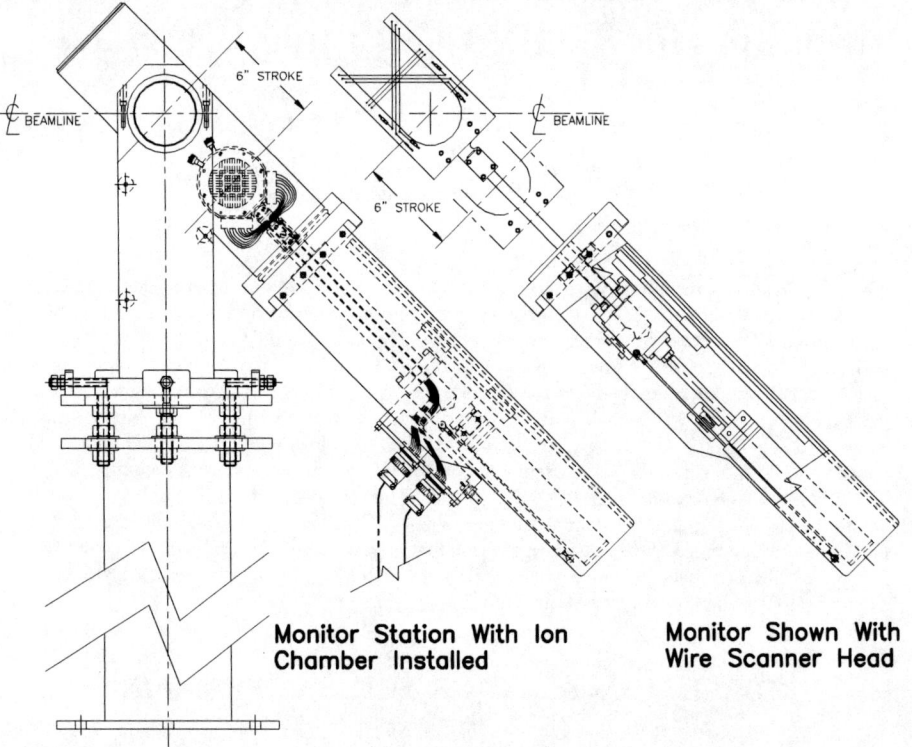

FIGURE 1. Standard 2A beamline profile monitor station.

Wire Scanner Head

Three signal blades (0.125 in. wide x 0.002 in. thick Be-Cu) pass through the beam at a 45° angle. Signal blades are installed so that the beam sees the 0.002 in. edge. Horizontal and vertical blades provide X and Y information. A third blade is perpendicular to the scan direction. A tomograph program generates hexagonal contours of the beam density. Each blade has two 0.005 in. diam Mo, Au plated, bias wires installed 0.2 in. away from the 0.125 in. wide faces. The bias voltage is +100V through a 10MΩ current protection resistor. The signal blades and bias wires are spring tensioned. During a scan, sensing of the head position is provided by a 7 in. stroke, precision wirewound 10kΩ linear potentiometer. The scan speed is set to approximately 0.3 m/s using air cylinder speed control valves. Beam current limits are set at 10 µA due to heating of the blades or from beam trips caused by scattering.

GAS PACK ION CHAMBERS HEAD

The standard 2A ion chambers gas packs use only 5 basic parts, two outer windows, two wire boards, and one center mounting frame Fig. 2. The chamber must be gas tight with one atmosphere of 90% Ar/10% CO_2 inside and vacuum at 10^{-7} Torr outside.

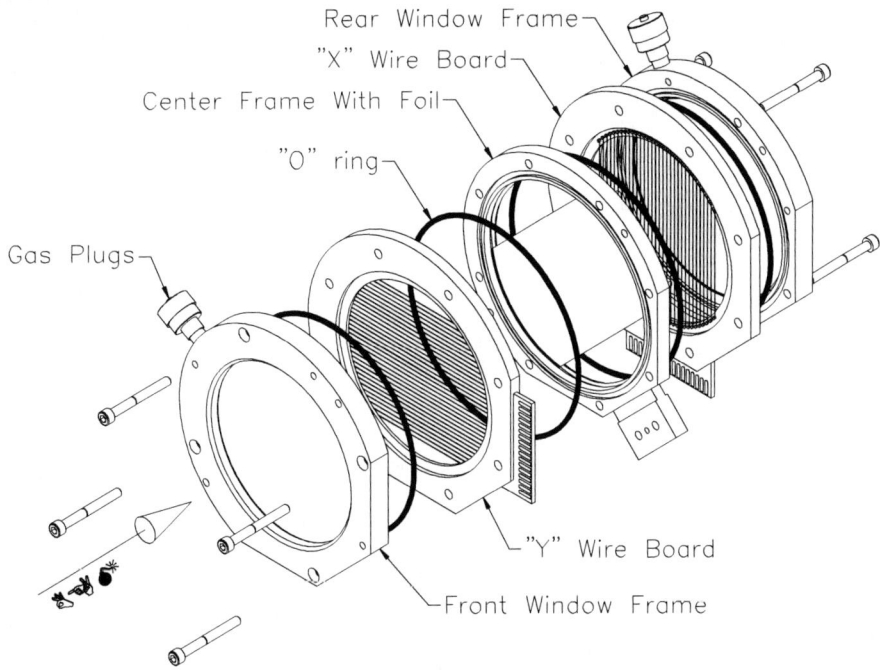

FIGURE 2. Exploded view of 2A gas pack ion chamber.

The main component of the chamber is the multi-layer wire boards made from G10 Fig. 3A. The same board layout is used for X and Y; one board is reversed and rotated 90°. The wire board is actually three ring shaped 0.062 in. thick circuit boards bonded together to form a 3/16 in. sandwich. A cross section of the board is shown in Fig. 3B. Embedded in this center board are Au plated signal traces that run from 0.015 in. diam vias (plated through holes) located around the center opening in the board to the perimeter of the board. The vias are spaced to permit the installation of 32 signal wires at 2 mm spacing. Vias are used for ease of assembly and the ability to change a single wire should it become damaged. Vias also retain the wire should the solder joint soften with heat, although this is not a consideration in the 2A chambers. Signal wires used are 0.005 in. diam Mo with 5% Au plating. Au plating permits easy soldering. Traces at the perimeter of the board permit the use of a readily available edge card connector (part # 345 034 500 202). Edac Inc. manufactures the connectors from green diallyl phthalate plastic that has good radiation properties [3]. Two outer circuit boards are bonded to either side of the center board by the manufacturer using A11-108 prepregs [4]. The wire boards must be bonded together with the ability to hold vacuum along their edges. The outer boards have ground planes embedded within them that reduce bias leakage to the signal traces. They have exposed tabs that permit connecting the ground plane to an edge card connector trace on the center board. Exposed faces of the outer boards act as an 'O'-ring sealing surface and must be free of defects.

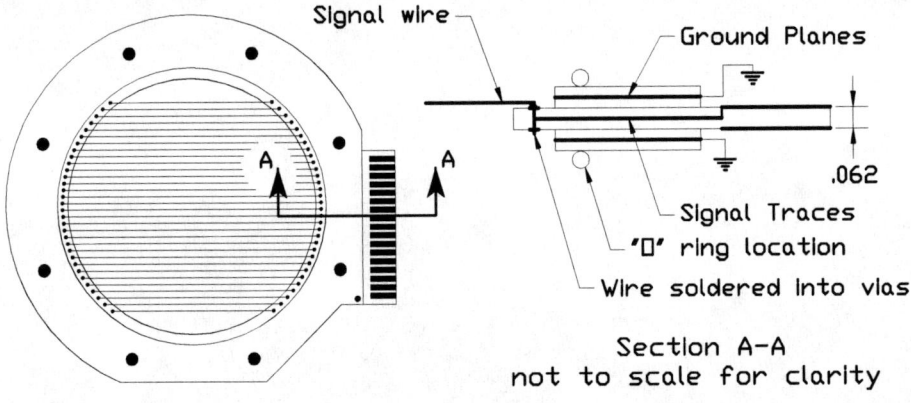

FIGURE 3A. Plan view of wire board with wires installed.

FIGURE 3B. Cross section A-A showing multiple layers.

Two outer window frames are machined from 5086 aluminum and have a 0.010 in. thick Al window electron beam welded into the center. It is important to match the alloys correctly for welding. The welding is a critical operation as it must be vacuum tight and mechanically strong. After calculating the maximum deflection of the window, 0.060 in. [5], the maximum stress on the window at the center is calculated at 15660 psi [5]. The first time the chamber is pressurized, the window plastically deforms to the 0.06 in. deflection; this increases the yield strength of the window.

An 'O'-ring groove is machined into the sealing face of the frame. Each window frame has a screw-type gas purging plug set in epoxy. Care was taken when machining the gas plug holes in the frame. The holes were machined with a flat bottom that matches the end of the plug fitting. This minimizes the surface area of the epoxy exposed to the gas, reducing the out-gassing from the epoxy into the chamber. Gas plugs used are Cajon Ultra-Torr tube fittings (#SS-2-UT-A-4), with a small machined plug inserted where the 1/8 in. tube would normally be used.

In the center of the chamber is a frame machined with two "O" ring grooves to enable vacuum sealing with the circuit boards. The center frame has a tab for mounting the chamber to the drive mechanism through insulating bushings. The insulating bushings enables the chamber to be biased at –300 V. In the middle of the center frame is a 0.001 in. thick 2 in. wide Al foil held in place by an Al wire ring. This foil, and the two outer windows, act as high voltage bias planes.

Ion Chamber "Gas Pack" Assembly and Testing

Using fixtures the windows and wire boards are leak checked before assembly. After leak checking, the signal wires are tensioned to approximately 5 N and soldered into the vias. After assembly using good vacuum practices, a final leak check is performed on the entire chamber. The assembled package is placed in a vacuum chamber connected to a leak detector. Two tubes are installed into the gas plug fittings. The tubes run from the gas pack to the outside of the vacuum chamber. The vacuum chamber is evacuated and a flow of helium probe gas is passed through the inside of the gas package. A leak rate Q of 10^{-6} Torr l/s or better is required. The volume V_c and the pressure P_1 inside the chamber is 0.136 liters and 760 Torr respectively. If the leak rate is considered to be linear, then from equation (1), the package would take $t_2 \approx 3.2$ years to leak all its gas to the vacuum space [6]. After 6 months the pressure would be approximately 633 Torr. This meets the design specification that gas loss is less than approximately 20% in 6 months. Since gas quantity and signal gain are proportional, the monitor would still give profiles but with a 20% signal loss. Low gas loss also lessens the reverse stress on the window when the beamline is vented to atmosphere. When in service, the chambers are routinely refilled with gas at about a 6 month interval, corresponding to TRIUMF's shutdown maintenance schedule. The gas used is 10% CO_2 / 90% Ar; this ratio is not critical therefore a standard welding gas, "Praxair, Mig Mix Gold" [7], is used.

$$Q = \frac{P_1 V_c}{t_2} \Rightarrow t_2 = \frac{P_1 V_c}{Q} \qquad (1)$$

Electronics and Operation

The signal electronics is beyond the scope of this article but a simple block diagram is shown in Fig. 4.

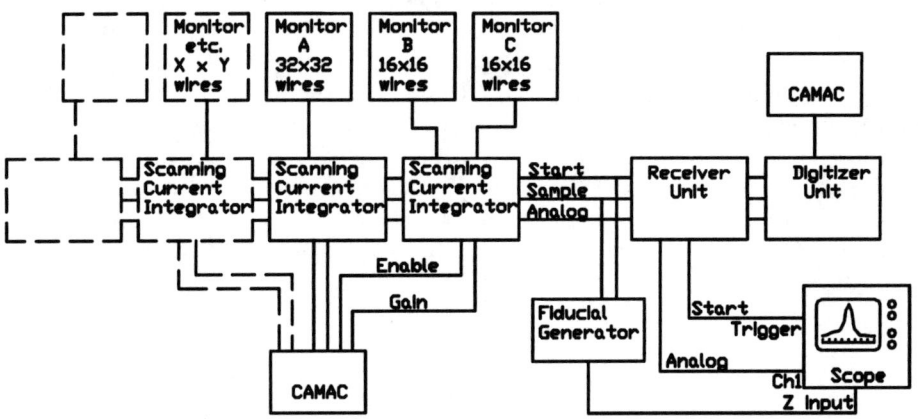

FIGURE 4. Block diagram of ion chamber electronics

A scanning current integrator collects signals from one 32x32 wire monitor or two 16x16 wire monitors. A CAMAC module controls the integrators gain and enable. A fiducial generator indicates wire position on an oscilloscope. A beam profile is shown in Fig. 5. Current limits are set at 100 nA to avoid beam trips due to excessive scattering, typical use is ≤ 1 nA p^+.

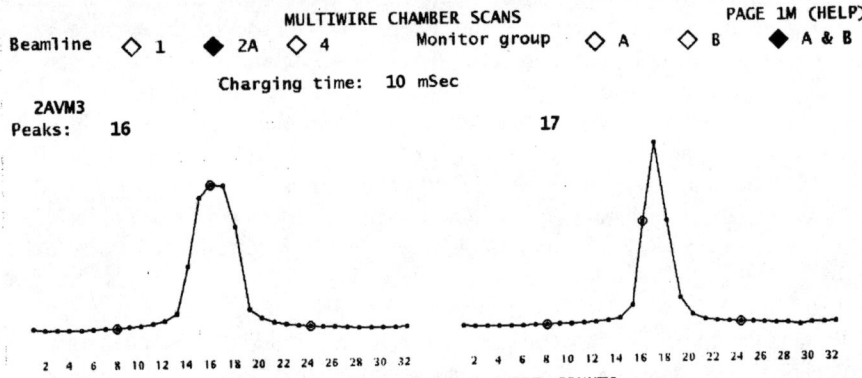

FIGURE 5. Beam profile from ion chamber

ADAPTATIONS OF THE GAS PACK DESIGN

Entrance Module Profile Monitors

In addition to the 13 standard profile monitor stations, there is also a profile monitor in each of the Entrance Modules at the ISAC target stations. The modules have been designed with the ability to change the profile monitor sensing head to either an ion chamber or a wire scanner, using a remote handling hot cell. Sensing heads are similar to the ones used in the 2A beamline, however, they are installed onto a frame which permits interchangeability on the same monitor. The gas pack is hard wired to a plug attached to the frame. During the initial start up and low current commissioning, the first target station was fitted with an ion chamber Fig 6B. The chamber was subsequently changed to a wire scanner Fig 6A. Currently the entrance modules to both target stations are fitted with wire scanner heads. Due to design constraints the sensing heads move in and out of the beam in a vertical direction. The vertical profile is measured directly but the horizontal profile is derived from a tomograph program.

FIGURE 6A. Entrance module profile monitor, fitted with wire scanner head.

FIGURE 6B. Entrance module profile monitor, fitted with gas pack ion chamber.

Upgrades to existing Monitors

It was recently decided to upgrade the original swing style ion chamber monitors in other proton beamlines at TRIUMF. These old chambers used circuit boards with signal traces on the surface. These boards then required additional G10 boards to be epoxied to each face to provide an "O" ring sealing surface [8]. This assembly led to a high failure rate, due to gas leakage into the vacuum space. The leaks usually occurred from gaps between the epoxy and traces. The wire boards were hard wired to multiple 9 pin feed-throughs (FT) making service difficult. The chambers swing 90° in and out of the beam via an in-house fabricated ferro-fluid FT. After years of service these FT's became prone to vacuum leakage. They would have to be periodically "topped up" with ferro-fluidic fluid.

The upgraded monitors have been fitted with a gas package style ion chamber of a similar design to those used in the 2A beamline. The upgraded chamber is made with circuit boards of either 16 wires at 3 mm spacing or 16 wires at 5 mm spacing. Unlike the 2A boards, a unique board is required for the X and Y. However different wire spacing can be used for X and Y in the same chamber.

The drive mechanism was also upgraded. A new off the shelf ferro-fluid FT was installed and new radiation tolerant 41 pin signal FT's were also installed. The original drive motor and electronics were maintained. A prototype of the ion chamber is shown in Fig. 7A.

1AM8 an ion chamber with remote gas flow

A special "gas pack" ion chamber is used on the 1AM8 monitor Fig. 7B. This monitor is under several layers of concrete shielding blocks, making regular service difficult. For this reason the monitor was fitted with remote gas flow. On this installation the gas plugs were replaced with tube fittings. Metal tubes were piped to allow gas to flow through the chamber continuously. Gas flow is approximately 1 cc/min. Polyimide insulating connectors were spliced into the lines using epoxy. This allows the gas package to be electrically isolated at −300 V bias.

FIGURE 7A. Prototype swing style ion chamber

FIGURE 7B. 1AM8 special ion chamber with remote gas flow

ACKNOWLEDGMENTS

The authors would like to thank the following, G. MacKenzie, for his many years of knowledge relating to beam instrumentation, A. Hurst, for his encouragement and support, W. Rawnsley, for the electronics, J. Yandon, for vacuum related topics. G. Dennison, for the layout of circuit boards, D. Ross and the TRIUMF design office for help in preparing drawings, R. Roper and TRIUMF's machine shop, for the welded windows, T. Ries, for stress calculations.

REFERENCES

1. G. M. Stinson, TRIUMF report TRI-DNA-96-05, TRIUMF, 1996. (Internal report)
2. W.R. Rawnsley, "Beam Diagnostics at TRIUMF", Beam Instrumentation Workshop, AIP Conf. Proc. 333 1994 page 125
3. NASA, SP-8053 (June 1970), Nuclear and Space Radiation Effects on Materials, Page 11
4. B. Devonald, BH Devonald and Associates, West Vancouver B.C. Canada, Email communication, April 4 2002
5. Roark's Formulas For Stress & Strain, Warren C. Young, 6th edition, MacGraw Hill, page 457 and page 477.
6. John Yandon, verbal communication.
7. Praxair Inc., Email communication, April 16 2002
8. G. Mackenzie, IEEE Trans. On Nuclear Science, NS-26, 1979, page 2316.

An Update of The Diagnostic Systems Proposed for The New Third Generation UK Light Source, DIAMOND

Stephen R Buckley, Michael J Dufau, Robert J Smith

Daresbury Laboratory, Keckwick Lane, Daresbury, Warrington, WA4 4AD. UK

Abstract. This paper describes the currently proposed systems for electron beam position monitoring (EBPM) and diagnostics for the DIAMOND synchrotron. Although the basic requirements have remained unaltered, the philosophy of implementation has been subject to change, influenced by the experiences of other national light sources, and the emerging availability of commercial equipment, suited to the needs of DIAMOND. This paper focuses in greatest detail on the storage ring systems, including data acquisition and control. Details of Total Current Monitor (TCM) systems, and an active, beam position based interlock system for protecting ID vessels against thermal damage, by beam mis-steer, are also included.

INTRODUCTION

Diagnostic systems for DIAMOND, the UK's proposed 3GeV 3rd Generation Light source, will be essential for the rapid commissioning and successful operation of the new facility. This paper proposes a specification for the expected requirements. The options to meet these requirements are discussed in turn and technical solutions are indicated to allow costs to be estimated and to inform design work in other DIAMOND systems.

Electron beam diagnostics has undergone a period of rapid technological development. A lot of this technology is now maturing and has been assessed by many long standing and newer 3rd generation light sources. This now provides the opportunity to specify and recommend, commercially available and hence more cost effective diagnostic implementations. It should be borne in mind however that the development of systems continues apace, and over the period of construction of DIAMOND, it will be necessary to review and refine some proposals at the onset of project procurement to capitalise fully on new technology and the experience of others.

Synchrotron radiation experiments depend increasingly for success on excellent source properties in terms of the stability of source dimensions and position. It has been assumed in the following that position stability to 1 micron or better will be required on all beam lines in both planes measured on time scales from milliseconds to hours for both local and global feedbacks. In addition and on the microsecond scale, individual bunch instabilities must also be considered to further reduce the beam size and suppress beam blowup, though such instability feedback is not considered here. Although high quality diagnostic systems will play a major role in achieving the required performance, beam stability specifications must be considered at the earliest design stages of all DIAMOND systems. The general problem of achieving stability at the user's sample is discussed, leading to some recommendations for consideration in the design of the facility as a whole.

STORAGE RING ELECTRON BEAM POSITION MONITORS (EBPMS)

The proposed storage ring EBPM system is the most extensive and complex of all DIAMOND diagnostics. Each of the 24 cells will be equipped with 7 two plane monitors. At least one extra EBPM position within the ring will be earmarked for installation as a test station and another for the instability feedback signal source. The mechanically defined positional layout within the cell is shown in Figure 1. It is proposed to add further EBPMs within the insertion device (ID) vessels as they are installed, to provide facilities for stand alone vessel protection systems for vertical beam misalignment, and to provide additional high precision beam position measurement to supplement local position servos.

All results presented are by mathematical electrostatic simulations from theory [1]. Further tests using an electroststic finite element analysis package will be carried out to refine absolute button pickup positions.

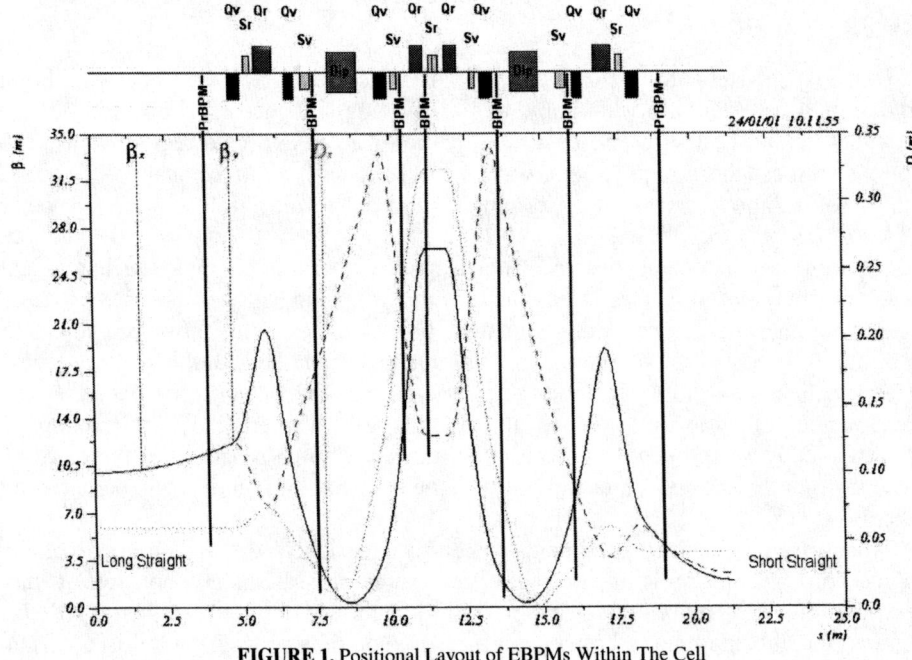

FIGURE 1. Positional Layout of EBPMs Within The Cell

Primary Electron Beam Position Monitor (PEBPM) Pickups

Of the seven EBPMs in each cell, the two devices at either end of insertion straights will be of the high performance, high stability type. Mounted in such a way as to minimise any mechanical movement, their positions will be constantly measured in respect to machine survey monuments, incorporated with these devices. To facilitate this, the PBPMs will be isolated by bellows from the rest of the cell and fixed to individual rigid stands. These devices will provide a reference beam position and can also be used to determine position and angle through the insertion device for local correction. Continuous physical position measurements on these PBPM vessels will also be made to discern beam from vessel movements.

The PBPM pickup head will utilise the standard Daresbury (i.e. modified ESRF) type button feedthroughs [2] to detect the internal electric field generated by the beam. The buttons will be arranged in an overlapping pattern of four buttons, mounted on the top and bottom faces of the vessel, with sufficient separation to allow installation welding. An additional pair of on axis vertical buttons will be added to allow complete de-coupling of the vertical measurement signals detection from the horizontal. These will provide an independent vertical position measurement if required. These vessels will be of a narrow vertical aperture, to give enhanced performance, with significantly improved signal to noise ratios. Although it is desirable to minimise the vertical apertures still further for the EBPM response improvement this gives, it is important to balance this against the lifetime effect that this will have. An adequate response however is still obtainable for a PBPM with a larger than ideal vertical aperture of 20mm, and a button separation of 12mm centre to centre, configured in a longitudinally displaced overlapping configuration as described. The resultant vertical and horizontal electrical calibration factors for this geometry can be seen in Figure 2. They are approximately the same in both planes around the central region of interest.

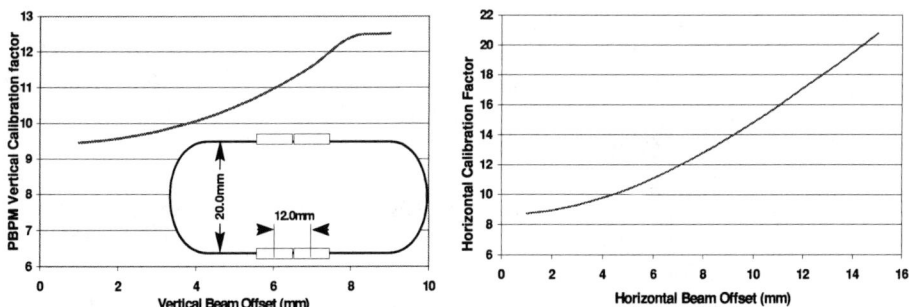

FIGURE 2. Central 'on-axis' Electrical Calibration Factor Variation for the DIAMOND PBPMs

As is shown a multiplying calibration factor of around 9 is found in both planes around the central region of interest, which will give adequate signal to noise ratios for the desired positional resolution of less than 1 micron, in the vertical plane. Further refining of the button positions, taking into account the installation of the on axis pair of buttons can result in a further improvement in calibration factors in one plane only. Using the 4 button BPM only for the horizontal plane, and adjusting the button spacing to 24.0mm can produce calibration factors as low as 7. Values below even this (~6) are obtained from the single on axis vertical buttons. These results should be balanced however against the increased cost of the detector electronics with which the PBPM devices will be instrumented.

The extent to which this high accuracy and resolution measurement is useful will depend on the relationship between the reference in which the user is working or in which his optics is located and the reference established by the local EBPMs. This in turn will depend on the details of the building and floor design and the proximity of the user or his equipment to the source. The same argument can be made for the relevance of independent white beam photon position monitors, which may be remote from the user and some distance from his optics. The fast global feedback loop will however be capable of operating much faster, and in conjunction with fast steering elements in the ID straights will allow the removal of motions due to ambient vibration and electrical interference up to ≈ 100Hz.

Standard 'In Vessel' EBPM Pickups

It will not be practical to bellows decouple the remaining 5 EBPMs in the cell from the vacuum vessel, but they will be kinematically mounted to ensure that thermal motion is well constrained and repeatable. Global correction regimes lock the beam to a reference defined by the location of all the EBPM vessels. These in turn will be referenced to the position of the PBPMs, which define a 'standard' position. However, in the context of global correction, the reference defined by all the EBPMs is unlikely to be any more relevant to the facility users than the reference defined by the locations of the magnet systems and any benefits from this to users will have to be assessed. One solution adopted by some 3rd generation light sources, which lock the EBPMs to adjacent quadrupoles will only be effective if the mechanical engineering problems associated with differential expansion of the vacuum vessels and the magnets under the influence of varying ambient temperatures and varying stored beam power loading are solved. Substantial development work continues in this area.

The EBPM pickup head will again utilise the standard Daresbury (i.e. modified ESRF) type button feedthroughs to detect the internal electric field generated by the beam. The buttons will be arranged within the standard vessel geometry to give a reasonable response in both planes. A partially optimised example of such a vessel response is shown in Figure 3.

The offsets of the electrical centres of the EBPMs will be measured on calibration rigs before final assembly. However, experience dictates that systems allowing in situ measurement of the relationship between the electrical centres should be provided. The simplest way to make this measurement is to modulate the excitation of individual quadrupoles and look for a minimum response on the beam at the modulation frequency. The present plan for DIAMOND includes the use of individually programmed power supplies for each quadrupole which will allow simple implementation of this measurement technique.

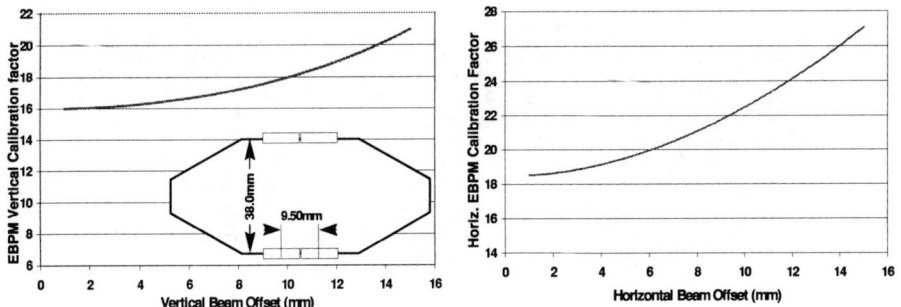

FIGURE 3. Central 'on-axis' Electrical Calibration Factor Variation for the DIAMOND in vessel EBPMs

DETECTION ELECTRONIC SYSTEMS

The installed EBPM detection systems must, in combination, provide position measurement with sufficient accuracy and range in several situations. These include the storage ring, where capabilities of instantaneous single/first turn position is required, and also during beam transportation along the flight paths, and during its time within the booster accelerator. They must in addition provide the storage ring with very high quality, stable position information at rates sufficient to allow the operation of global beam feedback servos, at rates up to 100Hz. This will meet the requirements, of both long and short-term stability of less than 10% of beam size.

Until recently, such high precision detection electronics had to be built in house. Today however, commercial systems are available to tackle such problems. Two systems are compared and a suitable implementation presented.

Commercial Systems

The two systems discussed have used very different approaches to the task of EBPM signal detection, and both have merits and drawbacks. An overview of the operational techniques is given below.

The Bergoz Instruments, Multiplexed BPM (MBPM)

This device is the more mature of the two commercial types, and its function is based around analogue circuitry to produce DC values, relating to X and Y beam position within a normal type four button pickup EBPM head [3]. A schematic showing the detector hardware can be seen in Figure 4.

The key to the operation of this device is that each button pickup from the EBPM head, is processed through the same electronic detector. This is achieved by a switching multiplexer, that passes all signals in turn through a series of filters and gain control, to the detector chip, based around a standard television technology detector, running at 21MHz. The resultant voltages are held in sample and hold devices that allow the appropriate summing and differences to result in X and Y positions being produced as DC Voltages. These voltages are then digitised by a separate ADC to interface them to the control system.

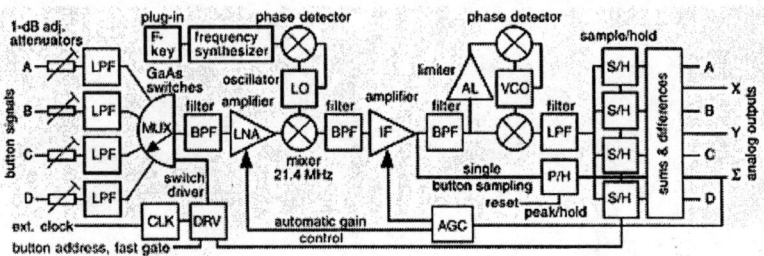

FIGURE 4. Bergoz Instrumentation Detector Electronics Schematic (MBPM)

Since this detector uses a multiplexer on its input, it uses the same detector for each button signal, and produces excellent results, with compensated electronic drift, however the multiplexer means that it has limitations with regard to its position reading update rate. Bergoz MBPM cards have been used on the SRS for over two years for BPM measurements and active machine protection for insertion devices.

The Instrumentation Technologies (I-Tech) Digital BPM (DBPM)

This device is relatively new to the market, and its detection method is based around the technique of processing all four button pickups down identical separate channels. The results are then digitised and the summing and difference required for beam position readings carried out within the VME controller, or other higher level computer [4]. A block diagram of its implementation within the Swiss Light Source (SLS) is shown in Figure 5.

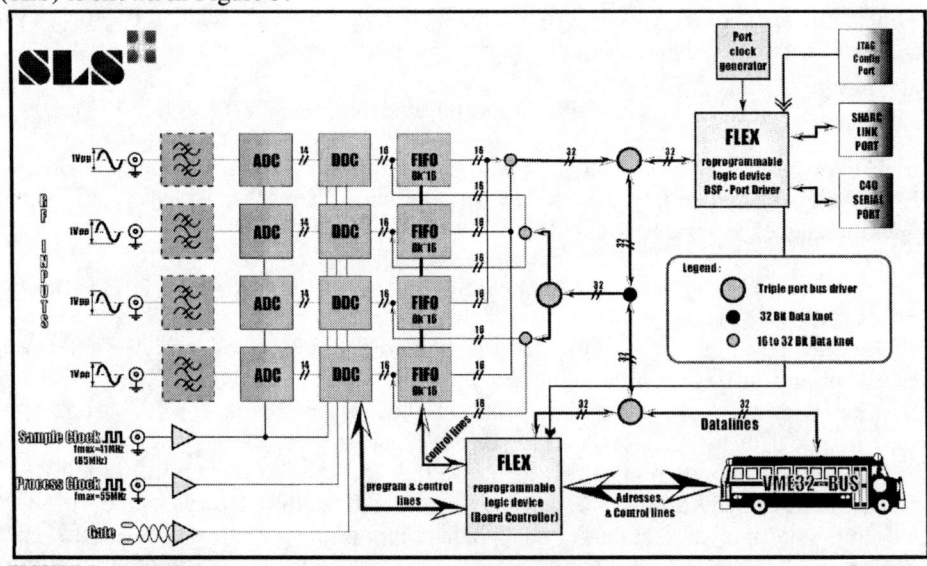

FIGURE 5. I-Tech four channel, 'Quad-Receiver' Schematic/Overview installed in the SLS EBPM Configuration

I-Tech themselves are becoming established with this and other products based around BPM technology, including tune measurement and multibunch feedback system front end detectors.

The basic methodology used here is to detect each input signal down its own detector channel, in order to provide very high speed (single/first turn) Beam position signals. Each single input channel is mixed down to an intermediate frequency of 36MHz and gain equalised against a separate pilot frequency, applied to all channels, generated on board before final filtering. This pilot frequency is essential to help keep all channel gains close together, to maintain the reading accuracy. These four 36MHz signals are digitised by an adjacent dedicated 14 bit ADC, with 16 bit resolution being achieved by oversampling. Hence the 14 bit mode is available for the single/first turn mode, and the 16 bit data at much slower rates for high precision read back for use with global servos, typically at frequencies up to 100Hz. Further enhancements to the system include a dedicated DSP module (installed one per VME crate), to fully exploit the single turn capabilities. A major feature of this system is its ability to store up to 16,000 turns of data, allowing a 'black box' mode to investigate beam losses and some instabilities exhibiting sufficient positional change to be observed at single turn.

COMPARISON OF KEY SPECIFICATIONS

Table 1 reviews and compares the major technical specification, features and performance of the two systems.

TABLE 1. Key Specification Comparison

	I-Tech	*Bergoz*
Dynamic Range (Beam Current)	>100dB	>90dB
High Speed Resolution (Turn by Turn/'Pulsed')	±20µm (500kHz Bandwidth)	Not Supported
Memory Depth (No. of selectable turns)	1k to 16k	Not Supported
Low Speed Resolution (Closed Orbit)	<1µm (2kHz Bandwidth)	<1µm (9kHz Bandwidth)
Output X, Y, Σ	14 bits Digital (16 bits -oversampling)	Analogue +/-10V
Wide Range Beam Current Dependence	25µm (-65dBm to –5dBm)	1µm (SLAC Test) (-50dBm to -3dBm)
Narrow Range Beam Current Dependence	2.5µm (in any 14dB range)	>1.0µm (in any 14dB range)
Long Term Stability	2.5µm	>1µm (LEP Test)
Form Factor	VME 64X Compatible	3U Eurocard

Detection Electronics Summary

Commercial systems are available for DIAMOND EBPM application, thus avoiding the considerable time/cost outlay in the development of further in-house systems, which invariably are more expensive. One is a mature, established system, the other becoming so and still under considerable development.

The system offered by Bergoz Electronics is a simple fully analogue design but offers resolution, accuracy and stability compatible with the requirements of a third generation light source. Being an analogue system digitising of the output is required prior to any software interface and unlike its I-Tech counterpart the Bergoz system is unsuitable for pulsed beams. It is however well established, it has a proven reliability record and is in commission in most of the light sources around the world.

The relative performances of the two systems are naturally reflected in their respective costs. BPM channel for BPM channel the I-Tech system is several times as expensive as the Bergoz system, and thus may possibly only be justified for applications that will truly exploit the full potential of the system. The I-Tech DBPM detector is specified elsewhere for all EBPM locations within the DIAMOND pre-injector. Its transient beam capabilities will allow instantaneous beam position within all transport lines and also provide position/tune measurement for the Booster synchrotron. The dynamic range of this system is also adequate for top up mode, low level electron beams.

Within the storage ring, a combination of the two systems following the examples given in Figure 6, will provide DIAMOND with first class EBPM systems, for the best value for money.

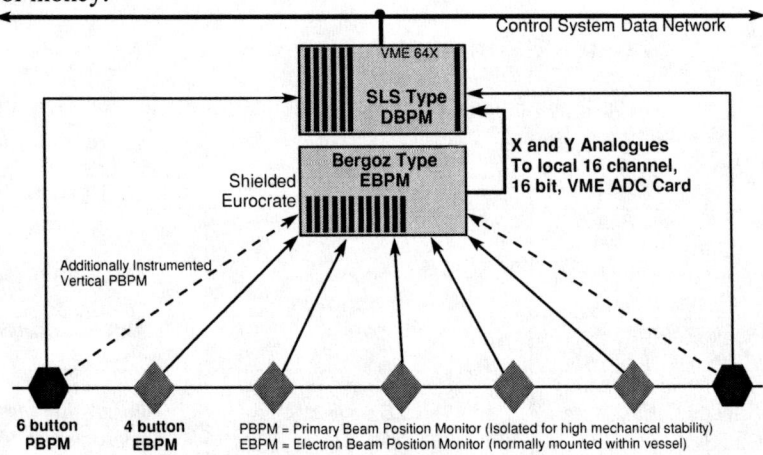

FIGURE 6. Proposed Storage Ring Detector Implementation

MACHINE PROTECTION SYSTEMS

Beam mis-steer damage protection, of narrow-gap insertion device vessels, is a fringe area of beam diagnostic systems, thus it is therefore appropriate to detail the system in this section.

DIAMOND is intended to be essentially an insertion device machine so eventually most straights will be fitted with narrow-gap beam vessels introducing attendant beam steering limitations. Theoretical analysis has shown that mis-steer of the electron beam, could cause catastrophic thermal damage to the vessel in less than one second. Front line protection systems have been developed for the SRS narrow gap vessels to help prevent such damage [5], and are suitable, with minor modification, for installation on DIAMOND insertion devices.

Machine Protection Hardware Installation

When dangerous mis-steering conditions prevail, the electron beam will be switched off by tripping the RF power to the beam. The main interlock signals to achieve this will have been generated by excessive vertical beam displacement (shown as Beam Position Upstream and Downstream), and excessive rise in vessel wall temperature (Thermal Trip). The single Thermal Trip interlock will be generated from a strategically placed array of thermocouples, affixed to the vessel wall and monitored by a Programmable Logic Controller (PLC).

Since initial injection at low current is deemed to be a safe operating area, a current sensitive bypass will be included as shown. This will facilitate beam steering through the narrow gap of the vessel at low currents, by permitting a wider tolerance on position.

Because confidence in the reliability of beam position measurement and interlock operation is paramount, secondary interlock signals will be required to become active when the integrity of electronic hardware and associated support signals is suspect. These are shown in Figure 7, an expanded organizational diagram, as BPM Electronics Failure Upstream and Downstream, Total Current Monitor Failure, Power Supplies Failure and Thermal Monitoring (PLC) Failure. A keyswitch controlled override of Beam Position interlocks is included for possible Accelerator Physics application, to permit wider beam positioning at high current levels, or unrestricted control under special user conditions (e.g. ultra low current running).

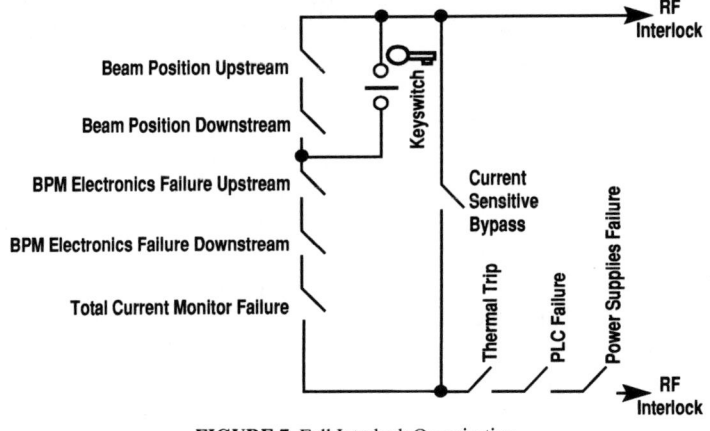

FIGURE 7. Full Interlock Organisation

There will be provision for full status monitoring interface to the DIAMOND control system, and local (Control Console) real time display of beam positions and current sensitive bypass status.

Because experience has shown that there is interaction between the vessel protection system and beam loss when an unrelated fault occurs, i.e. beam loss creates numerous vessel protection interlock failures, Event Sequence Discrimination circuitry will be installed to distinguish cause from effect. The result of the discrimination process will be displayed through the Control System and visually at the Control Console.

When the time for installation of vessel protection systems arrives, all development from both a design philosophy and engineering aspect will be complete and installation should be modular on demand.

TOTAL CURRENT AND BUNCH CHARGE MONITORS (TCMS AND BCMS)

Accurate and stable measurement of the DC component of the stored charged DIAMOND beam is an important requirement, since it will be a measure of the light intensity and the parameter from which lifetime will be calculated. The parameter of critical importance, for monitors which are designed to non-intrusively measure charged particle beam currents, is resolution, a specification that is dictated by zero drift. For a third generation machine such as DIAMOND a zero drift figure in the order of 5.0 microamps is required.

Zero drift of a monitor is defined primarily by the thermal properties of the core material that is employed for the toroidal monitor head. In this area the current generation of monitors use amorphous magnetic materials for the cores, materials that demand extensive expertise and resources to process. Consequently the manufacture of modern monitors has tended to remain within the province of the commercial sector.

In the commercial field a leading manufacturer of high specification Total Current Monitors TCMs (now more commonly referred to as Parametric Current Transformers, PCTs) is the French firm Bergoz Electronics. This company has supplied PCTs to most of the leading synchrotron radiation laboratories world wide; their products have a good reputation for performance and reliability and the company has demonstrated its long term commitment to product support.

Thus in line with the cost effective policy to be applied to DIAMOND of implementing systems by use of commercial instrumentation, it is intended to equip DIAMOND with a Bergoz Electronics PCT, with the assurance that the potential for obsolescence is minor.

A standard production model from the Bergoz Company offers zero drift of 5.0 microamps. This specification satisfies both the measurement and lifetime calculation demands of DIAMOND, but to achieve the quoted performance figure, careful consideration will have to be given to the magnetic and electrostatic screening requirements of the toroidal monitor head. To implement the full theoretical monitor head screening required to guarantee the demanded performance, a free beam pipe length of approximately 0.6 metres is required.

Finally, since DIAMOND is to operate in a top-up, mode it is recommended that two PCTs be installed in the DIAMOND Storage Ring to crosscheck beam current measurement by comparison. This is important if the top-up operation is to be performed at a defined and repeatable level of beam decay. Additionally a PCT is required for the DIAMOND Booster ; this device can be of a lower specification than that for the Storage Ring, but there would be little economic gain in purchasing an inferior instrument.

The potential user community for DIAMOND have expressed a desire for customised bunch charge fill patterns on request, a facility that will require selection and charge fill of individual beam bunches. This in turn will require the capability of measuring the individual bunch charge content.

Selection of an individual bunch can be achieved through the DIAMOND timing system by variable delay from the Orbit Clock ; measurement of the bunch charge content will require the installation of a Beam Charge Monitor.

Again in order to adhere to the policy of using commercial instrumentation, a Beam Charge Monitor manufactured by Bergoz Electronics will be installed. This device has an integrate - hold - reset capability on an individual pulse basis, that offers a maximum sensitivity < 60 pico Coulombs per volt and a resolution of approximately 3×10^6 particles.

These devices will also be installed as part of the beam loss monitoring throughout the DIAMOND to assist with radiation protection system, in order to assess and interlock if necessary, the effective beam injection/transport efficiency.

REFERENCES

1. T. Ring, "Beam Position Monitors for the High Brightness Lattice", Daresbury Report, DL/SCI/TM41A (1985).
2. M J Dufau, D M Dykes, R J Smith, "Electron Beam Position Monitor (EBPM) Diagnostics for DIAMOND", Proceedings, PAC99, New-York (1999)
3. J. A. Hinkson, K. B. Unser., "Precision Analogue Signal Processor for Beam Position Measurements in Electron Storage Rings", Proceedings, 2nd DIPAC, Travemunde, Germany, (May 1995)
4. Brown, M. P., and Austin, K., M. Dehler, A. Jaggi, P. Pollet, T. Schilcher, V. Schlott, R. Ursic, "New Digital BPM System for the Swiss Light Source", DIPAC99, Chester, UK (May 1999).
5. M J Dufau, R J Smith, "a fast protection system for narrow-gap insertion Device vessels", DIPAC99, Chester, UK (May 1999)

Study and Design of a New Over-damped Cavity Kicker for the PEP II Longitudinal Feedback System

F. Marcellini°, M. Tobiyama*, P. MacIntosh*, J. Fox*, H. Schwarz*, D. Teytelman*, A. Young*

°*Laboratori Nazionali di Frascati-INFN, Via E. Fermi 40, Frascati (Rome) Italy*
*Stanford Linear Accelerator Center, 2575 Sand Hill Road - Menlo Park Ca 94025 USA

Abstract. PEP-II has been running for several years using drift-tube style longitudinal kickers. They have functioned well at the design current in the HER and LER. Machine upgrade plans for PEP-II have encouraged the analysis and design of cavity kickers for the longitudinal feedback systems in PEP-II. The cavity kicker design is based on the use of an extremely low Q cavity, where the Q of the system is determined primarily by ridged waveguides coupling to external loads. This kicker design has originally developed at LNF-INFN, and is attractive for use at PEP- II to reduce the kicker impedance at frequencies outside the working bandwidth and consequently reduce the strong beam-heating of the structure and the feedthroughs. The cavity-style kicker is also better suited to external cooling, as it is without internal elements which must be cooled through either radiation or conduction out through some path. The design options, including the choice of operating frequency (9/4*RF vs. 13/4*RF), the kicker shunt impedance, the number of external coupling ports (4 vs. 8) and the selection of the kicker bandwidth, are briefly described and three different solutions are proposed. Results are presented estimating the shunt impedance, bandwidth and HOM impedances via the use of the Ansoft HFSS code.

INTRODUCTION

Both the High Energy and Low Energy rings of PEP-II at SLAC utilize two-element drift tube type kickers in the longitudinal feedback system [1]. They have worked with success since machine commissioning and the machine is currently running above the design luminosity. Future plans for PEP-II include operation at higher currents, and we are considering the use of cavity type kickers for future use. These cavity style kickers have limited impedance outside the operating bandwidth (BW), and are potentially less subject to problems of beam-induced heating. This paper reports on a design study for a cavity type kicker suitable for PEP-II requirements.

GENERAL CONSIDERATIONS ABOUT THE KICKER DESIGN

The cavity kicker design is based on the idea of using the voltage seen by the beam when it crosses a cavity gap, to provide the bunch with the proper longitudinal kick needed to damp longitudinal coupled bunch motion. As the coupled bunch modes cover a wide frequency range (up to $1/2\ f_{RF}$ in case of every bucket filled), for good efficiency (and uniform gain across the full bandwidth of coupled-bunch modes) the cavity has to be a broadband resonator, with a Q much lower than typical values for conventional cavities. Since $Q = 2\pi f_0 U/P$, where f_0 is the central frequency of resonance, U is the energy stored in the cavity volume and P is the total power dissipated in the cavity walls and in external loads, the cavity Q can be lowered by increasing the power losses.

The overdamped cavity kicker design uses strong coupling to external loads to set the value of Q at the desired value (typically in the range from 4 to 10). In this approch the external Q of the system is really defining the system bandwidth, and the external loads are available to dissipate beam-induced power.

In detail, the kicker consists of a pill box cavity with a number of waveguides (WGs) attached symmetrically on both cavity sides.

Figure1 shows a drawing of the first cavity type kicker. It was designed for the DAFNE Φ-Factory at Frascati, Italy [2].

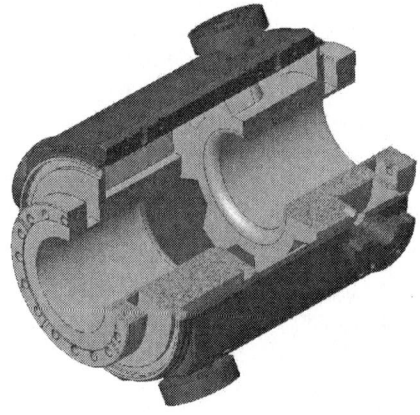

FIGURE 1. Cut view of the DAFNE longitudinal feedback kicker.

Each WG is followed by a transition to a standard coaxial line via coaxial vacuum insulators to facilitate the connection towards external devices (driving amplifiers and dummy loads). Special ridged WGs are designed to lower the cut off frequency below the frequencies of the fundamental mode band. By extending the transmission response of the transitions up to the beam pipe cut off, effective damping of all HOMs of the cavity can be obtained.

As the power losses are strongly dominated by the dissipation in the external loads, driving half the ports of the structure in phase and connecting the remaining ports to dissipative loads results in the cavity structure being perfectly matched at its central frequency. However the power reflected towards amplifiers at the edge of the working frequency band and the power released in the cavity by the beam for interaction with the fundamental and HOM impedances require the use of circulators to protect the driving amplifiers. This is in contrast to the existing drift-tube kickers, for which absorbtive low-pass filters are used to protect the amplifiers from the beam-induced HOM power, while the directivity of the drift-tube structure dissipates the majority of the fundamental beam-induced power in the terminations of the drift-tube kicker, so that the in-band power reflected back into the driving amplifiers is within the power dissipation rating of the amplifiers.

THE CHOICE OF OPERATING FREQUENCY

We have explored two possible operating frequencies for the cavity kicker: $9/4\ f_{RF} = 1.071$ GHz and $13/4\ f_{RF} = 1.547$ GHz (odd multiples of $1/4\ f_{RF}$ allow the minimization of beam-induced power and help provide uniform gain across the operating band [3]. As part of this initial design study the preliminary design of a cavity- kicker at 13/4 RF was performed to better estimate the likely shunt impedance and HOM impedances, and better understand the trade-offs in choice of operating frequency [4]. The higher frequency option presents some advantages, such as the existing power amplifiers have more power output in this range of frequencies and the HOM content of the cavity is reduced to only two dipoles because of the shift of the other HOM resonant frequencies beyond beam pipe cut off. On the other hand, for the $9/4\ f_{RF}$ solution, the increased ratio between cavity and beam pipe radii results in higher R/Q. Consequently more shunt impedance is obtainable in spite of a reduction of Q by a factor 13/9 for a fixed BW. Moreover, the current system is operating at $9/4\ f_{RF}$ and there is no need to modify many back-end components. Finally the timing of the system is less critical at lower frequencies.

Weighting pros and cons of each option, it has been decided to design the cavity kicker at $9/4\ f_{RF}$.

DESIGN OF THE THREE DIFFERENT PROPOSALS

For this basic design there are two important parameters available to specify the over-all design: the number of coupling ports and the coupling strengths, which determine the BW of the kicker.

The system BW requirements are depending on the bucket filling patterns. In case of by 2 filling, all the possible coupled bunch modes oscillate at frequencies contained in a span of 120 MHz (476 MHz/4). With the rough assumption that the cavity fundamental mode R/Q is constant when the cavity body profile is not changed, the BW of the cavity kicker is inverse proportional to the maximum shunt impedance. In other words it seems convenient to keep the BW as narrow as possible to maximize the shunt impedance R_s. But this dependence is less sensitive at the edge of the BW, so a BW larger than 120 MHz presents the advantage to have a better gain flatness across the operating band and a better phase linearity of the system. The narrower bandwidth structures have higher shunt impedance, and at first glance one might assume that the voltage seen by the beam would be higher due to the higher shunt impedance. However, there is an additional issue with the QPSK modulation scheme, in which the QPSK "carrier" is resynchronized to the RF every 2.25 cycles, so that the voltage seen by the beam for for the various cases studied is roughly identical [5].

In our initial studies we decided that a device having a BW of about 160 MHz would be a good compromise between the opposite demands. The first two solutions proposed have been designed with this BW value as target. A third possible design has investigated the possibility to have a wider bandwidth cavity kicker with a BW closer to the drift-tube design. The BW obtained for the third solution is 224 MHz.

For the number of coupling ports the two more feasible options are 4 or 8. Just 2 ports could not yield the required coupling for mode damping. Moreover the WGs, transitions, feedthroughs, cables, circulators and loads must be designed to be able to manage each one half the whole flowing power. More than 8 WGs cannot be allocated on the cavity surface for lack of space, and a 6 port solution does not convienently fit the current scheme with 4 amplifiers mounted on 2 kickers. We also considered the possibility to feed all ports instead of half of them, obtaining a $\sqrt{2}$ kick voltage gain. In this approach ports on opposite sides of the cavity gap must be powered 180 degrees out of phase, and the number of circulators must be doubled as every port is connected to an amplifier. But as the circulator BW does not extend up to higher frequency modes that could remain trapped, this idea has been rejected.

An additional operation constraint has been considered. If we want to minimize the necessary changes to the installed cable plant and high-powered terminations, the alternative is between two 4 port kickers per beamline vs. a single 8 port structure.

Since, for a fixed BW, the shunt impedance does not depend on the number of ports, the total kick voltage obtainable from a pair of 4 port devices results $\sqrt{2}$ times the voltage given by a single 8 port kicker. On the other hand the power per feedthrough is higher in case of a 4 port geometry and the contribution to beam coupling impedance is also doubled. Moreover a wide BW is easier to achieve with an increased number of WGs.

The first of the designs proposed has been designed with 8 ports, while the second and the third options are 4 port structures.

The first option

As a first step we considered an 8 port structure. It was not obvious to arrange 4 WGs in the space of each cavity side wall, as the WGs cross section should be large enough to keep the frequency cut off well below the operating frequencies. Figure 2 shows the HFSS model (1/8 of the whole structure, for the geometric and electromagnetic symmetry of fundamental mode) used as input geometry for simulations.

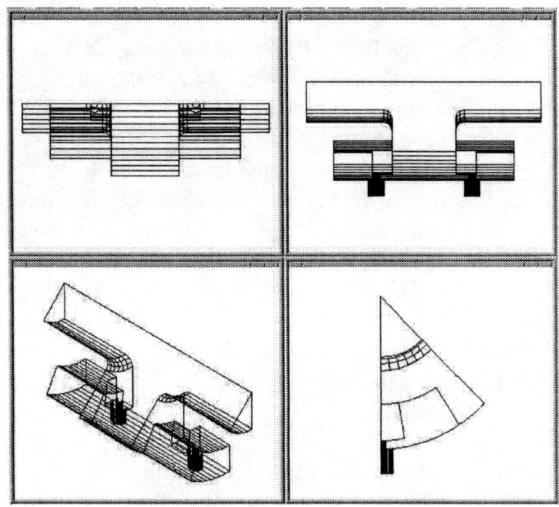

FIGURE 2. HFSS input geometry.

In Fig. 3a the transmission response of the WG and transition to 7/8" coaxial standard designed for this case is reported. To simplify the mechanical construction the WG profile has been obtained as a slice (65 degrees wide) of a hollow cylinder, which main radius is the same as the cavity radius. The ridge profile is also a cylinder slice and it is truncated in proximity of the transition giving rise to a back cavity. The length of this cavity can be tuned to center the frequency band of the transition response.

Figure 3b shows the transmission coefficient S_{21} between the 4 input ports and the 4 output ports for the cavity fundamental mode. A little adjustment of central frequency should be done since the estimated value has been 1.068 GHz. The 3dB BW is 153 MHz and the calculated shunt impedance peak value is about 950 Ω.

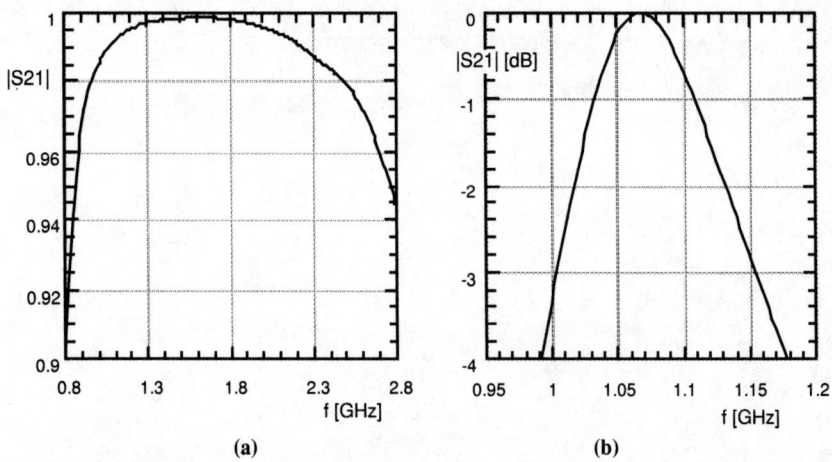

FIGURE 3. Transmission frequency response of kicker (b) and of its transition (a).

The HOM characterization has revealed 4 parasitic modes (2 dipoles and 2 monopoles) remain trapped into the cavity, but they are quite damped by external loading through the coupling WGs. Table 1 summarizes what has been found in terms of frequencies, Qs and impedances.

TABLE 1. Parameter of HOM characterization.

Mode	Freq. [GHz]	Q	R
TM_{110}	1.5616	15	15.3 kΩ/m
TE_{111}	1.6819	541	60.3 kΩ/m
TM_{011}	2.2203	12	70 Ω
TM_{020}	2.4266	35	50 Ω

Various filling patterns have been considered to estimate the power released by the beam on the kicker total impedance (both of the operating mode and of HOMs) [6].

In any case the power per port never reaches 1 kW, that seems a very conservative value compared with the performances of feedthroughs, cables and loads. For this reason, the following designs concentrate on kickers with only 4 ports.

The second option

The design feasibility of a kicker having the same 160 MHz BW but just half port number of the previous one has been investigated.

In this case, each single WG has to provide a doubled loading of the fundamental mode. So the transverse dimensions of the WG have been increased (the height is 4 mm more and the width is 100 degrees instead of 65).

The gap between the ridge and the WG upper wall has been increased too (from 6 to 8 mm), to reduce the field intensity in this region. The capability to sustain the flowing power results increased as well.

Figure 4 shows the input geometry for HFSS. WGs on opposite side of the cavity are 90 degrees rotated, to allow a better coupling with high polarity modes (particularly dipoles). A quarter of geometry had to be modeled in this case.

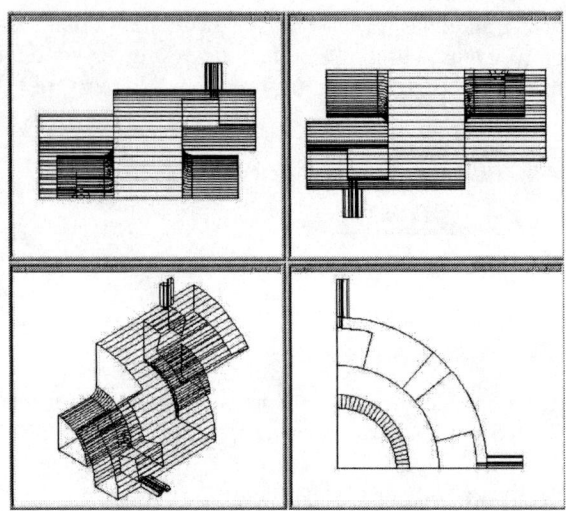

FIGURE 4. HFSS input geometry.

Figure 5 shows the transmission response of both the transition (left) and the kicker between its input and output ports (right). The transition band extends up to the beam pipe frequency cut off (approximately 2.6 GHz). The kicker BW is 165 MHz centered at 1.071 GHz, while the shunt impedance peak value is now about 890 Ω.

In Table 2 results of HOM characterization are listed.

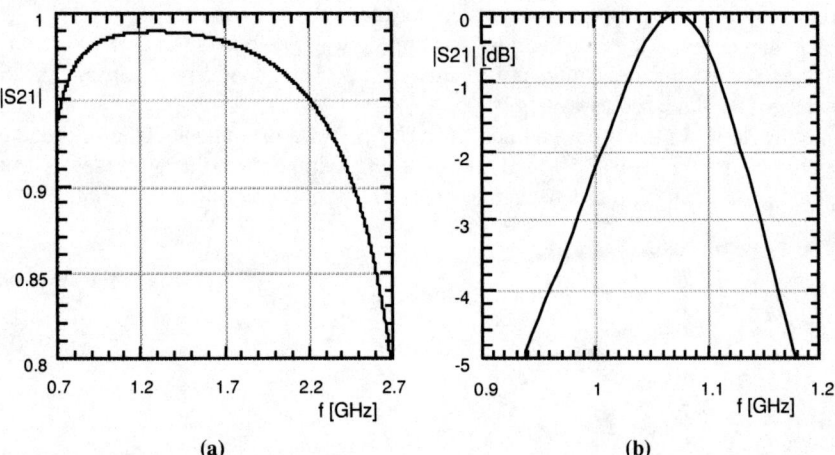

FIGURE 5. Transmission frequency response of kicker (b) and of its transition (a).

TABLE 2. Parameter of HOM characterization.

Mode	Freq. [GHz]	Q	R
TM_{110}	1.5264	25	18.2 kΩ/m
TE_{111}	1.6802	539	56 kΩ/m
TM_{011}	2.0568	12	61 Ω
TM_{020}	2.3848	60	69 Ω

The third option

Finally a 4 port device with a wider BW (comparable to the drift tube structure one) has been considered. The main difference respect to the design of the previous option is in the WG cross section, that has been further on increased (the height and the width being now respectively 41 mm and 120 degrees).

Simulation results for the transition and the kicker frequency response are shown in Fig.6. Due to the WG larger size, its transmission response presents a notch around 2.6 GHz. However, the HOM damping is not compromised as the higher frequency of the trapped HOMs is about 2.4 GHz.

The 3dB kicker BW is 224 MHz with a shunt impedance of 626 Ω estimated at the center frequency (1.071 GHz).

Table 3 summarizes HOM parameters calculated for this structure.

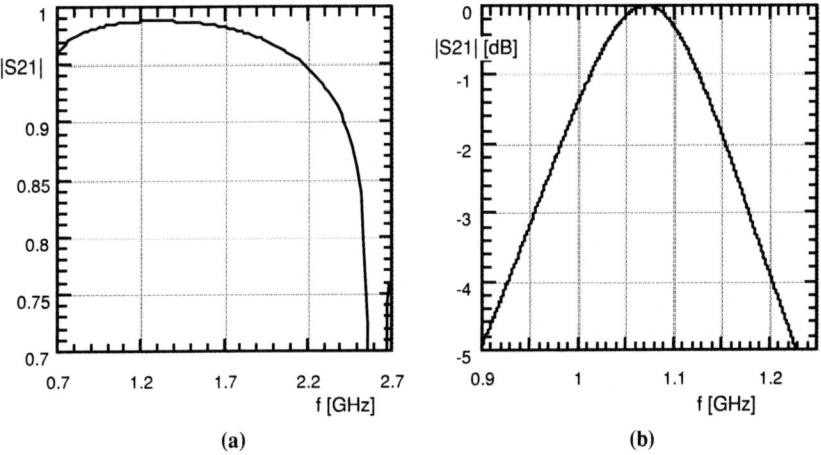

FIGURE 6. Transmission frequency response of kicker (b) and of its transition (a).

TABLE 3. Parameter of HOM characterization.

Mode	Freq. [GHz]	Q	R
TM_{110}	1.5151	22	16.7 kΩ/m
TE_{111}	1.7114	210	19.2 kΩ/m
TM_{011}	2.0507	13	65 Ω
TM_{020}	2.4117	116	76 Ω

SUMMARY

The design of a cavity kicker involves many details of electrical, thermal and vacuum requirements. Our group is proceeding with the detailed design of the wideband 4 port structure. Ongoing efforts are centered on fine optimization of the cavity bandwidth and shunt impedance, and the consideration of the heating of the structure and feedthroughs. A study is underway to consider the possibility of multipactor effects in the structure to help select the cavity material and surface finish [7]. We plan on completing the final design and moving to the fabrication phase in summer 2002.

ACKNOWLEDGEMENTS

The authors thank Uli Weinands and John Seeman of SLAC for their support of this design, and thank Mario Serio for his consistent and supportive advice in all phases of our collaboration on longitudinal feedback.

REFERENCES

1. J.N. Corlett, J. Johnson, G. Lambertson, F. Voelker: Longitudinal and Transverse Feedback Kickers for the Als, Presented at 4th European Particle Accelerator Conference (EPAC 94), London, England, 27 Jun – 1 Jul 1994.
2. R. Boni, A. Gallo, A. Ghigo, F. Marcellini, M. Serio, M. Zobov: "A Waveguide Overloaded Cavity as Longitudinal Kicker for the DAFNE Bunch-by-Bunch Feedback System", Particle Accelerators, 1996, Vol. 52, pp. 95-113, 1996.
3. A. Gallo, A. Ghigo, F. Marcellini, M. Miglioratti, L. Palumbo, M. Serio : "Simulations of the Bunch By Bunch Feedback Operation with a Broadband RF Cavity as Longitudinal Kicker", DAFNE Technical Note G-31 April 1995.
4. M. Tobiyama, private communication, summarized in the December 2001 Cavity Kicker Design Review.
5. D. Teytelman, private communication, summarized in the December 2001 Cavity Kicker Design Review.
6. J. Fox, private communication, summarized in the December 2001 Cavity Kicker Design Review.
7. P. MacIntosh, H. Schwarz, private communications.

In-Situ Calibration: Migrating Control System IP Module Calibration from the Bench to the Storage Ring

J. M. Weber, M. J. Chin

Lawrence Berkeley National Laboratory, 1 Cyclotron Road, Berkeley, CA 94720

Abstract. The Control System for the Advanced Light Source (ALS) at Lawrence Berkeley National Lab (LBNL) uses in-house designed IndustryPack® (IP) modules contained in compact PCI (cPCI) crates with 16-bit analog I/O to control instrumentation. To make the IP modules interchangeable, each module is calibrated for gain and offset compensation. We initially developed a method of verifying and calibrating the IP modules in a lab bench test environment using a PC with LabVIEW. The subsequent discovery that the ADCs have significant drift characteristics over periods of days of installed operation prompted development of an "in-situ" calibration process--one in which the IP modules can be calibrated without removing them from the cPCI crates in the storage ring. This paper discusses the original LabVIEW PC calibration and the migration to the proposed in-situ EPICS control system calibration.

INTRODUCTION

The ALS Instrumentation Group designed custom IP Modules for the ALS Control System. Housed in cPCI crates, these modules control instruments such as magnet power supplies and beam position monitors (BPMs). Each IP Module consists of two 16-bit analog control channels (DACs) and four 16-bit analog monitor channels (ADCs). To make modules interchangeable and increase channel accuracy, each channel is calibrated for gain and offset compensation. We developed LabVIEW calibration software (IPCal) to calculate calibration constants for each channel and compile this data in a file. The file contains slope, offset, and mean-square-error (MSE) constants for each of the six channels on the IP module. Each IP module stores its calibration file in an onboard EEPROM. The ALS control system database uses the calibration constants to adjust data read from and written to each channel.

Originally, the test bench setup was sufficient to perform an initial calibration on each module before installing the modules into cPCI crates in the storage ring. We then discovered that many of the ADC channel offsets drifted several mV over weeks of installed operation. Further investigation revealed that the ADC channels exhibit an exponential drift from their initial offset as illustrated in Figure 1.

The ADC (ADS7825U) data sheet specifies a maximum zero offset error of ±10mV with a typical drift of ±2ppm/°C . ALS instrumentation and control requires a maximum drift of 10mV/year. These specifications coupled with the measured drift prompted a recalibration of each installed module. To perform the recalibration, each module would have to be removed during a storage ring shutdown, recalibrated in the test bench environment, and reinstalled during the same or a subsequent shutdown. Moreover, future recalibration, verification, and troubleshooting necessitate access to the installed and operating IP modules.

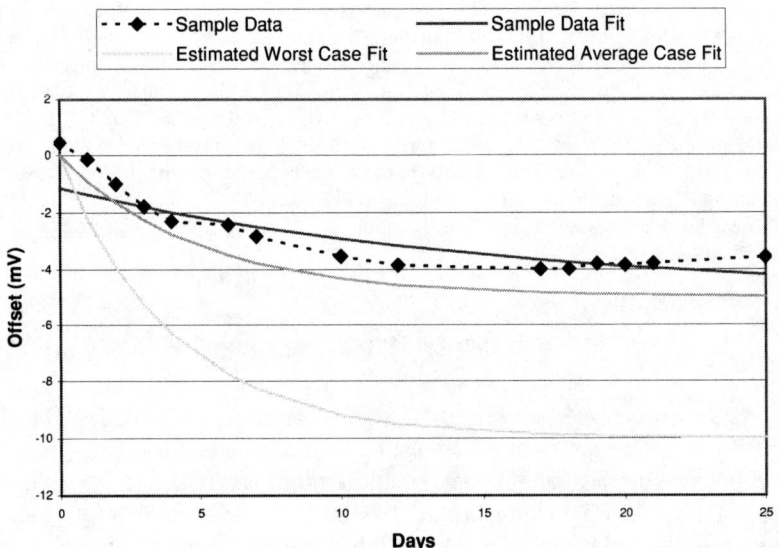

FIGURE 1. Plots show measured and estimated ADC offset drift. Estimated fits are based on testing a small sample of IP modules under varying conditions. Actual average and worst case offset drift of installed operating modules may differ. Assume initial offset of zero.

To improve the recalibration process, we are developing an in-situ calibration, wherein each module can be calibrated quickly and efficiently without being removed from the storage ring. Instead of powering down the cPCI crate, removing the IP modules, and disabling the corresponding instruments all at once, IP modules can be individually disabled to perform the in-situ calibration. This makes the in-situ process flexible, fast, and minimally disruptive to storage ring instrumentation.

CALIBRATION METHOD

To calibrate an IP module, each DAC output is connected to two ADC inputs and to an analog input channel on an Agilent (HP) 34970a Data Acquisition Switch Unit. The calibration software (IPCal) configures the 34970 for 20 bit resolution on each analog scan. The PC reads the 34970 digital output via RS-232 after each scan.

The full scale range of each DAC and ADC is 16 bits representing -10 to +10 Volts. IPCal writes to the DAC values from 0x1000 to 0xF000 (-8.75 to +8.75 Volts) to avoid nonlinear behavior at the limits of the range. The DAC values are written in steps of 160 (0xA0) for a total of 358 steps.

To begin the calibration sequence, IPCal writes a value to the DAC. The output of each ADC connected to the DAC is sampled 10 times and averaged. Each average is stored in an array of ADC output values. Then the 34970 scans the analog (DAC output and ADC input) value which is also stored in an array. This cycle is repeated until the sample range has been exercised. After generating the arrays, IPCal calculates the calibration constants. This entire process is repeated for the other channel set (DAC and 2 ADCs). When calibration constants have been generated for both channel sets, IPCal writes these values to a calibration file stored locally on the PC.

After IPCal generates the calibration file, a simple verification is performed to ensure the IP module will operate within tolerance. When IPCal generates calibration constants for a channel set, it displays several plots used for manual verification. These plots show the following: the averaged and raw ADC output values; the difference between input and output for the DACs and ADCs at each sample point; the step size between each sample point. These plots allow the tester to visually check gain, offset, and linearity for each channel.

In a separate window, IPCal uses the calibration parameters to simulate the gain and offset compensation. The software allows the tester to write to the DAC in volts and monitor the bit patterns for the DAC input and the ADC output in both bits and calibrated volts. As a coarse gain and offset test, the tester can exercise the DAC input in one volt steps over the ±10 Volt range and visually inspect the calibrated ADC. The software also allows access to the individual bits on the two 8-bit digital I/O ports. Each pin on port A is connected to port B across low value resistors (120Ω). The tester sets one port as input, the other as output and toggles each bit. This tests the data and control lines which could affect the results of the calibration if they are not working properly.

LABVIEW PC BENCH CALIBRATION

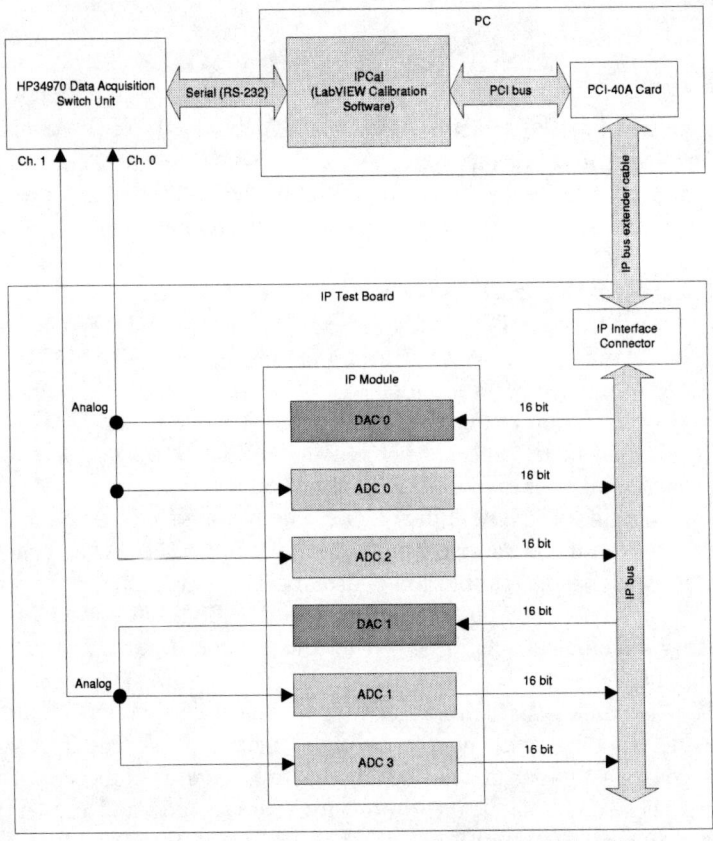

FIGURE 2. Block Diagram of the Test Bench Calibration.

Hardware

As shown in Figure 2, the bench calibration requires a PC with a PCI interface to the IP module and an RS-232 serial interface to the 34970. The PC contains an SBS Greenspring PCI-40A card with 4 slots for IP modules. For improved accessibility to the IP module and increased speed during the calibration process, we developed an in-house general purpose IP test board with hot swap capability and optional calibration configuration. This board allows us to probe both sides of the IP module during operation and swap IP modules without power cycling the PC. This board connects to the PCI card through an IP extender cable. The HP34970 accesses the IP module analog channels through BNC connectors on the IP test board.

Software

The IP module calibration is controlled at the highest level by LabVIEW software (IPCal). LabVIEW provides both PCI bus access and RS-232 serial control. LabVIEW interfaces to the PCI bus through BSquare (formerly Bluewater Systems) WinRT, a hardware driver that controls the PCI-40A card. This driver enables LabVIEW to access the digital signals on the IP module. LabVIEW controls the HP34970 with VISA functions via the serial port. VISA functions allow the user to control devices with uniform function calls via GPIB, RS-232, or Ethernet.

IN-SITU CALIBRATION

Hardware

Figure 3 shows the proposed in-situ calibration setup. The components of the in-situ calibration include a PC to run IPCal, an HP34970 Switching Unit, a breakout box for access to analog signals on the IP module, and a cPCI crate with a carrier board containing the IP module to be calibrated. The cPCI crate is located inside the storage ring and the PC and the 34970 are located on a mobile cart close to the cPCI crate. As in the bench calibration, the PC connects to the 34970 through a serial port. The PC communicates with the cPCI crate via Ethernet.

The ALS uses an in-house custom cPCI rear I/O card to interface IP modules to instruments. A test/breakout box with calibration configuration capability is being developed to interface the cPCI rear I/O connectors to the 34970. This breakout box will output the IP module analog signals via BNC connectors, enabling the 34970 to access these channels as in the bench calibration.

To perform an in-situ calibration, the cables from the cPCI crate to the controlled instrument are disconnected and the breakout box is inserted in series. This causes a temporary loss of operation for the controlled instrument. As such, an in-situ calibration may only be performed while the ALS is not operating with beam in the storage ring.

Software

The in-situ LabVIEW calibration software has the same user interface (IPCal) as the bench calibration software. Since the 34970 is still controlled by the PC through a serial interface, the VISA functions that control the 34970 remain the same. However, the in-situ software uses an ActiveX component to communicate with the cPCI chassis over the network.

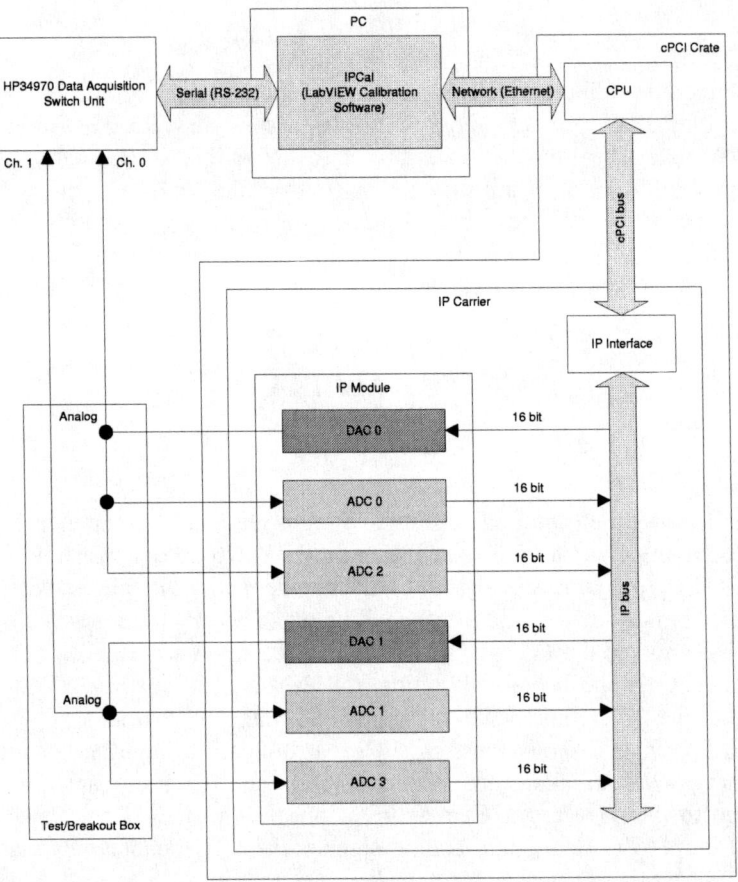

FIGURE 3. Block Diagram of the In-Situ Calibration.

The ALS control system EPICS records typically report values in engineering units (i.e. amps). Since the calibration software requires raw values (i.e. ±32768 bits), the cPCI crate must be rebooted to replace the operating records with test records that report raw values.

LabVIEW uses ActiveX through Simple Channel Access (SCA), an EPICS tool used for network communication. SCA allows LabVIEW to invoke methods in the ActiveX component. The ALS Controls Group has developed custom methods to read and write values in EPICS records. With these methods, the calibration software can access the digital signals on the IP module.

SYSTEM RESULTS AND PERFORMANCE

The current test bench calibration software and hardware configuration will be used to benchmark system performance. This includes using a 400 MHz PC with 128MB RAM to run the calibration software. Since PC processor speed and memory impact LabVIEW application overhead, we base our comparison of the calibration systems on this PC. Table 1 gives a timing comparison for each calibration setup.

Table 1. Calibration Timing Comparison.

Operation	Calibration Setup	
	Bench (PC)	In-Situ (cPCI)
ADC Read	0.38 ms	2.32 ms
DAC Write	0.26 ms	4.46 ms
HP34970 Read Setup Delay	100 ms	100 ms
HP34970 Read	100 ms	100 ms
Single Calibration Cycle (1 Write, 10 Reads, 34970)	210 ms	390 ms
Single Channel Set Calibration	1.30 min	2.40 min
Generate Calibration File	2.70 min	4.90 min

In the test bench environment, there are two significant timing factors: the 100ms HP34970 read setup delay, and the 100ms HP34970 read. These factors account for the majority of the calibration cycle time. In the in-situ calibration, the significant timing factors are the HP34970 read setup delay, the HP34970 read, the ADC read, and the DAC write. The HP34970 read time remains the same because the default resolution is used in both setups. The most significant timing differences between the two setups is in the ADC Read and DAC Write. This difference stems from communicating over the network instead of over the PCI bus in the PC. The ALS uses high speed 100Mbps Ethernet, which is still slower than the PCI bus in the PC.

ACKNOWLEDGEMENTS

The authors would like to thank the ALS Instrumentation Group and Control Systems Group for supporting this project, S. Jacobson and C. Timossi for EPICS, cPCI, and ActiveX support and development, and A. Geyer for building the breakout box. This work was supported by the Director, Office of Science, U.S. Department of Energy under Contract No. DE-AC03-76SF00098.

Set-up of PEP-II Longitudinal Feedback Systems for Even/Odd Bunch Spacings

D. Teytelman* and J. Fox*

*Stanford Linear Accelerator Center[1]
P.O. Box 4349, Stanford, CA 94309

Abstract. Feedback systems installed for control of coupled-bunch longitudinal instabilities in PEP-II collider have been designed to process bunch data at one half of the ring RF frequency. As a result these systems are ideally suited for controlling ring fills where only even or only odd RF buckets are populated (even bunch spacings). However in the operation of PEP-II per bunch charge considerations require fill patterns that alternately populate even and odd buckets. In this note we present a technique that allows to use existing hardware to provide feedback control of all bunches in such fills.

INTRODUCTION

PEP-II is a collider comprised of two circular accelerators of differing energies, Low Energy Ring (LER) and High Energy Ring (HER) [1, 2]. Both rings operate above coupled-bunch instability thresholds at the design currents. Feedback systems are required to stabilize the beams in transverse and longitudinal planes [3, 4]. Both rings have a 476 MHz RF frequency which corresponds to a bucket spacing of 2.1 ns. The operation was planned around filled bucket spacings of 4.2 ns or every other RF bucket. In order to reduce the digital signal processing load and analog bandwidth the longitudinal feedback systems were designed to operate at a 238 MHz sampling frequency (one half of the RF).

The operation of a collider is driven by luminosity considerations. Among many factors affecting the luminosity is per bunch charge. During the operation of PEP-II the combination of operating current and optimal per bunch charge made fill patterns with 6.3 ns (every third bucket) and 10.5 ns (every fifth) bunch spacings desirable. In their initial configurations the longitudinal feedback systems could not control fill patterns where both even and odd RF buckets were populated (mixed fills). A modification of the design has been made to address that problem. Here we will describe the design changes required, present corrected beam timing procedures and provide experimental results for the new mode of operation.

There are three signal processing components involved in controlling unstable motion: front-end detection and sampling, filtering by digital signal processors (DSPs), and kick signal generation in the back-end. These components are illustrated in Fig. 1. In order to

---

[1] Work supported by Department of Energy contract No. DE-AC03-76SF00515

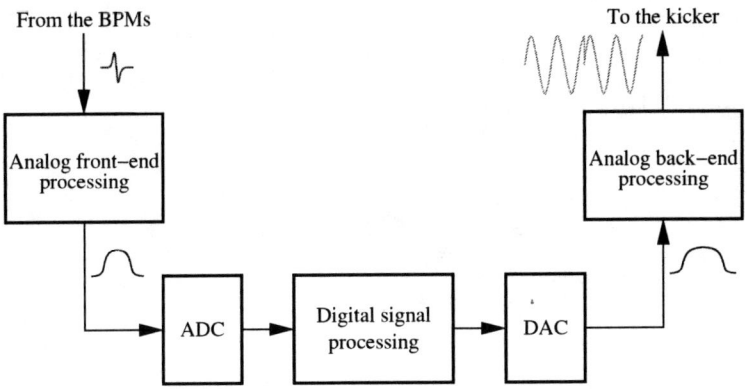

FIGURE 1. Block diagram of the PEP-II longitudinal feedback system

support mixed fills signals from both even and odd buckets must go through the signal processing chain and the kicker signal has to be correctly aligned with the passing bunch. Let's start the discussion from the front-end.

FRONT-END SIGNAL PROCESSING

Analog front-end processing of the longitudinal feedback system is designed to measure time of arrival of each bunch relative to the RF master oscillator clock. This is done using a phase detector operating with the master oscillator derived carrier at $6 \times f_{rf}$. The bunch-induced signal on the BPM output is a very short differentiated pulse. If such a pulse is used as an input to the phase detector, the output signal will be also a short baseband pulse. As a result sampled signal will be very sensitive to both pulse and clock timing. In order to avoid this problem in PEP-II the longitudinal feedback systems use a comb generator filter. In response to a BPM output pulse this filter produces a burst of pulses spaced at $T_{rf}/6$. When such a burst is phase detected baseband signal has a rectangular envelope with duration equal to that of the input burst [5]. Figure 2 shows the front-end block diagram.

Originally both PEP-II longitudinal systems were equipped with 4-tap comb generators that resulted in a baseband pulse of 1.4 ns duration. A timing diagram using such comb generators is given in Fig. 3. By comparing the baseband phase signal with the ADC sampling clock we determine that only even or only odd buckets can be sampled for a fixed clock.

The comb generator design allows one to stretch the detected pulse. If we lengthen the baseband pulse to span 2 RF buckets (2.1 $ns < T_{pulse} <$ 4.2 ns), the ADC clock can be set up to sample both even and odd buckets. This arrangement is illustrated in Fig. 4 for a 10-tap comb generator. In this case the pulse is 3.5 ns long. If for even buckets we put the sampling clock within $T_{offset} <$ 1.4 ns of either rising or falling edges of the pulse, odd buckets will be sampled within $T_{pulse} - T_{rf} - T_{offset} = 1.4\ ns - T_{offset}$ of the

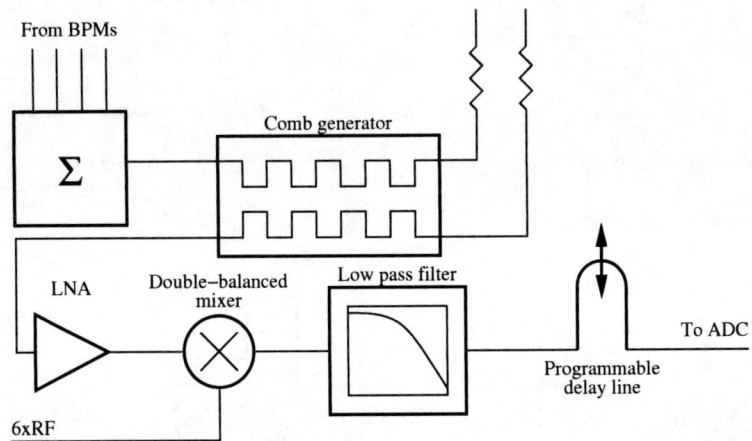

FIGURE 2. Block diagram of longitudinal front-end analog processing

opposite edge. To minimize timing jitter sensitivity we set $T_{offset} = 700$ ps.

We can arbitrarily select whether to sample filled even buckets near the front or the tail of the pulse. However that choice, as we will see later, affects set up of the back-end. In addition, examining Fig. 4 we see, that for the timing shown bucket 3 is sampled by clock pulse 1. If even buckets are sampled near the front of the pulse, bucket 3 gets sampled by clock pulse 2. This sample shift would not affect feedback processing in these bunch-by-bunch feedback systems with independent digital signal processing for each bunch. However analysis of beam data acquired by the DSPs is dependent on which sampling model (pulse front or tail) is chosen. In order to eliminate ambiguity we've chosen in

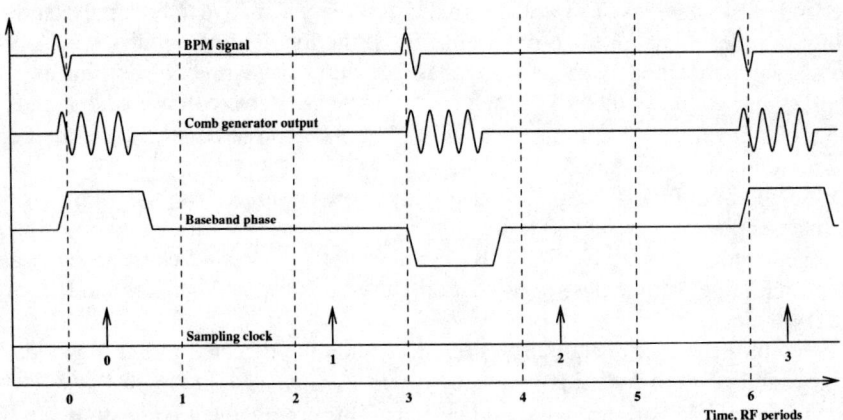

FIGURE 3. Timing with 4-tap comb generator. The top trace shows BPM signals for 3 bunches in the every third bucket fill pattern. Dashed lines show ideal synchronous times, bunches 0 and 6 arrive early while bunch 3 is late.

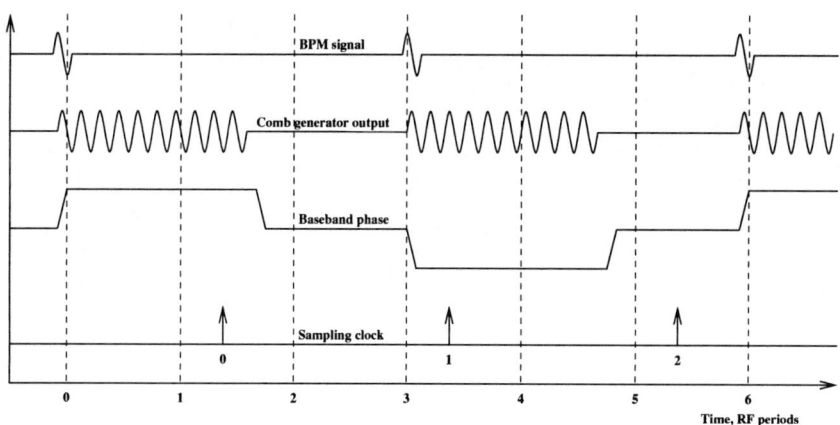

FIGURE 4. Timing with 10-tap comb generator.

PEP-II to sample even buckets at the tail of the pulse.

There are several trade-offs in mixed fill pattern sampling with a long comb generator. First of all, the longer baseband pulse increases coupling between the neighboring bunches. In an ideal system with fast pulse rise and fall times there would be no coupling for pulse lengths below 4.2 ns. However due to ringing in RF components (such as the low-pass filter) baseband pulse can last longer than the length of the comb generator burst. The coupling is worst at the minimal bunch spacing of 4.2 ns. Since the pulse is lengthened by a ringing tail, coupling is most significant for all odd bucket fill patterns. Another negative effect comes from tap spacing errors in the comb generator. If tap spacing is different from nominal $T_{rf}/6$, frequency of the resulting burst is offset from $6 \times f_{rf}$. Consequently detected phase pulse will ride on a slope. Since for mixed fill patterns we sample the pulses at different points for even and odd buckets, the bunch signals will have differing DC offsets. Even though overall DC offset is rejected by the phase servo loop in the feedback front-end, these even to odd bucket offsets cannot be easily compensated.

DSP FILTERS AND POST-PROCESSING

The feedback processing is done on a bunch-by-bunch basis. The harmonic number of PEP-II is 3492 and so the feedback system acquires $3492/2 = 1746$ bunch samples per revolution. These bunch signals are downsampled and processed independently from each other. Thus, as long as the bunch phase signal is properly sampled, it is processed in the same manner within any of the 1746 processing channels.

However post-processing of the data acquired by the feedback system does depend on the actual fill pattern. As shown in Fig. 4 the signal from filled bucket 0 or 1 appears in processing channel 0. In order to accurately extract modal information from the bunch-by-bunch time-domain record we need to know whether even or odd bucket is filled. This

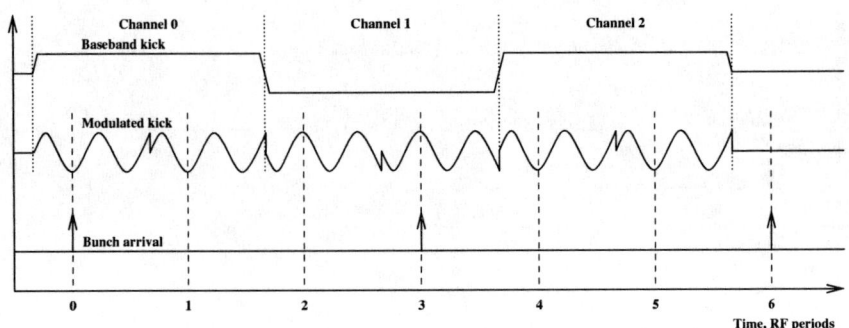

FIGURE 5. Back-end kick signals and timing to the bunch arrivals

information is not present in the data, therefore knowledge of the fill pattern is needed for proper signal reconstruction. Once the fill pattern is known it is easy to determine RF bucket to feedback processing channel correspondence. For front-end timing setup described in section bucket and channel numbers are related as follows:

$$N_{channel} = [N_{bucket}/2] \qquad (1)$$

BACK-END SIGNAL PROCESSING

The correction values computed by the DSPs are converted into a baseband analog signal using a D/A clocked at $f_{rf}/2$. The D/A output is then a series of 4.2 ns long steps. In the frequency domain most of the power is in the DC to $f_{rf}/4$ band. The longitudinal feedback kicker for PEP-II is designed with center frequency of $9f_{rf}/4$. Therefore the baseband output of the D/A is mixed with a carrier frequency to place the power within kicker bandwidth. A carrier signal at $9f_{rf}/4$ undergoes a 90 degree phase shift over one RF period. If bunch 0 is timed for maximum positive kick, bunches 1 and 3 will arrive at zero crossings while bunch 2 will see negative kick. To avoid this in PEP-II longitudinal feedbacks phase of the carrier is shifted by -90 degrees every RF period using quadrature phase shift keying (QPSK).

In this discussion we will consider the bunch passage time through the longitudinal kicker to be short relative to the carrier wavelength. PEP-II kickers have three field gaps spaced at 1/4 carrier wavelength. However we can model these as a single short (broadband) gap in combination with a bandpass kicker transfer function. When the bunch passes through the gap it samples the kick voltage. In order to maximize the effect, the kick has to be timed so that a bunch samples the peak of the waveform. As the single-bunch kick is 4.2 ns long there are multiple peaks that can be synchronized with the beam. However, the requirement to affect both even and odd RF buckets restricts the timing. In Fig. 5 waveforms of the back-end signals are shown. Our goal is to chose a kick timing relative to the beam so that both buckets 0 and 1 would sample peaks of the channel 0 kick waveform. There are 5 possible timing settings. However, in order to

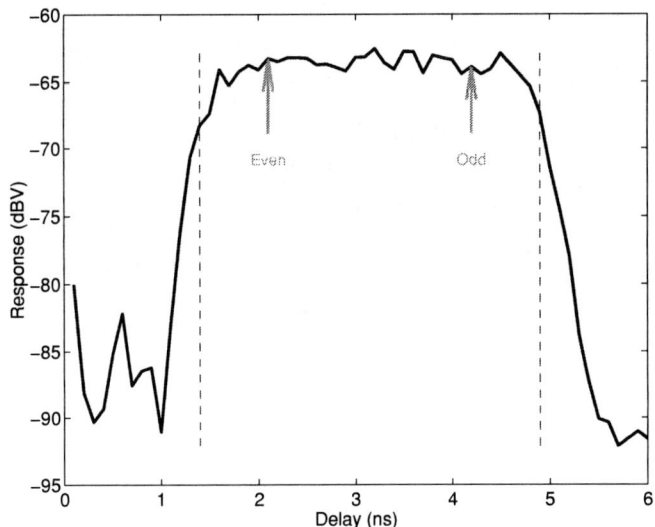

FIGURE 6. PEP-II LER front-end timing sweep. Approximate extent of the phase detector output pulse is indicated by dashed vertical lines. The "even" arrow indicates where the pulse from an even filled bucket is sampled by the ADC, while "odd" pointer shows the sampling for the odd bucket.

minimize interbunch coupling we try to use the peaks closest to the burst center point. For the carrier phasing shown in Fig. 5 there are two timing settings which place either the even or the odd bunch closer to the center point. The choice between the two settings is arbitrary.

It is interesting to note that both in the front-end and the back-end we are dealing with sampling of a finite-duration pulse. In the front-end, the ADC clock samples the baseband phase while in the back-end the bunch samples an amplitude-modulated QPSK carrier. However for the front-end setup where ADC samples the tail of the pulse we have to configure the back-end so that the bunch samples the head of the kick pulse.

EXPERIMENTAL RESULTS

In this section we will describe the procedures for setting up the correct front-end and back-end timings and present experimental data for feedback control of even/odd filling patterns. The front-end timing is controlled by by the setting of the programmable delay line. The delay line allows one to adjust the placement of the phase detector output pulse relative to the ADC clock. The clock signal is locked to the ring master oscillator and for fixed RF voltage phase remains in a constant relation to the beam phase.

In order to determine optimal setting for the front-end delay line an automated timing utility is used. First a single bucket in the ring is filled to nominal per bunch charge. The feedback system is set up to deliver a back-end correction signal for only one channel - the channel that must sample the filled bucket. Then under program control the delay

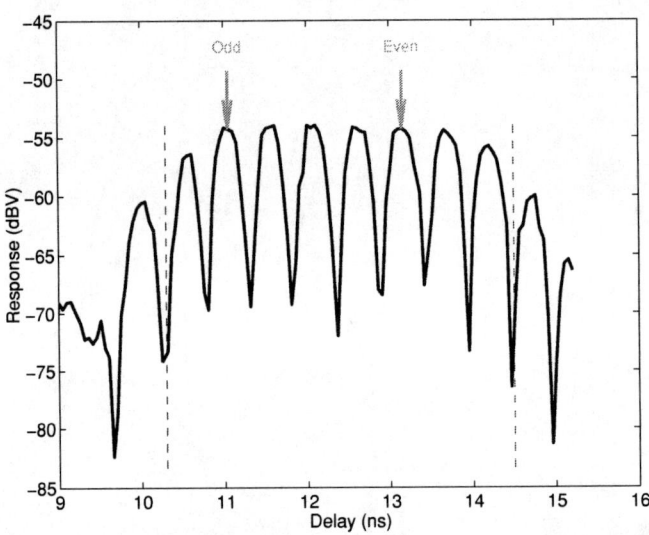

FIGURE 7. PEP-II LER back-end timing sweep. The dashed lines show the DAC pulse length of 4.2 ns. This pulse is stretched due to the bandlimiting in the kicker. The "even" arrow indicates the point where the even bucket samples the kick waveform.

line is swept through a range of values and at each setting the baseband spectrum of the DAC output is measured. Due to the RF noise excitation the beam oscillates at the synchrotron frequency. That motion is amplified by the feedback filter and allows the determination of the amplitude sampled by the ADC. By selecting a spectral component at the synchrotron resonance peak we obtain sampled signal amplitude versus delay as shown in Fig. 6. Note that the delay axis goes in the opposite direction from the time axis, that is larger delay setting corresponds to sampling the earlier part of the pulse. Since we decided to sample the even buckets towards the tail of the pulse the "even" timing arrow is placed closer to the left edge of the delay sweep.

A similar automated procedure is used to set the back-end delay line. In this case with a single bunch in the ring we program the DSPs to produce sinusoidal excitation at the synchrotron frequency in a single channel. As the delay line is swept through the region of interest at each setting we record the spectrum of the beam via the phase-detector output. When the kicker burst is correctly aligned with the beam the driven motion is maximal. By plotting the signal amplitude at the synchrotron frequency against the delay setting we get the magnitude response of the kicker as illustrated in Fig. 7. Here two response lobes 2.1 ns apart are selected for even and odd bucket timing. As expected the positions of even and odd buckets are reversed relative to the front-end sweep.

After the optimal timing settings are determined we are able to fill the rings in the even/odd fill patterns, for example every third bucket. After filling the LER to 800 mA we recorded beam data in the closed-loop feedback configuration to verify the system stability. In Fig. 8 the RMS amplitudes of the recorded data are show for the first 12 DSP channels. Empty channels show RMS motion of 1 ADC count while the

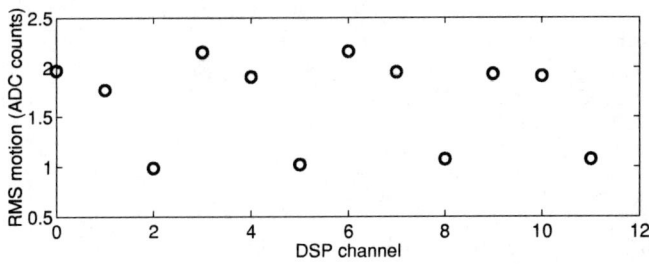

FIGURE 8. RMS of the beam data recorded by the DSPs. Bucket 0 is sampled by channel 0 while bucket 3 signal is in channel 1.

channels sampling filled buckets show 2 counts. Since the front-end is timed to sample even buckets near the tail of the pulse we can reconstruct the underlying fill pattern unambiguously.

In Fig. 9 average amplitudes of the coupled-bunch eigenmodes are shown. These values are obtained by transforming the abovementioned data set to the eigenmodal basis. Since the measurement was made in the closed-loop configuration we obtain the steady-state modal amplitudes. Most of the eigenmodes are damped by the feedback to the noise floor with the notable exception of eigenmode 0 oscillating at 0.09 degrees. This lowest-frequency mode is excited by the RF noise from the klystron high-voltage power supplies. The fill pattern has periodic minigaps which act to alias mode 0 every 72 revolution harmonics.

SUMMARY

Modification of the comb generator length in combination with proper front-end and back-end timing procedures described here have enabled PEP-II to use the mixed even/odd ring filling patterns required to optimize luminosity. Under these conditions we have demonstrated feedback stabilization of longitudinal coupled-bunch instabilities at the design beam currents.

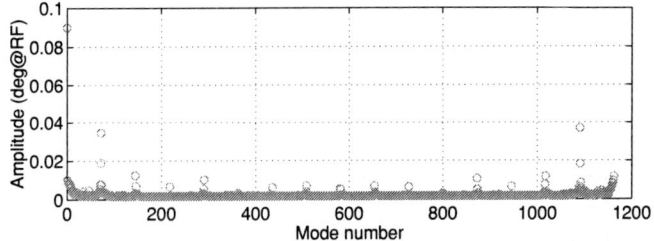

FIGURE 9. Average modal amplitudes in the LER at 800 mA

REFERENCES

1. PEP-II: An asymmetric B factory. conceptual design report (1993).
2. Seeman, J., et al., "Status Report of PEP-II Performance," in *EPAC 2000: Proceedings*, IOP Publishing, Philadelphia, PA, USA, 2001, pp. 38–42.
3. Barry, W., et al., "Design of the PEP-II transverse coupled bunch feedback system," in *Proceedings of the 1995 Particle Accelerator Conference and International Conference on High - Energy Accelerators*, IEEE, Piscataway, NJ, USA, 1996, pp. 2681–2683.
4. Oxoby, G., et al., "Bunch-by-bunch longitudinal feedback system for PEP-II," in *EPAC 94: proceedings*, World Scientific, River Edge, NJ, USA, 1994, pp. 1616–1618.
5. Young, A., Fox, J., and Teytelman, D., "VXI based multibunch detector and QPSK modulator for the PEP-II/ALS/DAPHNE longitudinal feedback system," in *1997 IEEE Particle Accelerator Conference: Proceedings*, IEEE, Piscataway, NJ, USA, 1998, pp. 2368–2370.

Operation of the Beam Diagnostics System for Tevatron Electron Lens

X.Zhang, K.Bishofberger*, J.Fitzgerald, G.Kuznetsov, M.Olson, A.Semenov, V.Shiltsev, N.Solyak

PO Box 500, FNAL, Batavia, IL 60510
*University of California at Los Angeles, P.O. Box 951547, Los Angeles, CA 90095-1547

Abstract. The first Tevatron Electron Lens (TEL) has been installed and commissioned successfully as part of the Beam-Beam Compensation project at Fermilab[1]. Currently it is operated routinely for DC beam cleaning during Tevatron luminosity stores and for advanced beam-beam studies. This paper reviews the electron and proton (antiproton) beam diagnostics, which allow us to measure beam intensity, waveform, losses, position, timing and profile. In addition, other proton (antiproton) diagnostics, available from the Tevatron control system, which are used for tuning beam parameters in the TEL (tune-shift, orbit, emittances, lifetime measurements, etc) are also described. We also present the results of measurements of the beam parameters and discussions for future upgrades.

1 BEAM POSITION MONITORING

We have installed two pairs of pickup electrodes in the TEL system near each end of the interaction region that enables us to measure horizontal and vertical positions of electron, proton and antiproton beams entering and exiting the TEL. Each pickup electrode pair is made of a stainless steel cylinder with a diameter of 70mm and cut diagonally in half. The TEL layout and the BPM pickup electrodes are shown below in Figure 1.

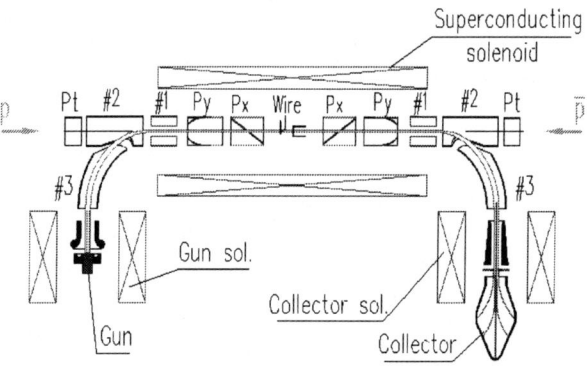

FIGURE 1. Layout of the TEL (*Px, Py are the beam position pickup electrodes*)

We have a broadband three-beam BPM system, which is used to measure proton, antiproton, and electron beam positions. The requirement for beam position accuracy is 50μm rms with an electron-(anti)proton position difference less than 100μm.

The three-beam BPM system[2] is composed of a LabVIEW application program operating on a Macintosh computer utilizing a digital oscilloscope for data acquisition. The four position detectors are sequentially connected to the oscilloscope's inputs through the RF Multiplexer. The computer communicates with the oscilloscope and the multiplexer and links with Tevatron Accelerator Controls Network(ACNET). A Beam Synchronous pulse generated by a Camac 279 module triggers the scope's main sweep. The oscilloscope must be operated in the delay trigger mode to obtain the finer timing resolution required to capture the 20 ns bunch signal. The delay must be adjusted properly for each position detector to compensate for differences in cable lengths and beam flight times. The default delays for the oscilloscope trigger are automatically set from a look-up table, depending upon the selected bunch and BPM detector, and have been empirically determined to trigger the oscilloscope about 10 ns before a given bunch arrives.

When the beam traverses the detector, it generates a doublet current signal similar to Figure 2 on each plate of the same detector. Then both signals are digitized to 500 points by the oscilloscope and transferred to the computer through the GPIB interface. The vertical scale of the oscilloscope can also be changed to improve the performance according to the bunch intensity.

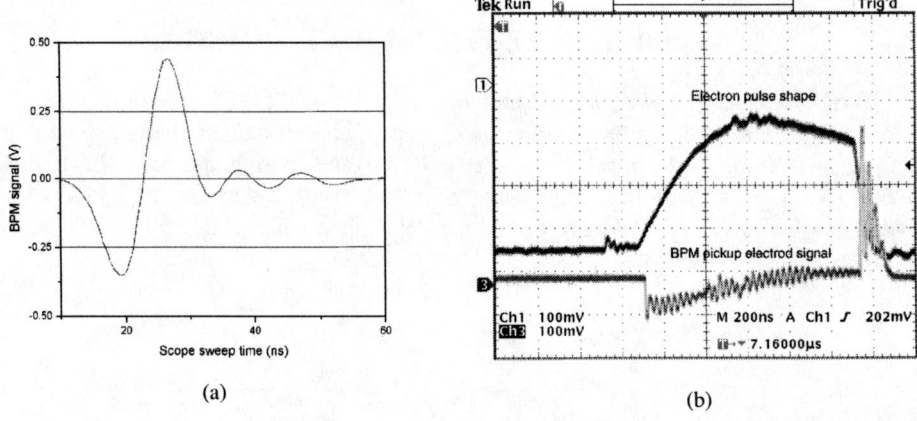

FIGURE 2. Doublet signal from pickup electrode (a) antiproton pulse and (b) electron pulse

To calculate the beam position, the digitized doublet signal of each pickup plate is first averaged 128~512 times by the scope to minimize noise. Next the signal strength of the two plates is determined by digitally integrating the signals individually. Finally, a difference over sum calculation is performed to obtain a position reading. The above process repeats until all four proton and all four antiproton beam positions have been obtained.

To obtain the integral of the signal strength, several methods were tested. We first tried a zero-crossing method, where the most positive and most negative points of the averaged traces are found, and a cubic polynomial is fit to the data between them to find the zero-crossing point. Then the signal is rectified by multiplying all points after the zero-crossing with -1. The advantage of the above procedure over a simple addition is that the effect of offset, noise, ringing, or satellite bunch signals outside the central bunch is greatly reduced. The signal strength of each plate is then determined by digitally integrating the rectified signals individually. In addition, a software Band-Pass filter can be applied to optimize the signal-to-noise ratio. However, the distorted waveform creates an error in the zero crossing and hence an error of the reported beam position. This method works poorly for the electron beam, since the electron beam pulse has a 30MHz modulation coming from the high-voltage modulator circuits, and its shape also makes it very difficult to find the correct zero crossing. (Figure 2(b)).

The second method is trying to find the averaged peak value of the traces. Unfortunately the peak has a small signal-to-noise ratio, giving a larger error in beam position readings. This is especially true for electron beams.

The third method is taking the absolute value of the averaged traces and integrating. The software Band-Pass filter can still be applied to optimize the signal-to-noise ratio. This method is relatively easy and fast and gives a satisfying answer. We use it mostly.

Typically, we calibrate the BPM readings by moving the electron beam transversely with magnetic steering coils. Figure 3(a) shows the measurement of the linearity of one pair of electrodes made with the electron beam, which is quite satisfactory in the range of ±6 mm around centerline (the BPM pickup pipe diameter is 70 mm). By this way, we can also get the calibrated coefficient of the BPM system.

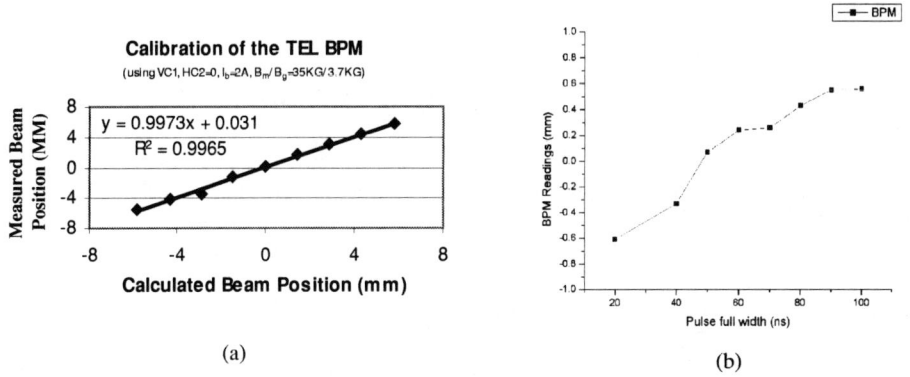

FIGURE 3. (a) The linearity measurement of the BPM. (b) Measured beam position vs. beam pulse width on a BPM test stand

A lot of effort has been taken to reach our goal of high resolution and small error for beam position. Besides increasing the number of averages, we improved the vacuum feedthroughs with SMA connectors instead of simple solder-on pins, which led to a substantial reduction of mismatch and ringing. Using 128 averages and full scope bandwidth, a beam resolution of about 30μm rms was achieved for a 2A peak current of 800ns-long electron pulse. An rms resolution of 50μm was obtained for a single proton bunch with the intensity of 9.4×10^{10}, and the antiproton resolution was about 90μm rms, which was over ten times weaker.

However, the most serious problem comes from the discrepancy between the electron beam position and proton beam position readings. One source of the offset comes from the difference in the channels of the oscilloscope. To solve this, we use the same channel to read both plates of the same BPM pair by switching inputs via the multiplexer. We also improved the beam synchrotron signal to scope trigger to reduce the mis-triggering. By doing these, we eliminated an offset of 0.3mm. However, the major source of the offset comes from the different BPM impedances for electron beam and proton beam signals, since for proton-like signal the main frequency component is about 53MHz while for electron beam the main frequency component is less than 2MHz. The capacitance between the two plates and the cross-talking between pairs of electrodes might also contribute to the offset. The maximum offset between electron beam and proton beam is over 1mm (see Figure 3(b)). This makes it very difficult to align the electron beam exactly to the proton beam orbit.

To minimize the offset, various software and analog filters were tested. In the end, a 5MHz software low pass filter was used, which decreased the average position difference to 0.3mm. Thereafter during the electron beam colliding with the proton bunch, a fine-tuning of the beam alignment was carried out ad hoc by maximizing tune shift and minimizing the proton beam loss.

As a part of the TEL upgrading plan, new stripline electrodes will replace the current arrangement in summer of 2002. This will decrease cross-talking, flatten frequency response, increase the sensitivity and be better calibrated initially. We hope we can achieve a better beam position resolution and lower the offsets to less than 100μm in future.

2 ELECTRON BEAM QUALITY MONITORS

We have installed electron beam quality monitors which measures the electron beam current, beam profile, current stability, and beam pulse shape.

2.1 Electron Beam Profile Monitor

Beam profile is a crucial characteristic of the electron lens. For linear beam-beam compensation the electron beam should have a profile with uniform charge distribution. For future non-linear beam-beam compensation, the electron beam is required to have a charge distribution closer to the Gaussian distribution.

To measure the electron beam profile, two wire scanners, one horizontal and one vertical, are installed in the TEL near the center of the main solenoid. Wires can be moved in or out of the beam pipe by remotely controlled stepping motors. In normal Tevatron operation, they are moved completely out of the beam orbit in order not to disturb the beams. The geometry of the wire is shaped like a "fork". The distance between the fork claws is 15mm, from the wire to top edge is 22mm, the wire diameter is 100μm, and the beam pipe diameter is 70mm. Moreover, the dimensions give us a good scale for calibration of steering strength of correctors for the electron beam and in turn, to calibrate the pickup BPM systems.

By steering the electron beam vertically or horizontally by a known amount we can scan the beam across the wire. The portion of the beam intercepted by the wire gives a sliced X (or Y) beam profile as shown in Figure 4(a). Then the radial beam profile can be restored assuming almost radial symmetry. This restored beam profile, also in Figure 4(a), indicates a mostly flattop profile with somewhat less charge in the center than around the edge. This is caused by the electron space charge effect (the electrons in the center are moving slower than those in the edge). The beam diameter is about 3.5mm. The restored profile is in good agreement with the two-dimensional electron current profile (Figure 4(b)), previously measured by a "pinhole" collector scanner on a testbed[3]. This collector simply has a very small hole that allows a small fraction of the current to pass and be measured. Scanning in two dimensions allows for detailed profile measurements.

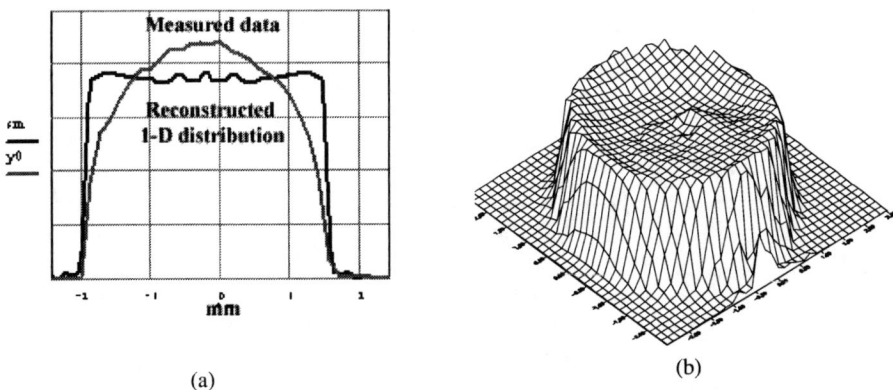

FIGURE 4. (a) 1-D and (b) 2-D beam profiles (I_{peak}=2A).

Unfortunately, both wires in the TEL have burnt out recently. Due to a major upgrade in summer 2002, there will be no room for the wire apparatus. Instead, right-angled "knives" will be installed near the beam pipe wall. The electron beam will be scanned through the knives to measure the profile.

2.2 Electron Beam Current and Charge Monitors

The electron beam current is measured by wideband current transformers both at the cathode and collector. They monitor the electron beam pulse shapes as well as the beam loss through the TEL system. Typical output waveforms are shown in Figure 2(b) and 5.

The electron beam charge monitor uses common BPM pickup electrodes. An RF switch is used to switch the signals between the BPM electronics and the charge monitor. There are four sets of identical electronics, each for one of the four pairs of BPM detectors. The charge amplifiers are used to get the electron beam charge distribution along the beam pulse. The lower curve of Figure 5 shows a typical signal from one channel of the electron charge monitor, which is dependent on the beam position. By adding up both channels from the same pair of pickup electrodes, we can get the total charge of the electron beam independent of the beam position.

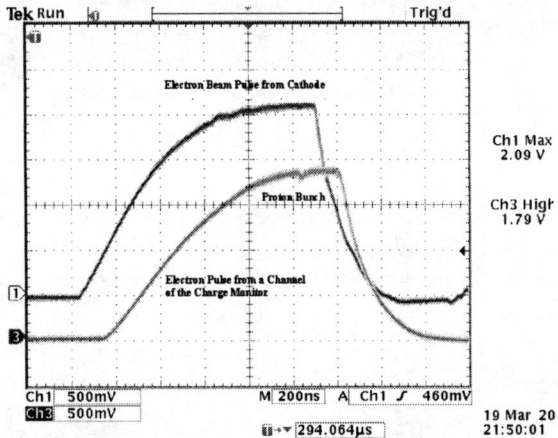

FIGURE 5. The lower trace taken from one channel of the charge monitor, the small dent in the lower waveform is the proton bunch signal, which is timed during the flat part of the electron pulse.

The charge amplifiers have a saturation threshold and we have to be careful with the input signal. Fortunately, in most cases, the electron current is not so high as to overfeed the system. This system can also be used to measure the electron beam position. It will be further upgraded by fine synchronizing signals from both pickup electrodes and digitizing to enable the data to be shared over ACNET.

We are also prototyping an electron-current stability-measuring system. It has two channels, with a 14-bit fast ADC (AD6644) and a 262Kx18-bit high-speed FIFO memory (IDT72V2105) each channel. This allows the system to sample and acquire data at speeds up to 66MHz and up to data record lengths of 512K. There are two modes of operation: continuous mode and gated mode. In continuous mode, the digitizer samples and writes continuously at a rate of 66MSPS to FIFO memory until full. In gated mode, the digitizer only samples signals that we want and writes to the FIFO. In each case, 7.6ms of data will be recorded with a time resolution 15ns. In addition, the threshold, delay, and window parameters for the gating can be preprogrammed and the gating rate can be as low as 1Hz. Using the cathode current or charge intensity monitor as the input, preliminary results show that the stability of the electron beam current from pulse to pulse was about 3.6×10^{-3} at 2.5A peak current.

3 (ANTI)PROTON DIAGNOSTICS

Besides the BPM system of the TEL, we also use the beam diagnostics of the Tevatron to monitor proton and antiproton parameters, which include intensity, emittance, orbit, lifetime, and tune[4]. We are also able to monitor the luminosity and the proton losses bunch-by-bunch, which is disseminated by the CDF detector. The tunes are measured by Shottky spectra analyzers, whose output appears in Figure 6(a). In this example, there were only two proton bunches in the Tevatron. The spectrum in the left part shows the tune of the typical bunch that is not colliding with the electron beam, but the right part shows the shifted tune of the other bunch colliding with the electron beam. We rely on these spectra heavily together with the proton loss monitor in order to fine-tune the electron beam onto the proton beam orbit. Our goals are to maximize the tuneshift while minimizing any proton losses.

A new bunch-by-bunch tune meter[5] is currently being commissioned. The beam emittances are measured by the flying wire systems[6], which have errors of about 10%. We expect that the recently upgraded synchrotron light monitor will give us a double-check and allow us to monitor the proton or antiproton emittance variations continuously during beam-beam compensation studies[7]. Also the beam lifetime is monitored by the Fast Beam Integrator (FBI), which relies on the wall current monitor[4].

A very accurate method of aligning the electron beam with the proton bunch is by means of 'tickling' the proton beam orbit. We do this by modulating the electron beam current[1]. That method provides us the information for precise centering of the electron beam onto the proton (or antiproton) beam. The Tevatron orbit measurement system has a resolution of 150 micrometers, which helps us to double check the our BPM measurement.

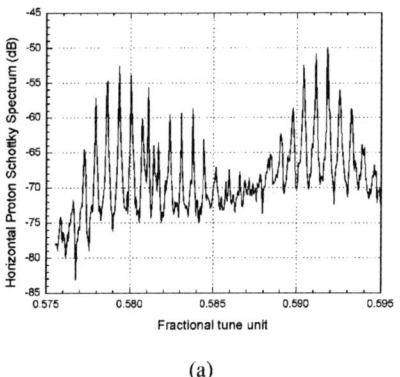
(a)

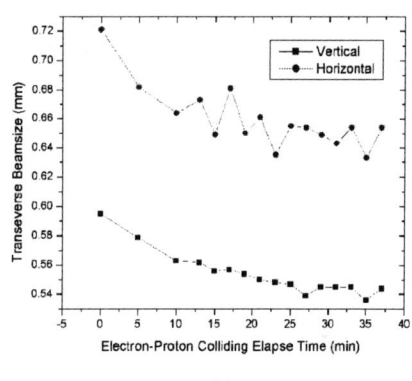
(b)

FIGURE 6. (a) Tuneshift due to colliding with electron beam
(b) Proton beamsize during the collision with electron beam

Figure 6(b) shows the proton beam size measured over time while colliding with the electron beam by Tevatron flying wires. Since the electron beam had a hollow shape, the non-linearity of the electron beam edge is very strong. Therefore, the tail of the proton bunch was scraped off, leaving the major part of the proton beam which can fit into the electron beam comfortably. In this measurement, the initial proton beam sizes were larger than the design value. But the equilibrium beam sizes correspond to the electron beam acceptance of 20~25π mm•mrad, which is what the TEL is designed to accommodate. The coming upgrade of the TEL will decrease the bend angle of the electron beam from 90° degrees to about 60° degrees, and more solenoids will be added in the bend to minimize the electron beam size in the bends. Also a new electron gun with a parabolic beam profile will be installed in order to eliminate the sharp edge and compensate the hollow in the electron beam profile. By doing these, we hope that we will be able to vary the beam size by a factor of two and also increase the electron beam acceptance significantly.

4 CONCLUSION

The TEL is a very important project for Tevatron upgrading. It is not only working as a setup for advanced beam-beam studies, but also being operated routinely as the Tevatron DC beam cleaner. Occasionally a troublesome DC beam is generated in the Tevatron, which causes a spiky background in the CDF and sometimes quenches during aborting. The beam diagnostic systems for the TEL electron lens played a crucial role in commissioning. We were able to measure the electron beam parameters. We also have successfully aligned the electron beam to the proton beam and obtained excellent beam tune shift. The BPM offset issues between the electron beam and the proton beam have not allowed us to quickly align the electron beam to the proton beam, but a new system will be installed in the coming months that will deliver huge improvements. At the same time, substantial efforts are being put into further improvements on other beam diagnostics in order to secure the successful operation of the TEL in future.

ACKNOWLEDGMENTS

We thank Jim Crisp, Jim Steimel, Dan Wolff, Dave McGinnis, Howard Pfeffer and David Peterson for their helpful advice and discussions on BPM issues. We also thank Stephen Pordes and Wim Blockland for their help using their flying wire system for proton size measurement, Dean Still with the Tevatron Schottky tune measurement and C.Y. Tan for his advices on the tune-meter system. Finally we thank the Accelerator Controls Department for their help for the TEL control system and the Tevatron operation crew and during TEL studying shifts.

REFERENCES

1. V.Shiltsev *et al*, submitted to PAC2001
2. M. Olson, A. A. Hahn, AIP conference proceedings 390, pp. 468-475, Argonne Il., May 1996
3. A. Shemyakin, *et al*, Proc.of EPAC 2000, p.1271
4. Tevatron Run II Handbook, Chapter 6.12, http://www-runii.fnal.gov/
5. C.Y. Tan, FERMILAB-TM-2078, 2000
6. J. Gannon et al, FERMILAB-CONF-89/64, 1989
7. A.A. Hahn, HEACC'92, pp. 248-250

Techniques for Electro-Optic Bunch Length Measurement at the Femtosecond Level[†]

P. Bolton, D. Dowell, P. Krejcik[*], J. Rifkin

Stanford Linear Accelerator Center
2575 Sand Hill Rd, Menlo Park CA 94025

Abstract. Electro optic methods to modulate ultra-short laser pulses using the electric field of a relativistic electron bunch have been demonstrated by several groups to obtain information about the electron bunch length charge distribution. We discuss the merits of different approaches of transforming the temporal coordinate of the electron bunch into either the spatial or frequency domains. The requirements for achieving femtosecond resolution with this technique are discussed. These techniques are being applied to the Linac Coherent Light Source (LCLS) and the Sub-Picosecond Photon Source (SPPS) currently under construction at SLAC.

INTRODUCTION

New generations of accelerator-based x-ray laser sources will utilize extremely short electron bunches. The bunches are considerably shorter than can be measured with existing streak camera technology and will require new, innovative techniques to diagnose and tune them. The Linac Coherent Light Source (LCLS) [1] to be built at SLAC utilizes electron bunches as short as 80 femtoseconds rms to generate self-amplified stimulated emission (SASE) X-ray radiation in a FEL. The new Sub-Picosecond Photon Source (SPPS)[2] at SLAC offers a near term opportunity to test and compare these different diagnostic techniques with bunches as short as 30 fs rms, far shorter than anything so far produced in a high energy electron accelerator.

Conventional laser technology has succeeded in producing and characterizing visible light pulses with the time structure that is well matched to these electron beam requirements. The bridge between the two systems exists in the form of electro optic (EO) crystals whose birefringence properties are modulated by the electric field of the electron bunch to be measured. The transmission of polarized light through an EO crystal is in turn modulated and the problem of measuring an electron bunch length is thereby transformed into one of measuring the duration of a light pulse.

This paper discusses first the choices of configurations between laser and electron beam geometry. For a given geometry the sensitivity and bandwidth requirements are then presented. The light pulse detection techniques discussed show the evolution from interferometric techniques to nonlinear autocorrelation methods and the advantages they offer. Finally, the limits to resolution are discussed and the latest developments towards pushing this limit.

---

[*] corresponding author, pkr@slac.stanford.edu   [†]This work is supported under the DOE contract DE-AC03-76SF00515.

ELECTRO OPTIC PROBE GEOMETRY

Two basic configurations are possible. One is where the probe laser pulse is transverse to the electron beam and the second where the two beams are parallel and co-propagating. Each of these geometries has implications for: laser power in the crystal; whether the pulse sampling should be performed in the spatial or frequency domain; and on the ultimate resolution that could be achieved. An illustration of the transverse geometry is shown in figure 1, similar to that proposed in reference [3].

In the transverse scheme a cylindrical lens brings the polarized laser beam to a ribbon focus at the EO crystal where it overlaps the electric field of the electron bunch. The length of the ribbon focus determines the duration of the timing gate in which the electron bunch must be coincident. A cross polarizer normally extinguishes the light reaching the CCD detector. However, the light will be modulated in proportion to the change in birefringence induced by the electric field of the bunch. The image recorded on the detector is therefore a convolution of the electron bunch charge distribution and the temporal profile of the laser pulse. In order to achieve good resolution a laser pulse length much shorter than the electron bunch length must be used. For example, a Ti:sapphire laser with 20 fs would resolve sub-picosecond electron bunches.

A 100 fs electron bunch is 30 µm in length so some magnifying optics must be used in front of the detector to overcome the pixel resolution limit of approximately 9 µm in a CCD camera. Furthermore, the retardation in the crystal is wavelength dependant and since a short pulse laser has by nature a very large bandwidth, care must be given to calibrating the system. Wavelength calibration is difficult in the transverse geometry where the extent of the bunch is to be measured in the spatial domain.

This problem is partially overcome in the second transverse geometry scheme shown in figure 1. The laser pulse is chirped and the modulation of the electron bunch now acts to gate a portion of the stretched laser pulse. A spectrometer detects the modulation in the frequency domain thereby avoiding limitations due to pixel resolution and beam size. The wavelength calibration of the optical elements is readily done with the electron beam off and using quarter-wave plates.

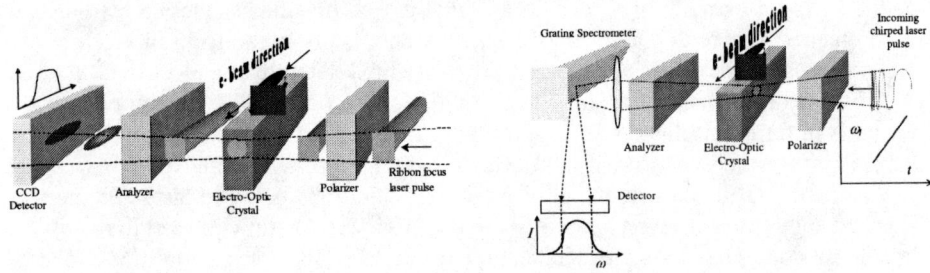

FIGURE 1. Transverse, line-focus probe geometry can produce a spatial image of the bunch[3] (left), or gating of a chirped laser pulse (right).

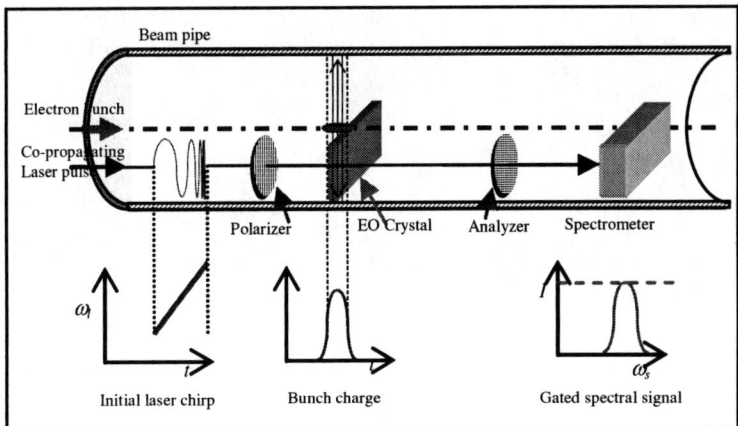

FIGURE 2. In the longitudinal geometry the electric field from a bunch modulates the polarization in an electro optic crystal and gates the transmission of a co-propagating, chirped laser pulse.

In order to achieve good resolution the laser must be focused to a spot at the EO crystal much smaller than the bunch length. Spot sizes of a few microns would be adequate for a 30 micron (100 fs) bunch but the technique now suffers from excessive power density in the EO crystal.

Transverse probe geometries were used successfully in early experiments[4] with picosecond beams, but for femtosecond resolution the longitudinal probe geometry shown in figure 2 has distinct advantages. In the longitudinal geometry the laser pulse co-propagates with the electron bunch. The chirped laser pulse is stretched to around 10 ps to ensure timing overlap with the electron bunch. The chirp provides a correlation between the time domain and the frequency spread of the pulse which can be measured with great precision[5] with a spectrometer. The laser pulse need not be focused to small dimensions in the EO crystal so there is no limitation due to laser power density in the crystal. To understand the limiting resolution of this technique we first look at some of the properties of the EO interaction.

ELECTRO OPTIC BASICS

The electric field of the bunch alters the optical retardation of the birefringent crystal. The polarization, P, of the probe laser in the crystal changes with the electric field, E, in proportion to the susceptibility according to

$$P = \varepsilon_0 \left[X_1 E + X_2 E^2 + X_3 E^3 + \ldots \right] \quad (1)$$

The susceptibility, X, has the same symmetry as the crystal so that all crystals except those with inversion symmetry exhibit some EO properties. The first term in the expansion in equation (1) is the linear, isotropic term. It is the second-order term that is exploited here, namely the linear EO effect, or Pockels effect, in which the polarization changes under the influence of the external, applied electric field of the bunch. A third-order term also exists which drives the second-order EO effect, referred to as the Kerr effect. Birefringence is the anisotropy of the crystal's refractive indices and results in differing propagation velocities along the different crystal axes. In isotropic media the index of refraction, n, is related to the susceptibility by

$$X_1 = n_1^2 - 1 \tag{2}$$

It is convenient to define the change in index through the EO crystal, following the methodology of Yariv[6], by defining an index ellipsoid with principal axes

$$\frac{x^2}{n_x^2} + \frac{y^2}{n_y^2} + \frac{z^2}{n_z^2} = 1 \tag{3}$$

The change in index with applied field is then given by the coefficients r_{ij} for a specific crystal

$$\Delta\left(\frac{1}{n^2}\right)_i = \sum_{j=1}^{3} r_{ij} E_j \tag{4}$$

The electric field of an ultra-relativistic bunch of N_e electrons at a distance r, from the crystal is [7]

$$E_r = 9\times 10^9 \frac{2N_e e}{r} \frac{e^{-\frac{s^2}{2\sigma_z^2}}}{\sqrt{2\pi}\sigma_z} \quad \text{in m.k.s units} \tag{5}$$

$$\text{and} \quad E_{r_{max}} = 9\times 10^9 \frac{2N_e e}{r} \frac{1}{\sqrt{2\pi}\sigma_z}$$

measured at the center of a Gaussian bunch of length σ_z.

The opening angle, $1/\gamma$, of the electric field lines dictates that the crystal should be within a couple of millimeters of the bunch for low energy bunch length measurements at the electron gun. At GeV energies the crystal can be a centimeter away without loss of resolution or sensitivity.

The retardation, Γ, of the light transmitted by a crystal with trigonal symmetry such as LiTaO$_3$ or LiNbO$_3$ is

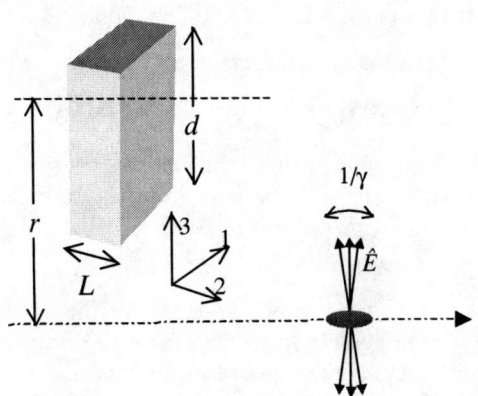

FIGURE 3. A relativistic bunch at a distance r from the crystal and laser pulse.

$$\Gamma = \frac{\pi L}{\lambda_0} E \left(n_e^3 r_{33} - n_o^3 r_{13} \right) \quad (6)$$

and is determined by the electric field strength, E, and the thickness of the crystal along the light path, L,

The electric field corresponding to a retardation of $\lambda/2$ (when the polarized light changes from zero to 100% transmission) is therefore

$$E_\pi = \frac{\lambda_0}{L} \frac{1}{\left(n_e^3 r_{33} - n_o^3 r_{13} \right)}$$

$$= 9 \times 10^9 \frac{2 N_e e}{r \sqrt{2\pi} \sigma_z} \quad (7)$$

from which the required thickness of the crystal can be determined.

AUTOCORRELATION AND DETECTION

The light pulse transmitted by the EO crystal is characterized by its amplitude A, carrier frequency ω_0, carrier phase φ_0 and chirp $\varphi(t)$,

$$E(t) = A(t) \cos\left(\omega_0 t + \varphi(t) + \varphi_0 \right) \quad (8)$$

Its temporal profile, $C(t)$, can be measured by interferometric means, shown in figure 4a, which is an autocorrelation technique

$$C(t) = \int_{-\infty}^{\infty} \overline{E}(\tau) E(t+\tau) d\tau = \overline{E}(-t) \otimes E(t) \quad (9)$$

However, by the Wiener-Khinchin theorem this is equal to the Fourier transform of the magnitude squared, $F\left[|E_v|^2 \right]$, and consequently does not preserve phase or asymmetry information about the bunch.

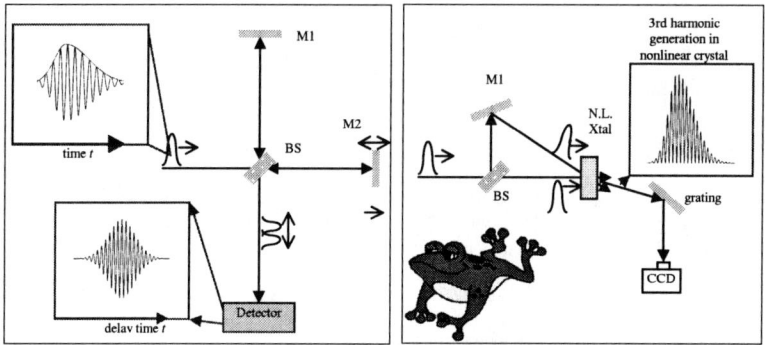

FIGURE 4. Autocorrelation (left) with an interferometer, and a nonlinear autocorrelator (right) for Frequency Resolved Optical Gating.

Alternatively, if the light rays cross at an angle inside a nonlinear crystal, as illustrated in figure 4b, higher harmonics are generated and the unbalanced, third-order correlation terms preserve pulse asymmetry information. Combining the nonlinear autocorrelator with a grating spectrometer results in a powerful diagnostic to Frequency Resolve the Optically Gated (FROG) pulses.

LIMITS TO RESOLUTION

The EO detection method is resolution limited by the phase slippage that occurs from the slight difference in propagation velocity for the sub-millimeter radiation from the bunch and the 800 nm laser radiation in the crystal arising from the difference in refractive index, Δn,

$$\Delta t = \frac{L}{c}\Delta n \qquad (10)$$

This effect is minimized by choosing a crystal with small Δn and small thickness, L. For example, in $LiTaO_3$ $\Delta n = 0.1$ so that a 100 µm thick crystal has a slippage of 30 fs.

As an alternative we are investigating a new technique based on measuring the wavelength shift that occurs at the entry and exit of the crystal where there is a time-dependant change in refractive index

$$\frac{\Delta \lambda}{\lambda} = \frac{L}{c}\frac{dn}{dt} = \frac{L}{c}\frac{1}{2}\left(n_e^3 r_{33} - n_o^3 r_{13}\right) \qquad (11)$$

This effect has been observed in laser-plasma interactions[8] and is proposed as a means to surpass the resolution limits set by phase slippage. For the crystal parameters above, equation (11) predicts a 3% wavelength shift at the edges of the crystal.

In conclusion, it is expected that improvements in EO detection techniques will go hand-in-hand with progress in the generation of ultra-short electron bunches over the next few years.

REFERENCES

1. "LCLS CDR", SLAC-R-593, April 2002.
2. M. Cornacchia et al., "A Subpicosecond Photon Pulse Facility for SLAC", SLAC-PUB-8950, LCLS-TN-01-7, Aug 2001. 28pp.
3. T. Srinivasan-Rao et al., "Novel Single Shot Scheme to Measure Submillimeter Electron Bunch Lengths Using Electro-Optic Technique", Phys. Rev. ST Accel. Beams 5, 042801 (2002).
4. M.J. Fitch et al., "Electro-optic Measurement of the Wake Fields of a Relativistic Electron Beam", Phys. Rev. Letters, Vol. 87, Num. 3, 034801(2001).
5 I. Wilke et al., "Single-shot Electron-beam Bunch Length Measurements", Physical Review. Letters 88, 124801, 2002.
6. A. Yariv, "Quantum Electronics", Springer-Verlag (1967).
7. A.W. Chao, "Physics of Collective Beam Instabilities", John Wiley & Sons, New York, 1993.
8. P. Bolton et al., "Propagation of Intense, Ultrashort Laser Pulses Through Metal Vapor", J.O.S.A. B, Volume 13, Issue 2, p. 336 (Feb. 1996).

Laser-Compton Scattering as a Potential Electron Beam Monitor

K. Chouffani[1,*], D. Wells[1], F. Harmon[1], G. Lancaster[2], J. Jones[2]

[1]*Idaho Accelerator Center, 1500 Alvin Ricken DR., Pocatello, ID 83209, USA*
[2]*Idaho National Engineering and Environmental Laboratory, P.O. Box 1625, MS 2802, Idaho Falls, ID 83415-2802, USA*

Abstract. LCS experiments were carried out at the Idaho Accelerator Center (IAC); sharp monochromatic x-ray lines were observed. These are produced using the so-called inverse Compton effect, whereby optical laser photons are collided with a relativistic electron beam. The back-scattered photons are then kinematically boosted to keV x-ray energies. We have first demonstrated these beams using a 20 MeV electron beam collided with a 100 MW, 7 ns Nd:YAG laser. We observed narrow LCS x-ray spectral peaks resulting from the interaction of the electron beam with the Nd:YAG laser second harmonic (532 nm). The LCS x-ray energy lines and energy deviations were measured as a function of the electron beam energy and energy-spread respectively. The results showed good agreement with the predicted values. LCS could provide an excellent probe of electron beam energy, energy spread, transverse and longitudinal distribution and direction.

1 INTRODUCTION

Laser-Compton scattering (LCS) was originally viewed as a potential bright and tunable short pulse x-ray source. Recent results and publications have shown that LCS can also be used for beam diagnostics and have proved to be excellent tools to measure the electron beam divergence, longitudinal electron bunch distribution and electron beam spot sizes as small as 10 nm [1,2]. LCS is the exchange of energy between relativistic electron and laser beams. When photons interact with high energy moving electrons (in the MeV region), the electrons scatter low energy photons to a higher energy at the expense of the electrons' kinetic energy. This interaction results in the emission of highly directed (peaked in the direction of the incident electron beam), mono-energetic (see below), highly polarized and tunable x-ray beams with a divergence on the order of $1/\gamma$, where γ is the energy of the electrons relative to the electron rest energy. LCS experimental observations of bright photons generated by backscattered laser photons from relativistic electrons were reported in literature [3-5] and more recently by K. Chouffani et al. [6].

2 KINEMATICS OF LCS

Figure 1 shows a schematic illustration of LCS geometry. When an incoming photon strikes an electron of energy E_B and velocity β with a relative angle α, the

---

[*] Corresponding author. Tel:1-208-282-5874; Fax:1-208-282-5878.
E-mail address: khalid@physics.isu.edu (K. Chouffani)

energy of the scattered photon in the plane of incidence (defined by the electron beam and incoming photon directions) is given by:

$$E_\gamma = \frac{E_L(1+\beta\cos\alpha)}{1-\beta\cos\theta+E_L(1+\cos(\alpha-\theta))/E_B} \quad (1).$$

E_γ and E_L are the photon and laser quantum energy respectively, E_B is the electron beam energy, α and θ are the angle between the laser and electron beam and emission angle respectively. LCS can occur at any

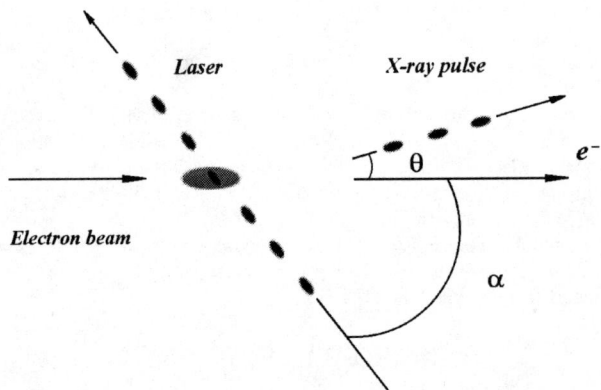

Figure 1. Laser Compton scattering (LCS) mechanism.

crossing angle between the electron and laser beam; however, two basic configurations are typically used in LCS. Head-on collision where the laser photon acquires the highest gain in energy (most energy efficient geometry). In this geometry, the x-ray pulse duration τ_x is determined primarily by the electron bunch length τ_b [5] and is given by $\tau_x = \tau_b + \tau_{Laser}/4\gamma^2$, where τ_{Laser} is the laser pulse length. In this configuration the photon energy is given by:

$$E_\gamma \approx \frac{4\gamma^2 E_L}{1+\gamma^2\theta^2+4E_L E_B/m^2c^4} \quad (2),$$

where m is the electron mass and c is the speed of light. The maximum photon energy occurs for $\theta = 0$ and is equal to $E_\gamma^{Max} \approx 4\gamma^2 E_L$. The other configuration is the 90° geometry where the electron beam and laser beam are orthogonal to each other. In our experiments the 90° geometry was not attempted and we will discuss only data taken from the head-on collisions. LCS spectra, when summed over θ, are relatively broad, with a range of energy of $\Delta E_\gamma/E_\gamma \leq 1$. The spectral width is limited by the electron beam emittance and energy spread [7]. Monochromaticity can, however, be achieved by collimating the forward directed x-ray beam, i.e. decreasing the detector solid angle. This will confine the LCS x-rays into a narrow energy window ranging from a cut-off energy $E_\gamma^{cut-off}$ to E_γ^{Max}.

3 EXPERIMENT

The laser-Compton scattering experiments were carried out at the Idaho Accelerator Center (IAC). The 1.3 GHz RF linac produces a 20-22 MeV electron beam that is brought to a head-on collision with a pulsed 7 ns long, 10 Hz repetition rate and 100 MW peak power Nd:YAG laser. The Nd:YAG laser fundamental wavelength is equal to 1064 nm and the second harmonic is equal to 532 nm. Interaction with a 20 MeV electron beam generates two x-ray lines at 7.5 and 15 keV, respectively, when the observation direction coincide with the electron beam direction. The laser-pulse energies of 1064 and 532 were 750 and 250 mJ respectively. During the experiment, the electron macrobunch length was varied between 2 and 20 ns and the electron bunch charge ranged from 1.4 to 8nC. A pair of slits located between two 22.5° bending magnets [6] enabled us to change the electron beam energy distribution as well as the electron beam current. The electron beam spot size and emittance were determined from optical transition radiation (OTR) measurements. During the OTR measurements, the electron beam slits were open to a maximum ($\approx$ 5 cm). The electron beam angular spread was consistent with the previous measurements made by Fiorito on the same LINAC [6,8].

Figure 2 shows the experimental and optical setup. The 9 mm diameter Nd:YAG laser beam is first expanded to a 45 mm diameter beam and then focused by a 5 m focal lens. The lens focal point coincides with the center of the interaction chamber. An off axis 45° broadband mirror, whose center is located 2.4 cm from the beam line axis, steers the laser beam toward the center of the interaction chamber. The laser beam spot size was equal to 0.12 mm for the 532 nm lines. The HeNe laser on the optics table follows the same path as the Nd:YAG laser, it used only for alignment purposes. Two fluorescent screens were placed on motorized actuators, one located at the center of the interaction chamber and the other positioned 38.1 cm upstream, to enable the LINAC operator to position the electron beam along their axis. The centers of the fluorescent screens were aligned when they were intercepted by another HeNe laser collinear with the electron beam's drift-line axis. The angle between the Nd:YAG laser beam and the beam line axis was equal to 4.65 mrad. The angle between the laser-beam and electron beam was 1.6 mrad (see below).

The x-rays from the LCS passed through a 51 μm thick Kapton window and traveled 1.8 m in air before reaching the liquid nitrogen cooled high resolution Si(Li) detector placed at 0° with respect to the beam line axis. The distance between the center of the interaction chamber and detector was equal to 6.83 m, and the solid angle sustended by the detector was about 0.68 μsr.

A scintillator coupled to a photomupliplier (PMT) tube was placed closed to the x-ray detector in order to monitor the bremsstrahlung background during beam tuning. A fast photomutiplier positioned on the optics table and a fast electron beam toroid placed at the exit of the 45° bending magnet were used for timing purposes.

The lowest acceptable bremsstrahlung signal recorded was obtained when the angle between the electron beam and the beam line axis was approximately 3mrad. The electron beam was focused on the first actuator (upstream from the interaction chamber) and the laser beam was steered toward its center. At this position the laser beam spot size is approximately 3.4 mm. From this interaction geometry, the crossing

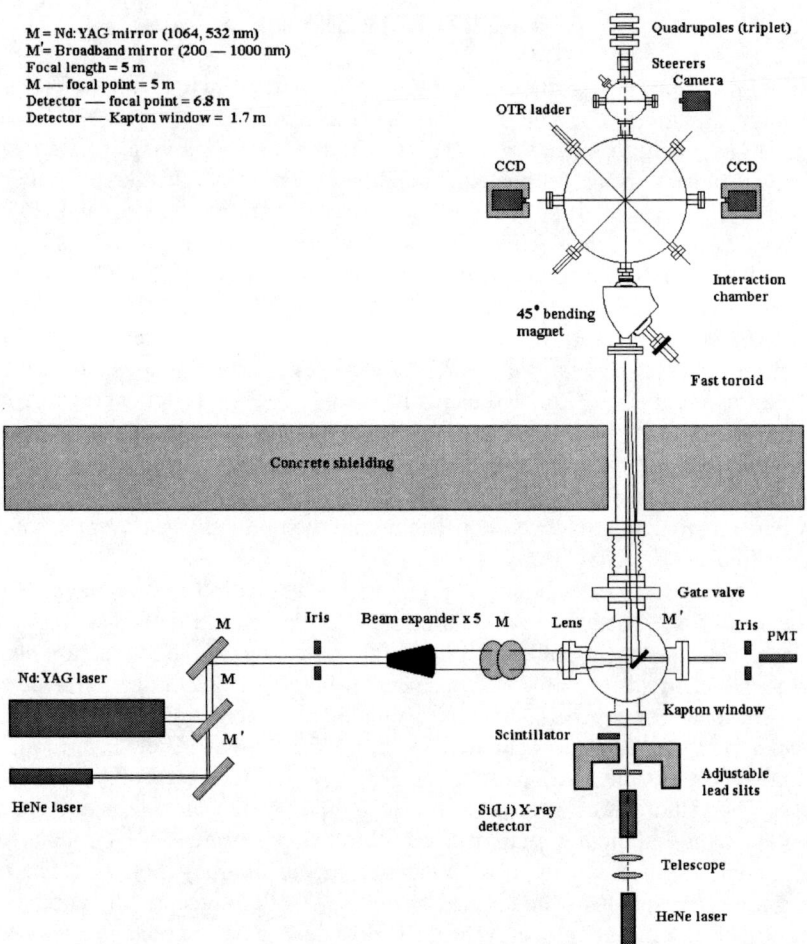

Figure 2. LCS experimental setup.

angle was of the order of 1.6 mrad. The energy calibration of the Si-(Li) detector was determined by x-ray fluorescence from the bremsstrahlung x-ray beam hitting a 343 μm thick Zr target. The detector energy resolution both with and without linac operation was determined by measuring the widths of the x-ray peaks from calibrated radioactive sources, and the bremsstrahlung induced Zr k_α x-ray emission line. The LINAC-on resolution is given by:

$$\Gamma^2_{fwhm} = 0.03956 + 0.00262 E_\gamma \quad (3),$$

where Γ_{fwhm} is the x-ray fwhm in keV, and E_γ is the x-ray energy in keV. The contribution from the RF noise was found to be small. For $E_\gamma = 15$ keV and $E_\gamma = 7.5$ keV, the detector resolution is 0.2808 and 0.243 keV respectively. The detector efficiency was determined using calibrated Am^{241} x-ray sources. For x-ray energies ranging from 6 to 20 keV, the efficiency approaches unity.

4 EXPERIMENTAL RESULTS

1. Laser-Compton spectra

Once a clear LCS signal was detected, a time scan was performed. The goal of this time scan was to optimize the LCS x-ray yield as a function of the delay between the electron beam and laser pulses. In order to remove possible pile up from the 7.5 keV x-ray resulting from the interaction of the electron beam with the 1064 nm laser line, a 25.4 µm thick stainless steel (SS) foil was placed in front of the Si(Li) detector. The x-ray transmission in air and SS foil for the 7.5 keV line was about 0.01% compared to 20% for the 15 keV line. Figure 3 shows the LCS signal as a function of the time delay between the electron beam and laser bunches. In these measurements the laser pulse length was equal to 7 ns, each point corresponds to an integrated yield from a LCS spectrum and the data collection time for each point was equal to 600 s. Figure 3 gives information on the electron bunch longitudinal distribution. One can also deduct from figure 3 an approximate value of the jitter between the electron beam and laser, for a 7ns long laser pulse and a 5 ns electron beam burst, the jitter is small and is of the order a ns. In the head-on collision geometry, this value of the jitter does not appear to be a severe limitation to observe LCS.

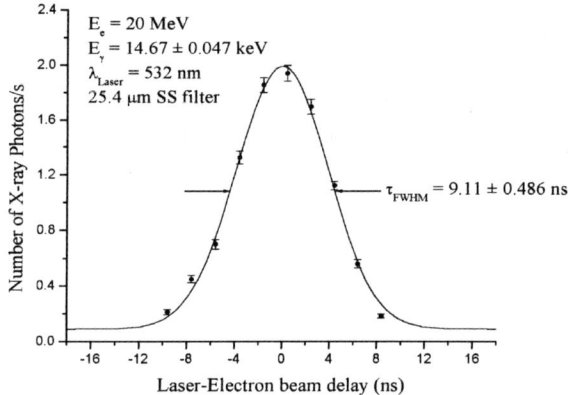

Figure 3. Time delay between laser and electron beam pulses. The curve was obtained using the x-rays generated from the interaction of the electron bunch and the 532 line of the Nd:YAG laser pulse.

Figure 4 shows the LCS spectrum for an observation angle equal to 3 mrad. The collection time for this spectrum was 600 s. The spectrum shows a clear sharp distinct monochromatic x-ray peak resulting from the interaction of the 20 MeV electron beam with the 532 nm laser line on top of a low bremsstrahlung background. The additional higher energy peak is due to pile up from the major line. The FWHM of the peak at 29.35 keV is about $\sqrt{2}$ larger than the width of the main peak, which is consistent with pileup.

Figure 5 shows the photon energies as a function of the electron beam energy for the second order LCS together with Eq. 2 for an emission angle of 3 mrad. There is good agreement between the data collected and the theoretical predictions. The discrepancy observed for the electron energy equal to 22 MeV results from a change in the electron beam direction during beam tuning and therefore a change in the observation angle of the order of 5 mrad.

2. Laser-Compton energy spread

LCS energy spread depends on the energy spread of the laser, the electron beam energy and angular spread and the spread of the scattering angle subtended by the detector. The natural LCS spectral width is be given by:

$$(\Delta E_\gamma / E_\gamma)^2 = (\Delta E_L / E_L)^2 + (\Delta E_b^e / E_b^e)^2 + (2\Delta E_{Beam}/E_{Beam})^2 + (\Delta E_\gamma^\theta / E_\gamma^\theta)^2 \quad (4),$$

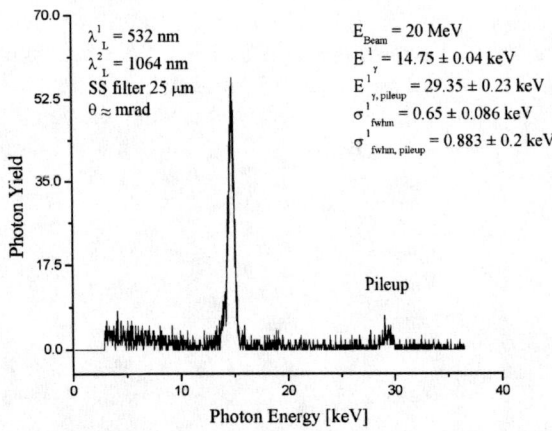

Figure 4. LCS spectrum using the Nd:YAG laser 532 line.

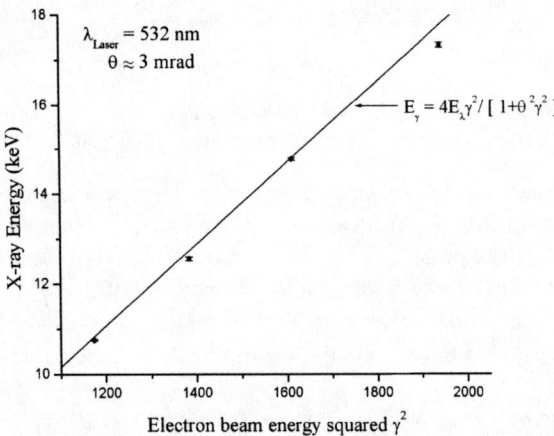

Figure 5. Second order LCS x-ray energy as a function of electron energy square.

where ΔE_L, ΔE_{Beam} correspond to the energy deviation of the laser and electron beam respectively and $\Delta E_L/E_L = 2.0\ 10^{-4}$ is negligible compared to $\Delta E_{Beam}/E_{Beam}$. ΔE_b^ε takes into account the electron beam angular spread. Because of the small solid angle, we have ignored the contribution to the energy resolution due to the finite collimation ΔE_γ^θ [9] in the LCS natural spectral width. The observed LCS energy spread ΔE_{Obs} is the quadradure sum of the LCS width and detector energy resolution given in Eq. 4. The LCS spectral width ΔE_γ is of the form:

$$\Delta E_\gamma^2 = \Delta E_{Obs}^2 - \Gamma_{fwhm}^2 \quad (5a),$$

The electron beam energy deviation is dominated by the width of the analyzing slits. The maximum electron beam current is recorded when the width of the slits is larger or equal to 3.5 cm. Figure 6 shows the variation of LCS energy spread (Eq. 5a) as a function of electron beam energy deviation for an observation angle θ equal to 3 mrad and x-ray energy equal to 14.75 keV, together with Eq. 5b given by:

$$\Delta E_\gamma = \frac{2E_\gamma \Delta E_{Beam}}{[(E_{Beam} + mc^2)(1+\gamma^2\theta^2)]} \quad (5b).$$

The upper x-axis in Fig. 6 is the analyzing slits width and the lower x-axis is the corresponding electron beam energy spread. From this measurement, we see that the LCS width is solely determined by the electron beam energy deviation and that the contribution of electron beam angular spread is negligible.

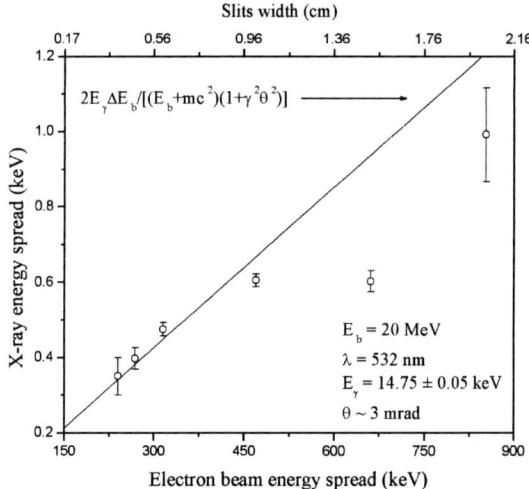

Figure 6. LCS natural energy spread as a function of the electron beam energy deviation. The electron beam energy spread was varied by changing the width of the 90° port slits [6]. The maximum slits width was about 5 cm.

There is good agreement between experimental and predicted values. As the widths of the electron beam slits are increased (larger than 0.9 cm) LCS energy spread reaches a maximum and a constant value as can be seen from figure 6.

There are two possible explanations for the constant value of LCS energy spread. Because the width of the slits also changes the size of the electron beam, we can conclude that the electron beam spot size becomes comparable to that of the laser when the maximum LCS energy spread is reached. Also, if we assume that the electron beam divergence is negligible because of the narrow slits' width, one sees that the maximum LCS energy spread equal to 0.6 keV corresponds (using eq. 5b) to a maximum electron energy spread equal to 420 keV. This value remains constant as the width of the slits is increased. Therefore, only electrons with an energy deviation $\Delta E_{beam} \leq 420$ keV are bent toward the interaction area [6]. Increasing the slits width above 0.9 cm would not affect the width of the Compton peaks but increase the electron beam current and spot size.

When the width of the slits is increased beyond 1.5 cm, the extra broadening of LCS width is mainly due the increase in bremsstrahlung background. The narrowest LCS width obtained is shown on figure 7. The width of the slit was set to optimize the signal to noise ratio and corresponds to an electron energy spread of 384 keV ($\Delta E_b / E_b \approx 2\%$). The electron beam was then tuned to reduce the electron beam phase angle spread. The width of the observed LCS spectral peak was equal to 0.397 keV. This corresponds, once the detector resolution is subtracted, to a width equal to 0.28 keV. From Eq. 5b, the electron beam energy spread was equal to 198.2 keV ($\Delta E_b / E_b \approx 1\%$).

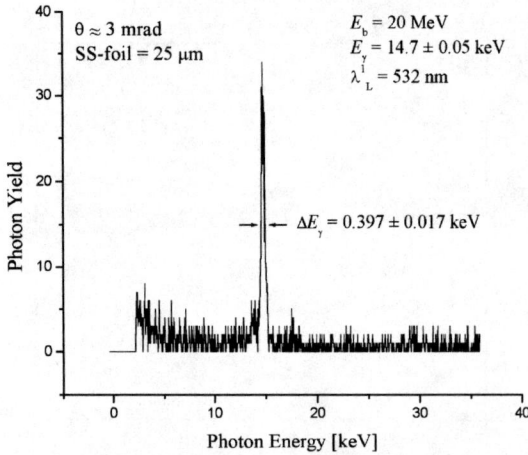

Figure 7. Narrowest line observed during the LCS experiment.

3. Angular distribution

During this experiment the electron beam was equal to 22 MeV and as before, only the Nd:YAG laser second harmonic was used. The x-rays travel through a 50.8 μm thick, 4 inches in diameter stainless steel (SS) window and 0.5 m of air before they reach a Si-PIN Photodiode x-ray detector. The detector has an opening area equal to 7mm^2 and was placed on translation stages for horizontal (x) and vertical (y) scans. Once the center of the x-ray cone is located (maximum x-ray energy in x and y directions) a finer scan is performed. Because the broadband mirror was close to the beam line axis, only results from the vertical scan are shown. Figure 8a shows the variation of the x-ray energy as a function of the vertical observation angle together with a fit using Eq. 1.

The cut-off angles correspond to the limit of the scan. The detector is aligned with the beam line axis when $\theta = 0$ mrad. Several observations can be made from Fig. 8a. One can see that the electron beam makes a vertical angle equal to 2.15 mrad with respect to the beam line axis and therefore by steering the electron beam down vertically the curve should shift to have its maximum at $\theta = 0$ mrad. The width of the curve, for a specific x-ray energy, gives the value of the distance between the detector and the interaction point, and its height is a direct measure of the electron beam energy. A fit to the data shows that electron beam energy is equal to 22.087 MeV and that the interaction point is located about 7.34 m from the x-ray detector. This will place the interaction point at the exit of the focusing quadrupoles. This value is consistent with

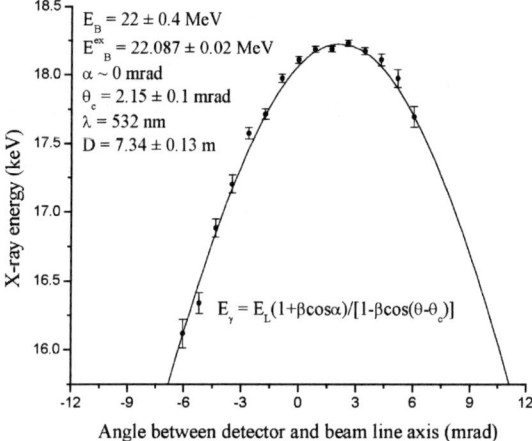

Figure 8a. Variation of LCS x-ray energy as a function of vertical observation angle. The SS windows blocks x-rays with angles greater and smaller than 6 and –6 mrad with respect to the beam line axis. Fit to the data provides values of the electron beam energy, distance from interaction point to detector and electron beam direction with respect to the beam line axis.

the time delay between the signal from the photomultiplier and the fast pickup beam monitor (Fig. 2).

The total number of LCS x-rays $N(\theta)$, as a function of the observation angle θ is given by:

$$N(\theta) = \iint_{\Delta\Omega_{det}} Abs(E_\gamma) \frac{d^2\sigma^{cv}}{d\theta_x d\theta_y} d\Omega \quad (6a),$$

$$\frac{d^2\sigma^{cv}}{d\theta_x d\theta_y} = \frac{d^2\sigma}{d\theta_x d\theta_y} \otimes g(\sigma_x, \sigma_y) \quad (6b).$$

$\Delta\Omega_{det}$ is the solid angle subtended by the detector. θ_x, θ_y are the horizontal and vertical photon directions respectively. $Abs(E_\gamma)$ is a function that takes into account the detector efficiency and the x-ray absorption in different media. $d^2\sigma^{cv}/d\theta x d\theta y$ is the convolution of the single photon differential cross section $d^2\sigma/d\theta x d\theta y$ for linearly polarized incident radiation [10], with the electron beam horizontal and vertical angular distribution $g(\sigma_x, \sigma_y)$ of RMS widths σ_x and σ_y respectively. Because of the large distance between the detector and interaction region, the stainless steal window as well as the distance in air enabled us the narrow the x-ray cone by absorption of the lower energy x-rays. The solid angle subtended by the detector was equal to $d\Omega_\gamma \approx 1.3 \times 10^{-7}$ Sr, the width of the analyzing slits was equal to 1 cm and therefore about 25% of the total electron beam current was transmitted. At this slit's width, we believe that at the interaction, the electron beam spot size is larger than that of the laser and that the electron beam divergence is not negligible. Figure 8b shows the variation of the x-ray intensity as a function of the vertical observation angle (angle between the beam line axis and detector), together with the fit to the data with a reduced $\chi^2 \approx 3.25$. The normalization of the spectra was done by measuring the current from the fast toroid. With the current number and quality of the data collected (due to beam current and laser power fluctuations), it is difficult to obtain an accurate measurement of the electron beam divergence.

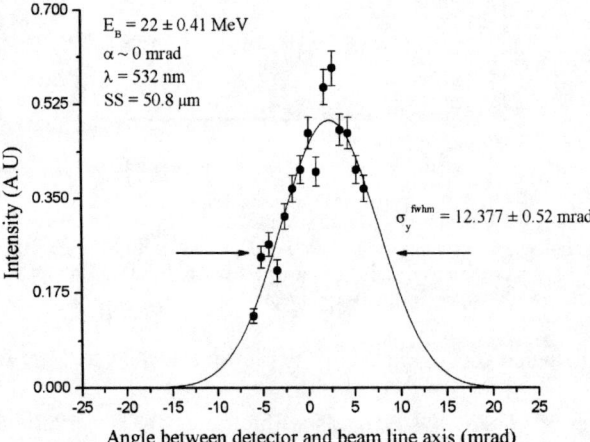

Figure 8b. Intensity as a function of observation angle, together with the fit to the data using Eq. 6a. The x-rays travel through a 50.8 µm thick window and 0.5 m in air before they reach the detector.

The vertical RMS σ_y = 3.54 mrad, previously measured using OTR [6] was obtained at a position located 1.2 m downstream from the interaction point and for a slits width greater than 3.5 cm.

5 CONCLUSION

LCS x-ray peaks resulting from the head-on interaction of a Nd:YAG laser and a 20-22 MeV electron beam were observed during our experiments at the IAC. We have shown that there is a good agreement between experiment and theoretical predictions. We have also demonstrated the direct relation between measured LCS width and electron beam energy deviation. This suggests that LCS could provide an excellent probe of electron beam energy and energy spread. We have also shown that although LCS can be viewed as a potential bright x-ray source, it can also provide valuable information on the longitudinal electron beam distribution, beam energy and direction with respect the electron beam line. By placing an array or arrays of Si-PIN detectors at the x-ray port and an appropriate data acquisition system, it is possible to have an online beam monitoring system.

ACKNOWLEDGMENT

This work partially funded by US Department of Energy under DOE Idaho Operations Office Contract Number DE-AC07-99ID13727.

REFERENCES

[1] W. P. Leemans et al., Phys. Rev. Lett. 77 (1996) 4182.
[2] T. Shintake, Nucl. Instr. and Meth A 311 (1992) 453.
[3] I. C. Hsu et al. Phys. Rev. E 54 (1996) 5657.
[4] G. Ya. Kezerashvili et al. Nucl. Instr. and Meth. B 145 (1998) 43.
[5] I. V. Pogorelsky et al., Nucl. Instr. and Meth. A 455 (2000) 176.
[6] K. Chouffani et al., to be published in Nucl. Instr. And Meth. A
[7] E. Esarey et al., Nucl. Instr. and Meth. A 331 (1993) 545.
[8] R. Fiorito, private communication.
[9] G. Matone et al. in *"Lecture Notes in Physics"*, S. Costa and C. Schaerf (Eds), Springer-Verlag, Berlin, 1976.
[10] J. D. Jakson, Classical Electrodynamics, 2nd ed. (Wiley, New York, 1975)

DESIGN OF A MULTI-BUNCH BPM FOR THE NEXT LINEAR COLLIDER[1]

Andrew Young, Douglas McCormick, Marc Ross, Stephen R. Smith

SLAC, Stanford, CA. USA

H. Hayano, T. Naito, N. Terunuma, S. Araki

KEK, Ibaraki, Japan

Abstract. The Next Linear Collider (NLC) will collide 180-bunch trains of electrons and positrons with bunch spacing of 1.4 ns. The small spot size ($\sigma_y < 3$ nm) at the interaction point requires precise control of emittance, which in turn requires the alignment of individual bunches in the train to within a fraction of a micron. Multi-bunch beam position monitors (BPMs) are to determine the bunch-to-bunch misalignment on each machine pulse. High bandwidth kickers will then be programmed to bring the train into better alignment on the next machine cycle. A prototype multi-bunch BPM system with bandwidth (350 MHz) sufficient to distinguish adjacent bunches has been built at SLAC. It is based on 5 G sample/s digitization of analog sum and difference channels. Calibration tone injection and logging of the single bunch impulse response provide the kernel for deconvolution of bunch-by-bunch position from the sum and difference waveforms. These multi-bunch BPMs have been tested in the Accelerator Test Facility at KEK and in the PEP-II ring at SLAC. The results of these measurements are presented in this paper.

OVERVIEW

The multi-bunch (MB) BPMs were designed to operate over a wide range of conditions (Table 1) allowing for testing to be performed at SLAC and KEK. The MB BPMs are used by the sub-train feedback, which applies a shaped pulse to a set of stripline kickers to straighten out a bunch train. These are qualitatively different from the quad (Q) and feedback (FB) BPMs due to their high bandwidth and relatively relaxed stability requirements. The primary requirement on the MB BPMs is a bunch train that generates a BPM signal, which is straight.

TABLE 1. BPM Specifications

Parameters	Value	Comments
Resolution	300 nm rms at 0.6×10^{10} e⁻/bunch	For bunch-bunch displacements freq. Below 300 MHz
Position range	±2mm	
Bunch spacing	1.4 ns	
No. of bunches	1-190	1.4ns
Beam current	$1 \times 10^9 - 1.4 \times 10^{10}$	Particles per bunch
No. of BPMs	278	

---

[1] Work supported by the Department of Energy, contract DE-AC03-76SF00515

Implementation

Figure 1 shows a simplified block diagram of the multi-bunch front-end prototype chassis. The BPM chassis contains directional couplers, sum and difference hybrid, bandpass filters for noise rejection, and sold-state amplifies. The BPM chassis takes the four inputs ("Up" "Down") from the BPM buttons that can be translated into X and Y positions and takes the sum and difference. The BPM signal is then amplified in order to run it on a long cable to a digitizer outside the radiation area. The front-end has a feature where a single tone can be injected into the inputs of the sum and difference hybrid for calibration. Thus allowing the operators to perform a transfer function measurement.

Figure 1. Block diagram of the BPM front-electronics

To meet the bandwidth and performance requirements, a four-tap directional coupler that is shown in figure 2 and allows one to inject a single-tone into the Hybrid and perform a transfer function calibration. The four-tap configuration allows the bandwidth to be broadband. This has the effect of a constant coupling ratio across the frequency band. This allows one to remove the phase variation in the hybrid and cable losses and insertion losses of other components. Also the calibration will allow one to remove any non-linearities in the digitizer that will be discussed later in the paper.

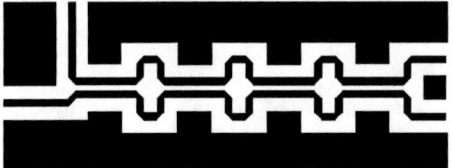

Figure 2. Stripline directional coupler

Other RF custom components such as amplifiers and switches were bought from manufactures and then integrated into a circuit board. Special care was taken to lower insertion losses by using low-loss dielectric materials. The most critical component is the sum and difference hybrid. Instead of a stripline $5/4\lambda$ sum-difference hybrid design, a toroid design was used. This improved the sum to difference isolation from 20dB to >30dB and the bandwidth improved from 150MHz to 300MHz.

The digitizer that was chosen for this experiment is a Tektronix 684..This is a 5GSa/s digital scope that has a maximum record length of 15000 points. The waveforms were read out using a MATLAB script that communicated through the GPIB. To insure that the scope was triggered away from the injection a trigger was timed to the middle of the beam train. Seven turns were recorded and then analyzed.

RESULTS

Three Y-position BPMs chassis were installed in the KEK Accelerator Test Facility (ATF). The toroid current was recorded with each data set. The data presented in this paper is with toroid current of $3e^{10}$ particles per bunch. Figure 3 shows the results of a frequency spectrum of a single-bunch beam stimulus.

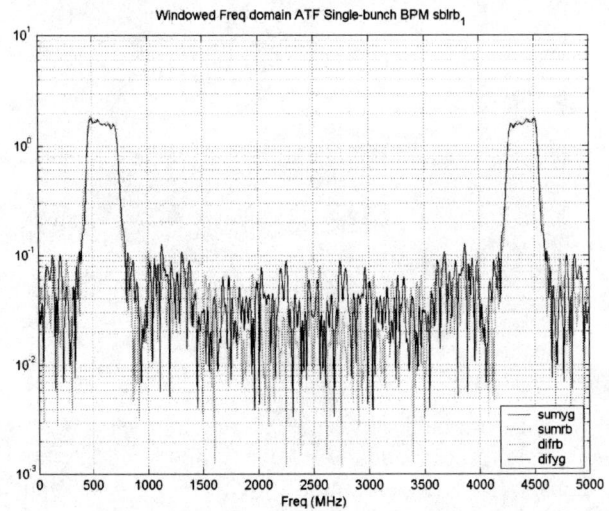

Figure 3. Frequency spectrum of a single bunch beam

An elaborate calibration procedure was used to normalize the front-end. The normalization was needed because the multi-bunch data will have overlapping bunches. Therefore, the single bunch data can be used to decimate the individual bunches of the multi-bunch data thus, transforming the multi-bunch data into charge and position of real bunches.

First the front end was calibrated using a single tone at 600MHz, the center frequency of the hybrid and RF coupler. Four gain and phase coefficients were derived (Up to sum, Up to difference, Down to sum, Down to difference). These coefficients were used in the single bunch data (see equation 1).

$$\text{Cor\_sb}=\text{FFT(sb)}*(\text{FFT(tone)}/\text{FFT(600MHz)})) \qquad (1)$$

The next calibration procedure was to inject a single bunch of charge into the machine and measure the response of the front-end. This also generated four coefficients over the complete bandwidth of the front-end. Because the single bunch beam calibration data, was digitized at separate times a phase error was injected into the signals. The corrected signals are illustrated in Figure 4. This phase error was corrected by writing a MATLAB script that ensured the digitizers started digitizing at the same time.

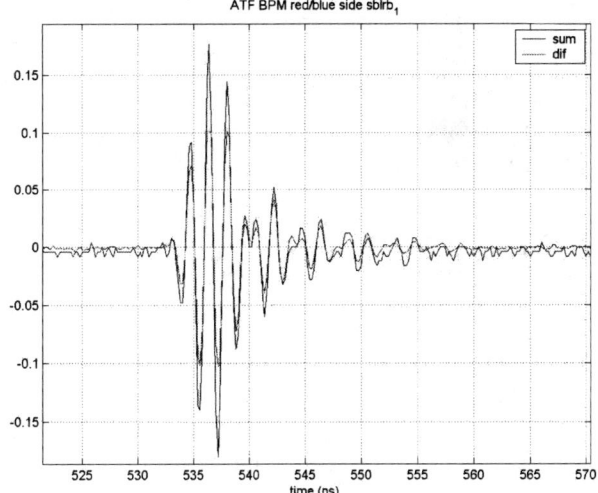

Figure 4. Single bunch raw beam data

Using the corrected single bunch data, an inverse matrix was generated that was used to correct the multi-bunch data (see equation 2 and 3)

$$\begin{bmatrix} U \\ D \end{bmatrix} = \begin{bmatrix} sb_{11} & sb_{12} \\ sb_{21} & sb_{22} \end{bmatrix}^{-1} * \begin{bmatrix} \Sigma_{meas} \\ \Delta_{meas} \end{bmatrix} \qquad (2)$$

$$\begin{bmatrix} \Sigma_{cor} \\ \Delta_{cor} \end{bmatrix} = \begin{bmatrix} U \\ D \end{bmatrix} * \begin{bmatrix} 1 & 1 \\ 1 & -1 \end{bmatrix} \qquad (3)$$

Two procedures were used in computing the corrected sum and difference signals. The first procedure was to compute the corrected sum and difference using a MATLAB script that simplifies the raw multibunch data by windowing the data around the turn frequency and then taking an FFT of the data. The second procedure, was to window the data around the turn frequency and then convert the data to baseband. Then data was decimated as a function of the bucket spacing. The windowed multi-bunch raw data is shown in Figure 5. In this figure, both the sum and difference signals are displayed. Examination of this data determined that there is eighteen bunches in the accelerator. The data shows that there is greater than –30dB of isolation between the sum and difference ports.

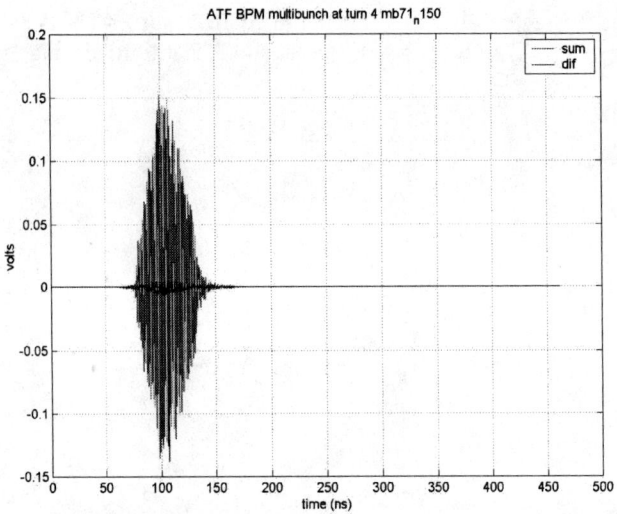

Figure 5. Multi-bunch raw beam data

Using the first procedure and Equation 2 and 3 the corrected multi-bunch data is computed. Figure 6 shows the envelope of the corrected multi-bunch sum data and the raw sum signal The overall eveelope matches the structure of the raw BPM signal. Using this signal along with the corrected difference signal the y-position and resolution can be computed.

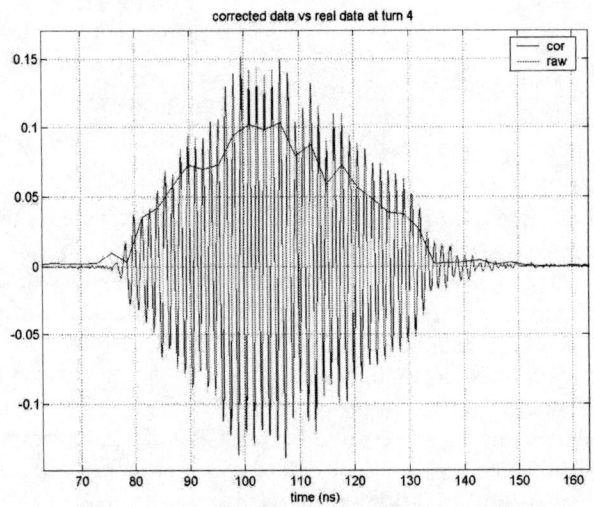

Figure 6. Multi-bunch corrected sum and raw sum data

Figure 7 shows the average beam position over the 18 bunches using the first procedure. As one can see the position and resolution begins to get large at the edge of the train. This effect is due to the corrected sum signal increasing and the corrected difference signal is approaching zero.

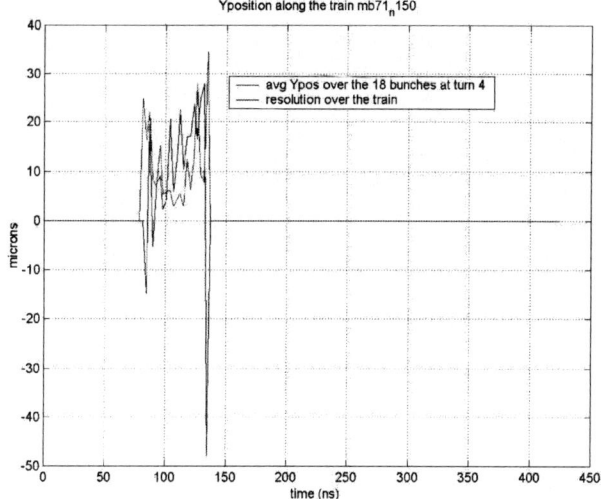

Figure 7. Multi-bunch Y-position over seven turns data

Using the second procedure and Equation 2 and 3 the corrected multi-bunch data is computed. Figure 8 shows the corrected multi-bunch sum data down-converted to baseband and low-pass filtered at turn 4. Also illustrated is the raw sum signal

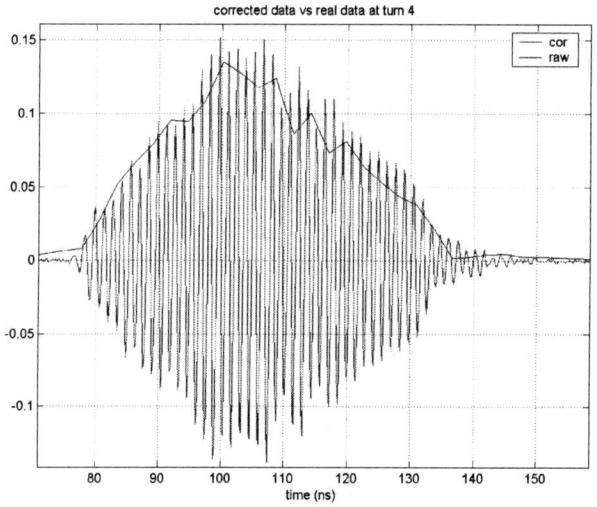

Figure 8. Multi-bunch baseband corrected sum and raw sum data

By sampling the maximum signal on the sum and difference ports, the position of the beam as a function of turns can be calculated. The equation for the Y-position is defined as Ypos=R/2(Δ/Σ) where, R is the radius of the beam pipe, R/2=3000microns, Δ is the corrected difference signal, Σ is the corrected sum signal Figure 9 shows that the beam position varies 15 microns over seven turns with a single shot resolution of 3 microns.

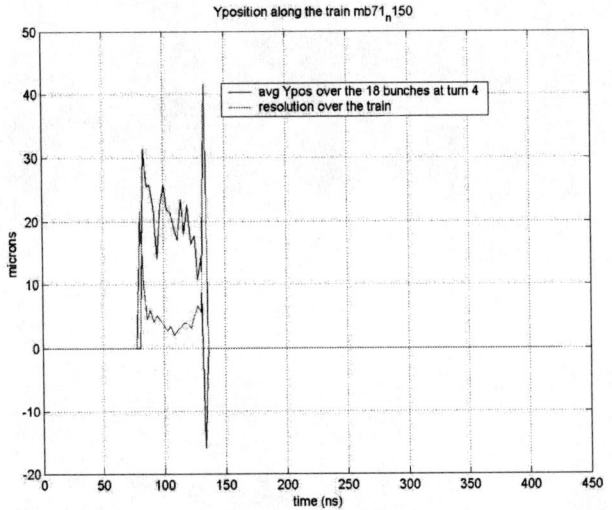

Figure 9. Multi-bunch Y-position over seven turns data

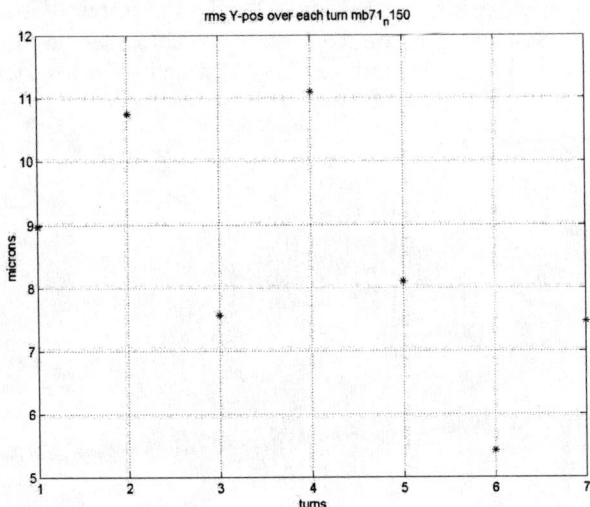

Figure 10. RMS Y-Position over each turn

Figure 10 illustrates that the position resolution over each turn and varies from 5.4-11.1 microns. This will need to be improved to meet the specification for the NLC.

SUMMARY

A multi-bunch BPM was built at SLAC and tested at KEK. The BPM electronics can resolve both single and multi-bunch fills. The data clearly indicates that the multi-bunch BPM electronics can measure the beam to within 15 microns and 6 microns over seven turns. However, the design goal of measuring the beam position within 1 micron was not achieved. The goals were not accomplished due to possible problems in the bandwidth of the output filters. In the next design broadband filters will be used to help with clock recovery and bunch spacing determination. More care will also be needed in setting up a trigger such that every BPM signal will be insensitive to timing jitters in the digitizer lead to the larger position resolution.

Another solution to solve this problem of position resolution is to use a higher frequency 1428MHz and perform hardware downconversion thus operating at an IF frequency of 200-MHz. The advantage of this solution is a better signal to noise ratio, smaller components, and less sensitivity to phase noise in the digitizers.

ACKNOWLEDGMENTS

The authors wish to acknowledge the work that Boni Cordova-Grimaldi and Dave Anderson did to insure the chassis were manufactured and tested properly. The authors are extremely appreciative of the ATF collaboration at KEK for allowing the testing of this device at their facility and especially thankful to the operators that were on staff during the beam testing of these devices.

Global Orbit Feedback In SRRC

C. H. Kuo, Jenny Chen, K. H. Hu, K. T. Hsu

Synchrotron Radiation Research Center
No. 1 R&D Road VI, Hsinchu Science-Based Industrial Park, Hsinchu 30077, Taiwan, R.O.C.

Abstract. The global orbit feedback system plays a critical role in the operation of the third generation light source in SRRC. This article addresses various issues concerning the orbit feedback system to optimize performance. The orbit feedback system has been recently upgraded to meet to user demands. Following upon operational experiences in recent years, SRRC has designed a new system to be more easily maintained, with a better diagnostic environment, robustness and flexibility controller. The new hardware structure is based on a general purpose CPU with a real-time operation system to reduce the upgrade time and the investment. Performance analysis tools are also developed to monitor the system and analyze data. Commercial Off-The Shelf (COTS) products, supporting effective integration, are applied to support various beam studies and accesses.

INTRODUCTION

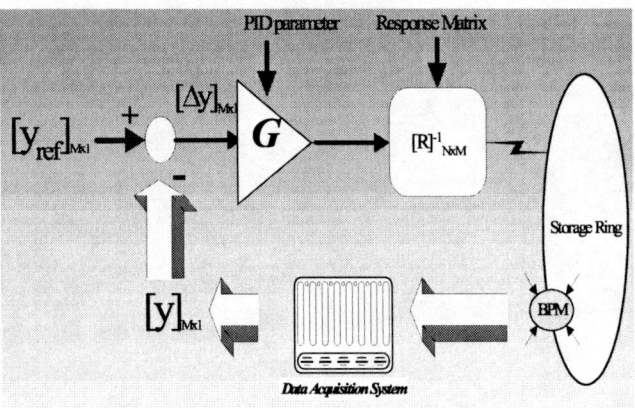

FIGURE 1. Basic concept of the orbit feedback system.

The orbit feedback system eliminates orbit irregularities that are caused by perturbation sources. Work to improve the stability of the orbit began in 1995, with the installation of the orbit feedback system. This orbit feedback system has been equipped with insertion devices, including undulators (U5 and U9) and, in particular, an elliptically polarized undulator (EPU5.6). Orbit drift and low frequency oscillations have also been reduced. The orbit feedback system in the storage ring of SRRC is now being upgraded. This upgrade includes increasing feedback bandwidth, increasing sampling rate, compensating for the eddy current effect of the vacuum chamber with a filter, and enhancing the performance and robustness of the control rules. This report summarizes the status of the orbit feedback development in SRRC.

EXISTING ORBIT FEEDBACK SYSTEM

A digital orbit feedback system [1,2] had been developed to suppress orbit disturbances such as long-term drift, low frequency oscillation and perturbation due to the operation of insertion devices. The main component of the system is the response matrix. The linear response matrix is measured by taking an electron beam position monitor (BPM) reading when each corrector is individually perturbed. This response matrix is inverted by a single value decomposition method for compensation calculation of system. The feedback controller is based on the PID algorithm. Digital filtering techniques were used to remove noise in the electron beam position reading, to compensate for the eddy current effect of the vacuum chamber, and to increase the bandwidth of the orbit feedback loop. The infrastructure of the digital orbit feedback system consists of an orbit acquisition system, gigabit fiber links, digital signal processing software, and high precision digital-to-analog converters. The orbit feedback constitutes a typical multiple input, multiple output problem. Figure 1 depicts the basic concept of the orbit feedback system. Implementing an analog matrix operation with many BPMs and correctors is technically difficult. However, a digital based feedback system is a better way to implement the multiple input/output system.

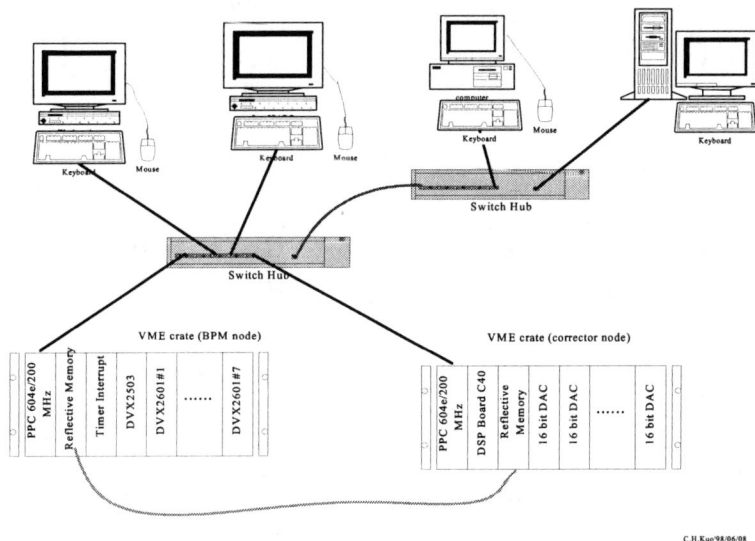

FIGURE 2. Hardware structure of system.

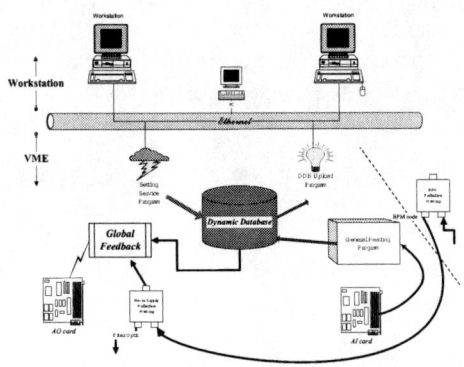

FIGURE 3. Software diagram of system.

HARDWARE STRUCTURE

Figure 2 illustrates the hardware configuration of the corrector control system in SRRC. The low layer is a VME crate system and includes a PowerPC 604e CPU board and I/O interface cards. The CPU board consists of a PowerPC microprocessor, 32 megabytes on-board memory, RS-232, PMC sites and Ethernet ports. The front-end devices are connected to this system via VME interfaces for analog I/O and digital I/O. A PowerPC-based server system is used as the TFTP file server for OS downloading and the network file server (NFS) for disk mounting. All application programs are stored on the server disk. These programs are developed and debugged on the client node to relieve the server loading. The real-time multi-tasking kernel is embedded in the single board computer of the VME bus. It provides satisfactory performance, reliability, and a rich set of system services. A new device is easily created by a modification in the device table file, just like editing a word file on line. The system can automatically boot and execute various applications in each VME node with the same operation system environment after rebooting or restarting the system. The upload task will be reduced to a low priority when the global feedback is on, avoiding interference with the latency time of the feedback system. This task handles the analog input, digital input and database service when it receives the broadcast upload message from the Ethernet. The cycle time of the system was improved to 1 ms by the VME interrupt. The present hardware system consists of two VME crates, an orbit server VME crate, and a corrector and computation VME crate. The bus adapter is inserted in slot one of the VME crate as a system controller. All programs were developed and debugged on the PC and downloaded to a DSP board. The DSP board, carrying the TMS320C40 module, handles all signal processing. This includes a digital low pass filter (LPF) and a PID controller. Completing the feedback processes takes 1 ms, including the operation of the PID, digital low pass filtering, matrix operation, BPMs data reading from reflective memory, and the corrector compensation. The corrector setting was upgraded to a 16 bit DAC to achieve sub-µrad steering resolution. All parameters can be remotely adjusted from graphical interfaces of the control system.

SOFTWARE STRUCTURE OF SYSTEM

The corrector node is attached to the corrector controller via a PowerPC, multiple 16 bit D/A and A/D cards and a DSP card. Figure 4 shows the structure of the application software. Some process tasks are performed in the PowerPC. The setting process involves a corrector setting when a setting command arrives from the database. It spawns child tasks to process command requests to save the PowerPC loading. The reading is triggered by the external 10 Hz clock signal from the network when the broadcasted upload message is received from the network. An event is sent from the timing server to wake up the data acquisition process. The data acquisition process of the PowerPC directly controls the I/O interface cards of crate. The DDB process handles communicating with the shared memory between the reading and setting process. It also provides data access for the external request to survey the feedback performance.

OPERATIONAL PERFORMANCE

Operational performance of the orbit feedback system was obtained by varing the gap of the insertion devices and the externally applied perturbation source in the corrector. Figures 4, 5 and 6 show the results. The cutoff frequency of LPF is 60 Hz, and the combination of PID parameters was selected to fulfill control goals to minimize the orbit variation due to any perturbation source in the devices. The PID parameters were modified while increasing the bandwidth of the feedback. The following description is based on parameters, $K_p = 0.8$, $K_i = 0.03$ and $K_d = 0$.

Perturbation with U5 Gap Variation

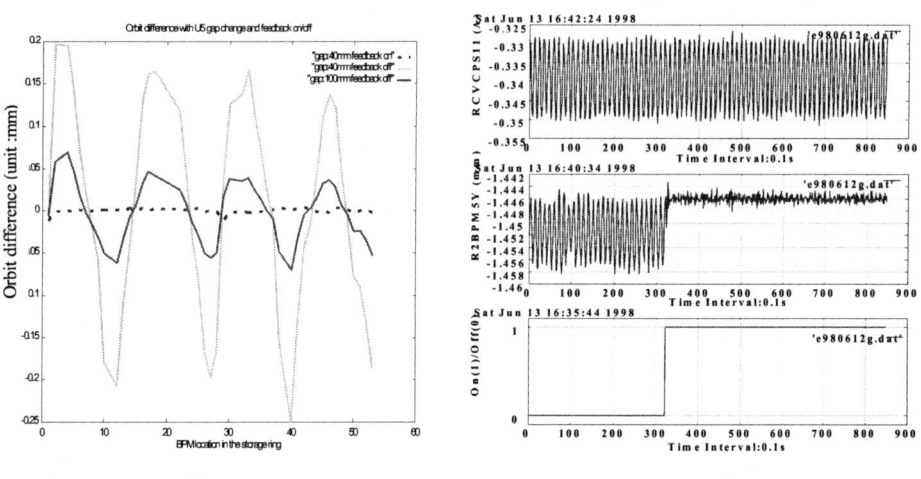

FIGURE 4. (a) Orbit displacement with various gaps of U5 and feedback on/off. (b) Orbit displacement with corrector perturbation and feedback on/off.

A new insertion device has been installed in the storage ring. It is a 4-meter long prototype undulator with a 5 cm period (U5). The orbit is changed due to the beta beating and field error of the insertion device. Figure 3 presents the orbit variation with and without DGFB, while adjusting the U5 gap. The difference orbit is defined from the gap variation of ID. The gap of ID is to 100mm and 40mm from the 219mm. The displacement of the orbit was much smaller when the digital global feedback was turned on than when it was off.

Perturbation Source due to Corrector

The displacement of the orbit due to a perturbation source at the corrector is much smaller when the orbit feedback is turned on than when it is off. Figure 4(b ii) shows the results of the performance testing. The beam position monitor, R2BPM5Y, is included in the feedback loop. An external perturbation source is applied in the corrector. The source is a 0.9 Hz sine wave with an amplitude of 0.02 A peak-to-peak. Figure 4(b i) shows the perturbation source, RCVCPS11. Figure 4 (b iii) shows the orbit feedback control status.

UPGRADE ISSUES

The new hardware system is based on a new model DSP board and is compatible with the new development environment. This Windows-based environment, with an Ethernet network, supports maintenance and trouble-shooting. The host of the corrector node handles interrupt requests and generates two signals to the DSP. One is for vertical orbit feedback; the other is for horizontal orbit feedback. The DSP, Texas Instruments TMS320C40, was also upgraded to 50 MHz from 40MHz.

This upgrade will be insufficient in the future, when many insertion devices are installed. The number of BPM signals is continually increasing and the corrector must be included in the feedback loops, especially since the insertion devices are operated at the same time. The orbit feedback system is currently being upgraded. The new system is planned to become operational in 2002. Several reasons for the upgrade exist. First, to improve system maintainability and stability. The DSP board of the original feedback system is embedded in the corrector control of VME crate, which is inconvenient for the development of the feedback system. The hindrance of feedback loop R&D due to machine operations is troublesome. Secondly, the system was implemented in 1995 with a slow DSP board; the functionality of the feedback loop is limited. Computing power is insufficient to handle the demand for more BPMs and correctors. The selection of a control algorithm is also limited. In the new implementation, the feedback calculation will be located in a separate VME crate. An Ethernet-based processor was selected to provide remote access. The corrector node is loosely coupled to. The upgraded system includes three VME crates, the BPM node, corrector node, and feedback node as shown in Figure 5. These three nodes are connected by reflective memory. Several fiber link reflective memory cards are held tightly together by one dedicated reflective hub that simplifies the wiring of the fiber link. The corrector node handles power-supply control. The feedback node is upgraded from the DSP to the PowerPC 7410 (G4) that handles feedback control, the control algorithm calculation and corrector compensation data conversion from orbit information. The correction results are sent to the host processor of the corrector node to do compensation processing.

DSP and PowerPC

The calculation power of the processor can handle more and more multi-input and multi-output controller loops in each millisecond. The new generation DSPs, such as TMS320C6201 and TMS320C6701 from Texas Instruments, suffice. However, DSP maintenance and stocking represent problems that are expensive to solve. The general purpose CPU, a PowerPC with a real time OS, is preferred over DSP for this. The manufacturer supports driver and Business Service Provider (BSP)) in the operation system to reduce the development time for this upgrade.

The PowerPC is based on Motorola's AltiVec technology, and, specifically, the fourth generation MPC74xx. The digital signal processing applications are beginning to migrate from a traditional DSP environment to a RISC environment. At the same time, Motorola has been increasing the processing power of the PowerPC, increasing the speed and flexibility of an already impressive portfolio of DSPs, and Texas Instruments has been introducing new parts for the C6000 family. Table 1 specifies some of the processors available from Texas Instruments and Motorola, these will satisfy requirements in the future.

TABLE 1. Processor lists for orbit feedback.

Manufacturer	Processor	Part Number
Texas Instruments	C6415	TMS320C6415
Texas Instruments	C6203	TMS320C6203
Texas Instruments	C6701	TMS320C6701
Motorola	7410	MPC7410

TABLE 2. Speed and benchmarks for DSP and PowerPC. The 7410 can perform up to 16 parallel integer calculations on 8-bit data every cycle, so the MIPS number would be 16 multiplied by 500 MHz or 8000 MIPS. 2) Some 7410 instructions can be performed at eight calculations per cycle.

	C6415	C6203	C6701	7410
Clock(MHz)	600	300	167	500
Instruction Cycle (ns)	1.67	3.33	6	2
Instruction Per Cycle	1~8	1~8	1~8	1~3
Peak MIPS	4800	2400	1336	917[1]
Floating-Point Operations Per Cycle	-	-	1~6	4[2]
Peak MFLOPS	-	-	1000	2000

Compare the clock speeds and peak processing power in the Table2. The performance of PowerPC is close to the DSP.

CONCLUSIONS

The feedback node is now a separate VME crate from the original corrector node. The sub-component of the feedback node is now being tested. These components include the VME interrupt request, VME bus access, signal processing and the controller loop. The task of corrector node will be simplified, has been modified, and will be tested during a shutdown of the storage ring. Some slots will be vacated in this crate due to removal of the DSP. It will reserve slot of VME to increase numbers of corrector control in the future. Figure 5 presents the new system structure.

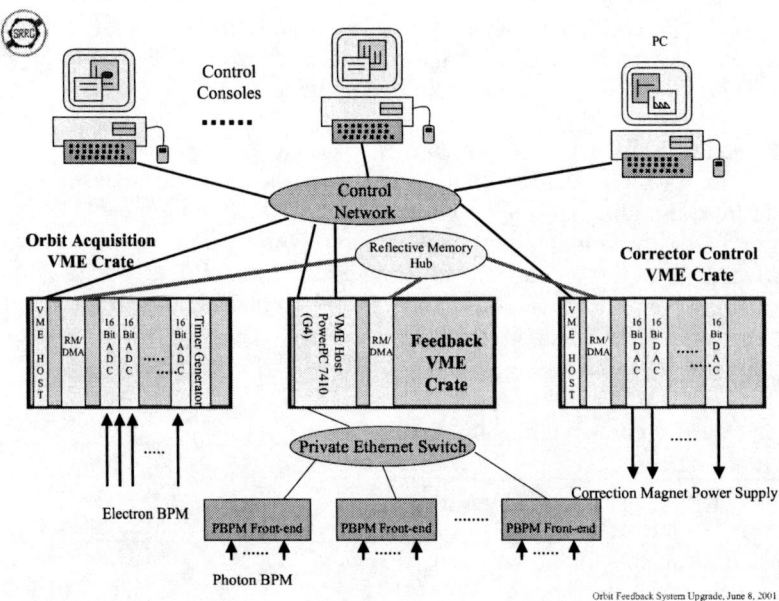

FIGURE 5. The block diagram of new system.

REFERENCES

1. C. H. Kuo, et al., Local Feedback Experiment in the Taiwan Light Source, Proceedings of 1997 IEEE Particle Accelerator Conference, Vancouver, 1997.
2. C. H. Kuo, et al., Digital Global Orbit Feedback System Developing in SRRC, Proceedings of 1997 IEEE Particle Accelerator Conference, Vancouver, 1997.

Simple Amplitude and Phase Detector for Accelerator Instrumentation

K. H. Hu, Jenny Chen, C. H. Kuo, Demi Lee, K. T. Hsu

Synchrotron Radiation Research Center
No. 1 R&D Road VI, Hsinchu Science-Based Industrial Park, Hsinchu, Taiwan.

Abstract. Amplitude and phase are important parameters of an RF system and beam properties. Commercial RF/IF gain and phase detector chip can be simply used to extract amplitude and phase information. This detector can measure the amplitude log ratio and phase difference between OR of two RF signals simultaneously. The method is simpler to implement than other methods. Design consideration and implementation details will be discussed in this report. Preliminary test results and possible applications will be summary in this report.

INTRODUCTION

Amplitude and phase properties of RF system and charge particle beam are important parameters of an accelerator system. Rich information is embedded in these two parameters. The standard method to measure phase and amplitude are done by two-separated detectors, an amplitude detector and a phase detector [1]. Quadrature signal processing by using I/Q detection techniques has been widely used in sophisticated applications[2] recently. This method can provide precision measurement, but with complicated circuitry and occupying much space. Most precision phase detectors do not work at the RF frequency. It needs an RF down-conversion or direct RF sampling process. It is too complicated for some applications.

The newly released commercial RF/IF gain and phase detector chip, AD8302 provides a simple way to measure amplitude log ratio and the phase difference of two RF signals simultaneously [3]. Its features include very few components and a wide frequency working range. The log amplifier has been reported to work with S band for BPM applications [4,5]. Direct working at an RF frequency up to the S band is possible for the detector. It provides a convenient way to implement simple amplitude and phase detection circuitry for accelerator applications.

AMPLITUDE AND PHASE DETECTOR

Commercial chip AD8302 is adopted to implement amplitude and phase detector, working at the RF frequency directly. Figure 1 shows a functional block diagram of the detector. The chip includes a closely matched pair of demodulator logarithmic amplifiers and a phase detector. The amplitude demodulator consists of seven stages of 60dB log-ratio amplifiers. The phase detector is of the multiplier type. It requires a few external components, has an input range from −60dBm to 0dBm, measures gain and phase up to 2.7GHz, normal gain sensitivity of 30mV/dB, a typical gain nonlinearly < 0.5 dB, normal phase sensitivity of 10mV/Degree, a typical phase nonlinearly <1 Degree, and a fast response time. These characteristics support the easy detection of amplitude and phase information.

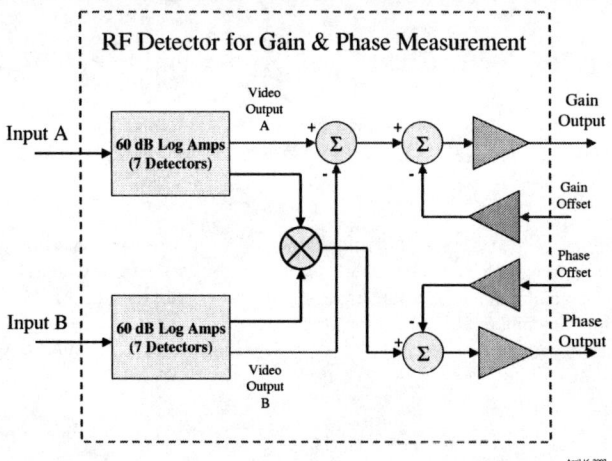

FIGURE 1. The functional block diagram of AD8302.

A measurement system was setup as show in Figure 2 to determine the performance of the detector. The system consists of reference and test signal input arms. The coherent output of the RF signal generator is attenuated to provide a −30dBm reference signal that is connected to the reference arm input of the detector. The output of the signal generator is then connected to the signal arm of the detector. Next, the signal generator is operated in the external pulse modulation mode. Additionally, a voltage controlled phase shifter is inserted between the signal generator output and signal input arm of the detector for phase detector functionality testing. This phase shifter can be removed or maintained at a constant phase position to test gain characteristics.

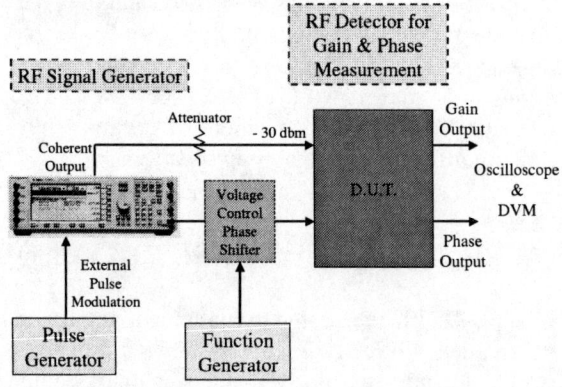

FIGURE 2. AD8302 amplitude and phase detector function test block.

Figure 3 displays the amplitude function test results. When this detector is operated at 500MHz, the amplitude ac performance is shown in Figure 3(a). Figure 3(b) illustrates that the amplitude sensitivity is 30mV at a reference power of -30 dBm and an input signal power sweep range from –60dBm to 0dBm. Phase function test results are shown in Figure 4. The phase ac performance is shown in Figure 4(a). Figure 4(b) illustrates that the phase sensitivity is 10 mV/Degree.

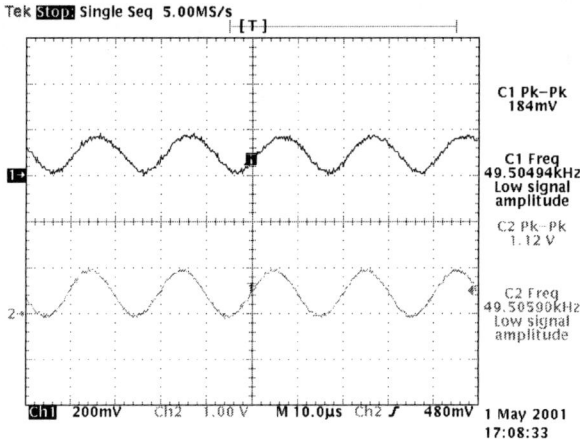

3 (a) Amplitude modulation test results.

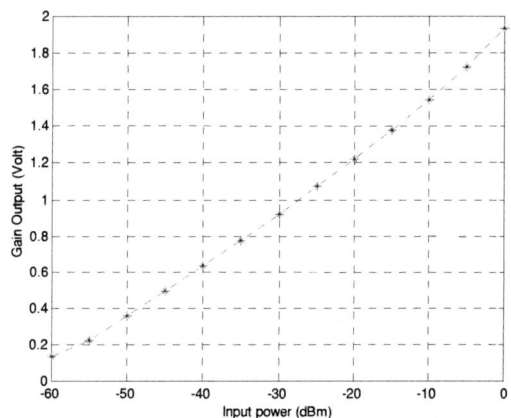

3(b) The amplitude sensitivity of detector.
FIGURE 3. Amplitude sensitivity of detector when input frequency is 500MHz.

PRELIMINARY BEAM TEST RESULTS

This amplitude and phase detector can be used in various applications involving machine physical experiments. Beam transfer function measurement was used to demonstrate the functionality of the detector. Figure 5 describes the experimental setup for beam transfer function measurement. In this experiment, the RF phase was

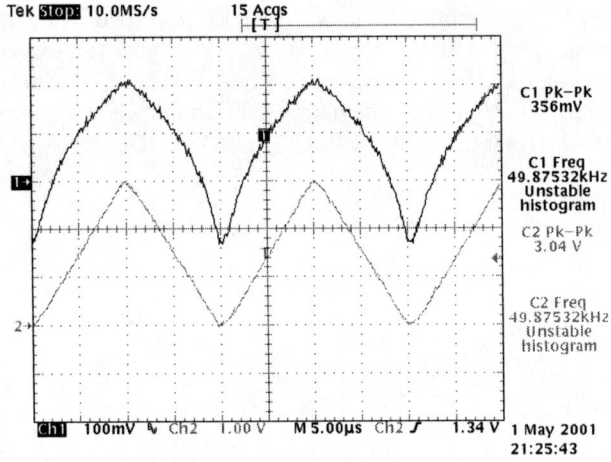

4 (a) The phase modulation test result.

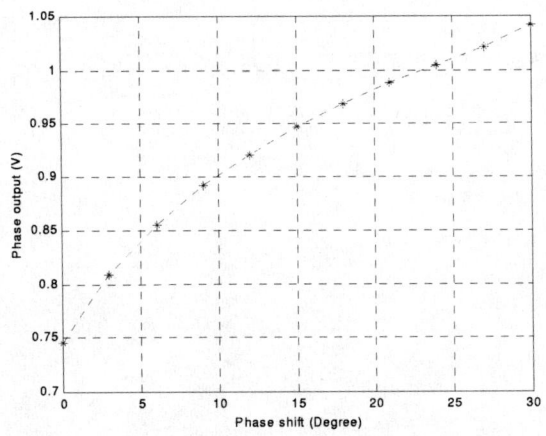

4 (b) The phase sensitivity of detector.
FIGURE 4. Phase sensitivity of detector when input frequency is 500MHz.

modulated near 1fs (synchrotron oscillation frequency) and the RF gap voltage modulation was approximately 2fs. The RF reference signal was connected to the reference arm of the detector to provide a reference for gain and phase measurement. The beam signal from the BPM output was passed through a 500 MHz narrow bandwidth band pass filter and connected to the signal arm of the detector. The beam transfer function was measured using a dynamic signal analyzer (DSA), working in the DC~100 KHz frequency range. This DSA was connected to a control network via NI's GPIB/ENET 100 bus interface. The experiment process was controlled by a MATLAB script, running on a control console. The phase modulation was applied to the phase shifter on low level RF electronics (LLRF) and the gap voltage modulation was also applied to the voltage-controlled attenuator on the LLRF system. The detector simultaneously outputs amplitude and the phase difference between beam signals.

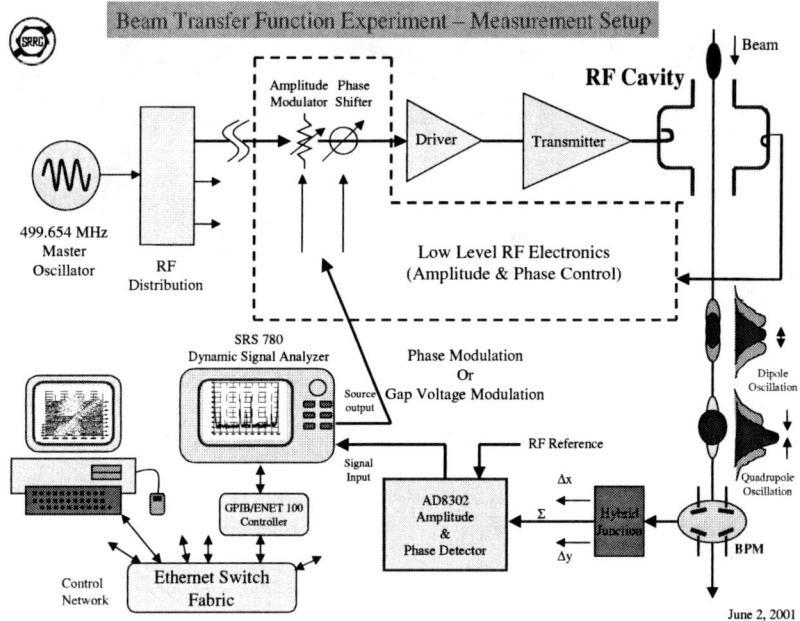

FIGURE 5. Beam amplitude and phase measurement block diagram.

Figure 6 displays the magnitude response of the beam transfer function with phase modulation. Several modulation amplitudes were set. Upward and downward sweeps were done in the experiment. Figures 6(a) and 6(b) summarize the preliminary results. The phase modulated beam transfer function shows strong hysteresis, which has been interpreted in several reports [6,7]. In large amplitude modulation, the beam splits into two beamlets, which oscillate out of phase. The dip in the amplitude spectrum is the overall response to this effect.

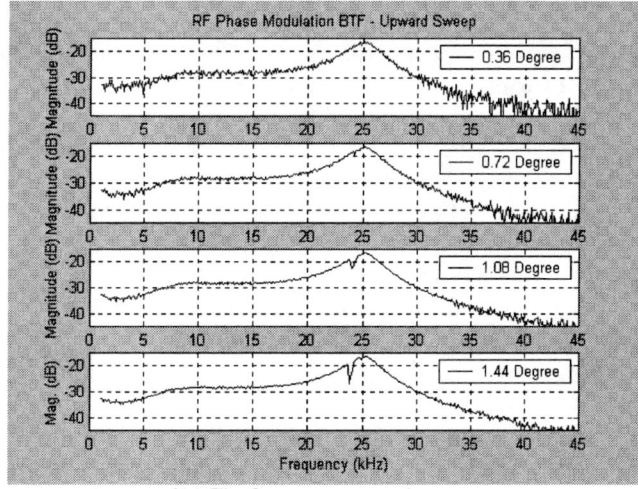

6(a) The frequency upward sweep.

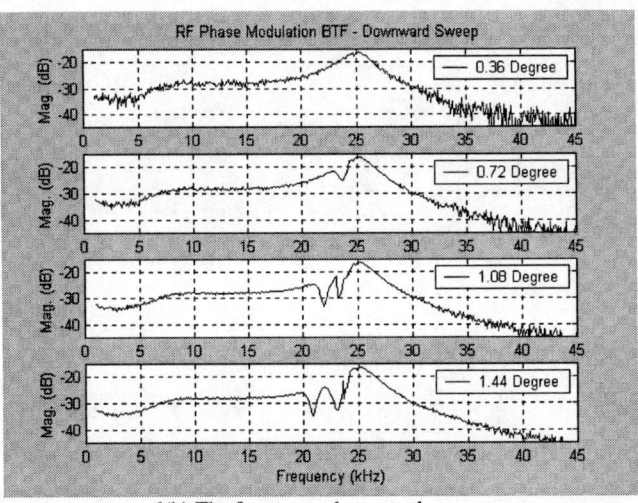

6(b) The frequency downward sweep.

FIGURE 6. Phase modulated magnitude response of beam transfer function in multi-bunch operation mode.

The storage ring of SRRC is hindered by strong coupled-bunch instabilities. Two existing DORIS cavities are the major sources of the instability. The RF system will be upgraded into a superconducting RF (SRF) system in 2003. The project will adopt CESR's nearly high order mode (HOM) free SRF cavity to eliminate longitudinal coupled bunch instability and double the stored beam current (400 mA). Before the upgrade, the storage ring uses RF gap voltage modulation to combat the longitudinal instability [8] in routine operation. Figure 7 shows the effect of the RF gap voltage modulation. The figure shows the 2fs upper sideband of the 200th revolution harmonics. The modulation amplitude is kept constant with a 10% gap voltage. The frequency is scanned from 47.5 kHz to 51.5 kHz. The sharp line spectrum is the modulated source and the broad spectrum is due to the instabilities. When the modulation frequency is near 2fs (49.5 kHz), the instability is suppressed. The motion of the gap voltage modulated beam near 2fs follows a dumbbell like shape rotated in phase space with a frequency of 1fs. Amplitude modulation is introduced and possibly detect by the gain measurement circuitry in AD8302. The gain measurement capability of AD8302 enables it to detect the amplitude variation. Figure 8 illustrates the magnitude response of the gap voltage modulation beam transfer function. The gap voltage response modulation amplitude varies from 5% to 7 %. Figure 8(a) shows the DSA output frequency upward sweep show and 8(b) shows the downward sweep. Strong hysteresis is observed. The hysteresis increases with the excitation amplitude. The beam stable region is located the valley near the crest and near 2fs. Further study is required to obtain clearly the gap voltage modulation beam transfer function.

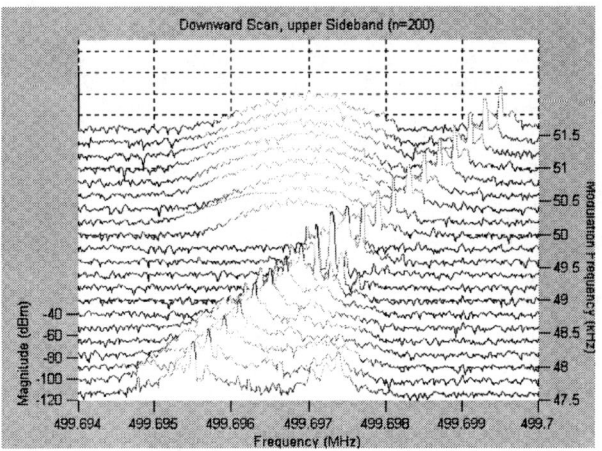

FIGURE 7. Gap voltage modulation spectrums as a function of modulation frequency scan.

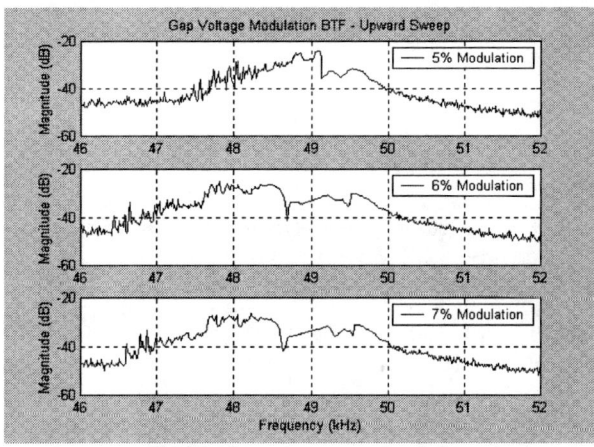

(8a) The frequency upward sweep.

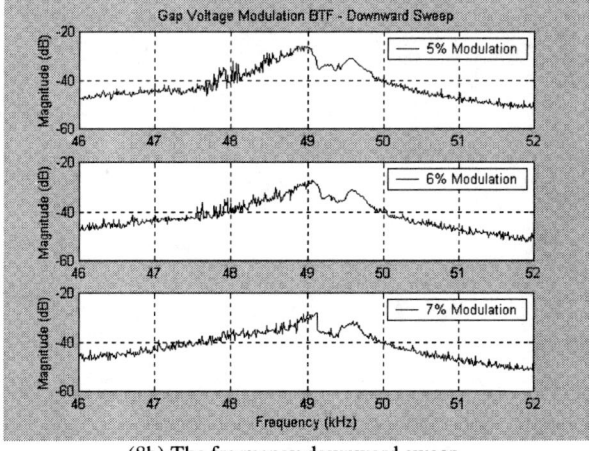

(8b) The frequency downward sweep.

FIGURE 8. Amplitude detector responses during perform gap voltage modulation.

SUMMARY

The AD8302 based detector is simple and easy to implement. The detector combines two functions in one unit. Thus, it doesn't need complex circuitry. The detection frequency is up to the S band. The detection bandwidth is approximately 30MHz, so amplitude and phase variation can be measured on a time scale of 50 nsec. The amplitude sensitivity is 30 mV/dB and the phase sensitivity is 10 mV/Degree. These features of the detector satisfy meet general purpose measurement requirements.

This simple amplitude and phase detector have several possible applications, including (1) studying beam transfer functions; (2) monitoring low precision RF parameters; (3) measuring S-band LINAC RF parameters; (4) acting as a simple log-ratio processor for BPM applications, working at the RF frequency. This report implement a simple detector, based upon the AD8302 chip, to measure a beam transfer function. The performance of the detector module is still being improved to for use in various applications.

REFERENCE

[1]. D. Cheever, et al., "The Bates Phase and Amplitude Monitor System", ICALEPCS'2001 proceeding, Nov 27-30. San 'Jose.
[2] M. Satoh, et al., "First Beam Test of $\Delta\Phi$-A Initial Beam Loading Compensation for Electron Linacs", Proceedings of the 2001 Particle Accelerator Conference, Chicago. p. 3954, 2001.
[3]. AD8302 datasheet, data sheet of Analog Device Incorporation.
[4]. K. Yanagida, et al., "Signal Processor for Spring-8 Linac BPM", DIPAC 2001 Proceedings, ESRF, Grenoble, May 5-9, 2001.
[5] K. Yanagida et al., " A BPM System for the Spring-8 Linac", Proc. Of the 20th Int. Linac Conf., Monterey USA, Aug. 2000, pp 190-192.
[6] J. M. Byrd, et al., "Nonlinear Effects of Phase Modulation in an Electron Storage Ring", Phys. Rev. E57, 4706 (1998).
[7] J. M. Byrd, "Longitudinal Beam Transfer Function Diagnostics in an Electron Storage Ring", Particle Accelerator, 1997, Vol. 57, pp. 159-173..
[8] D. Li, et al., "Effects of rf voltage modulation on particle motion", Nucl. Instrum. and Meth., NIM A364, 205 (1995).

Appendices

10th Beam Instrumentation Workshop
May 6-9, 2002

Workshop Participants

Saeed Assadi
Oak Ridge National Laboratory
701 Scarboro Road
Oak Ridge, TN 37830
USA
saeed@sns.gov

Mei Bai
Brookhaven National Laboratory
Bldg. 911B, C-AD Department
P.O. Box 5000
Upton, NY 11973
USA
mbai@bnl.gov

Julien Bergoz
Bergoz Instrumentation
Espace Allondon Quest
St. Genis Povilly, 01630
FRANCE
bergoz@bergoz.com

Willem Blokland
Oak Ridge National Laboratory
SNS
701 Scarboro Road
Oak Ridge, TN 37830
USA
blokland@sns.gov

J. Michael Brennan
Brookhaven National Laboratory
Bldg. 911B, C-AD Department
P.O. Box 5000
Upton, NY 11973
USA
brennan@bnl.gov

Stephen R. Buckley
Daresbury Laboratory
Keckwick Lane
Warrington, Cheshire WA44AD
United Kingdom
s.r.buckley@dl.ac.uk

Peter Cameron
Brookhaven National Laboratory
Bldg. 817
35 Lawrence Avenue
Upton, NY 11973
USA
cameron@bnl.gov

Karel Capek
TJNAF
12000 Jefferson Avenue, MS 59
Newport News, VA 23606
USA
capek@jlab.org

Brian E. Chase
Fermilab
240 LeGrande Blvd.
Aurora, IL 60506
USA
chase@bnal.gov

Daniel H. Cheever
MIT-Bates
21 Manning Road
Middleton, MA 01949
USA
Cheever@mit.edu

Khalid Chouffani
Idaho Accelerator Center
1500 Alvin Ricken Drive
Pocatello, ID 83209
USA
khalid@physics.isu.edu

Gerald W. Codner
Cornell University – CESR
Wilson Synchrotron Lab
Ithaca, NY 14853
USA
gcodner@lns.cornell.edu

Roger Connolly
Brookhaven National Laboratory
Bldg. 817
35 Lawrence Avenue
Upton, NY 11973
USA
Connolly@bnl.gov

James L. Crisp
Fermilab
MS 308
P.O. Box 500
Batavia, IL 60510
USA
crisp@fnal.gov

Craig Dawson
Brookhaven National Laboratory
Bldg. 817
Upton, NY 11973
USA
dawsonw@bnl.gov

Bernd Dehning
CERN
Meyrin
Genever 23
CH-1213 Switzerland
dehning@cern.ch

Alfred J. Della Penna
Brookhaven National Laboratory
Bldg. 817
Upton, NY 11973
USA
dellapenna@bnl.gov

Joseph H. DeLong
Brookhaven National Laboratory
Collider-Accelerator Department
Bldg. 911A
Upton, NY 11973
USA
Delong@bnl.gov

Jean-Claude Denard
Synchrotron Soleil
Centre Universitaire Paris Sud
- Bât. 209H
Orsay 91898
FRANCE
Jean-claude.denard@soleil.u-psud.fr

Juergen Dietrich
Forschungszentrum Juelich GmbH
Leo-Brandt_str.
Juelich, NRW D-52425
GERMANY
j.dietrich@fz-juelich.de

Rudolf Dölling
Paul Scherrer Institute
Villigen-PSI
Villigen, CH-5232
SWITZERLAND
Rudolf.doelling@psi.ch

Angelika Drees
Brookhaven National Laboratory
Collider-Accelerator Department
Bldg. 911B
Upton, NY 11973
USA
drees@bnl.gov

Michael J. Dufau
Daresbury Laboratory
Keckwick Lane
Daresbury, Warrington WA4 4AD
Cheshire, UK
m.j.dufau@dl.ac.uk

Pierre-André Duperrex
Paul Scherrer Institute
Villigen-PSI
Villigen AG 5232
UK
Pierre-andre.duperrex@psi.ch

Yury I. Eidelman
Brookhaven National Laboratory
Bldg. 830
Upton, NY 11937
eidelman@bnl.gov

Michael P. Fahmie
Lawrence Berkeley National Laboratory
1 Cyclotron Road
Berkeley, CA 94702
mpfahmie@lbl.gov

Mario Ferianis
Sincrotrone Trieste
Strada Statale 14, Km 163.5
Trieste, 34012
ITALY
Mario.ferianis@elettra.trieste.it

Mike Fettig
Texas Instruments/Arrow
220 Rabro Drive East
Hauppauge, NY 11788
USA
mfettig@arrow.com

Ralph Fiorito
TR Research Inc./Univ. of Maryland
3 Lauer Terrace
Silver Spring, MD 20901
USA
rfiorito@rocketmail.com

Alan S. Fisher
Stanford Linear Accelerator Center
MS 18
2575 Sand Hill Road
Menlo Park, CA 94025
AFisher@SLAC.Stanford.edu

Josef C. Frisch
Stanford Linear Accelerator Center
MS 66
2575 Sand Hill Road
Menlo Park, CA 94025
frisch@slac.stanford.edu

David M. Gassner
Brookhaven National Laboratory
C-A Department, Bldg. 911A
Upton, NY 11973
gassner@bnl.gov

James Gibney
Texas Instruments/Arrow
220 Rabro Drive East
Hauppauge, NY 11788
USA
jgibney@arrow.com

John D. Gilpatrick
Los Alamos National Laboratory
MS H808
Los Alamos, NM 87545
USA
gilpatrick@lanl.gov

Daniel T. Gray
TRIUMF
4004 Wesbrook Mall
Vancouver, British Columbia V6T 2A3
CANADA
gray@triumf.ca

Justin D. Gullotta
Brookhaven National Laboratory
C-A Department, Bldg. 911A
Upton, NY 11973
USA
gullotta@bnl.gov

Lee R. Hammons
Brookhaven National Laboratory
C-A Department, Bldg. 911A
Upton, NY 11973
USA
hammons@bnl.gov

Robert A. Hardekopf
Los Alamos National Laboratory
MS H824
Los Alamos, NM 87545
USA
hardekopf@lanl.gov

Robert O. Hettel
SSRL/SLAC
2575 Sand Hill Road
Menlo Park, CA 94025-7015
USA
hettel@slac.stanford.edu

James A. Hinkson
GMW Associates
5805 Robin Hood Drive
El Sobrante, CA 94803
USA
jhinkson@aol.com

Tobias Hoffmann
GSI Darmstadt
Strahldiagnose, Planckstrasse 1
Darmstadt, D-64291
GERMANY
t.hoffmann@gsi.de

John Holst
Texas Instruments/ERA
354 Vets Highway
Commack, NY 11725
USA
jholst@erareps.com

Rodger H. Hosking
Pentek, Inc.
One Park Way
Upper Saddle River, NJ 07458
USA
rodger@pentek.com

Ken Jacobs
Univ. Wisc. Synchrotron Radiation Center
3731 Schneider Drive
Stoughton, WI 53589
USA
kjacobs@src.wisc.edu

Andreas Jansson
CERN
Route de Meyrin
Geneva 1211
SWITZERLAND
Andreas.jansson@cern.ch

Fred Joerger
Joerger Enterprises, Inc.
166 Laurel Road
East Northport, NY 11731
USA
joerger@joergerinc.com

Ronald G. Johnson
Stanford Linear Accelerator Center
2575 Sand Hill Road
Menlo Park, CA 94025
USA
ron_johnson@slac.stanford.edu

Alan A. Jones
SNS/ORNL
701 Scarboro Road
Oak Ridge, TN 37830
USA
jonesaa@sns.gov

Kevin Jordan
Jefferson Lab FEL
12000 Jefferson Avenue
Newport News, VA 23606
USA
Jordan@jlab.org

Roland R. Jung
CERN
Geneva CH1211
SWITZERLAND
Roland.jung@cern.ch

Alexandre Kalinine
Bergoz Instrumentation
Espace Allondon Quest
Saint Genis Pouilly 01630
FRANCE
kalinine@bergoz.com

Tony Kershaw
CLRC, Rutherford Appleton Laboratory
Chilton
Didcot, Oxon OX11 0QX
UK
a.h.kershaw@rl.ac.uk

Sabine K. Kessler
American Institute of Physics
2 Huntington Quadrangle, Suite 1NO1
Melville, NY 11747-4502
USA
skessler@aip.org

Martin Kesselman
Brookhaven National Laboratory
Bldg. 817
Upton, NY 11973
USA
kesselman@bnl.gov

Ralph C. Kimball
Echotek Corporation
555 Sparkman Drive, Suite 400
Huntsville, AL 35816
USA
Ralph@echotek.com

Matthias J. Kirsch
Struck Innovative Systeme
Harksheider Str. 102A
Hamburg 22399
GERMANY
matthias.kirsch@struck.de

Geoffrey A. Krafft
TJNAF
MS-7A
12000 Jefferson Avenue
Newport News, VA 23606
krafft@jlab.org

Patrick Krejcik
SLAC
MS 18
2575 Sand Hill Road
Menlo Park, CA 94025
pkr@slac.stanford.edu

Samuel Krinsky
Brookhaven National Laboratory
Bldg. 725B
Upton, NY 11973
USA
krinsky@bnl.gov

Changhou Kuo
SRRC
No. 1, R&D Rd 6
Science-Based Industrial Park
Hsinchu, Taiwan 30077
TAIWAN
chkuo@srrc.gov.tw

Saul Kupferburg
Contech/Kepco Inc.
131-38 Sanford Avenue
Flushing, NY 11352
USA
skupferberg@kepcopower.com

James F. Kurtz
Echotek Corporation
555 Sparkman Drive, Suite 400
Huntsville, AL 35816
jim@echotek.com

Yongbin Leng
Brookhaven National Laboratory
SNS, Bldg. 817
Upton, NY 11973
USA
leng@bnl.gov

Dmitri Liakine
Institute for Theoretical & Exp. Physics
Bol. Cheremushkinskaja 25.
Moscow 117259
RUSSIA
d.liakin@gsi.de

Robert M. Lill
Argonne National Laboratory
9700 S. Cass Avenue
Argonne, IL 60439
USA
blill@aps.anl.gov

Ping Lu
NSRL of USTC
P.O. Box 6022
Hefei, Anhui 230029
CHINA
lup@ustc.edu.cn

Alex H. Lumpkin
APS/ANL
9700 S. Cass Avenue
Argonne, IL 60439
USA
lumpkin@aps.anl.gov

Christian Magne
CEA-Saclay
SACM, Gif-sur-Yvette cedex 91191
FRANCE
cmagne@cea.fr

Fabio Marcellini
INFN – LNF
Via E. Fermi 40
00044 Frascati, Rome
ITALY
Fabio.marcellini@inf.infn.it

Jerry C. Melcher
GMW Associates
955 Industrial Road
San Carlos, CA 94070
USA
jerry@gmw.com

Thomas S. Meyer
Fermilab
P.O. Box 500
MS 308
Batavia, IL 60510
USA
tsmeyer@fnal.gov

Robert J. Michnoff
Brookhaven National Laboratory
C-A Department
Bldg. 911C
Upton, NY 11973
USA
michnoff@bnl.gov

Brian M. Minato
TRIUMF
4004 Wesbrook Mall
Vancouver, B.C. V6T 2A3
CANADA
minato@triumf.ca

Istvan Mohos
Forschungszentrum Juelich GmbH
Leo Brand Str. 1
Juelich, NRW D-52425
GERMANY
i.mohos@fz-juelich.de

Christoph H. Montag
Brookhaven National Laboratory
C-A Department
Bldg. 911B
Upton, NY 11973
Montag@bnl.gov

Gerhard Mrotzek
CLRC Daresbury Laboratory
Keckwick Lane
Warrington, Cheshire WA4 5HH
UK
g.mrotzek@dl.ac.uk

John Musson
TJNAF
MS 58
12000 Jefferson Avenue
Newport News, VA 23606
USA
musson@jlab.org

Peter H. Oddo
Brookhaven National Laboratory
C-A Department
Bldg. 911A
Upton, NY 11973
USA
poddo@bnl.gov

James F. O'Hara
Los Alamos National Laboratory
MS H817
Los Alamos, NM 87544
USA
ohara@lanl.gov

Ralph J. Pasquinelli
Fermilab
P.O. Box 500, MS 340
Batavia, IL 60510
USA
pasquin@fnal.gov

Michael A. Plum
LANL
MS H838
P.O. Box 1663
Los Alamos, NM 87544
USA
plum@lanl.gov

Tom Powers
TJNAF
MS 58
12000 Jefferson Avenue
Newport News, VA 23606
powers@jlab.org

Peter S. Prieto
Fermilab
Kirk and Wilson
RF&I MS 308
Batavia, IL 60510
prieto@fnal.gov

Bryan S. Quinn
Institute for Research in Electronics and
Applied Physics, Unv. Of Maryland
Energy Research Facility Bldg. 223
College Park, Maryland 20742
bquinn@glue.umd.edu

William R. Rawnsley
TRIUMF
4004 Wesbrook Mall
Vancouver, BC V6T 2A3
CANADA
rawnsley@triumf.ca

Thomas Russo
Brookhaven National Laboratory
C-A Department
Bldg. 911A
Upton, NY 11973
USA
trusso@bnl.gov

Surajit Sarkar
Massachusetts General Hospital
North East Proton Therapy Center
30 Fruit Street, Room 111
Boston, MA 02114
USA
sarkar@hadron.mgh.harvard.edu

Shigeki Saski
SPpring-8
Kouto 1-1-1, Mikadzuki-cho
Sayo-gun
Hyogo-ken
JAPAN
saski@spring8.or.jp

Christof L. Schneider
Forschungszentrum Rossendorf (FZR)
Radiation Source ELBE
Bautzner Landstrasse 128
P.O. Box 51 01 19
01314 Dresden
GERMANY
Christof.Schneider@fz-rossendorf.de

Rico Schurig
Forschungszentrum Rossendorf (FZR)
Postfach 510119
Dresden 01328
GERMANY
r.schurig@fz-rossendorf.de

William C. Sellyey
Los Alamos National Laboratory
LANSCE-1, MS –H808
Los Alamos, NM 87545
USA
sellyey@lanl.gov

Robert E. Shafer
SNS/LANL
MS H808
Los Alamos, NM 87545
USA
Rshafer174@aol.com

Coles Sibley
SNS/ORNL
701 Scarboro Drive
Oak Ridge, TN 37830
USA
sibley@sns.gov

Om V. Singh
Argonne National Laboratory
9700 S. Cass Avenue
Bldg. 401
Argonne, IL 60439
USA
singh@aps.anl.gov

Gary A. Smith
Brookhaven National Laboratory
C-A Department
Bldg. 911C
Upton, NY 11973
USA
gasmith@bnl.gov

Robert J. Smith
Daresbury Laboratory
Keckwick Lane
Daresbury, Warrington WA4 4AD
Cheshire, UK
r.j.smith@dl.ac.uk

Stephen R. Smith
SLAC
2575 Sand Hill Road
Stanford, CA 94019
USA
ssmith@slac.stanford.edu

Vern Smith
SLAC
2575 Sand Hill Road, MS 50
Menlo Park, CA 94025
USA
vrs@slac.stanford.edu

Lars Soby
CERN, Div. PS
Rue de Meyrin
Geneva, 23
1211
SWITZERLAND
Lars.soby@cern.ch

Angelo Stella
INFN – LNF
Via E. Fermi, 40
00044 Frascati Italy
Frascati, Rome
ITALY
Angelo.stella@inf.infn.it

Jocelyn Tan
CERN
PS Division
Geneva 1211
SWITZERLAND
Jocelyn.tan@cern.ch

Gianni R. Tassotto
Fermilab
P.O. Box 500
Batavia, IL 60510
USA
tassotto@fnal.gov

Dmitry Teytelman
SLAC
2575 Sand Hill Road
Menlo Park, CA 94025
USA
dim@slac.stanford.edu

Frank Toich
Contech/Kepco Inc.
131-31 Sanford Avenue
Flushing, NY 11352
USA
ftoich@kepcopower.com

Ed Tomicich
Contech Marketing Associates
7 Lincoln Highway, Suite 103
Edison, NJ 08820
USA
edt@contech-rti.com

Gary J. Turchetta
Echotek Corporation
555 Sparkman Drive, Suite 400
Huntsville, AL 35816
USA
gary@echotek.com

Klaus B. Unser
Bergoz Instrumentation
Espace Allondon Quest
Saint Genis Pouilly 01630
FRANCE
unser@bergoz.com

Rok Ursic
Instrumentation Technologies
Srebrnicev trg 4a
Solkan SI-5250
Slovenia
rok@I-tech.si

Ian J. Walker
GMW Associates
955 Industrial Road
San Carlos, CA 94070
USA
ian@gmw.com

Richard L. Walker
TJNAF
12000 Jefferson Avenue
Newport News, VA 23606
USA
Richard@jlab.org

Junhua Wang
NSRL of University of Science &
Technology of China
Hefei, Anhui 230029
P.R. CHINA
wjhua@ustc.edu.cn

Robert C. Webber
Fermilab
MS 341
P.O. Box 500
Batavia, IL 60510
USA
webber@fnal.gov

Jonah M. Weber
Lawrence Berkeley National Laboratory
1 Cycltron Road
Berkeley, CA 94702
USA
jmweber@lbl.gov

Manfred Wendt
DESY
Notkestr.85
Hamburg 22607
GERMANY
wendt@us-bello.desy.de

Michelle Wilinski
Brookhaven National Laboratory
C-A Department
Bldg. 911A
Upton, NY 11973
USA
wilinski@bnl.gov

Mark Wissmann
TJNAF
12000 Jefferson Avenue, MS59
Newport News, VA 23606
USA
wissmann@jlab.org

Richard L. Witkover
TechSource Inc./ORNL/BNL
Brookhaven National Laboratory
Bldg. 817
Upton, NY 11973
USA
witkover@bnl.gov

Kay Wittenburg
DESY
Notkestr. 85
Hamburg, Germany D-22607
GERMANY
Kay.wittenburg@desy.de

Bingxin Yang
Argonne National Laboratory
9700 South Cass Avenue
APS/AOD
Argonne, IL 60439
USA
bxyang@aps.anl.gov

Andrew Young
Stanford Linear Accelerator Center
2575 Sand Hill Road
MS 50
Menlo Park, CA 94309
USA
ayoung@slac.stanford.edu

Adim J. Yousif
Fermilab
731 S. Batavia Avenue
Batavia, IL 60510
USA
yousif@fnal.gov

James Zagel
Fermilab
MS 308
P.O. Box 500
Batavia, IL 60510
USA
zagel@fnal.gov

Xiaolong Zhang
Fermilab
MS 307
P.O. Box 500
Batavia, IL 60510-0500
USA
zhangxl@fnal.gov

Participating Vendors

Advanced Design Consulting, Inc.
126 Ridge Road
P.O. Box 187
Lansing, NY 14882
USA
Tel: (607) 533-3531
Fax: (607) 533-3618
http://www.adc9001.com

Bergoz Instrumentation, Inc.
Espace Allondon Quest
Saint Genis Pouilly
01630 France
Tel: +33-450.426.642
Fax: +33-450.426.643
http://www.bergoz.com

Contech Marketing Associates
7 Lincoln Highway
Suite 103
Edison, NJ 08820
Tel: (732) 744-2500
Fax: (732) 744-2505
Toll Free: (800) 219-9417
http://www.contech-rti.com

Drivesoft, LLC
4813 Wayside Drive
Old Hickory, TN 37138-1349
Tel: (615) 758-3718
Fax: (615) 758-3973
http://www.drivesoft.com

Echotek Corporation
555 Sparkman Drive
Suite 400
Huntsville, AL 35816
Tel: (256) 721-1911
Fax: (256) 721-9266
http://www.echotek.com

GMW Associates
955 Industrial Road
San Carlos, CA 94070
Tel: (650) 802-8292
Fax: (650) 802-8298
http://www.gmw.com

Joerger Enterprises, Inc.
166 Laurel Road
East Northport, NY 11731
Tel: (631) 757-6200
Fax: (631) 757-6201
http://www.joergerinc.com

MicroGraphics, Inc.
P.O. Box 125
Waterloo, WI 53594
Tel: (920) 478-2889
Fax: (920) 478-3689
http://www.MicroG.com

Pentek, Inc.
One Park Way
Upper Saddle River, NJ 07458
Tel: (201) 818-5900
Fax: (201) 818-5904
http://www.pentek.com

Texas Instruments
354 Veterans Memorial Highway
Commack, NY 11725
Tel: (631) 543-0510 x311
Fax: (631) 543-0758
http://www.ti.com

AUTHOR INDEX

A

Allison, S., 267
Anderson, S., 237
Araki, S., 508
Arnold, N., 401

B

Bellavia, S., 353
Bentley, B. E., 305
Benz, S., 361
Berg, W. J., 212, 433
Biedron, S., 212
Bishofberger, K., 483
Bolton, P., 491
Borland, M., 212
Brennan, M., 134, 409
Brown, K. A., 353
Buckley, S. R., 447

C

Cameron, P., 134, 134, 376, 409
Cerniglia, P., 134, 376
Chae, Y. C., 212
Chase, B. E., 368
Chen, J., 516, 523
Chiang, I. H., 353
Chin, M. J., 467
Chouffani, K., 497
Connolly, R., 134, 134, 409
Crisp, J. L., 103
Cupolo, J., 134, 134, 376

D

Danailov, M., 203
Dawson, W. C., 134, 376, 417
Day, L., 195
Degen, C., 134, 376
Dehning, B., 229
Dejus, R., 212
DellaPenna, A., 134, 376

DeLong, J., 134
Denard, J.-C., 118
Dölling, R., 361
Dowell, D., 491
Drees, A., 134, 289, 409
Dufau, M. J., 447
Duperrex, P.-A., 361

E

Ellis, S., 195
Erdmann, M., 212
Evtushenko, P., 313

F

Ferianis, M., 203
Ferioli, G., 229
Fiorito, R. B., 187
Fisher, A. S., 267
Fitzgerald, J., 483
Forck, P., 329
Fox, J., 458, 474
Frei, U., 361
Friesenbichler, W., 229
Frisch, J., 237

G

Gassner, D., 134, 134, 337, 345, 353
Gilpatrick, J. D., 297, 305
Grau, M., 134
Gray, D., 439
Gruchalla, M., 297
Gschwendtner, E., 229

H

Hardekopf, R., 195
Harmon, F., 497
Hayano, H., 237, 508
He, D., 283
Hoffmann, T., 329

Hong, J., 283
Hosking, R. H., 248
Hsu, K. T., 516, 523
Hu, K. H., 516, 523
Huang, Z., 212
Huhn, A., 376
Humbel, M., 361

J

Jansson, A., 3
Jobe, K., 237
Johnson, R., 321
Jones, J., 497
Jung, R., 220

K

Kalinin, A., 384
Kamperschroer, J., 297
Kesselman, M., 134, 134, 376, 417
Kim, K.-J., 212
Kim, S., 393
Kimball, R. C., 79
Komorowski, P., 220
Koopman, J., 229
Krafft, G. A., 118
Krejcik, P., 162, 491
Krinsky, S., 23
Kuo, C. H., 516, 523
Kuznetsov, G., 483

L

Lancaster, G., 497
Laznovsky, M., 267
Ledford, J. E., 305
Lee, D., 523
Lee, R., 134
Lehnert, U., 313
Lewellen, J., 212
Li, W. M., 259, 283
Li, Y., 212
Li, Z., 321
Liakin, D. A., 329
Lill, R., 393, 401
Liu, J. H., 259

Liu, Z. P., 259, 283
Lu, P., 259, 283
Lumpkin, A. H., 212, 433

M

MacIntosh, P., 458
Marcellini, F., 458
Marusic, A., 134
McCormick, D., 237, 508
McKee, B., 237
Mead, J., 134, 376
Meisner, K. G., 368
Meyer Sr., R., 195
Michel, P., 313
Michnoff, R., 134, 289, 409
Milton, S. V., 212
Minato, B., 439
Montag, C., 409
Moog, E., 212

N

Naito, T., 237, 321, 508
Nelson, J., 237

O

O'Hara, J. F., 195, 275, 297, 305
Olson, M., 483
O'Shea, P. G., 187

P

Pei, Y. J., 283
Peng, S., 134
Petree, M., 267
Pile, P., 353
Plum, M. A., 195
Ponce, L., 220
Power, J., 195
Prigl, R., 353

R

Rezzonico, L., 361
Rifkin, J., 321, 491
Robin, J., 267
Rose, C., 195
Roser, T., 289
Ross, M., 237, 508
Rotela, E., 393
Roybal, R. J., 305
Rule, D. W., 212

S

Sajaev, V., 212
Sasaki, S., 425
Schneider, C., 313
Schultheiss, C., 134
Schurig, R., 313
Schwarz, H., 458
Seeman, M., 267
Sellyey, W. C., 275
Semenov, A., 483
Shafer, R. E., 44, 195
Sharma, S., 393
Shiltsev, V., 483
Shkvarunets, A. G., 187
Shurter, R. B., 305
Sikora, R., 134, 134, 376
Singh, O., 401
Smith, G. A., 289
Smith, R. J., 447
Smith, S. R., 321, 508
Smith, T., 237
Smith, V., 321
Solyak, N., 483
Soutome, K., 425
Stettler, M., 195
Stovall, J., 195
Sun, B. G., 259, 283

T

Tanaka, H., 425
Teichert, J., 313
Terunuma, N., 237, 508

Teytelman, D., 458, 474
Thanos, S. N., 248
Tobiyama, M., 458
Tommasini, D., 220

U

Uršič, R., 179

V

Van Zeijts, J., 134

W

Wang, J. H., 259, 283
Weber, J. M., 467
Wells, D., 497
Wienands, U., 267
Wilinski, M., 289
Witkover, R. L., 337, 345

X

Xu, H., 283

Y

Yang, B. X., 59, 212, 393, 433
Young, A., 458, 508

Z

Zhang, X., 483
Zhang, Z. Z., 259

Previous Proceedings in the Series of Beam Instrumentation Workshops

	Year	Held in	Publisher	ISBN
9th	2000	Cambridge, Massachusetts	AIP Conf. Proceedings vol. 546	1-56396-975-0
8th	1998	Stanford, California	AIP Conf. Proceedings vol. 451	1-56396-794-4
7th	1996	Argonne, Illinois	AIP Conf. Proceedings vol. 390	1-56396-612-3
6th	1994	Vancouver, B.C, Can.	AIP Conf. Proceedings vol. 333	1-56396-352-3
5th	1993	Santa Fe, New Mexico	AIP Conf. Proceedings vol. 319	1-56396-389-2

Other Related Titles from AIP Conference Proceedings

647 Advanced Accelerator Concepts: Tenth Workshop
Edited by Christopher E. Clayton and Patrick Muggli, December 2002, 0-7354-0102-0

642 High Intensity and High Brightness Hadron Beams: 20th ICFA Advanced Beam Dynamics Workshop on High Intensity and High Brightness Hadron Beams; ICFA-HB2002
Edited by Weiren Chou, Yoshiharu Mori, David Neuffer, and Jean-François Ostiguy, November 2002, 0-7354-0097-0

639 Production and Neutralization of Negative Ions and Beams:
Ninth International Symposium
Edited by Martin P. Stockli, November 2002, 0-7354-0094-6

592 High Quality Beams: Joint US-CERN-JAPAN-RUSSIA Accelerator School
Edited by S. I. Kurokawa, S. Y. Lee, J. Miles, E. A. Perevedentsev, November 2001, 0-7354-0034-2

581 Physics of, and Science with, the X-Ray Free-Electron Laser: 19th Advanced ICFA Beam Dynamics Workshop
Edited by C. Pellegini, S. Chattopadhyay, M. Cornacchia, and I. Lindau, August 2001, 0-7354-0022-9

572 Electron Beam Ion Sources and Traps and Their Applications:
8th International Symposium
Edited by Krsto Prelec, June 2001, 0-7354-0011-3

536 Instrumentation in Elementary Particle Physics: VIII ICFA School
Edited by Sehban Kartal, September 2000, 1-56396-960-2

520 Bates 25: Celebrating 25 Years of Beam to Experiment
Edited by T. W. Donnelly and W. Turchinetz, June 2000, 1-56396-949-1

496 Workshop on Instabilities of High Intensity Hadron Beams in Rings
Edited by T. Roser and S. Y. Zhang, December 1999, 1-56396-910-6

To learn more about these titles, or the AIP Conference Proceedings Series, please visit the webpage **http://proceedings.aip.org/proceedings**